COMPUTATIONAL INTE
IN CANCER DIAGN

COMPUTATIONAL INTELLIGENCE IN CANCER DIAGNOSIS

Progress and Challenges

Edited by

JANMENJOY NAYAK
P.G. Department of Computer Science, Maharaja Sriram Chandra Bhanja Deo University, Baripada, Odisha, India

DANILO PELUSI
Faculty of Communication Sciences, University of Teramo, Teramo, Italy

BIGHNARAJ NAIK
Department of Computer Application, Veer Surrendra Sai University of Technology, Burla, Odisha, India

MANOHAR MISHRA
Electrical and Electronics Engineering Department, Institute of Technical Education and Research, Bhubaneswar, Odisha, India

KHAN MUHAMMAD
Department of Applied Artificial Intelligence, Sungkyunkwan University, Seoul, South Korea

DAVID AL-DABASS
Department of Computer Science, Nottingham Trent University, Nottingham, United Kingdom

ELSEVIER

ACADEMIC PRESS
An imprint of Elsevier

Academic Press is an imprint of Elsevier
125 London Wall, London EC2Y 5AS, United Kingdom
525 B Street, Suite 1650, San Diego, CA 92101, United States
50 Hampshire Street, 5th Floor, Cambridge, MA 02139, United States
The Boulevard, Langford Lane, Kidlington, Oxford OX5 1GB, United Kingdom

ISBN 978-0-323-85240-1

For information on all Academic Press publications
visit our website at https://www.elsevier.com/books-and-journals

Publisher: Stacy Masucci
Acquisitions Editor: Rafael E. Teixeira
Editorial Project Manager: Kristi Anderson
Production Project Manager: Niranjan Bhaskaran
Cover Designer: Christian Bilbow

Typeset by STRAIVE, India

Contents

6. Machine learning for thyroid cancer diagnosis

Feyzullah Temurtas, Kutlucan Gorur, Onursal Cetin, and Ilyas Ozer

2

Prediction of cancer susceptibility

7. Machine learning-based detection and classification of lung cancer

Shubham Dodia and B. Annappa

8. Deep learning techniques for oral cancer diagnosis

Ijaz Ul Haq, Fath U Min Ullah, Khan Muhammad, and Sung Wook Baik

9. An intelligent deep learning approach for colon cancer diagnosis

Pemmada Suresh Kumar, K. Anisha Kumari, and Uttam Ghosh

10. Effect of COVID-19 on cancer patients: Issues and future challenges

H. Swapnarekha and Janmenjoy Nayak

11. Empirical wavelet transform-based fast deep convolutional neural network for detection and classification of melanoma

Bhanja Kishor Swain, Mrutyunjaya Sahani, and Renu Sharma

3

Advance computational intelligence paradigms

12. Convolutional neural networks and stacked generalization ensemble method in breast cancer prognosis

Tahmina Akter Tisha, Mir Moynuddin Ahmed Shibly, Kowshik Ahmed, and Shamim H. Ripon

13. Light-gradient boosting machine for identification of osteosarcoma cell type from histological features

Etuari Oram, Pandit Byomakesha Dash, and Bighnaraj Naik

14. Deep learning-based computer-aided cervical cancer diagnosis in digital histopathology images

Pandia Rajan Jeyaraj, Edward Rajan Samuel Nadar, and Bijaya Ketan Panigrahi

15. Deep learning techniques for hepatocellular carcinoma diagnosis

Dillip Kumar Bishi, Priyadarshini Padhi, Chhabi Rani Panigrahi, Bibudhendu Pati, and Chandi Charan Rath

16. Issues and future challenges in cancer prognosis: (Prostate cancer: A case study)

Dukka Karun Kumar Reddy, H. Swapnarekha, H.S. Behera, S. Vimal, Asit Kumar Das, and Danilo Pelusi

Contributors

Soutrik Acharya Department of Radiophysics and Electronics, University of Calcutta, Kolkata, West Bengal, India

Kowshik Ahmed Department of Computer Science and Engineering, East West University, Dhaka, Bangladesh

David Al-Dabass School of Computing & Informatics, Nottingham Trent University, Nottingham, United Kingdom

K. Anisha Kumari Department of Artificial intelligence, Northeastern University, Boston, MA, United States

B. Annappa Department of Computer Science and Engineering, National Institute of Technology, Mangaluru, Karnataka, India

Sung Wook Baik Sejong University, Seoul, South Korea

H.S. Behera Department of Information Technology, Veer Surendra Sai University of Technology, Burla, Odisha, India

Dillip Kumar Bishi Department of Biotechnology, Rama Devi Women's University, Bhubaneswar, Odisha, India

Roberto Cannella Department of Biomedicine, Neuroscience and Advanced Diagnostics (BIND), University Hospital of Palermo; Department of Health Promotion, Mother and Child Care, Internal Medicine and Medical Specialties (PROMISE), University of Palermo, Palermo, Italy

Onursal Cetin Electrical and Electronics Engineering, Bandirma Onyedi Eylul University, Balikesir, Turkey

Asit Kumar Das IIEST Shibpur, Howrah, West Bengal, India

Pandit Byomakesha Dash Department of Information Technology, Aditya Institute of Technology and Management (AITAM), Tekkali, Andhra Pradesh, India

Shubham Dodia Department of Computer Science and Engineering, National Institute of Technology, Mangaluru, Karnataka, India

Dyuti Ghosh Department of Radiophysics and Electronics, University of Calcutta, Kolkata, West Bengal, India

Uttam Ghosh Department of CS & DS, Meharry Medical College, Nashville, TN, United States

Kutlucan Gorur Electrical and Electronics Engineering, Bandirma Onyedi Eylul University, Balikesir, Turkey

R. Gowri Department of Computer Applications, Sona College of Arts and Science, Salem, Tamilnadu, India

Ijaz Ul Haq Department of Computer Science & Engineering, School of Engineering and Computer Science, Oakland University, Rochester, MI, United States

Pandia Rajan Jeyaraj Department of Electrical and Electronics Engineering, Mepco Schlenk Engineering College, Sivakasi, Tamilnadu, India

Pemmada Suresh Kumar Department of Computer Science and Engineering, GITAM School of Technology, GITAM (Deemed to be University), Visakhapatnam, Andhra Pradesh, India

Roberto Lagalla Department of Biomedicine, Neuroscience and Advanced Diagnostics (BIND), University Hospital of Palermo, Palermo, Italy

Shuai Liu Hunan Provincial Key Laboratory of Intelligent Computing and Language Information Processing; College of Information Science and Engineering, Hunan Normal University; Xiangjiang Institute of Artificial

Intelligence, Changsha, People's Republic of China

Xiangbin Liu Hunan Provincial Key Laboratory of Intelligent Computing and Language Information Processing; College of Information Science and Engineering, Hunan Normal University; Xiangjiang Institute of Artificial Intelligence, Changsha, People's Republic of China

Massimo Midiri Department of Biomedicine, Neuroscience and Advanced Diagnostics (BIND), University Hospital of Palermo, Palermo, Italy

Fath U Min Ullah Department of Electronic and Electrical Engineering, University of Sheffield, Sheffield, South Yorkshire, United Kingdom

Manohar Mishra Siksha 'O' Anusandhan Deemed to be University, Bhubaneswar, Odisha, India

Subhashree Mohapatra Department of Computer Science & Engineering, Institute of Technical Education and Research, Siksha 'O' Anusandhan Deemed to be University, Bhubaneswar, Odisha, India

Khan Muhammad Department of Applied Artificial Intelligence, Sungkyunkwan University, Seoul, South Korea

Edward Rajan Samuel Nadar Department of Electrical and Electronics Engineering, Mepco Schlenk Engineering College, Sivakasi, Tamilnadu, India

Bighnaraj Naik Department of Computer Application, Veer Surendra Sai University of Technology, Burla, Odisha, India

Debasish Swapnesh Kumar Nayak Department of Computer Science & Engineering, Institute of Technical Education and Research, Siksha 'O' Anusandhan Deemed to be University, Bhubaneswar, Odisha, India

Janmenjoy Nayak P.G. Department of Computer Science, Maharaja Sriram Chandra Bhanja Deo University, Baripada, Odisha, India

SoumyaRanjan Nayak KIIT (Deemed-to-be) University, Bhubaneswar, Odisha, India

Etuari Oram Department of Computer Application, Veer Surendra Sai University of Technology, Burla, Odisha, India

Ilyas Ozer Computer Engineering, Bandirma Onyedi Eylul University, Balikesir, Turkey

Priyadarshini Padhi Department of Biotechnology, Rama Devi Women's University, Bhubaneswar, Odisha, India

Bijaya Ketan Panigrahi Department of Electrical Engineering, Indian Institute of Technology Delhi, New Delhi, Delhi, India

Chhabi Rani Panigrahi Department of Computer Science, Rama Devi Women's University, Bhubaneswar, Odisha, India

Bibudhendu Pati Department of Computer Science, Rama Devi Women's University, Bhubaneswar, Odisha, India

Danilo Pelusi University of Teramo, Teramo, Italy

Chandi Charan Rath Department of Life Science, Rama Devi Women's University, Bhubaneswar, Odisha, India

R. Rathipriya Department of Computer Science, Periyar University, Salem, Tamilnadu, India

Dukka Karun Kumar Reddy Dr. Lankapalli Bullayya College of Engineering, Visakhapatnam, Andhra Pradesh, India

Shamim H. Ripon Department of Computer Science and Engineering, East West University, Dhaka, Bangladesh

Mrutyunjaya Sahani Siksha 'O' Anusandhan Deemed to be University, Bhubaneswar, Odisha, India

Renu Sharma Siksha 'O' Anusandhan Deemed to be University, Bhubaneswar, Odisha, India

Mir Moynuddin Ahmed Shibly Department of Computer Science and Engineering, United International University, Dhaka, Bangladesh

Liping Song Hunan Provincial Key Laboratory of Intelligent Computing and Language Information Processing; College of Information Science and Engineering, Hunan Normal University; Xiangjiang Institute of Artificial Intelligence, Changsha, People's Republic of China

Bhanja Kishor Swain Siksha 'O' Anusandhan Deemed to be University, Bhubaneswar, Odisha, India

H. Swapnarekha Department of Information Technology, Aditya Institute of Technology and Management (AITAM), Tekkali, Andhra Pradesh, India

Tripti Swarnkar Department of Computer Application, Institute of Technical Education and Research, Siksha 'O' Anusandhan Deemed to be University, Bhubaneswar, Odisha, India

Feyzullah Temurtas Electrical and Electronics Engineering, Bandirma Onyedi Eylul University, Balikesir, Turkey

Tahmina Akter Tisha Department of Computer Science and Engineering, East West University, Dhaka, Bangladesh

Federica Vernuccio Institute of Radiology, Department of Medicine-DIMED, University of Padova, Padova, Italy

S. Vimal Ramco Institute of Technology, Rajapalayam, Tamilnadu, India

About the editors

Janmenjoy Nayak is an Assistant Professor, P.G. Department of Computer Science, Maharaja Sriram Chandra Bhanja Deo University, Baripada, Odisha, India. He has been a Gold Medallist in Computer Science twice in his career, and has been awarded the "Innovation in Science Pursuit for Inspired Research" (INSPIRE) Research Fellowship from the Department of Science & Technology, Government of India (at both Junior Research Fellow and Senior Research Fellow level) and Best Researcher Award from Jawaharlal Nehru University of Technology, Kakinada, Andhra Pradesh for the academic year 2018–19. He has received many other awards from national and international academic agencies. Dr. Nayak has edited 19 books and 8 special issues on the applications of computational intelligence, soft computing, and pattern recognition, which have been published by Springer and Inderscience. He has published more than 190 refereed articles in various book chapters, conferences, and peer-reviewed journals of Elsevier, Inderscience, Springer, the Institute of Electrical and Electronics Engineers (IEEE), and others. He has also served as a reviewer for more than 100 journals and conferences produced by the IEEE, the Association for Computing Machinery (ACM), Springer, Elsevier, Wiley, and Inderscience. He has 11 years of experience in both teaching and research. Dr. Nayak is a senior member of the IEEE and a life member of societies such as the Soft Computing Research Society (SCRS), the Computer Society of India (CSI India), the Orissa Information Technology Society (OITS), the Orissa Mathematical Society (OMS), and the International Association of Engineers (IAENG), Hong Kong. He has successfully conducted and is associated with 14 internationally renowned series conferences such as ICCIDM, HIS, ARIAM, CIPR, and SCDA. His areas of interest include data mining, nature-inspired algorithms, and soft computing.

Danilo Pelusi is an Associate Professor in the Department of Communication Sciences, University of Teramo, where he received his PhD in Computational Astrophysics. He is an Editor of books for Springer and Elsevier, and an Associate Editor of *IEEE Transactions on Emerging Topics in Computational Intelligence and IEEE Access*, and was an Associate Editor of *International Journal of Machine Learning and Cybernetics*. He is a Guest Editor for Elsevier, Springer, and Inderscience journals and keynote speaker in several IEEE conferences; he is also an editorial board member of many journals. His research interests include fuzzy logic, neural networks, information theory, machine learning, and evolutionary algorithms.

Bighnaraj Naik is an Assistant Professor in the Department of Computer Application, Veer Surendra Sai University of Technology (formerly UCE Burla), Odisha, India. He has published more than 100 research articles in various peer-reviewed international journals, conferences, and book chapters. He has edited 10 books for publishers including Elsevier, Springer, and IGI Global. At present, he has more than 10 years of teaching experience in the field of computer science and information technology. He is a member of the Institute of Electrical and Electronics Engineers (IEEE) and his areas of interest include data science, data mining, machine learning, deep learning, computational intelligence (CI), and CI's applications in science and engineering. He has served as Guest Editor of various special issues of journals such as *Information Fusion* (Elsevier), *Neural Computing and Applications* (Springer), *Evolutionary Intelligence* (Springer), *International Journal of Computational Intelligence Studies* (Inderscience), and *International Journal of Swarm Intelligence* (Inderscience). He is an active reviewer of various journals from publishers including IEEE Transactions, Elsevier, Springer, and Inderscience. Currently, he is undertaking a major research project as Principal Investigator, which is funded by the Science and Engineering Research Board (SERB), Department of Science and Technology (DST), Government of India.

Manohar Mishra is an Associate Professor in the Department of Electronics and Electrical Engineering, Faculty of Engineering and Technology, Siksha 'O' Anusandhan Deemed to be University, Bhubaneswar, Odisha, India. He received his PhD in Electrical Engineering, MTech in Power Electronics and Drives, and BTech in Electrical Engineering in 2017, 2012, and 2008, respectively. He has published more than 50 research papers in various peer-reviewed international journals and conferences. He has edited three research books, and has served as a reviewer for journal publishers such as the Institute of Electrical and Electronics Engineers (IEEE), Springer, Elsevier, and Inderscience. At present, he has more than 10 years of teaching experience in the field of electrical engineering. He is a senior member of the IEEE, and is currently guiding four PhD scholars. His areas of interest include power system analysis, power system protection, signal processing, power quality, distribution generation systems, and micro-grids. He has served as Convener and Volume Editor of the International Conference on Innovation in Electrical Power Engineering, Communication and Computing Technology (IEPCCT-2019, IEPCCT-2021) and the International Conference on Green Technology for Smart City and Society (GTSCS-2020). Currently, he is serving as Guest Editor for different journals such as *International Journal of Power Electronics* (Inderscience), *International Journal of Innovative Computing and Application* (Inderscience), and *Neural Computing and Application* (Springer).

Khan Muhammad received his PhD in Digital Contents from Sejong University, South Korea in February 2019. He was an Assistant Professor in the Department of Software, Sejong University from March 2019 to February 2022. He is currently the director of Visual Analytics for Knowledge Laboratory (VIS2KNOW Lab) and an Assistant Professor (Tenure-Track) in the Department of Applied AI, School of Convergence, College of Computing and Informatics, Sungkyunkwan University, Seoul, South Korea. His research interests include intelligent video surveillance, medical image analysis, information security, video summarization, multimedia data analysis, computer vision, and smart cities. He has registered 10 patents and contributed more than 220 papers in peer-reviewed journals and conference proceedings in his research areas. He is an Associate Editor/Editorial Board Member for more than 15 journals. He was among the most highly cited researchers in 2021 and 2022, according to the Web of Science (Clarivate).

David Al-Dabass is Professor Emeritus and holds the personal Chair of Intelligent Systems in the Department of Computer Science, Nottingham Trent University, United Kingdom. He is a graduate of Imperial College, London University, holds a PhD, and has held postdoctoral and advanced research fellowships at the Control Systems Centre, University of Manchester Institute of Science and Technology (UMIST), Manchester University. He is a member of the Institute of Electrical and Electronics Engineers (IEEE) and a Fellow of the Institution of Engineering and Technology (IET), Institute of Mathematics and its Applications (IMA), and British Computer Society (BCS). He is the founder and Editor-in-Chief of *International Journal of Simulation: Systems, Science and Technology*, currently serves as President of the UK Simulation Society, and has previously served on the European Council for Modeling and Simulation as Director of the European Simulation Multiconference series. He has authored or coauthored more than 180 scientific publications on intelligent systems, modeling, and simulation, and has served as general Chair for more than 50 international conferences.

Foreword

Currently, medical systems development is moving away from tools that only enable a physician to examine organs and toward systems of a new type, involving the integration of the results of automatic image analysis with the results obtained by the doctor. Combining these results ensures more accurate and timely diagnosis of diseases, which significantly increases the effectiveness of treatment; this in turn reduces the cost of treatment and lessens the burden on medical personnel.

The development of clinical decision support systems (CDSS) is especially important in the case of cancer diagnostics. On the one hand, the worldwide increase in life expectancy has led to an increase in the relative and absolute incidence of cancer. On the other hand, it is axiomatic to say that the earlier the oncological changes are detected, the more effective the treatment is. For example, more than 700,000 people die from stomach cancer in the world every year. Timely treatment reduces the risk of death dramatically: complete cure for stage I gastric cancer occurs in 70% of cases, while for stage III, this figure is 26%. Thus, early and accuracy diagnosis provided by CDSS is essential today.

One of the most effective approaches to analyzing medical data in CDSS involves deep learning technologies, and in particular, convolutional neural networks. The field of medicine is undoubtedly one of the most promising in terms of effective implementation of deep learning technologies. This is due to the fact that this technology is focused on the analysis of data with significant variability and, of course, medical images are among the data in which variability is pronounced.

It should be noted that when developing CDSS, in addition to automatic analysis, an important consideration is increasing the diagnostic value of images offered to the doctor.

This book contains modern methods of image processing, realizing both automatic analysis and an increase in diagnostic value to cancer prediction and prognosis. Topics run the gamut of research in this field: cancer data analysis using machine learning approaches, with a focus on different real tasks and applications such as neural networks for lung cancer diagnosis, oral cancer diagnosis, and—a hot topic at the time of writing—the effect of COVID-19 on cancer patients.

Due to the accelerating progress in machine learning and its employment in cancer detection in all its complexity, this book's publication is not only timely but also much needed. In my opinion, this book is a valuable resource for graduate students, engineers, and researchers interested in understanding and investigating this important field of science.

Nataliia Obukhova

Head of Television and Video Engineering Department, St. Petersburg Electrotechnical University "LETI," St. Petersburg, Russia

Preface

In the last two decades, the progress of artificial intelligence has reshaped personalized clinical practices and medical applications in a successful way. The exploration of high-dimensional medical data and the development of intelligent methods have led to the implementation of many successful healthcare practices. In particular, the convergence of big data and advancements of computational intelligence (CI) techniques has paved the way to manage many cancer studies. At any stage of human life, cancer is commonly considered as worse than any other disease. Although scientists, researchers, and oncologists are aware that correct early stage diagnosis of such deadly diseases may increase a patient's chances of survival, accurate prognosis with proper treatment is essential for any type of cancer. Cancer is a multifaceted disease that affects tissues comprising thousands of body cells and results in a number of direct and indirect complications. CI techniques hold great promise to combat disorders at an early stage for better treatment. CI-based smart assistance to pathologists and physicians could offer an immense leap forward for risk prediction, disease classification, diagnosis, prognosis, and treatment.

Computational intelligence (CI) is progressively strengthening its impact on many healthcare problems and is believed to have a dominant influence on effective prognosis of cancer treatment. Oncologists have become more interested in the use of CI for high-level supervision as well as numerous problem-solving ideas to handle various complexities associated with cancer. CI research is witnessing unprecedented innovations for dealing with cancer issues from the histological level to the molecular level, along with a sustained impact on treatment and healthcare practices. Many advanced techniques such as deep neural networks, ensemble methods, fuzzy logic, and optimization methods are pushing the frontiers of healthcare and clinical practices. Although many of these techniques are being leveraged to achieve solutions in terms of biomedical big data, the way to clinical success (e.g., efficient prevention and care, improvised treatment methods, better clinical trial matching) will involve using the insights gained from previous experience to resolve future problems. The methodological effectiveness of CI is that it has the ability to characterize the interdisciplinary and miscellaneous nature of healthcare problems to a great extent. The prominent convergence of CI and big healthcare data could be helpful in interlacing the challenges of the multifarious landscape of cancer problems, which are solved according to high-level methodological standards.

The book will explore a few advanced intelligence-based solutions relating to the growing plethora and complexity of medical data in cancer research. A special collection of advanced CI methodologies with the usefulness of automated tools, procedures to handle complex problems, and the directions to control a disease's harm is focused in a detailed way. This book will also be helpful in addressing the numerous research challenges and new solutions for practitioners, oncologists, and researchers.

This book contains in-depth research contributions that are both fundamental and advanced; working from a methodological perspective, the chapters deal with various applications of CI methods for cancer problems. The current challenges and difficulties in handling complex oncology problems are some of the major highlights of the book. The motive and outcome of the book may help in improving the solutions and rationality among different CI methods and could augment the significance and utilization of various application domains for end-users. The main objective is to integrate the advances of CI and cancer research for effectual analysis and usage in attempts to solve some important problems related to various emerging issues. With the swift advances of precision medicine and intelligent applications in oncology research, CI presents several opportunities to combat the challenges in this area, facilitating the growth of independently customized preventive and therapeutic interventions. This book is aimed at oncologists, scientists, specific domain healthcare researchers, health professionals, professors, and research scholars. It will reveal diverse dimensions of CI applications with detailed illustrations of their benefits in various cancer studies. This volume comprises of three parts and 17 chapters, and is organized as follows.

Part 1: Introduction to computational intelligence approaches

In Chapter 1, a brief yet lucid explanation on the clinical perspective of adoption of computational intelligence in cancer diagnosis is reported. Federica Vernuccio et al. have aimed to provide an overview of the most significant breakthroughs in radiomics and AI for cancer diagnosis from a clinical perspective. In addition, the main challenges that prevent the routine clinical adoption of computational intelligence in cancer diagnosis are emphasized.

In Chapter 2, the authors give a detailed description of gene expression in digital as well as microarray form, and explain its use in disease diagnosis. A broad literature study is carried out that relates to the feature prediction of high-dimension microarray cancer data using various deep learning techniques. The chapter concludes with a critical analysis, identifying several challenges and future research scopes.

Chapter 3 discusses the importance of early prediction of cancer to provide effective treatment for patients diagnosed with ovarian cancer. Acharya et al. propose a deep convolutional neural network-based architecture for the diagnosis of ovarian carcinoma on a dataset of 349 patients obtained from Kaggle. Data cleaning and a combination of feature extraction techniques are applied to this dataset. The effectiveness of the proposed model is demonstrated by comparing it with seven conventional machine learning classifiers: random forest, multilayer perceptron, decision tree, linear discriminant analysis, adaptive boosting (AdaBoost), gradient boosting, and extreme gradient boosting (XGBoost).

Chapter 4 provides a simple and in-depth analysis of intelligent diagnosis methods. According to the classification of several common medical images, different cancer detection and diagnosis methods are considered. This chapter also introduces the widespread applications of convolutional neural networks (CNNs) for medical image processing in cancer diagnostic analysis.

In Chapter 5, Liu et al. summarize the application of neural networks in terms of the diagnosis of lung cancer, which includes the detection of lung nodules, the segmentation of lung tumors, and the classification of lung nodules. The chapter also discusses the technical route for lung cancer diagnosis in detail.

Chapter 6 focuses on the importance of the rapid advancement of probabilistic methods and machine learning techniques for the success of computer-aided diagnosis in thyroid cancer. Temurtas et al. revise several machine learning and deep learning techniques for early detection of thyroid cancer. The principles, types, advantages, and disadvantages of these techniques are explained briefly. This chapter also presents a comparative study of the most effective diagnostic methods using information from databases on thyroid cancer.

Part 2: Prediction of cancer susceptibility

Chapter 7 explores various machine learning classifiers for lung cancer classification. The authors discuss the diagnosis procedures of lung cancer in two steps: lung cancer nodule detection and classification. In nodule detection, from a given computed tomography (CT) scan, the nodules and nonnodules are identified. The next step is to classify the detected nodules as cancerous or noncancerous. A majority voting scheme is used to classify nodules. The chapter also offers an in-depth analysis of the performance of different machine learning algorithms.

Chapter 8 provides a short tutorial on the basics of deep learning as applied to medical images, and then describes recent advances in deep learning models for oral cancer diagnosis. Haq et al. also provide a brief discussion of the overall results achieved by these models on various datasets. They discuss the challenges and limitations of approaches in the current literature, and suggest some future research directions for open issues.

In Chapter 9, Kumar et al. propose a deep neural network trained by a genetic algorithm (DNN-GA) to identify colon cancer patients. This method is used in experiments on a microarray dataset that consists of symptoms of cancer patients. The performance of the proposed model is compared with performances of other machine learning approaches such as K-nearest neighbor, stochastic gradient descent, decision tree, random forest, multilayer perceptron, and deep neural network; the model is found to be superior for efficient identification of colon cancer.

Chapter 10 discusses some complications, issues, and adverse effects of COVID-19 on cancer patients. The COVID-19 pandemic has caused particular concern for the oncology community, as the disease has negative consequences for cancer patients because of their immune-suppressed status. In this chapter, Swapnarekha and Nayak present a systematic analysis of the impact of COVID-19 infection on various types of cancers. The study then focuses on case studies representing scenarios of cancer patients in various countries. Finally, the study delineates the major challenges and future directions for the efficient management of the cancer community during the COVID-19 pandemic.

Chapter 11 provides effective analysis of empirical wavelet transform-based fast deep convolutional neural networks for detection and classification of melanoma. This is a type of skin cancer that develops from melanocytes, which are responsible for skin color. The severity of melanoma cancer is defined on the basis of different stages, which depend upon the depth of penetration; early detection of melanoma at its prodromal stage is crucial to stop its advancement. In this chapter, a novel variant of deep convolutional neural networks (DCNNs) is developed, called a fast deep convolutional neural network (fast-DCNN), to perform binary classification of normal nevus and melanoma by using dermoscopic images of a PH^2 dataset.

Part 3: Advance computational intelligence paradigms

Chapter 12 aims to develop appropriate deep learning models that detect breast cancer from breast histopathology images of the BreakHis dataset. To build such deep learning models, seven popular convolutional neural networks—DenseNet, ResNetV2, Inception, InceptionResNetV2, VGG16, VGG19, and Xception—are used. These models are trained on the magnification factor-based subdatasets from the BreakHis dataset. The experimental results reveal the efficiency in prognosis of breast cancer using deep learning networks.

Chapter 13 discusses the early detection of common osteosarcoma cell malignancy (bone cancer), which is crucial for patient health and helpful to a physician, as appropriate medication benefits the future quality of life and survival condition of the patient. In this chapter, Oram et al. propose a model based on a light-boost gradient boosting machine (LightGBM) for the identification of osteosarcoma cell types from histological features. Experimentation and simulation results show that the proposed model is faster and lighter because of its use of exclusive feature bundling and gradient-based one-side sampling; it also maintains good accuracy compared to other machine learning approaches.

Chapter 14 introduces deep learning-based computer-aided cervical cancer diagnosis in digital histopathology images. Cervical cancer has high mortality and morbidity with a higher risk of death for women. Early diagnosis using a computer-assisted system is vital to increase the survival rate. Jeyaraj et al. propose a deep learning method to detect global features by utilizing patch features extracted with the feature fusion method. The proposed feature encoding is obtained at various times (15, 30, 45, and 60 s) using the acetic acid test of captured colposcopy images. Each patch-level information is obtained by feature encoding using the transfer learning approach. This proposed transfer learning fusion method is trained and validated by HERLEV with 560 patient images.

Chapter 15 focuses on the novel insights of deep learning techniques for improved diagnosis and evaluation of hepatocellular carcinoma (HCC). HCC stands second in the list of cancer-associated deaths and its increasing rate of incidence is a growing concern globally. Recent revolutions in computational intelligence tools have shown promising reliance of physicians on computer-assisted analysis of various test data as compared to human experts. The authors address relevant issues of precise diagnosis and prognosis, and measures taken by physicians on HCC biomarkers and gene expression studies. Furthermore, the roles of convolutional neural networks and artificial neural networks are discussed in terms of dealing with histopathological imaging, radiomics, and multiomics.

Chapter 16 conducts a systematic literature study to analyze the efficacy of different machine learning models on detecting clinically relevant prostate cancer disease from cytological features based on a digital scan. To handle the variability in real-world data, Reddy et al. explore the gradient boosting (GB) algorithm for prostate cancer. The chapter aims to analyze and differentiate between clinically significant and insignificant prostate cancer assessment through hyperparameter optimization for the GB algorithm by using the particle swarm optimization technique. With proper hyperparameter tuning, the proposed method is found to be suitable for detecting prostate cancer.

Chapter 17 proposes a computational approach in dealing with protein interactions. The application of an optimized multi-functional coclustering approach (MR-CoCVFO) for mining drug target modules on a human dataset is discussed in this chapter. This is designed as a multifunctional approach with optimal functional score selection using the Venus flytrap optimization (VFO) algorithm. The results are tested for protein complex coverage and compared with prior approaches. The biological significance of the results is analyzed and presented with the mapping of drug-target characteristics and the targeted cancerous protein. From the observed mapping, cancer drug target modules are suggested for the development of targeted therapies.

We are grateful to all the authors and reviewers for their contributions and dedicated efforts to ensure the publication of this book. In particular, we thank the editorial team of Elsevier for their valuable technical support and dedication. We hope that the work reported in this volume will motivate further research and development efforts in the evaluation of different types of cancer and its related fields.

Janmenjoy Nayak
Danilo Pelusi
Bighnaraj Naik
Manohar Mishra
Khan Muhammad
David Al-Dabass

PART 1

Introduction to computational intelligence approaches

CHAPTER

1

The roadmap to the adoption of computational intelligence in cancer diagnosis: The clinical-radiological perspective

Federica Vernuccio[a], *Roberto Cannella*[b,c], *Roberto Lagalla*[b], *and Massimo Midiri*[b]

[a]Institute of Radiology, Department of Medicine-DIMED, University of Padova, Padova, Italy [b]Department of Biomedicine, Neuroscience and Advanced Diagnostics (BIND), University Hospital of Palermo, Palermo, Italy [c]Department of Health Promotion, Mother and Child Care, Internal Medicine and Medical Specialties (PROMISE), University of Palermo, Palermo, Italy

Introduction

According to global cancer statistics, an estimated 19.3 million new cancer cases and almost 10.0 million cancer deaths occurred in 2020, and 28.4 million cases in 2040, a 47% rise from 2020 (Sung et al., 2021). In 2021, almost one million and nine hundred thousand new cancer cases are estimated to occur in the United States (Siegel et al., 2021). The leading cancer type is different in men and women, being prostate in men and breast in women (Siegel et al., 2021). However, lung and colorectal tumor are the second most common tumor in both genders (Siegel et al., 2021). It is interesting to notice that in the time frame between 1995 and 2017, there has been a drop of cancer incidence rate of about 2% in women and 20% in men, mainly related to differences in lung cancer trends (Siegel et al., 2021). An estimated 608,570 cancer deaths are projected to occur in the same year, corresponding to more than 1600 deaths per day with lung cancer being the main cause of death (22%) in both genders (Siegel et al., 2021).

https://doi.org/10.1016/B978-0-323-85240-1.00020-1

Therefore, the burden of cancer worldwide is expected to increase over time, and an early diagnosis of the tumor may help in improving the prognosis by establishing a prompt treatment. In this setting, radiologists are of utmost importance particularly in high-risk patients by performing screening programs with imaging diagnostic techniques. Screening programs are performed to find disease before symptoms onset with the aim of detecting a high-prevalent disease at its earliest and most treatable stage, thus reducing the number of deaths from the given disease. The relevance of imaging screening programs has been known for years in case of hepatocellular carcinoma in cirrhotic patients. European and American guidelines recommend ultrasound surveillance and CT or MRI for characterization of liver lesions in patients at risk for hepatocellular carcinoma, including cirrhotic patients, patients with hepatitis B infection, and patients with prior hepatocellular carcinoma (Elsayes et al., 2018; European Association For The Study Of The Liver & European Organisation For Research And Treatment Of Cancer, 2012). The adoption of this screening program for hepatocellular carcinoma has proven to be cost-effective and leads to better prognosis in different studies (Cadier et al., 2017; Cucchetti et al., 2012; Sherman, 2014; Yuen & Lai, 2003). Imaging screening programs have been also implemented for other tumors including breast cancer and lung cancer. Imaging screening program for lung cancer has been recently implemented in some countries. Indeed, in the USA, Canada, and in some European countries, screening programs for lung cancer with low-dose CT have been implemented because it may lead to a reduction in lung cancer mortality by 20% based on the results of the National Lung Screening Trial (https://www.cdc.gov/cancer/ncccp/pdf/lungcancerscreeningprograms.pdf, n.d.; National Lung Screening Trial Research Team et al., 2011).

The increasing breakthroughs in biomedical technologies for cancer diagnosis expose clinicians to an exponential amount of complex data (Claussnitzer et al., 2020; Mantini et al., 2021; Yachida et al., 2019). The integration of clinical, laboratory, genomic, and radiological data is nowadays the cornerstone of personalized precision medicine. Along with advances in the biomedical field, in the last decade, important breakthroughs took place in the clinical applications of computational intelligence in oncology, and these breakthroughs are likely very to be helpful for management and interpretation of the huge amount of data obtained thorough novel biomedical technologies, thus facilitating and improving patients' management (Momeni et al., 2020; Schork, 2019; Vernuccio et al., 2020).

The main breakthroughs in biomedical technologies for cancer diagnosis are related to genomic profiling and radiomics analysis, with their integration known as radiogenomics (Lo Gullo et al., 2020). Radiogenomics of tumors is a promising new strategy for obtaining cancer signatures and for predicting clinical outcomes (Jamshidi et al., 2016; Katabathina et al., 2020; Lv et al., 2018). However, clinical studies involving radiogenomics analysis include usually few patients and heterogeneous data sources. The application of artificial intelligence (AI) in cancers care is likely to improve the accuracy and speed of cancers diagnosis, but many limitations and challenges impair its clinical use (Schork, 2019; Vernuccio et al., 2020).

This chapter is aimed at providing an overview of the most significant breakthroughs in radiomics and AI for cancer diagnosis with a clinical perspective and to briefly resume the main challenges that still prevent the routine clinical adoption of computational intelligence in cancer diagnosis.

Radiomics and artificial intelligence for cancer diagnosis and treatment

Accurate diagnosis of cancers remains a crucial clinical demand in patients with focal lesions detected on imaging studies. This is especially relevant in the setting of screening populations or in incidentally detected focal lesions, and several approaches such as adoption of dual-energy CT have been recently investigated for this purpose (Al-Najami et al., 2017; Dai et al., 2013; Meyer et al., 2019; Patel et al., 2018). Criteria based on qualitative imaging features are still being investigated to yield high accuracy in tumor characterization (Pourvaziri et al., 2019; Vernuccio, Cannella, et al., 2019); however, qualitative imaging features are frequently affected by readers' experience and may be prone to inter-reader variability. Despite several efforts in standardization of imaging acquisitions and reporting systems, such as the widespread structured reporting, the application in clinical practice remains limited (An et al., 2019; Davenport et al., 2019; Dimarco et al., 2020; Eiber et al., 2018).

Radiomics and AI models have been constructed and tested to overcome some of these problems in cancers diagnosis, staging, and assessment of treatment response. In a recent large study, a deep learning-based model has been proposed to differentiate primary brain cancer—glioblastoma with very poor prognosis and limited therapy options—from single cerebral metastasis, achieving an excellent performance for the differential diagnosis in internal and external cohort of 589 and 143 patients, respectively (Shin et al., 2021). In patients undergoing low-dose CT for lung cancer screening, computational intelligence algorithms have provided high accuracy of malignancy risk estimation in detected pulmonary nodules, improving the diagnosis of experienced radiologists, and helping in to select the optimal management in lung cancer screening (Venkadesh et al., 2021). In breast cancer screening, the application of AI reduced the workload of breast cancer screening program without affecting the sensitivity for cancer detection, reducing the number of false-positive reassessments (Raya-Povedano et al., 2021). Cirrhotic patients are another population at risk of developing cancer, especially hepatocellular carcinoma (HCC). Surveillance is usually performed with abdominal ultrasound, while contrast-enhanced CT or MRI is used for characterization of focal liver lesions. However, several preneoplastic lesions (i.e., regenerative or dysplastic nodules) or benign lesions can be diagnosed in the setting of chronic liver disease and may mimic HCC (Vernuccio et al., 2021). The main AI applications in this context include lesion segmentation and differential diagnosis between HCC and other non-HCC benign and malignant lesions (Fig. 1.1) (Castaldo et al., 2021; Wels et al., 2019). In prostate cancer imaging, AI algorithms have been constructed and tested for improving the detection and characterization of prostate lesions, but also for improving the quality of imaging acquisition with noise reduction and for automatic segmentation of the prostate (Gassenmaier et al., 2021; Giambelluca et al., 2021).

Beside cancers detection and diagnosis, it should be noted patients' prognosis is also strictly related to histopathological tumor characteristics. In most patients, reliable histopathological information related to tumor biological aggressiveness can be obtained only with dedicated biopsy or after surgical resection. Nevertheless, biopsy results can be affected by the tumor internal heterogeneity and may turn out to be false-negative (Vernuccio, Rosenberg, et al., 2019), while not all patients are candidates to surgical resection due to the advanced cancers stage. Therefore, computational intelligent models have been proposed

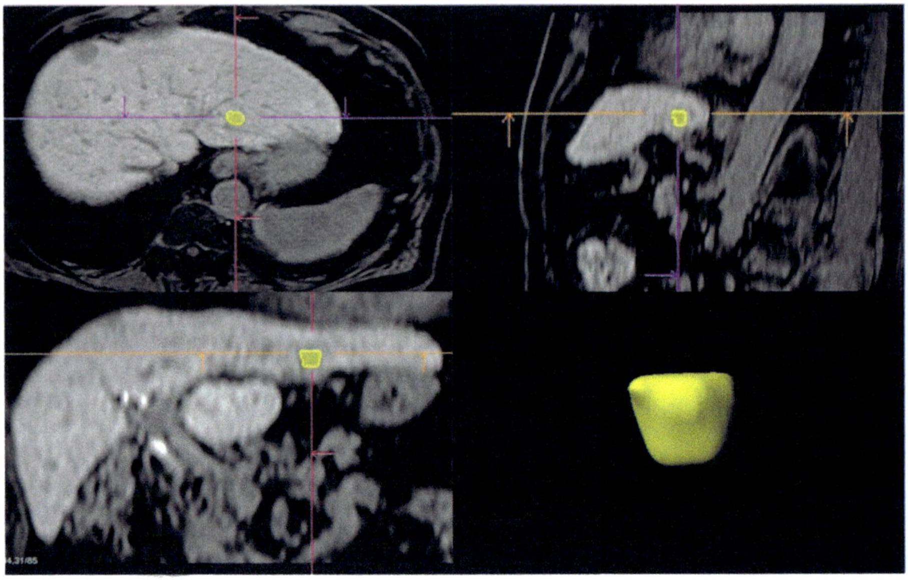

FIG. 1.1 76-year-old man with history of hepatitis C-related cirrhosis, under virologic suppression with Tenofovir. Hepatic lesion segmentation on hepatobiliary phase MRI with 3D volume rendering reconstruction performed using a prototype research software Radiomics, version 1.0.9 (Siemens Healthineers, Forchheim, Germany).

to predict histopathological characteristic based on preoperative imaging analysis, including tumor grade in glioblastoma and other cerebral neoplasms, tumor histology on non-small-cell lung cancer, and microvascular invasion in HCC (Chaunzwa et al., 2021; Marcus et al., 2020; Wilson et al., 2020).

The integration of genomic data into computational intelligent system has allowed to explore the potential to predict cancers underlying mutational status, which may be used to guide personalized treatments with target therapies (Cannella et al., 2020). In a multicentric study, a convolutional neural network was trained to predict tumor genomic subtype in patients with low-grade glioma based on the MRI of the tumor with fair performance (Buda et al., 2020). A systematic review evaluating radiogenomics correlation in patients with glioma concluded that radiomics could be a valid alternative to conventional biopsy (Bhandari et al., 2021).

AI is also emerging as a useful tool for predicting the outcome in patients with cancers after treatments, indicating the potential of AI for applying personalized medical care. Several treatment strategies are nowadays available in cancers patients, and the choice of optimal treatment is crucial to prolong patients' survival. In patients with HCC, a study evaluated the performance of an automated machine-learning algorithm to predict response after transarterial chemoembolization based on quantitative features on pretreatment contrast-enhanced CT, and achieved an accuracy of 72% for response prediction (Fig. 1.2) (Morshid et al., 2019). A reliable prediction of response using AI could be particularly helpful especially in patients treated with novel immunotherapy drugs, as these treatments are often characterized by delayed response on conventional imaging studies. In this setting, initial studies have explored the role of computational intelligence algorithms in patients with advanced lung cancers undergoing immunotherapy (Tunali et al., 2019).

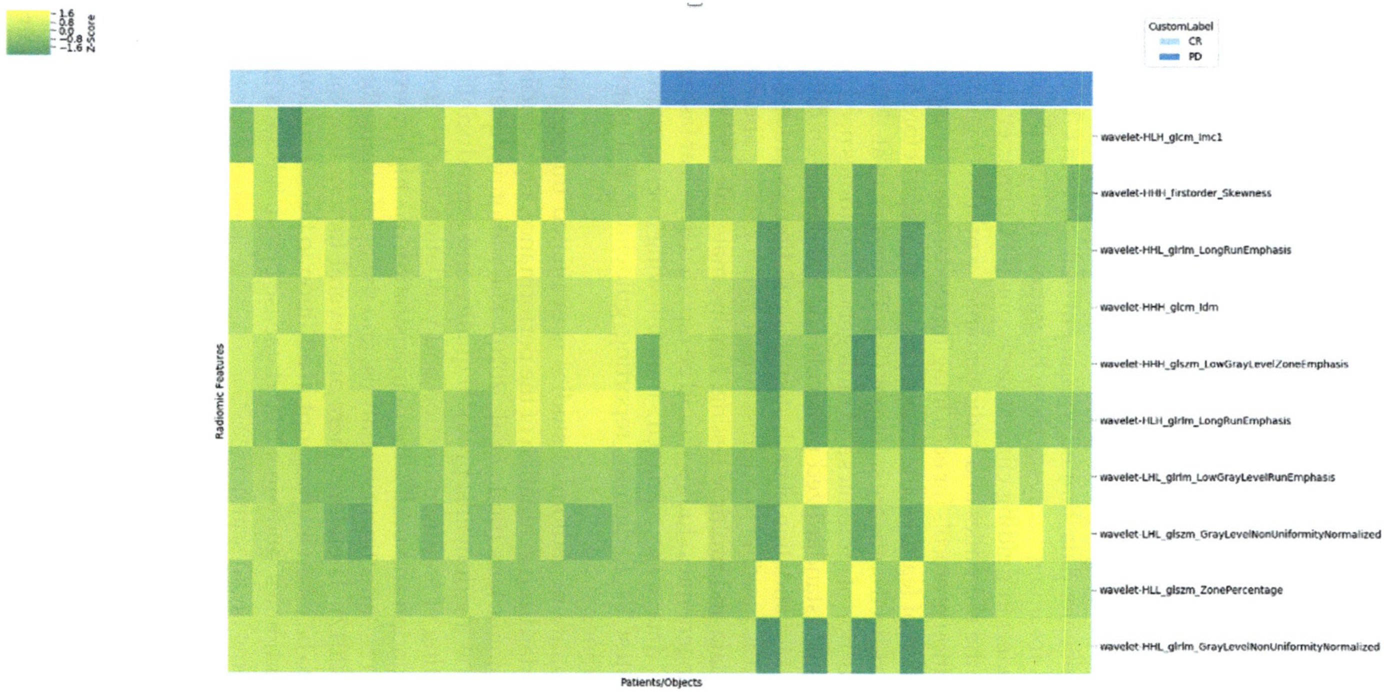

FIG. 1.2 Heatmap with selected wavelet features using the prototype research software Radiomics, version 1.0.9 (Siemens Healthineers, Forchheim, Germany) in a simulated study comparing progressive disease (PD) and complete response (CR) after locoregional therapy for hepatocellular carcinoma.

Challenges and perspectives

Radiomics and radiogenomics diagnostic and predictive models have the potential to assist oncologic multidisciplinary teams in the different steps of patient management, with a more efficient workflow from diagnosis to treatment. Before this can happen though, the significant gap between research study setting and application in clinical practice must be filled-in with most radiomics studies being of insufficient quality yet (Cannella et al., 2021; Park et al., 2020a, 2020b). Indeed, the clinical applicability of radiomics and radiogenomics models requires multicenter prospective studies that test their repeatability and reproducibility.

Many radiomics and radiogenomics research studies are exploratory or proof-of-concept in nature using single-center data, limited equipment (few types of CT/MRI scanners) as well as different softwares for image analysis; these limitations hamper the effective development of computational models in guiding clinical decisions (O'Connor et al., 2017).

The complexity of the radiomics and radiogenomics workflow mandates the need to standardize image acquisition protocols for each imaging modality and computational methods, including the computation of features, the image-processing steps before feature extraction, and the adoption of standardized software solutions (van Timmeren et al., 2020). On the other hand, the design and construction of radiomics and radiogenomics studies need prospective accrual of patients with multicenter standardized protocols and validation on different cohorts, with a large number of data for validation of each step of the radiomics process and the results. Only with large numbers, it is possible to address tumor heterogeneity and avoid statistical issues related to overfitted data and multiple testing. Finally, materials and methods and the results of radiomics and radiogenomics studies should be reported in a clear fashion that allows to have a clear idea of how data were managed, thus allowing for reproducibility (https://ibsi.readthedocs.io/en/latest/04_Radiomics_reporting_guidelines_and_nomenclature.html#reporting-guidelines, n.d.).

Nowadays, the quality of radiomics studies is evaluated through the radiomics quality score, which is a standardized evaluation of the performance, reproducibility, and/or clinical utility of radiomics biomarkers (Buda et al., 2020). The radiomics quality score includes the evaluation of the following domains: protocol quality and reproducibility in image and segmentation, the reporting of feature reduction and validation, the performance index, the biological/clinical validation and utility, the demonstration of a higher level of evidence with a prospective study or cost-effectiveness analysis, and the open availability of source code and data (Lambin et al., 2017). Therefore, there are several factors that increase the quality of radiomics and need to be taken into account in addition to those previously mentioned, such as registration of a radiomics study in a trial database before it begins and the calibration of features with phantoms on all scanners.

Conclusion

The implementation of computational analysis in clinical practice in the oncologic setting is expected to facilitate patient management, and this applies to different cancers and to the different steps of oncology. However, clinical implementation of computational analysis will be practically feasible only by formulating a standardized workflow and internationally agreed methods in large-scale multicenter prospective studies through international concerted efforts.

References

Al-Najami, I., Lahaye, M. J., Beets-Tan, R. G. H., & Baatrup, G. (2017). Dual-energy CT can detect malignant lymph nodes in rectal cancer. *European Journal of Radiology*, *90*, 81–88. https://doi.org/10.1016/j.ejrad.2017.02.005.

An, J. Y., Unsdorfer, K. M. L., & Weinreb, J. C. (2019). BI-RADS, C-RADS, CAD-RADS, LI-RADS, Lung-RADS, NI-RADS, O-RADS, PI-RADS, TI-RADS: Reporting and data systems. *Radiographics*, *39*, 1435–1436.

Bhandari, A. P., Liong, R., Koppen, J., Murthy, S. V., & Lasocki, A. (2021). Noninvasive determination of IDH and 1p19q status of lower-grade gliomas using MRI radiomics: A systematic review. *AJNR. American Journal of Neuroradiology*, *42*, 94–101.

Buda, M., AlBadawy, E. A., Saha, A., & Mazurowski, M. A. (2020). Deep radiogenomics of lower-grade gliomas: Convolutional neural networks predict tumor genomic subtypes using MR images. *Radiology. Artificial Intelligence*, *2*, e180050.

Cadier, B., Bulsei, J., Nahon, P., Seror, O., Laurent, A., Rosa, I., Layese, R., Costentin, C., Cagnot, C., Durand-Zaleski, I., Chevreul, K., & ANRS CO12 CirVir and CHANGH groups. (2017). Early detection and curative treatment of hepatocellular carcinoma: A cost-effectiveness analysis in France and in the United States. *Hepatology*, *65*(4), 1237–1248. https://doi.org/10.1002/hep.28961.

Cannella, R., La Grutta, L., Midiri, M., & Bartolotta, T. V. (2020). New advances in radiomics of gastrointestinal stromal tumors. *World Journal of Gastroenterology*, *26*, 4729–4738.

Cannella, R., Sartoris, R., Grégory, J., et al. (2021). Quantitative magnetic resonance imaging for focal liver lesions: Bridging the gap between research and clinical practice. *The British Journal of Radiology*, *94*(1122), 20210220. https://doi.org/10.1259/bjr.20210220.

Castaldo, A., De Lucia, D. R., Pontillo, G., et al. (2021). State of the art in artificial intelligence and radiomics in hepatocellular carcinoma. *Diagnostics (Basel)*, *11*, 1194.

Chaunzwa, T. L., Hosny, A., Xu, Y., et al. (2021). Deep learning classification of lung cancer histology using CT images. *Scientific Reports*, *11*, 5471.

Claussnitzer, M., Cho, J. H., Collins, R., et al. (2020). A brief history of human disease genetics. *Nature*, *577*(7789), 179–189. https://doi.org/10.1038/s41586-019-1879-7.

Cucchetti, A., Trevisani, F., Cescon, M., Ercolani, G., Farinati, F., Poggio, P. D., Rapaccini, G., Nolfo, M. A. D., Benvegnù, L., Zoli, M., Borzio, F., Giannini, E. G., Caturelli, E., Chiaramonte, M., Pinna, A. D., & Italian Liver Cancer (ITA.LI.CA) Group. (2012). Cost-effectiveness of semi-annual surveillance for hepatocellular carcinoma in cirrhotic patients of the Italian Liver Cancer population. *Journal of Hepatology*, *56*(5), 1089–1096. https://doi.org/10.1016/j.jhep.2011.11.022.

Dai, X., Schlemmer, H. P., Schmidt, B., Höh, K., Xu, K., Ganten, T. M., & Ganten, M. K. (2013). Quantitative therapy response assessment by volumetric iodine-uptake measurement: Initial experience in patients with advanced hepatocellular carcinoma treated with sorafenib. *European Journal of Radiology*, *82*(2), 327–334. https://doi.org/10.1016/j.ejrad.2012.11.013.

Davenport, M. S., Hu, E. M., Zhang, A., SAR Disease-Focused Panel on RCC, et al. (2019). Standardized report template for indeterminate renal masses at CT and MRI: A collaborative product of the SAR Disease-Focused Panel on Renal Cell Carcinoma. *Abdominal Radiology*, *44*(4), 1423–1429. https://doi.org/10.1007/s00261-018-1851-2.

Dimarco, M., Cannella, R., Pellegrino, S., et al. (2020). Impact of structured report on the quality of preoperative CT staging of pancreatic ductal adenocarcinoma: Assessment of intra- and inter-reader variability. *Abdominal Radiology*, *45*, 437–448.

Eiber, M., Herrmann, K., Calais, J., et al. (2018). Prostate cancer molecular imaging standardized evaluation (PROMISE): Proposed miTNM classification for the interpretation of PSMA-ligand PET/CT. *Journal of Nuclear Medicine*, *59*(3), 469–478. https://doi.org/10.2967/jnumed.117.198119. Erratum in: J Nucl Med. 2018;59(6):992.

Elsayes, K. M., Kielar, A. Z., Elmohr, M. M., Chernyak, V., Masch, W. R., Furlan, A., Marks, R. M., Cruite, I., Fowler, K. J., Tang, A., Bashir, M. R., Hecht, E. M., Kamaya, A., Jambhekar, K., Kamath, A., Arora, S., Bijan, B., Ash, R., Kassam, Z., … Sirlin, C. B. (2018). White paper of the Society of Abdominal Radiology hepatocellular carcinoma diagnosis disease-focused panel on LI-RADS v2018 for CT and MRI. *Abdominal Radiology, 43*(10), 2625–2642. https://doi.org/10.1007/s00261-018-1744-4.

European Association For The Study Of The Liver, & European Organisation For Research And Treatment Of Cancer. (2012). EASL-EORTC clinical practice guidelines: Management of hepatocellular carcinoma. *Journal of Hepatology, 56*(4), 908–943. https://doi.org/10.1016/j.jhep.2011.12.001. Erratum in: J Hepatol. 2012 Jun; 56(6): 1430.

Gassenmaier, S., Afat, S., Nickel, D., et al. (2021). Deep learning-accelerated T2-weighted imaging of the prostate: Reduction of acquisition time and improvement of image quality. *European Journal of Radiology, 137*, 109600.

Giambelluca, D., Cannella, R., Vernuccio, F., et al. (2021). PI-RADS 3 lesions: Role of prostate MRI texture analysis in the identification of prostate cancer. *Current Problems in Diagnostic Radiology, 50*, 175–185.

https://ibsi.readthedocs.io/en/latest/04_Radiomics_reporting_guidelines_and_nomenclature.html#reporting-guidelines (Accessed on 07 August 2021).

https://www.cdc.gov/cancer/ncccp/pdf/lungcancerscreeningprograms.pdf (Accessed 03 September 2021).

Jamshidi, N., Jonasch, E., Zapala, M., et al. (2016). The radiogenomic risk score stratifies outcomes in a renal cell cancer phase 2 clinical trial. *European Radiology, 26*(8), 2798–2807. https://doi.org/10.1007/s00330-015-4082-8.

Katabathina, V. S., Marji, H., Khanna, L., et al. (2020). Decoding genes: Current update on radiogenomics of select abdominal malignancies. *Radiographics, 40*(6), 1600–1626. https://doi.org/10.1148/rg.2020200042.

Lambin, P., Leijenaar, R. T. H., Deist, T. M., et al. (2017). Radiomics: The bridge between medical imaging and personalized medicine. *Nature Reviews. Clinical Oncology, 14*(12), 749–762. https://doi.org/10.1038/nrclinonc.2017.141.

Lo Gullo, R., Daimiel, I., Morris, E. A., & Pinker, K. (2020). Combining molecular and imaging metrics in cancer: Radiogenomics. *Insights Into Imaging, 11*(1), 1. https://doi.org/10.1186/s13244-019-0795-6.

Lv, J., Zhang, H., Ma, J., et al. (2018). Comparison of CT radiogenomic and clinical characteristics between EGFR and KRAS mutations in lung adenocarcinomas. *Clinical Radiology, 73*(6), 590.e1–590.e8. https://doi.org/10.1016/j.crad.2018.01.009.

Mantini, G., Pham, T. V., Piersma, S. R., & Jimenez, C. R. (2021). Computational analysis of phosphoproteomics data in multi-omics cancer studies. *Proteomics, 21*(3–4), e1900312. https://doi.org/10.1002/pmic.201900312.

Marcus, A. P., Marcus, H. J., Camp, S. J., Nandi, D., Kitchen, N., & Thorne, L. (2020). Improved prediction of surgical resectability in patients with glioblastoma using an artificial neural network. *Scientific Reports, 10*, 5143.

Meyer, M., Nelson, R. C., Vernuccio, F., et al. (2019). Comparison of iodine quantification and conventional attenuation measurements for differentiating small, truly enhancing renal masses from high-attenuation nonenhancing renal lesions with dual-energy CT. *AJR. American Journal of Roentgenology, 213*(1), W26–W37. https://doi.org/10.2214/AJR.18.20547.

Momeni, Z., Hassanzadeh, E., Saniee Abadeh, M., & Bellazzi, R. (2020). A survey on single and multiomics data mining methods in cancer data classification. *Journal of Biomedical Informatics, 107*, 103466. https://doi.org/10.1016/j.jbi.2020.103466.

Morshid, A., Elsayes, K. M., Khalaf, A. M., et al. (2019). A machine learning model to predict hepatocellular carcinoma response to transcatheter arterial chemoembolization. *Radiology. Artificial Intelligence, 1*, e180021.

National Lung Screening Trial Research Team, Aberle, D. R., Adams, A. M., Berg, C. D., Black, W. C., Clapp, J. D., Fagerstrom, R. M., Gareen, I. F., Gatsonis, C., Marcus, P. M., & Sicks, J. D. (2011). Reduced lung-cancer mortality with low-dose computed tomographic screening. *The New England Journal of Medicine, 365*(5), 395–409.

O'Connor, J. P., Aboagye, E. O., Adams, J. E., et al. (2017). Imaging biomarker roadmap for cancer studies. *Nature Reviews. Clinical Oncology, 14*(3), 169–186. https://doi.org/10.1038/nrclinonc.2016.162.

Park, J. E., Kim, D., Kim, H. S., et al. (2020a). Quality of science and reporting of radiomics in oncologic studies: Room for improvement according to radiomics quality score and TRIPOD statement. *European Radiology, 30*(1), 523–536. https://doi.org/10.1007/s00330-019-06360-z.

Park, J. E., Kim, H. S., Kim, D., et al. (2020b). A systematic review reporting quality of radiomics research in neuro-oncology: Toward clinical utility and quality improvement using high-dimensional imaging features. *BMC Cancer, 20*(1), 29. https://doi.org/10.1186/s12885-019-6504-5.

Patel, B. N., Rosenberg, M., Vernuccio, F., et al. (2018). Characterization of small incidental indeterminate Hypoattenuating hepatic lesions: Added value of single-phase contrast-enhanced dual-energy CT material attenuation analysis. *AJR. American Journal of Roentgenology, 211*(3), 571–579. https://doi.org/10.2214/AJR.17.19170.

Pourvaziri, A., Parakh, A., Mojtahed, A., Kambadakone, A., & Sahani, D. V. (2019). Diagnostic performance of dual-energy CT and subtraction CT for renal lesion detection and characterization. *European Radiology, 29*(12), 6559–6570. https://doi.org/10.1007/s00330-019-06224-6. Erratum in: Eur Radiol. 2019 Jul 11.

Raya-Povedano, J. L., Romero-Martín, S., Elías-Cabot, E., et al. (2021). AI-based strategies to reduce workload in breast cancer screening with mammography and tomosynthesis: A retrospective evaluation. *Radiology, 300*, 57–65.

Schork, N. J. (2019). Artificial intelligence and personalized medicine. *Cancer Treatment and Research, 178*, 265–283. https://doi.org/10.1007/978-3-030-16391-4_11.

Sherman, M. (2014). Surveillance for hepatocellular carcinoma. *Best Practice & Research. Clinical Gastroenterology, 28*(5), 783–793. https://doi.org/10.1016/j.bpg.2014.08.008.

Shin, I., Kim, H., Ahn, S. S., et al. (2021). Development and validation of a deep learning-based model to distinguish glioblastoma from solitary brain metastasis using conventional MR images. *AJNR. American Journal of Neuroradiology, 42*, 838–844.

Siegel, R. L., Miller, K. D., Fuchs, H. E., & Jemal, A. (2021). Cancer statistics, 2021. *CA: A Cancer Journal for Clinicians, 71* (1), 7–33. https://doi.org/10.3322/caac.21654. Epub 2021 Jan 12. Erratum in: CA Cancer J Clin. 2021; 71(4): 359 33433946.

Sung, H., Ferlay, J., Siegel, R. L., Laversanne, M., Soerjomataram, I., Jemal, A., & Bray, F. (2021). Global Cancer Statistics 2020: GLOBOCAN estimates of incidence and mortality worldwide for 36 cancers in 185 countries. *CA: A Cancer Journal for Clinicians, 71*(3), 209–249. https://doi.org/10.3322/caac.21660.

Tunali, I., Gray, J. E., Qi, J., Abdalah, M., et al. (2019). Novel clinical and radiomic predictors of rapid disease progression phenotypes among lung cancer patients treated with immunotherapy: An early report. *Lung Cancer, 129*, 75–79.

van Timmeren, J. E., Cester, D., Tanadini-Lang, S., Alkadhi, H., & Baessler, B. (2020). Radiomics in medical imaging-"how-to" guide and critical reflection. *Insights Into Imaging, 11*(1), 91. https://doi.org/10.1186/s13244-020-00887-2.

Venkadesh, K. V., Setio, A. A. A., Schreuder, A., et al. (2021). Deep learning for malignancy risk estimation of pulmonary nodules detected at low-dose screening CT. *Radiology, 300*, 438–447.

Vernuccio, F., Cannella, R., Comelli, A., Salvaggio, G., Lagalla, R., & Midiri, M. (2020). Radiomica e intelligenza artificiale: nuove frontiere in medicina [Radiomics and artificial intelligence: New frontiers in medicine.]. *Recenti Progressi in Medicina, 111*, 130–135.

Vernuccio, F., Cannella, R., Meyer, M., et al. (2019). LI-RADS: Diagnostic performance of hepatobiliary phase hypointensity and major imaging features of LR-3 and LR-4 lesions measuring 10-19 mm with arterial phase hyperenhancement. *AJR. American Journal of Roentgenology, 213*(2), W57–W65. https://doi.org/10.2214/AJR.18.20979.

Vernuccio, F., Porrello, G., Cannella, R., et al. (2021). Benign and malignant mimickers of infiltrative hepatocellular carcinoma: Tips and tricks for differential diagnosis on CT and MRI. *Clinical Imaging, 70*, 33–45. https://doi.org/10.1016/j.clinimag.2020.10.011.

Vernuccio, F., Rosenberg, M. D., Meyer, M., et al. (2019). Negative biopsy of focal hepatic lesions: Decision tree model for patient management. *AJR. American Journal of Roentgenology, 212*(3), 677–685. https://doi.org/10.2214/AJR.18.20268.

Wels, M. G., Lades, F., Muehlberg, A., & Suehling, M. (2019). General purpose radiomics for multi-modal clinical research. In *Proc. SPIE 10950, Medical imaging 2019: Computer-aided diagnosis* (p. 1095046). 13 March https://doi.org/10.1117/12.2511856.

Wilson, G. C., Cannella, R., Fiorentini, G., et al. (2020). Texture analysis on preoperative contrast-enhanced magnetic resonance imaging identifies microvascular invasion in hepatocellular carcinoma. *HPB (Oxford), 22*, 1622–1630.

Yachida, S., Mizutani, S., Shiroma, H., et al. (2019). Metagenomic and metabolomic analyses reveal distinct stage-specific phenotypes of the gut microbiota in colorectal cancer. *Nature Medicine, 25*(6), 968–976. https://doi.org/10.1038/s41591-019-0458-7.

Yuen, M. F., & Lai, C. L. (2003). Screening for hepatocellular carcinoma: Survival benefit and cost-effectiveness. *Annals of Oncology, 14*(10), 1463–1467. https://doi.org/10.1093/annonc/mdg400.

CHAPTER

2

Deep learning approaches for high dimension cancer microarray data feature prediction: A review

Debasish Swapnesh Kumar Nayak[a], *Subhashree Mohapatra*[a], *David Al-Dabass*[b], *and Tripti Swarnkar*[c]

[a]Department of Computer Science & Engineering, Institute of Technical Education and Research, Siksha 'O' Anusandhan Deemed to be University, Bhubaneswar, Odisha, India [b]School of Computing & Informatics, Nottingham Trent University, Nottingham, United Kingdom [c]Department of Computer Application, Institute of Technical Education and Research, Siksha 'O' Anusandhan Deemed to be University, Bhubaneswar, Odisha, India

Introduction

One of the diseases known to be deadly in human being is cancer which is triggered by irregular growth of body cells giving rise to various abnormalities with diverse pathological characteristics (Varadhachary, 2007). Cancer classification rate in healthcare needs to be increased so as to improve patient survival rate. Fig. 2.1 shows the rise in new cancer cases from year 2010 to 2020 (CDC, n.d.). Fig. 2.2 shows the proportion of death due to cancer and other diseases as per a survey carried out by the Canadian health society (Canadian Cancer Society, n.d.). The revolution in biological research has given a new direction to medical science. It helps predict the diseases and also give direction for the preparation of the medicine as well. But, the major challenge in biological research is to characterize the gene expression patterns of cells in various biological states (Chen et al., 2016). The selection of significant genes is an important factor in biomedical research for identification and treatment of a particular disease. The traditional feature selection is based on individual genes, whereas the feature selection based on group of genes is more effective. The group feature selection puts an

https://doi.org/10.1016/B978-0-323-85240-1.00018-3

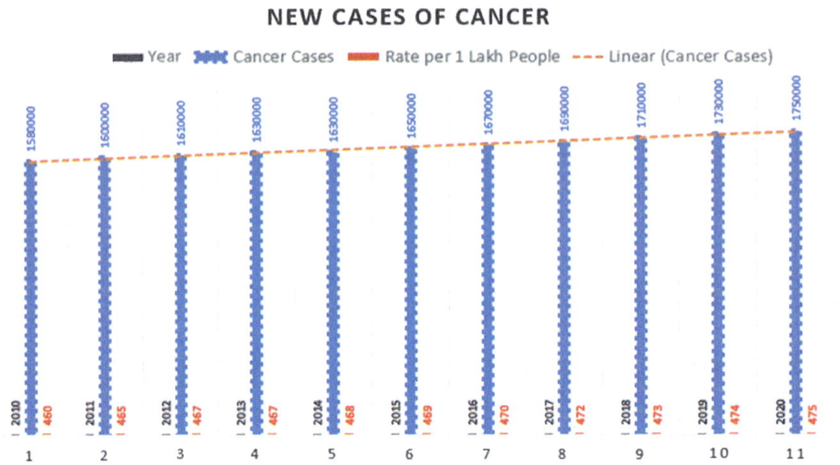

FIG. 2.1 Rate of increase in cancer cases from year 2010 to 2020 (CDC, n.d.).

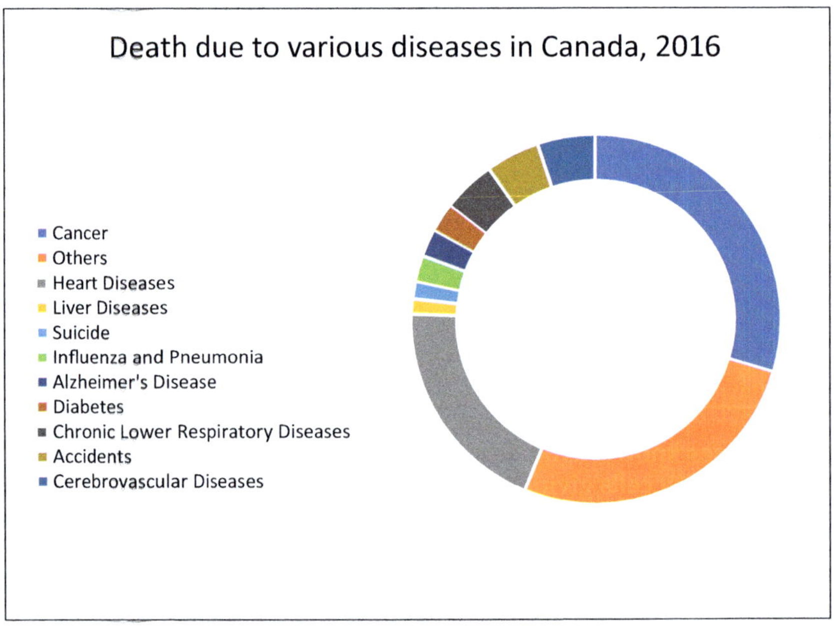

FIG. 2.2 Death due to different diseases in Canada, 2016 (Canadian Cancer Society, n.d.).

insight for better identification, accurate prediction, and treatment of disease (drug discovery) (Nayak et al., 2018).

The feature selection on gene expression data is a complex task in terms of time, accuracy, and cost. It is due to the reason that the gene expression dataset or the microarray data carry large number of genes but with less sample size. The high dimension of microarray leads to large p and small n problem. Thus, the analysis on gene expression data is complex, and to reduce this complexity, the feature extraction technique plays an important role. The main aim of the feature extraction method is to reduce the dimensionality of the gene expression data (Mahapatra & Swarnkar, 2021). So, before imposing the gene selection method, it is better to implement the feature extraction method for better clarity in gene expression data and to get more accurate selection results. The feature extraction does not remove the irrelevant genes, instead it shapes the expression data into a different presentation. While the information loss is taken into account, the feature extraction is considered better than gene selection. Thus, feature extraction helps solve the problem related with high dimensionality (Tan et al., 2014). However, due to this limitation of gene expression microarray data, effective and robust feature prediction/selection methods are highly essential.

The research scope in Bioinformatics and Machine Learning especially with microarray data has reached to a new level in the last decade. The emerging trend of machine learning with bioinformatics gives a new direction to research on microarray data. The microarray data are generally collecting the information from tissue and cell samples based on the differences in gene expression. This information after proper analysis is useful for diagnosis and identifying the particular disease. Even if the dataset for training and testing is very small (often less than 100 patients), the feature present in the raw data is having a range from 6000 to 60,000 as it is measuring the gene expression in large numbers (Bolon-Canedo et al., 2014). The classification techniques conduct the differentiation between healthy and diseased patients; this is the binary approach of a classifier. It is not a complex problem to separate the diseased from the healthy based on the particular disease. The complexity arises when we have to distinguish the diseased patient based on label of disease. The label of disease may be types of tumor, types of cancer, or stages of cancer. This problem is called as multi-class approach. There is a microarray dataset where we have to identify the previous said problem, which makes classification more complicated.

As a result of the above issues, microarray data create a complex ultimatum for machine learning researchers. The huge difference in gene samples and gene expression field creates a high probability of finding "false positives." Therefore, it is very important to validate the model by implementing various robust methods and find their feasibility. In addition to this, the other trial and error complications such as noise and variability eradication make the microarray analysis an interesting domain (Bolon-Canedo et al., 2014). The microarray data which contain the gene expression data are noisy in nature. In addition to this, various feature selection methods play a role to irradiate the irrelevant/inappropriate genes which are less correlated and may cause noise and enhance the prediction model in terms of execution time and accuracy (Gunavathi et al., 2017).

Background

The most fundamental unit of life in a living organism is the cell (Valafar, 2002). Some of the living beings are made up of only one cell, whereas living creatures like humans have millions of cells. Each cell performs some basic function in the body. The cell comprises various components, but the most important part is the nucleus of the cell. The cell nucleus consists of the genome which is considered the blueprint of each living organism. The genome is encoded by the deoxyribonucleic acid (DNA) present inside the nucleus of the human being. Human genome is evenly spread along 23 chromosome pairs, where one set of each pair is transmitted

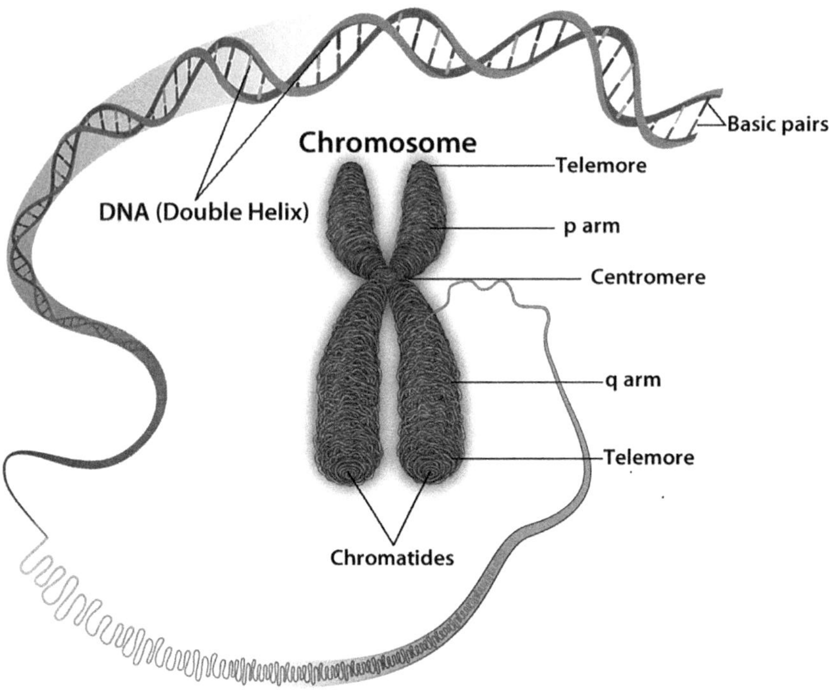

FIG. 2.3 Basic structure of chromosome.

from the maternal and another from the paternal side. The main composition of chromosomes is intertwined DNA. Fig. 2.3 shows the basic structure of a chromosome.

A DNA molecule is basically a double-stranded polymer that is built up in the form of double-helical structure. DNA have two segments, i.e., coding and non-coding segment, out of which the coding segment of DNA is called as genes. Genes specifies the shape of protein, which do the essential work in every organism (Swarnkar et al., 2015). The protein in cells is produced in two different steps. First the DNA is transcribed into messenger ribonucleic acid (mRNA), and then the mRNA is converted (translated) into proteins. The advancement in molecular genetics (DNA microarray) creates an opportunity to obtain a global view of the human cell. The molecular genetics technique enables the possibilities of measuring the simultaneous gene expression of thousands of genes (Piatetsky-Shapiro & Tamayo, 2003). The gene expression profiling data can be used in large-scale and hybrid data analysis techniques in order to obtain clarity in our understanding of normal and diseased state. The microarray basically consists of several genes with its features. These different genes are irrelevant in nature, thus causing the problem in accurate classification of the problem. Feature selection is the process of selecting the genes which are more relevant and removing the irrelevant genes. So, feature selection plays an important role in accurate classification of problem (Golub Todd et al., 1999).

The remaining part of the chapter is divided into multiple sections as mentioned. In the following section, a detailed concept of the microarray data is discussed along with the technology of microarray, gene expression profiling, the microarray dataset, microarray feature selection, and feature extraction techniques. In "Machine learning" section, the application of conventional-Machine Learning in microarray data analysis is discussed along with some related work. In "Deep learning" section, the application of Deep Learning in microarray data analysis is discussed along with some related work. In "Critical analysis: Challenges and future trends" section, different challenges and some future trends on microarray data analysis for cancer detection are discussed and finally concluding the chapter in "Conclusion" section.

Microarray data

In this section, the basic concept of the microarray technique along with the microarray gene expression profile is discussed. A brief explanation on the microarray dataset is also given, which can help future researchers to work on it.

Microarray technology

The Microarray technology came into picture in the late 90s. After the evolution of microarray data, feature gene selection was an important aspect for biomedical researchers. Based on microarray data, feature gene selection was started in early 2000s. In the early stage, the prediction was carried out by various statistical methods. In this approach, various factors like time and accuracy were a big challenge for researchers. In the current scenario, the microarray technology is the most popular tool in the research areas of biological science. It provides a significant procedure to collect the maximum number of gene expression data in the course of a single experiment with cost and time effectiveness. It also creates challenges for biologist, statisticians, and computer science researchers to develop appropriate methods and techniques to overcome the difficulties during Human DNA coding (Sabbar & Sulyman, 2014).

Microarray is the latest trend in experimental molecular biology, which allow to analyze the gene expression data which are very large in numbers (Hickey, 2009). The large number of features (2000–30,000) with very a smaller number of samples (<150) causes "curse of dimensionality" in the microarray (Arun Kumar et al., 2017). The large number of features in microarray makes it vulnerable to noise and variability. Apart from the advantages and disadvantages, microarray analysis nowadays is an interesting field of study for Machine Learning researchers, and the vulnerability of the microarray makes this domain even more exciting. It increases the possibility of cancer classification and diagnosis at the gene expression level (Pirooznia et al., 2008).

Microarray gene expression profile

Gene expression profiling is a process of measuring the expression of multiple genes at once, which helps in generating a clear picture of the cellular level function (Qaraad et al., 2019). The basic step includes the extraction of mRNA from a cell or tissue. The extracted mRNA is then labeled using reverse transcriptase enzyme (RT). In this step, a complementary DNA (cDNA) is generated with fluorescent nucleotides. Generating the cDNA helps in labeling the normal and tumor targets with different fluorescent dyes. These fluorescent cDNA are then put on the DNA microarrays. In the next step, the process of hybridization is carried out.

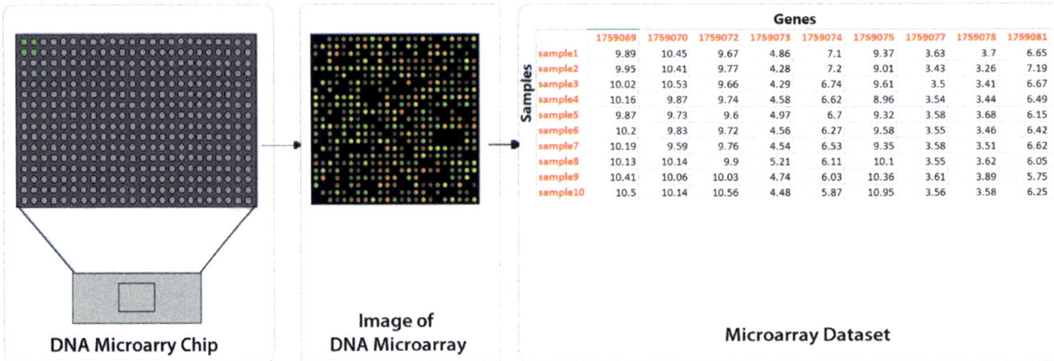

FIG. 2.4 Gene profiling process.

During hybridization, all the labeled cDNAs which signifies mRNA of a particular cell bind to the complimentary base pairs at an individual point on the microarray. Depending on the way in which the DNA bind together, the highly active gene generates several mRNAs. The more mRNA generated, more labeled cDNAs that hybridize with the DNA, thus producing a bright fluorescent portion. Along with very bright spots, some less bright spots are also formed, which indicates the presence of less active genes. On a microarray chip, the tumor and normal samples compete with each other for synthetic cDNA for which each spot is displayed in red, green, or yellow while it is scanned under laser. The significance of the red spot is that the particular gene is very much expressed in cancer cells than in normal cells. While a green area signifies that the gene is strongly suppressed in normal cells. And, the yellow point indicates that the gene is neither severely expressed nor suppressed in the cancer cell. In other words, yellow spot indicates equal expression of the gene in normal and cancer cells. Fig. 2.4 shows the microarray gene profiling process.

Microarray dataset

The microarray dataset is mostly in the form of a matrix where it has M number of rows and N columns. Fig. 2.5 (Qaraad et al., 2019) represents the basic structure of the microarray

'J' Genes

'I' Samples

	G_1	G_2	.....	G_J
S_1	V_{11}	V_{12}		V_{1J}
S_2	V_{21}	V_{22}		V_{2J}
S_3	V_{31}	V_{32}		V_{3J}
:				
S_I	V_{I1}	V_{I2}		V_{IJ}

FIG. 2.5 Structure of microarray matrix.

TABLE 2.1 Detail on dataset used in literature.

Name of the dataset	No. of classes	No. of features	No. of samples
Breast	2	24,481	97
Colon	2	2000	62
DLBCL	2	5469	77
Leukemia	2	7129	72
Lung	5	12,600	203
Lymphoma	2	4026	62
Ovarian	2	15,154	253
Prostrate	2	12,600	102
SRBCT	4	2308	83

data. In the matrix, the rows denote the number of samples and columns represent the features of the given samples (Zhang et al., 2019). Gene microarray data are usually of very high dimensions. So, analysis of such a dataset is a challenging task. And, it is even more challenging for the classification of the data due to presence of multiple classes, sometimes even more than five. Generally, the micro array data consist of very few numbers of samples and high number of features. Another challenge with the microarray data is that the data are imbalanced and common in application areas such as gene expression, bioinformatics, and image classification (Othman, 2020). Unbalanced data are the ones in which the number of samples for the individual classes is not the same: one class has very high number of instances, whereas the other classes may have very few numbers of samples. Table 2.1 gives a detailed information on the different microarray dataset used in various research works.

Microarray feature selection technique

Feature selection is one of the main steps in machine learning techniques. There are broadly three different groups of machine learning techniques for feature selection depending on the available class information (Manikandan & Abirami, 2018). One of them is the supervised feature selection method in which the discriminative features are chosen depending on the ranking of the feature relating to the class. The second type is the unsupervised feature selection methods in which the feature is selected depending on the internal assembling of the data with respect to data variance, separability, and data distribution. The third type is the semi-supervised feature selection techniques which consider both the labeled and unlabeled data for selecting the discriminate features. In Yu and Liu (2004), the features are categorized into broad four groups. The first group of features are the totally irrelevant features with noise, second one is partially relevant and redundant features, third type consists of partially relevant and non-redundant features, and the fourth one consists of the most relevant features. Among these group of features, the most relevant features play an important role in the classification of the samples, whereas the irrelevant features to not contribute in achieving the accuracy of the predictive model. So, during the feature prediction process, these irrelevant features need to be found out and then removed for constructing a good model. There are various standards that are set to calculate the relevance of a feature which is the main purpose of the feature prediction step. Depending on the different search parameters and techniques, the feature prediction is categorized into five groups which are discussed in details as follows.

Filter method

The filter method is used to filter out features that are constructed based on its internal characteristics with regard to the classification job. The internal characteristics of the features are measured with the help of different evolution criteria such as information, consistency, dependency, and distance. There are several statistical methods that are considered for scoring the features, and then considering the threshold value, different discriminative features are selected from the huge set. The filter method does not consider any classification algorithm for feature evaluation. Basically, in this method, there are two steps where in the first step the features are ranked depending on some evaluation standards and then in the second step the highest scoring features are selected to start machine learning classification. The evaluation criteria of the features could be either univariate or multivariate method. The univariate method considers ranking each of the features individually, and the multivariate method considers batch wise evaluation. The multivariate method has the ability to handle redundant and duplicate features.

There are various traditional filter methods that are implemented on the microarray data such as consistency-based filter, information gain, chi-square, consistency-based feature selection, correlation-based feature selection, INTERACT, relief, fast correlation-based feature selection, and minimum redundancy and maximum relevance. The filter technique is again grouped into two methods, namely, ranking and subset search methods (Ambusaidi et al., 2016). The ranking method chooses the perfect features taking into consideration the rank that are assigned to individual features. The gene having the maximum rank value is opted for classification purpose and rest are eliminated. The major disadvantage of this approach is that it does not check for the dependency of the features among themselves. This drawback is overcome by the subset search method as they evaluate the features in batch. On one hand, the filter method is fast, efficient, and scalable, whereas on the other hand, it does not check the indirect association between the features, which leads to fluctuating predictability (Nayak et al., 2018).

Wrapper method

Basically, in this method, the module related to feature search produces the feature set followed by the evaluation of features through performance estimation of the chosen subset using the classifier. For this method, the search space of "N" features is O(2N), and therefore the search is not feasible unless the "N" size is small (Manikandan & Abirami, 2018). To resolve such issues, different techniques such as best first, branch and bound, and genetic algorithm hill climbing are used. The wrapper technique measures the subsets of features as per the usefulness of the classifier. The wrapper method considers the dependency of features by taking into account the communication between the subsets of feature search. Here, the induction algorithm is applied for choosing the feature set that is optimal. The induction algorithm is provided with training samples such that each sample may feature attribute and values. After choosing the feature set, all the subsets are evaluated, and the subset space is searched. To evaluate the accuracy of the system, an n-fold cross-validation is implemented.

In the n-fold process, the total set of samples are divided into n-parts, in the first round of execution, the n − 1 part is taken for training and the nth portion is kept for testing and so on. The algorithm is run for n-times. The accuracy obtained in each round is combined to find the final accuracy. Algorithms such as backward elimination or forward selection are implemented for better feature selection. Wrapper techniques also consider the rank search method that examines exactly "n" subsets having the first ranked variables. It also applies agglomerative-ranked subsets that alternates between constructing the available ranking subset and the growing heuristic procedure which finds all the blends of the first three subsets in the ranking with each of the remaining subsets. The wrapper technique also used various different methods such as linear forward selection, re-ranking-based feature selection, nonparametric density estimation method, and multiple filter and multiple wrapper method for efficient feature selection.

Embedded method

Feature prediction forms the basis of the learning algorithm in the embedded approach along with choosing the feature subset. There are various algorithms used in the embedded method which are sparse multinomial logistic regression (SMLR), Bayesian logistic regression (BLogReg), recursive feature elimination for support vector machines (SVM-RFE), sparse logistic regression (SLogReg), feature optimization (JCFO), and joint classifier. While undergoing the process of classification, the decision of feature selection is made by the algorithm. An example of the embedded approach is the decision trees. This technique selects the subset of features along with the classification technique so as to evaluate the efficiency with respect to wrong classification rate. The particular method that is used here is the sequential forward selection (SFS) (Nayak et al., 2018). The most popular embedded technique is the SVM-RFE in which it uses the main principal of the classifier to formulate standards to rank the features. This particular technique carries out the work of feature selection by repetitively training the SVM model by using the present feature set and eliminating the least important features specified by the SVM model. Following this, a new embedded technique is introduced in which the important features are selected concurrently during the process of classifier formation by penalizing each of the variables used in the construction of the SVM. This is the kernel-penalized SVM. One of the other methods of the embedded technique is the embedded gene selector which is applied to microarray datasets. In this method, the backward elimination technique is used. The drawback of this method is the generation of imbalanced microarray datasets. To overcome this drawback, a new technique is introduced based on the random forest algorithm. To deal with the imbalanced dataset, this method selects the best training error for each of the class in run.

Hybrid method

The hybrid technique takes in the combined advantage of the filter and wrapper method to attain better efficiency. This method lessens the complexity of the wrapper technique and minimizes overfitting. Apolloni et al. (2016) have proposed an effective hybrid model known as binary differential evolution (BDE) algorithm for selecting the important features. Wrapper techniques address the drawback of finding the association between the features in the filter method by optimizing the feature subset. BDE evaluates the fitness function on the basis of the number of times the features are present and stated that the feature subset found had an accuracy near to 100%. Chuang et al. (2011) used the combination of two approaches. One of the methods is the correlation-based feature selection which uses the ranking technique and the other method is the Taguchi genetic algorithm which uses the SNR to improve the performance in each step. The basic genetic algorithm does not take into account the number of genes needed during analysis of feature selection. To eliminate this issue, Chuang et al. (2011) performed the ranking of genes based on the number of times the genes appear. Zorarpacı and Özel (2016) stated a hybrid technique so as to establish a balance between the exploitation and exploration of searching. The hybrid techniques combined the ant bee colony and differential evolution for the feature prediction process, which reduces the high CPU time essential for implementing parallelization and using only differential evolution for exploration.

Ensemble method

The main aim of this method is to provide a knowledge about the diversity and to intensify the regularity of the process to select the feature. This helps consider the strength of each selector as well as overcome the disadvantages. The evaluation of the features can be carried out in two different ways (Manikandan & Abirami, 2018). One way is the individual examination where each feature is assigned a rank by evaluating the level of relevance of individual features. But, in this process, the redundant features are not removed as they may have the similar ranking. The second way is the evaluation of the subset in which a final subset of the features is generated by obtaining successive feature subsets by using a predefined search strategy. This procedure is efficient in removing the duplicate features. As it is seen, there is some short fall of both the methods, so to evade the problem, the ensemble method is brought into picture. Here, several machine learning techniques can be combined to resolve the problem. The most popular ensemble methods implemented in classification are the bagging and boosting method. The bagging method helps in generating an ensemble by training different classifiers produced by arbitrary selection by replacing "n" instances. On the other hand, in the boosting method, the sampling is carried out proportionate to the weight of an instance. Four different approaches are used by ensemble techniques, namely, ensemble forward search, hill climbing, ensemble backward search, and genetic search. Two ways to construct the ensemble are heterogeneous centralized ensemble and homogeneous distributed ensemble. In the first method, "n" different models are produced by the help of various feature selection techniques. But here, the training data taken are the same and the computing time is less than that of the homogeneous technique. The homogeneous distributed ensemble produces 'n' different models by taking into consideration the same feature selection technique but with different training samples for each model. As compared to the heterogeneous one, this has a high computational time requirement. The advantages and disadvantages of each technique are given in Table 2.2.

Microarray feature extraction techniques

Extracting the feature from microarray data is an important accepts for predicting the disease type. In ML, different feature extraction techniques are used for microarray data. These methods are CFS, FCBF, INTERACT, Information Gain, ReliefF, mRMR and SVM-RFE. All of them are available in the well-known Weka tool, except for the mRMR filter, whose implementation is available for MatLab. There are some more methods also used, but we took the above most popular methods extensively found in our literature.

TABLE 2.2 The advantages and disadvantages of each feature selection technique.

Feature selection method	Pros	Cons
Filter	This method is scalable, fast with good computational complexity, and independent to classifier	This method does not consider the association between the features, and there may be possibility of redundancy
Wrapper	This method has an interaction with the classifier and also takes into account the association between the features	This method has the issue of overfitting, requires expensive operation, and the computational cost is also high
Embedded	This method has a better computaional accuracy, less susceptible to overfitting, and also considers the relation between the features	This method depends on the classifier and the computational cost is also high
Hybrid	This method shows a better performance than the filter method along with an effective computational cost, and it is also less suseptible to overfitting	This method depends on the classifier
Ensemble	This method is also less suseptible to overfitting and is scalable for dataset with high dimension	This method is not suitable to understand the combination of classifiers

Correlation-based feature selection (CFS)

CFS is a multivariate algorithm. It ranks the feature subset just as a correlation-based heuristic evaluation function. In this method, the subset which contains highly correlated features with the class and uncorrelated among each other holds the bias of the evaluation function (Bolon-Canedo et al., 2014). The features which are irrelevant should be discarded as they have very low correlation with the class. Redundant features are highly correlated with the remaining features; thus, it should be screened out. Further analysis of the selected feature depends on their predicted classes which should not be predicted by other features.

Fast correlation-based filter (FCBF)

FCBF is a multivariate algorithm. This algorithm measures the correlation between feature and class and among feature and feature. The feature selection in this technique is in a classifier independent manner and the highly correlated feature to the target and uncorrelated with others is selected based on symmetrical uncertainty (SU). The primary work of this algorithm is to select features which are above the set threshold by the SU. SU is the ratio between the entropy of two features and information gain (Bolon-Canedo et al., 2014).

To discard the irrelevant feature, three heuristics are applied which keep the features that are more relevant to the class and eradicate the redundant features. FCBF is effective for removing irrelevant and redundant features for high-dimensionality datasets. There is a disadvantage to this algorithm that it is unable to map the interaction between different features.

INTERACT

The INTERACT algorithm somehow acts as a supplement to the FCBF filter algorithm to measure its effectiveness especially the SU by taking consistency contribution into account.

It takes consistency contribution to obtain the level of consistency by eliminating various noisy features. The operational mode of this algorithms is having two different parts. In the first part, the descending ranking of features are performed based on their SU value. The second part includes the evaluation of features one by one starting from end of the ranked feature list. The feature is removed which consistency contribution is less that the threshold, else is selected. This method has proved its effectiveness in feature interaction and the selection of relevant features for high dimension datasets.

Information gain

Information gain (IG) is a univariate filtration algorithm. This algorithm is operating on the evaluation of information gain for each variable (features), and the selection of feature is based on the maximum information gain which, leads minimized entropy and effectiveness classification of the dataset into groups.

ReliefF

The original relief algorithm is limited to binary classification problem. It is based on random sampling of the data and then targets its neighbor from the correlated and uncorrelated class. It is also incapable of handling the missing values.The extended relief (ReliefF) is the advanced version of relief. The relevance score of each attribute is updated with the comparison values of the attributes of the nearest neighbors to the sample instance. It strengthens the original relief by providing the ability to deal with multiclass problem. Moreover, it is robust and also capable of dealing with incomplete and noisy data.

Minimum redundancy maximum relevance (mRMR)

The mRMR algorithm is used for accurate identification of the characteristics of genes and phenotypes. It also includes the description of relevance in feature selection. The subset often contains features which are relevant but redundant in nature, and the mRMR acts on this issue and eradicates the redundant feature. The features which are mutually far away and having high correlation among them are selected in this method.

Machine learning

Introduction

Feature selection methods are promising for prediction of diseases and nowadays are very popular. Due to this reason, various machine learning methods are proposed for microarray data for feature extraction. In last few decades, various machine learning (ML) feature extraction approaches play a very important role in the field of biomedical science for disease classification and prediction. Different machine learning algorithms for feature selection build the models by taking the necessary features. The three different types of feature selection methods are Filter, Embedded, and Wrapper. Besides this, the feature selection may be univariate or multivariate. In the univariate method, the dependency among different features is not taken into consideration but, in case of multivariate method, it is counted and act on multiple independent variables (Zorarpacı & Özel, 2016).

The feature extraction methods discussed in the previous section are the widely used classical selection methods in ML. Apart from the previously discussed methods, there are other techniques in ML also reviewed which are extensively used for microarray feature extraction. These techniques are combinations of one or more ML methods/algorithms. These techniques/applications are Principal Component Analysis (PCA), Bayesian Criteria (BAC), Independent Component Analysis (ICA), Partial Least Squares (PLS), and Local Linear Embedding (LLE).

Principal component analysis (PCA)

Among the methods used for the analysis of microarray data, PCA is one of them. It is basically a mathematical algorithm, while we apply PCA to the microarray data for dimension reduction, and the beauty of this algorithm is it reduces the dimension without affecting the important variations of data. The eigenvalue and eigenvector of sample can be obtained from the covariance matrix generated by few numbers of relative variables rather than thousands of variable (Verónica et al., 2013).

PCA is used as a linear dimension reduction method in the preliminary stage of microarray data analysis. The main aim of PCA is to project the number of correlated variables into smaller group of non-correlated variable called Principal Component (Sabbar & Sulyman, 2014). It helps in speeding up the processing of microarray (matrix) in advance stage.

Independent component analysis (ICA)

Independent Component Analysis (ICA) is considered an effective mining tool for microarray gene expression data analysis. As ICA deals with higher-order statistical analysis, it can perform a signature work for gene selection from microarray dataset. However, it acts as the advanced version of PCA that has been established considering the blind separation of independent sources from their linear combination.

Partial least squares (PLS)

PLS uses the methodology for creation of experimental information and is determined by very few numbers of latent characteristics. It is useful for identification of uncorrelation in linear transformation of the initial indicator characteristics which have high covariance with the reaction characteristics. In the meantime, PLS predicts reaction characteristics y, the assignment of regression, and reproduce initial matrix X, the undertaking of data modeling. The purpose of building components in PS is to optimize the covariance among the variable y and the initial predictor variables (Tan et al., 2014).

Local linear embedding (LLE)

In LLE, the first step includes the identification of every data point (gene features) and their nearest neighbors. Then, it assigns a set of weights to each data point which clearly relates each data point with its neighbor as a linear combination. It then adopts an eigen vector-based optimization method to find the low-dimensional set of points, which relate the same linear combination of each point with their neighbor. Most importantly, in the microarray gene expression dataset, LLE studies each gene expression data with its neighbor rather than with the outlying gene expression data point (remote data). However, the supervised LLE takes class label information into account and helps in improving the accuracy rate of the classification performance.

Related work

All the feature selection machine learning techniques are followed by different classification methods. There are various works carried out related to microarray data using machine learning. Authors in Pirooznia et al. (2008) have carried out a comparative analysis taking into account the different studies present in literature.

The classification, clustering, and feature selection methods are applied to eight datasets mentioned in Table 2.3. Each dataset is publicly available in microarray repositories, caGEDA website of University of Pittsburgh (Pirooznia et al., 2008):

- The Lymphoma dataset is having 25 samples. It contains normal vs. malignant plasma cells and have 7129 genes.
- The Breast Cancer dataset is having 84 samples of normal vs. tumor subtypes. It includes 1753 genes.
- The Colon Cancer dataset is having 45 samples of epithelial normal cells vs. tumor cells. It contains 7464 genes.
- The Lung Cancer dataset is having 72 samples. It consists of normal vs. malignant cells and have 917 genes.
- The Adenocarcinoma dataset contains 86 samples and 5377 genes.

TABLE 2.3 Various gene expression datasets with number of genes and samples.

Sl no.	Dataset	Variables (genes)	Samples
1	Lymphoma	7129	25
2	Breast cancer	1753	84
3	Colon cancer	7464	45
4	Lung cancer	917	72
5	Adenocarcinoma	5377	86
6	Lymphoma	4027	96
7	Melanoma	8067	38
8	Ovarian cancer	7129	39
9	11-Tumors	5729	60
10	Leukemia-1	5327	72
11	Leukemia-2	11,225	72
12	Lung-cancer	12,600	203
13	Prostate-tumor	10,509	102
14	Brain-tumor-2	10,367	50

- The Lymphoma dataset is having 96 samples. It is from DLBCL1 vs. DLBCL2 cells which includes 4027 genes.
- The Melanoma dataset contains 38 samples of normal vs. malignant and includes 8067 genes.
- The Ovarian Cancer dataset consists of 39 samples of normal vs. malignant cells. It is having 7129 genes.

The effectiveness of feature selection

The performance of pairwise combinations of the feature selection methods was examined by the authors for different samples as it is shown in Table 3 and the different classification methods in Fig. 2.6.

The gene numbers in the above dataset taken for tests like SVM-RFE, Correlation-based, and Chi Squared are 50, 100, 200, and 500. The outcomes of these methods mostly consistent for gene number size 50, 100, and 200 were compared. The performance of gene number size 50 is more accurate, and most importantly, it required minimum memory configuration and processing time (Pirooznia et al., 2008).

In the above study, it is observed that when feature selection methods are applied to the datasets, the overall accuracy performance of the classifier is improved significantly. In all of the above-described cases, SVM-RFE performs much better while it uses the SVM classification method.

SVM-RFE with the SVM classification method performs 100% while applied to the lymphoma dataset. SVM-RFE with Bayesian classification method gives a significant result.

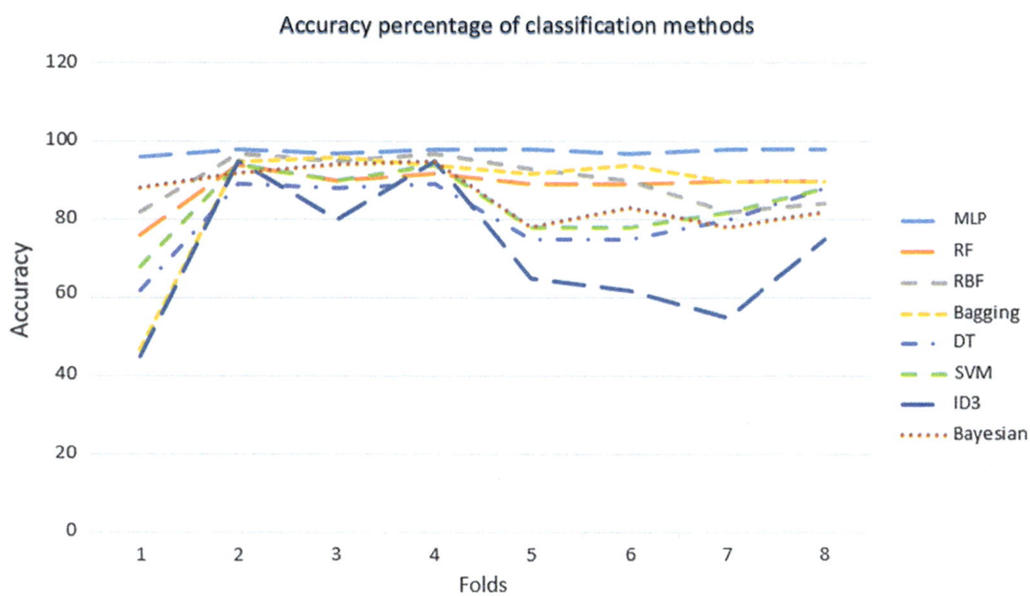

FIG. 2.6 Accuracy percentage of classification methods for all genes (un-processed data) using 10-fold cross validation.

The feature selection method Chi squared with SVM and Bayesian classification methods achieves 92% accuracy in both the cases.

In the context of classification accuracy, CFS and Chi squared perform much better. The Chi squared selection method while applied to the breast cancer dataset shows the least improvement.

The performance of all the feature selection methods imposed for the ovarian cancer dataset found very least difference among them. Whereas the SVM-RFE performs better compared to all other feature selection methods. It is found that the accuracy rate is 100% in the SVM-RFE feature selection method using SVM and RBF classifier.

In the lung cancer dataset, the classification accuracy of MLP with feature selection methods SVM-RFE, CFS and Chi squared is 94%, 92%, and 92%, respectively. The highest accuracy of feature selection method imposed to lung cancer dataset can be obtain using Decision Tree classification methods (both J48 and ID3) with all feature selection methods.

Feature selection from the leukemia dataset

Different classifier algorithm are applied to the Leukemia dataset before going for feature selection (unprocessed dataset) which is shown in Table 2.4.

The study on the Leukemia dataset is taken to observe the accuracy of performance of various ML methods and its effectiveness. The feature selection is studied using different ML methods. The studied methods and their results can be illustrated as follows:

Method-1: CF Subset evaluation method (Table 2.5).
Method-2: Classifier Subset evaluation method (Table 2.6).
Method-3: Consistency Subset evaluation method (Table 2.7).
Method-4: Fuzzy rough subset and fuzzy entropy as a measure method (Table 2.8).

Leukemia dataset result analysis

Based on the implementation of various Feature Selection Algorithms (FSA) on the Leukemia dataset, it is observed that one or more classifiers for the reduced feature subset produced 100% classification accuracy. For the unprocessed dataset, SVM achieves significant accuracy of 98.61%. Compared with the subset evaluation method, Best First and Greedy stepwise search algorithms perform much better. The Fuzzy Rough Subset (attribute evaluation and correlation-based attribute evaluation) method has achieved notable accuracy of 100% with

TABLE 2.4 Different classifiers and their accuracy rate for unprocessed dataset.

Sl no.	Classifier methods	Accuracy (%)
1	Support vector machine (SVM)	98.61
2	Naïve Bayes (NB)	97.20
3	Random forest (RF)	93.05
4	K-nearest neighbors (KNN)	84.72
5	Decision tree (DT)	79.20

TABLE 2.5 Feature selection using Method-1.

Sl no.	Search methods used	Time span in seconds	Reduced number of features	Accuracy of the classifier				
				SVM	KNN	DT	NB	RF
1	Best first	274.00	82	97.22	100	81.94	100	98.61
2	GA (genetic algorithm)	403.00	381	91.66	86.11	90.27	95.83	93.05
3	GS (greedy stepwise)	216.00	80	97.22	100	81.94	98.61	97.22

TABLE 2.6 Feature selection using Method-2.

Sl no.	Search methods used	Time span in seconds	Reduced number of features	Accuracy of the classifier				
				SVM	KNN	DT	NB	RF
1	Best first	196.00	04	97.22	95.83	81.94	88.88	91.66
2	GA (genetic algorithm)	20	74	81.94	83.33	90.27	61.11	77.77
3	GS (greedy stepwise)	101	04	97.22	95.83	81.94	88.88	91.66

TABLE 2.7 Feature selection using Method-3.

Sl no.	Search methods used	Time span in seconds	Reduced number of features	Accuracy of the classifier				
				SVM	KNN	DT	NB	RF
1	Best first	04.53	04	81.954	94.44	93.05	94.44	94.44
2	GA (genetic algorithm)	137.00	03	80.55	79.16	86.11	90.27	90.27
3	GS (greedy stepwise)	02.56	04	81.94	94.44	93.05	94.44	94.44

TABLE 2.8 Feature selection using Method-4.

Sl no.	Search methods used	Time span in seconds	Reduced number of features	Accuracy of the classifier				
				SVM	KNN	DT	NB	RF
1	Best first	65.00	04	97.22	98.61	94.44	95.83	95.83
2	GA (genetic algorithm)	02.30	32	73.61	77.77	62.50	76.38	79.16
3	GS (greedy stepwise)	29.19	04	97.22	98.61	94.44	95.83	95.83

the k-NN algorithm. Among the above-discussed methods, the Fuzzy Rough Subsets are considered best for feature selection in the Leukemia dataset.

Disadvantage

Although we applied several classification methods to different microarray gene expression datasets, still the best (hybrid) classification model is not yet developed which can improve the classification accuracy rate and helps analyze microarray gene expression data in a better way.

The process of gene selection methods is needed to be repeated for a particular microarray gene expression data with different classifiers to improve the accuracy, but each time, the result might vary. To obtain the best combination of classifier and feature selection method, it requires more processing time and memory usage, but still it is unable to identify the better combination. The most important drawbacks to various ML classifier methods and feature selection methods is that the results vary each time without any significant outcome (Pirooznia et al., 2008).

The issues like removing of irrelevant features from the dataset, sometime while applying various methods for analysis of microarray data, cause damage to the original dataset. The dimension reduction of the gene expression microarray dataset is a critical task for various classifiers, which reduces the accuracy of gene feature selection methods. It is also found that different classifiers and feature selection method takes more computational resources and slow down the feature selection procedure.

Deep learning

Introduction

Deep learning has gained popularity in recent years as an alternative to the above-mentioned feature prediction techniques. The central system of any deep learning models is the multi-layered neural networks (Welsh et al., 2001). These layers are much more efficient is extracting complex task-related features from the available data. Many deep neural network designs have been developed recently, which has helped researchers many difficult learning tasks mostly in the area of computer vision. Due to this, deep learning systems have succeeded in solving various biomedical image processing problems (Arvaniti et al., 2018; Esteva et al., 2017; Poplin et al., 2018; Rajpurkar et al., 2017). The idea of deep learning is basically initiated from the architecture of artificial neural networks. One best example of deep architecture is the multi-layer perceptron consisting of many hidden layers. Deep learning models have been implemented in several fields of research in the past few decades. In numerous applications, deep learning techniques have been proven to be performing better than the traditional techniques. Deep architectural model helps the computational system learn the hidden representation of data with different level of abstraction. This is possible by the help of multiple hidden processing layers in the network. The backpropagation technique helps the deep network learn the complicated design in large set of data. This helps the system change its core parameters that are used to evaluate the data in an individual layer from the data in the preceding layer. There are multiple procedures and techniques developed to find out group of genes which are highly associated with a particular type of cancer. Compared to this, deep learning implements machine learning techniques to different subsets of features and applies cross-validation to calculate the score of the individual subset of features. Deep learning takes the help of classifiers to evaluate the effectiveness of the feature subset for delivering a better accuracy for classification.

Related work

Authors in Angermueller et al. (2016) have proposed a deep CNN system in which ID gene expression data are converted into 2D images. The model was able to choose the optimal feature set from the total number of features of the genes which helped for better classification result. The proposed system was able to give an accuracy of 95.65% for classifying 33 different cancers. Authors in Joseph et al. (2019) several CNN-based systems presented efficient classification of the cancer genes with respect to individual cancer types. These models were also trained and tested for the gene expression data of 33 types of cancer. Different CNN models proposed were ID-CNN, 2D hybrid CNN, and 2D vanilla CNN. Authors in Mostavi et al. (2020) have proposed a hybrid model based on backpropagation neural network along with fast genetic algorithm. This hybrid technique was used to find the important features and to classify subtypes of cancer from the gene expression data on several microarray datasets. In Vimaladevi and Kalaavathi (2014), the CNN model was designed for efficient feature prediction and microarray data classification. Authors in Zeebaree et al. (2018) developed two deep learning techniques for microarray data classification. The two methods used were Sample Expansion-Based SAE (SESAE) and Sample Expansion-Based 1DCNN (SE1DCNN). These models were tested on three different cancer microarray datasets. Authors in Liu et al. (2017) have presented a critical analysis between the efficiency of machine learning and deep learning techniques for classification of cancer microarray data. The comparative study was carried out using 11 datasets related to cancer. Authors had obtained an accuracy of 90.0% for machine learning techniques and accuracy of 94.43% for deep learning techniques. The results obtained through the study concluded that deep learning techniques are more efficient in classification of cancer microarray data and is very helpful in early detection of cancer.

Authors in Tabares-Soto et al. (2020) developed a deep wavelet autoencoder with deep neural network for reducing the dimensionality of image. Authors in Mallick et al. (2019) developed a deep flexible neural forest network which was able to translate a multi-class problem into binary-class problem in individual forest. Authors in Xu et al. (2019) designed a model using a combination of PCA and autoencoder for better identification of features and selection. Authors in Zhang et al. (2018) designed a CNN for selecting and classifying the features. P.K. Ram et al. in Zeebaree et al. (2018) designed a model using a genetic algorithm for predicting the least number of features with higher accuracy from the microarray data. In Ram and Kuila (2019), a combination of tabu search and genetic algorithm was used for feature prediction along with SVM for classification. For the prediction of biomarkers in cancer data, authors in Bonilla-Huerta et al. (2016) have designed a hybrid system using semi supervised SVM and kernelized fuzzy rough set. W. Ke et al. in Chakraborty and Maulik (2014) have developed a model using score-based criteria fusion feature selection method.

In this work, a combination of symmetrical uncertainty and ReliefF technique was used. Here, the former method is used for evaluating the relevance of features, and the second method is used for characterizing the features. Authors in Ke et al. (2018) have designed a model based on the folded neural network autoencoder for reduction of dimension.

Advantage of DL over ML

The main characteristic of gene expression data is that it consists of quite large number of genes (Wang et al., 2012). Using this data, many researchers have evaluated and analyzed the classification of cancer microarray data by using statistical, machine learning, and data mining approaches. In literature, various works are available in which machine learning techniques have resulted in good classification accuracy and became popular in dealing with classification of cancer. Even then, there lie some drawbacks for which this task is considered as nontrivial (Shah et al., 2020; Wenyan et al., 2017). Much effort is required to carry out statistical analysis as there is systematic bias in the resultant microarray data. This is due to the high complexity feature of the data. Another issue with the machine learning method is that it needs pre-engineered structure of the raw data into structured form (Dang et al., 2020). The utility of the machine learning technique in microarray cancer data classification is its inability to process unstructured data. These drawbacks of machine learning are handled well by deep learning methodologies. Deep learning is a part of machine learning which is designed in layered structure to construct complex models with the capability of understanding complex data. Deep learning can easily process unstructured data (Luque-Baena et al., 2013) along with the property of automatic feature extraction which makes it the best choice to train on microarray data. Deep learning has ample potential in analyzing and classifying medical data and help assist professionals by reducing human error rate and help in early diagnosis of diseases especially different type of cancer.

Critical analysis: Challenges and future trends

In the research area of genetics, analysis of microarray data has proved to be a revolutionized technology. This technology helps in diagnosis and treatment of different diseases by help of gene expression profiles. The experiments carried out by the help of microarray data come along with number of issues. Techniques to validate gene-expression data and also improvising trail designs are few of the priorities to be handled in near future. After long success of microarrays, some visible limitations are identified, which makes technology like DNA sequencing come to play a more effective role for genome analysis. The relative concentrations of different DNA and RNA sequences are measured by microarray devices. In this technology, the arrays are used to establish the indirect measure of relative concentration among various DNA and RNA. The concentration value which is associated to an array is the assumption of a species at a particular location. It has been seen that the labels assigned to each array do not reflect the full proportional to the concentration of the species hybridizing. The common issues with this are in high concentration we obtained saturated array and no binding for low concentrations and thus leads to linear measure quantity. In microarray, a particular sequence is designed for a specific gene sequence, if the gene is having a homology of sequence, then it can create problems for gene families and multiple splice variant genes. Most important lacuna in microarray is it can only detect the sequence for which it is designed. Therefore, the genome is highly variable in nature while annotated by the microarray, and the resulted dataset contains a number of missing values and also not able to capture the most important genes.

Microarray data analysis: Challenges

Even though microarray data are very helpful in early detection of life-threatening diseases like cancer, it comes along with several challenges which needs to be figured out for better examination of the gene expression data. Some of the challenges are discussed in this section. The foremost challenge is the sample size. The central feature of the microarray is quite large data consisting of some thousands of genes, but the small sizes are very less (Kong & Yu, 2018). The major problem in this case is the proper estimation of error which is affected by small size and in turn, it leads to misclassification of samples. The sample size being small, there may be some noisy or inappropriate data or missing gene values which can further reduce the classification accuracy. Another challenge in handling microarray data is the cross-platform comparison of genes. As there are multiple platforms available for analyzing the gene expression data which differ in their designs, it becomes difficult to carry out cross-platform comparisons of microarray data. To handle this issue, minimal information on the microarray experiment (MIAME) was designed (Bhui et al., 2020) so as to enhance the sensitivity, reproducibility, and stability in the microarray data analysis. There is also possibility of confusion and bias that can occur due to some designing error, collection of data, or result analysis. So, for this, the designing phase of the microarray should be properly carried out also with proper collection of sample data. There are also chances of biological changes in the gene expression data, which can lead to wrong interpretation of the data. To overcome this issue, all the significant changes should be rightly identified and reported which can help professionals to more efficient system for diagnosis and treatment of cancer.

Microarray data analysis: Future trends

Number of samples

As discussed earlier, one of the shortfalls of microarray data analysis is the availability of data in a very small number. The reason behind the small sample size in the beginning years of research was the high sequencing cost restricting its numbers even less than 100. The small sample numbers make it difficult to efficiently examine the model performance. Researchers have opted for some alternatives like selecting an effective validation technique for evaluation of error during classification or combining multiple datasets for experimental work. But, these alternatives to small sample size did have their own cons and not very effective. This problem is gradually being addressed over the last few years, and the cost of sequencing has been reducing over time; hence, it is anticipated that in the coming years, the sample size would increase significantly. Increasing the sample size can open the door for the application of more sophisticated techniques such as deep learning for the micro array analysis. Most of the work carried out in the field of gene microarray data analysis is by using the machine learning techniques. But, the deep learning methods have produced promising output in extraction of high-level information from raw data of large, high-dimensional, and heterogeneous datasets (Brazma et al., 2001).

Feature selection methods

One of the major setbacks of microarray data is the small sample size but huge feature size. One way to tackle this problem is to reduce the number of features by selecting the important features. An important step in the DNA microarray feature prediction is a process of evaluating, selecting, and omitting extraneous features from the sample set. This process of feature prediction helps the model efficiently analyze the data and perform the classification task. As discussed in the previous sections, there are various methods for feature selection like filter, embedded, wrapper, hybrid, and ensemble method. As the cost of DNA sequencing is reducing, so it is estimated that the number of samples may increase in the near future, and this will require more advanced feature prediction techniques. A probable answer to this can be the use of distributed feature selection methods. The distributed methods process several subsets concurrently or in sequence (Mamoshina et al., 2016). One of the possible ways of doing this is dividing the set of features and then apply feature selection of each of the subset and then combine the results. Partitioning of the data can be carried out in two ways. One way is distributing the data vertically (by features) or horizontally (by instances). Authors in Bolón-Canedo et al. (2015) have carried out an in-depth study of consequences of the two ways of data partitioning on microarray data. An alternative to the above option is the use of graphical processing units (GPU) to partition and speed-up the processing of feature selection techniques. GPU abilities can be exploited to develop more efficient feature selection techniques to handle hundreds and thousands of genes. One more way to handle the huge number of features along with increased number of samples is using the concept of subsampling. The idea of subsampling is given by Politis and Romano (1994), which is an alternate process to the traditional resampling technique. This process has been implemented in recent years particularly for DNA microarray analysis. Authors in Politis and Romano (1994) have proposed a model by using the subsampling technique for normalizing the high-throughput sequence datasets.

Deep learning techniques

In past years, neural networks have been applied in microarray analysis, but those networks basically had very few hidden layers. Compared to the machine learning "shallow" networks, deep learning uses the neural networks with many hidden layers and complicated processing architecture. Removing the drawback of machine learning techniques of handcrafted feature extraction, deep learning has gained popularity in microarray data analysis by extracting important feature set automatically from the raw data. Even though deep learning has been implemented (de Cárcer et al., 2011; Leung et al., 2014) in the area of microarray data analysis, but the advantage of deep learning models having more computational power along with automatic feature extraction can be better utilized if the number of gene expression data is increased. Deep learning also has the capability of using multimodal algorithms which takes the benefit of cross-domain multimedia data. Authors in Xiong et al. (2015) have developed a model using sparse autoencoder which can perform a brief representation of features from unlabeled data. The unlabeled data are captured by combination of data from several tumor cells. But, the condition of doing so is that the data are generated by the help of same microarray platform, and this gives a advantage of more samples for the feature prediction process.

Interpretability and visualization

Black-box technology such as nonparametric deep learning methods have an issue with the interpretability of the outcomes attained. In other words, it is difficult to understand the biological interpretation of what the techniques are carrying out when there are hundreds of features available from the data samples. In this case, it is more effective for interpretation when some of the techniques use few important features selected from the hundreds of features available for processing. One of the possible ways is to select important features by the help of feature selection techniques based on the covariance significance test. But handling huge number of features becomes too difficult without the help of statistical methods, which justifies the false discovery rate. So, to regulate the false discovery rate, a combination of this techniques along with some existing methods has been applied. Authors in Benjamini and Hochberg (1995) and Fakoor et al. (2013) have proposed some efficient hybrid models which can be used for microarray analysis having a huge number of features. Another scope for future researchers is the development of new visualization methodologies for deep neural networks. Recently, the issue of understanding, interpreting, and visualizing the deep networks have gained lot of consideration. Unraveling the complicated structure of the convolutional network into symbolic and graphical structure for better visualization are also recent trends in research of microarray data analysis.

Software used

There is multiple software that is used in the microarray data analysis by different researchers. Table 2.9 gives a detailed knowledge on few of the software used in research works.

TABLE 2.9 Various software used for gene microarray data analysis.

Sl. no.	Software	Language	Operation performed
1	ExPosition (Benjamini & Yekutieli, 2001)	R	(i) PCA and Generalized Principal Component Analysis (ii) Worked with quantitative and qualitative data
2	ADE-4 (Beaton et al., 2013)	R	(i) PCA, CA, Principal Analysis Regression, PLS
3	MADE-4 (Thioulouse et al., 1997)	R	(i) Used for multivariate analysis (ii) Visualizing and plotting the results of analysis (includes 3D plots)

Continued

TABLE 2.9 Various software used for gene microarray data analysis—cont'd

Sl. no.	Software	Language	Operation performed
4	AMAP (Culhane et al., 2005)	R	(i) PCA, Generalized PCA, and Robust PCA (ii) Used for hierarchical clustering and *k*-means clustering (iii) Offers Plotting function to PCA (iv) Distance and dissimilarity matrices computation
5	ViSta (Lucas, 2013)	C++, Fortran	(i) PCA, Interactive Cluster Analysis, and Parallel Boxplots
6	FactoMineR (Young & Bann, 1992)	R	(ii) Dimension reduction methods like PCA, CA and MCA and function for PCA to handle missing values
7	Statistics Toolbox (Lê et al., 2008)	MATLAB	(i) Multivariate statistics like PCA, and Clustering (ii) Probability distributions (PD), linear models, nonlinear models for regression, statistical plots
8	Powered PLS Discriminant Analysis (The MathWorks, 2003)	R	(i) Deals with the information extraction for multivariate classification
9	SlimPLS (Liland & Indahl, 2009)	R	Used for implementing Multivariate feature extraction method
10	Sparse PLS (Karnanen, 2006)	R	This can help for feature extraction and dimension reduction simultaneously
11	Surrogate Variable Analysis PLS (Chung & Keles, 2010)	R	Extraction of the informative features with hidden confounders which are unaccounted for
12	PLS Path Modeling (Chakraborty & Datta, 2012)	R	A multivariate feature extraction analysis technique based on the cause-effect relationships of the unobserved and observed features
13	PLS Regression (Sanchez & Trinchera, 2013)	R	PLS regression is used to extract the predictive features from the generalized linear models

Critical analysis

Feature prediction on gene expression microarray data is really a complex procedure, and from literature, it is observed that,

- The data cleaning techniques applied over gene expression microarray dataset need to be more advanced in terms of time and result.
- The effectiveness of ML techniques for microarray feature prediction requires more improvisation to produce more accurate results.
- The proper identification of classifier and prediction methods is very important, and to overcome these issues, the communication gap between biological experts and ML researcher should be properly closed.

- Moreever, to obtain a perfect solution to the subjective analysis and prediction, the tie up between biological expert and research engineer is the need of the hour. The implementation of various ML techniques to medical biology (Computer vision) needs perfect training and testing dataset which only can feasible by choosing accurate classifier and prediction method.

However, in the research areas of gene feature extraction, DL plays a very important role due to its result accuracy and biological relevance still, the improvement on model development with selected and important feature need more attention. The pre-trained neural network model development required sufficient literature data to choose the number of layers for better result.

Conclusion

Feature prediction for gene expression microarray data is a complex task due to huge number of expression values. However, deep learning (deep neural networks) is capable of handling millions of data and produces state of the art result with high accuracy and more biological relevance. The feature of various deep learning techniques has a great impact on gene expression microarray data, which is because of their capability to make pre-trained model function in to feature extractors. The deep neural network is self-sufficient and may trained to handle (recognize) data from large datasets. The classifying accuracy also made DL more powerful. In the areas for gene feature extraction, DL may adopt transfer learning which generalizes the trained neural network for all domains. It helps the model with fast training and increase the adaptability to existing pipelines and better performance.

References

Ambusaidi, M. A., et al. (2016). Building an intrusion detection system using a filter-based feature selection algorithm. *IEEE Transactions on Computers*, *65*(10), 2986–2998.

Angermueller, C., Pärnamaa, T., Parts, L., & Stegle, O. (2016). Deep learning for computational biology. *Molecular Systems Biology*, *12*, 878.

Apolloni, J., Leguizamón, G., & Alba, E. (2016). Two hybrid wrapper-filter feature selection algorithms applied to high-dimensional microarray experiments. *Applied Soft Computing*, *38*, 922–932. 17.

Arun Kumar, C., Sooraj, M. P., & Ramakrishnan, S. (2017). A comparative performance evaluation of supervised feature selection algorithms on microarray datasets. *Procedia Computer Science*, *115*, 209–217. ISSN 1877-0509.

Arvaniti, E., Fricker, K. S., Moret, M., et al. (2018). Automated Gleason grading of prostate cancer tissue microarrays via deep learning. *Scientific Reports*, *8*, 12054. https://doi.org/10.1038/s41598-018-30535-1.

Beaton, D., Fatt, C. R. C., & Abdi, H. (2013). *Package 'ExPosition'*. http://cran.r-project.org/web/packages/ExPosition/ExPosition.pdf.

Benjamini, Y., & Hochberg, Y. (1995). Controlling the false discovery rate: A practical and powerful approach to multiple testing. *Journal of the Royal Statistical Society, Series B*, *57*, 289–300.

Benjamini, Y., & Yekutieli, D. (2001). The control of the false discovery rate in multiple testing under dependency. *The Annals of Statistics*, *29*, 1165–1188.

Bhui, N., Ram, P. K., & Kuila, P. (2020). *Feature selection from microarray data based on deep learning approach*. In *2020 11th International Conference on Computing, Communication and Networking Technologies (ICCCNT)* (pp. 1–5). IEEE.

Bolón-Canedo, V., Sánchez-Maroño, N., & Alonso-Betanzos, A. (2015). Recent advances and emerging challenges of feature selection in the context of big data. *Knowledge-Based Systems*, *86*, 33–45.

Bolon-Canedo, V., Sanchez-Marono, N., et al. (2014). A review of microarray datasets and applied feature selection methods. *Information Sciences*, *282*, 111–135.

Bonilla-Huerta, E., Hernandez-Montiel, A., Morales-Caporal, R., & Arjona-Liopez, M. (2016). Hybrid framework using multiple-filters and an embedded approach for an efficient selection and classification of microarray data. *IEEE/ACM Transactions on Computational Biology and Bioinformatics (TCBB)*, *13*(1), 12–26.

Brazma, A., et al. (2001). Minimum information about a microarray experiment (MIAME)—Toward standards for microarray data. *Nature Genetics*, *29*(4), 365.

Canadian Cancer Society, n.d., https://www.cancer.ca/en/cancer-information/cancer-101/cancer-statistics-at-a-glance/?region=on.

CDC, n.d., https://www.cdc.gov/cancer/dcpc/research/articles/cancer_2020.htm.

Chakraborty, S., & Datta, S. (2012). Surrogate variable analysis using partial least squares (SVA-PLS) in gene expression studies. *Bioinformatics*, *28*(6), 799–806.

Chakraborty, D., & Maulik, U. (2014). Identifying cancer biomarkers from microarray data using feature selection and semisupervised learning. *IEEE Journal of Translational Engineering in Health and Medicine*, *2*, 1–11.

Chen, Y., Li, Y., Narayan, R., et al. (2016). Gene expression inference with deep learning. *Bioinformatics*, *32*(12), 1832–1839.

Chuang, L.-Y., et al. (2011). A hybrid feature selection method for DNA microarray data. *Computers in Biology and Medicine*, *41*(4), 228–237.

Chung, D., & Keles, S. (2010). Sparse partial least squares classification for high dimensional data. *Statistical Applications in Genetics and Molecular Biology*, *9*(1).

Culhane, A. C., Thioulouse, J., Perriere, G., & Higgins, D. G. (2005). MADE4: An R package for multivariate analysis of gene expression data. *Bioinformatics*, *21*(11), 2789–2790.

Dang, S., Wen, M., Mumtaz, S., Li, J., & Li, C. (2020). Enabling multicarrier relay selection by sensing fusion and cascaded ANN for intelligent vehicular communications. *IEEE Sensors Journal*. https://doi.org/10.1109/JSEN.2020.2986322.

de Cárcer, D. A., Denman, S. E., McSweeney, C., & Morrison, M. (2011). Evaluation of subsampling-based normalization strategies for tagged high-throughput sequencing data sets from gut microbiomes. *Applied and Environmental Microbiology*, *77*(24), 8795–8798.

Esteva, A., et al. (2017). Dermatologist-level classification of skin cancer with deep neural networks. *Nature*, *542*, 115–118.

Fakoor, R., Ladhack, S., Nazi, A., & Huber, M. (2013). Using deep learning to enhance cancer diagnosis and classification. In: Proceedings of the 30th international conference on machine learning, ICML 2013. *Journal of Machine Learning Research: W&CP*, *28*.

Golub Todd, R., et al., Christine, H., Michelle, G., Mesirov, J. P., Hilary, C., Loh, M. L., Downing, J. R., & Caligiuri, M. A. (1999). Molecular classification of cancer: Class discovery and class prediction by gene expression monitoring. *Science. American Association for the Advancement of Science*, *286*(5439), 531–537.

Gunavathi, C., Premalatha, K., & Sivasubramanian, K. (2017). A survey on feature selection methods in microarray gene expression data for cancer classification. *Research Journal of Pharmacy and Technology*, *10*(5), 1395–1401.

Hickey, P. (2009). *X chromosome association testing in genome-wide association studies*. Ph.D thesis Department of Mathematics and Statistics, The University of Melbourne.

Joseph, M., Devaraj, M., & Leung, C. K. (2019). DeepGx: Deep learning using gene expression for cancer classification. In *2019 IEEE/ACM international conference on advances in social networks analysis and mining (ASONAM)* IEEE.

Karnanen, J. (2006). *Independent component analysis using score functions from the Pearson system*. http://cran.rproject.org/web/packages/PearsonICA/PearsonICA.pdf.

Ke, W., Wu, C., Wu, Y., & Xiong, N. N. (2018). A new filter feature selection based on criteria fusion for gene microarray data. *IEEE Access*, *6*, 61065–61076.

Kong, Y., & Yu, T. (2018). A deep neural network model using random forest to extract feature representation for gene expression data classification. *Scientific Reports*, *8*(1), 16477.

Lê, S., Josse, J., & Husson, F. (2008). FactoMineR: An R package for multivariate analysis. *Journal of Statistical Software*, *25*(1), 1–18.

Leung, M. K. K., Xiong, H. Y., Lee, L. J., & Frey, B. J. (2014). Deep learning of the tissue-regulated splicing code. *Bioinformatics*, *30*(12), i121–i129.

Liland, K. H., & Indahl, U. G. (2009). Powered partial least squares discriminant analysis. *Journal of Chemometrics*, *23*(1), 7–18.

Liu, J., Wang, X., Cheng, Y., & Zhang, L. (2017). Tumor gene expression data classification via sample expansion-based deep learning. *Oncotarget*, *8*(65), 109646.

Lucas, A. (2013). *Package 'amap'*. http://cran.r-project.org/web/packages/amap/vignettes/amap.pdf.

Luque-Baena, R., Urda, D., Subirats, J., Franco, L., & Jerez, J. (2013). Analysis of cancer microarray data using constructive neural networks and genetic algorithms. In *Proceedings of the IWBBIO, international work-conference on bioinformatics and biomedical engineering*.

Mahapatra, S., & Swarnkar, T. (2021). Gene selection using integrative analysis of multi-level omics data: A systematic review. In *Data analytics in bioinformatics: A machine learning perspective* (pp. 145–171). John Wiley & Sons, Inc.

Mallick, P. K., Ryu, S. H., Satapathy, S. K., Mishra, S., Nguyen, G. N., & Tiwari, P. (2019). Brain MRI image classification for cancer detection using deep wavelet autoencoder-based deep neural network. *IEEE Access*, *7*, 46278–46287.

Mamoshina, P., Vieira, A., Putin, E., & Zhavoronkov, A. (2016). Applications of deep learning in biomedicine. *Molecular Pharmaceutics*, *13*(5), 1445–1454.

Manikandan, G., & Abirami, S. (2018). A survey on feature selection and extraction techniques for high-dimensional microarray datasets. In A. S. Margret, & U. Wiil (Eds.), *Knowledge computing and its applications*. Singapore: Springer. https://doi.org/10.1007/978-981-10-8258-0_14.

Mostavi, M., Chiu, Y.-C., Huang, Y., & Chen, Y. (2020). Convolutional neural network models for cancer type prediction based on gene expression. *BMC Medical Genomics*, *13*, 1–13. 29.

Nayak, D. S. K., Mahapatra, S., & Swarnkar, T. (2018). Gene selection and enrichment for microarray data—A comparative network based approach. In K. Saeed, N. Chaki, B. Pati, S. Bakshi, & D. Mohapatra (Eds.), *vol. 564. Progress in advanced computing and intelligent engineering. Advances in intelligent systems and computing*. Singapore: Springer. https://doi.org/10.1007/978-981-10-6875-1_41.

Othman, S. M. (2020). *A survey on feature selection in microarray data: Methods, algorithms and challenges*. https://www.researchgate.net/publication/345393675.

Piatetsky-Shapiro, G., & Tamayo, P. (2003). Microarray data mining: Facing the challenges. *ACM SIGKDD Explorations Newsletter*, *5*(2), 1–5.

Pirooznia, M., et al. (2008). A comparative study of different machine learning methods on microarray gene expression data. *BMC Genomics*, *9*. SN-1471-2164.

Politis, D. N., & Romano, J. R. (1994). Large sample confidence regions based on subsamples under minimal assumptions. *The Annals of Statistics*, *22*, 2031–2050.

Poplin, R., et al. (2018). Prediction of cardiovascular risk factors from retinal fundus photographs via deep learning. *Nature Biomedical Engineering*, *2*, 158–164.

Qaraad, M., Amjad, S., Fathi, H., & Manhrawy, I. I. M. (2019). Feature selection techniques for cancer classification applied to microarray data: A survey. In *2019 International conference on intelligent systems and advanced computing sciences (ISACS), Taza, Morocco* (pp. 1–8). https://doi.org/10.1109/ISACS48493.2019.9068865.

Rajpurkar, P., et al. (2017). CheXNet: Radiologist-level pneumonia detection on chest X-rays with deep learning. *arXiv preprint arXiv*, *1711*, 05225.

Ram, P. K., & Kuila, P. (2019). Feature selection from microarray data: Genetic algorithm based approach. *Journal of Information and Optimization Sciences*, *40*(8), 1599–1610.

Sabbar, B. M., & Sulyman, M. R. (2014). Analysing of DNA microarray data using principle component analysis (PCA). *Journal of Theoretical and Applied Information Technology*, *70*(1).

Sanchez, G., & Trinchera, L. (2013). *Tools for partial least squares path modeling*. http://cran.r-project.org/web/packages/plspm/plspm.pdf.

Shah, S. H., Iqbal, M. J., Ahmad, I., Khan, S., & Rodrigues, J. J. (2020). Optimized gene selection and classification of cancer from microarray gene expression data using deep learning. *Neural Computing and Applications* (pp. 1–12).

Swarnkar, T., Simoes, S. N., et al. (2015). Identifying dense subgraphs in protein–protein interaction network for gene selection from microarray data. In *Network modeling analysis in health informatics and bioinformatics* (pp. 1–18). Springer Vienna.

Tabares-Soto, R., Orozco-Arias, S., Romero-Cano, V., Bucheli, V. S., Rodríguez-Sotelo, J. L., & Jiménez-Varón, C. F. (2020). A comparative study of machine learning and deep learning algorithms to classify cancer types based on microarray gene expression data. *PeerJ. Computer Science*, *6*, e270.

Tan, C. S., Ting, W. S., Mohamad, M. S., Chan, W. H., Deris, S., & Shah, Z. A. (2014). A review of feature extraction software for microarray gene expression data. *BioMed Research International, 2014*, 213656. 15 pages https://doi.org/10.1155/2014/213656.

The MathWorks. (2003). *Statistics toolbox for use with MATLAB, user guide version 4.* http://cda.psych.uiuc.edu/matlab_pdf/stats.pdf.

Thioulouse, J., Chessel, D., Dol´edec, S., & Olivier, J.-M. (1997). ADE-4: a multivariate analysis and graphical display software. *Journal of Statistics and Computing, 7*(1), 75–83.

Valafar, F. (2002). Pattern recognition techniques in microarray data analysis: A survey. *Annals of the New York Academy of Sciences, 980*, 41–64. https://doi.org/10.1111/j.1749-6632.2002.tb04888.x. 12594081.

Varadhachary, G. R. (2007). Carcinoma of unknown primary origin. *Gastrointestinal Cancer Research, 1*, 229–235.

Verónica, B.-C., et al. (2013). A review of feature selection methods on synthetic data. *Knowledge and Information Systems. Springer, 34*(3), 483–519.

Vimaladevi, M., & Kalaavathi, B. (2014). A microarray gene expression data classification using hybrid back propagation neural network. *Genetika, 46*(3), 1013–1026. 30.

Wang, J., He, H., & Prokhorov, D. V. (2012). A folded neural network autoencoder for dimensionality reduction. *Procedia Computer Science, 13*, 120–127.

Welsh, J. B., Zarrinkar, P. P., Sapinoso, L. M., Kern, S. G., Behling, C. A., Monk, B. J., Lockhart, D. J., Burger, R. A., & Hampton, G. M. (2001). Analysis of gene expression profiles in normal and neoplastic ovarian tissue samples identifies candidate molecular markers of epithelial ovarian cancer. *Proceedings of the National Academy of Sciences of the United States of America, 98*(3), 1176–1181.

Wenyan, Z., Xuewen, L., & Jingjing, W. (2017). Feature selection for cancer classification using microarray gene expression data. *Biostatistics and Biometrics Open Access Journal, 1*(2), 555557.

Xiong, H. Y., Alipanahi, B., Lee, L. J., Bretschneider, H., Merico, D., Yuen, R. K. C., Hua, Y., Gueroussov, S., Najafabadi, H. S., Hughes, T. R., et al. (2015). The human splicing code reveals new insights into the genetic determinants of disease. *Science, 347*(6218), 1254806.

Xu, J., Wu, P., Chen, Y., Meng, Q., Dawood, H., & Khan, M. M. (2019). A novel deep flexible neural forest model for classification of cancer subtypes based on gene expression data. *IEEE Access, 7*, 22086–22095.

Young, F. W., & Bann, C. M. (1992). ViSta: A visual statistics system. In R. A. Stine, & J. Fox (Eds.), *Statistical computing environments for social research* (pp. 207–235). Sage.

Yu, L., & Liu, H. (2004). Redundancy based feature selection for microarray data. In *Proceedings of the tenth ACM SIGKDD conference on knowledge discovery and data mining* (pp. 737–742).

Zeebaree, D. Q., Haron, H., & Abdulazeez, A. M. (2018). Gene selection and classification of microarray data using convolutional neural network. In *2018 International conference on advanced science and engineering (ICOASE)* IEEE.

Zhang, Y., Zhang, Q., Chen, Z., Shang, J., & Wei, H. (2019). Feature assessment and ranking for classification with nonlinear sparse representation and approximate dependence analysis. *Decision Support Systems, 122*, 113064.

Zhang, D., Zou, L., Zhou, X., & He, F. (2018). Integrating feature selection and feature extraction methods with deep learning to predict clinical outcome of breast cancer. *IEEE Access, 6*, 28936–28944.

Zorarpacı, E., & Özel, S. A. (2016). A hybrid approach of differential evolution and artificial bee colony for feature selection. *Expert Systems with Applications, 62*, 91–103.

Further reading

Alizadeh, A. A., Eisen, M. B., Davis, R. E., Ma, C., Lossos, I. S., Rosenwald, A., Boldrick, J. C., Sabet, H., Tran, T., Yu, X., et al. (2000). Distinct types of diffuse large B-cell lymphoma identified by gene expression profiling. *Nature, 403* (6769), 503–511.

Alon, U., Barkai, N., Notterman, D. A., Gish, K., Ybarra, S., Mack, D., & Levine, A. J. (1999). Broad patterns of gene expression revealed by clustering analysis of tumor and normal colon tissues probed by oligonucleotide arrays. *Proceedings of the National Academy of Sciences of the United States of America, 96*(12), 6745–6750.

Beer, D. G., Kardia, S. L., Huang, C. C., Giordano, T. J., Levin, A. M., Misek, D. E., Lin, L., Chen, G., Gharib, T. G., Thomas, D. G., et al. (2002). Gene-expression profiles predict survival of patients with lung adenocarcinoma. *Nature Medicine, 8*(8), 816–824.

Bertrand, F., Meyer, N., & Bertrand, M. M. *Partial least squares regression for generalized linear models.* http://cran.r-project.org/web/packages/plsRglm/plsRglm.pdf.

Bittner, M., Meltzer, P., Chen, Y., Jiang, Y., Seftor, E., Hendrix, M., Radmacher, M., Simon, R., Yakhini, Z., Ben-Dor, A., et al. (2000). Molecular classification of cutaneous malignant melanoma by gene expression profiling. *Nature, 406* (6795), 536–540.

De Vos, J., Thykjaer, T., Tarte, K., Ensslen, M., Raynaud, P., Requirand, G., Pellet, F., Pantesco, V., Reme, T., Jourdan, M., et al. (2002). Comparison of gene expression profiling between malignant and normal plasma cells with oligonucleotide arrays. *Oncogene, 21*(44), 6848–6857.

Garber, M. E., Troyanskaya, O. G., Schluens, K., Petersen, S., Thaesler, Z., Pacyna-Gengelbach, M., van de Rijn, M., Rosen, G. D., Perou, C. M., Whyte, R. I., et al. (2001). Diversity of gene expression in adenocarcinoma of the lung. *Proceedings of the National Academy of Sciences of the United States of America, 98*(24), 13784–13789.

Lee, C. P., & Leu, Y. (2011). A novel hybrid feature selection method for microarray data analysis. *Applied Soft Computing, 11*(1), 208–213. 19.

Mohapatra, P., Chakravarty, S., & Dash, P. K. (2016). Microarray medical data classification using kernel ridge regression and modified cat swarm optimization based gene selection system. *Swarm and Evolutionary Computation, 28,* 144–160.

Morán-Fernández, L., Bolón-Canedo, V., & Alonso-Betanzos, A. (2017). Centralized vs. distributed feature selection methods based on data complexity measures. *Knowledge-Based Systems, 117,* 27–45.

Peeters, J. K. (2008). Microarray bioinformatics and applications in Oncology. *Erasmus University Rotterdam.* Retrieved from http://hdl.handle.net/1765/12618.

Panigrahi, S., Nanda, A., & Swarnkar, T. (2018). Deep learning approach for image classification. In *2nd International conference on data science and business analytics (ICDSBA)* IEEE.

Pattnaik, P. A., & Swarnkar, T. (2018). Vision based malaria parasite image analysis: A systematic review. *International Journal of Bioinformatics Research and Applications, 15*(1).

Perou, C. M., Sorlie, T., Eisen, M. B., van de Rijn, M., Jeffrey, S. S., Rees, C. A., Pollack, J. R., Ross, D. T., Johnsen, H., Akslen, L. A., et al. (2000). Molecular portraits of human breast tumours. *Nature, 406*(6797), 747–752.

CHAPTER

3

Integrative data analysis and automated deep learning technique for ovary cancer detection

Soutrik Acharya[a], *Dyuti Ghosh*[a], *H. Swapnarekha*[b], *Manohar Mishra*[c], *and SoumyaRanjan Nayak*[d]

[a]Department of Radiophysics and Electronics, University of Calcutta, Kolkata, West Bengal, India [b]Department of Information Technology, Aditya Institute of Technology and Management (AITAM), Tekkali, Andhra Pradesh, India [c]Siksha 'O' Anusandhan Deemed to be University, Bhubaneswar, Odisha, India [d]KIIT (Deemed-to-be) University, Bhubaneswar, Odisha, India

Introduction

Cancer is considered the major health problem throughout the world as tens of millions of people are diagnosed with cancer every year and more than half of the patients eventually die from it (Fitzmaurice et al., 2013). Cancer is a kind of disease that happens mainly due to the unusual behavior of genes within the body of human beings. Normally, the cells in the body divide in a systematic way to produce new cells. However, the division of cells in the body in an abnormal way leads to cancer. Further, these cancerous cells inhibit the production of normal cells and are responsible for spreading the disease by producing new cancerous cells in the human body (Cancer Facts & Figures, 2021). Sometimes, these cancer cells spread to other organs from their origin, which is referred to as metastasis (Jin et al., 2017). According to the World Health Organization (WHO), carcinoma is the second leading source of deaths in both economically developed and developing countries throughout the world (WHO, The Global Health Estimates, 2021). From the global cancer statistics, it is revealed that an estimate of new cancer cases of 19.3 million and cancer deaths of approximately 10.0 million occurred throughout the world in 2020 (Sung et al., 2021). The most common cancers that often influence the health of women are breast, lung, endometrial, colorectal, skin, cervical, and ovarian

https://doi.org/10.1016/B978-0-323-85240-1.00007-9

cancer (Cancer Facts for Women, 2021). Among all types of cancer, it was found that ovarian carcinoma is the eleventh most prevalent cancer and the fifth predominant carcinoma responsible for cancer-related deaths in women. In the entire life span of a woman, the risk of getting ovarian carcinoma is approximately 1 in 78, and the risk of causing death due to ovarian cancer is approximately 1 in 108 (Ovarian Cancer Statistics, 2021). As per the global cancer statistics, an estimate of 3,13,959 new cases and 2,07,252 deaths are caused by ovarian cancer throughout the world in the year 2020 (Sung et al., 2021). As the early stage of ovarian cancer is typically asymptomatic, most of the cases are diagnosed at advanced stages, which results in a lower rate of long-term survival. Due to this, ovarian cancer is often known as silent cancer. After the identification of ovarian cancer in women, staging is carried out to find out the spread of cancer. If cancer is found at the part where it is initiated, then the stage is referred to as localized or stage 1. If cancer is found at different parts of the body, then the stage is referred to as distant or regional (National Cancer Institute SEER, 2021). Figs. 3.1 and 3.2 show the 5-year relative survival of ovarian cancer cases and percentage of cases by stage.

From the statistics of Fig. 3.2, it can be concluded that the localized stage has the highest five-year survival rate of 92.06% when compared with the other stages. As patients diagnosed with localized stage has the highest survival rate, effective treatment can be provided to patients diagnosed with stage 1, when compared with other stages. So, early diagnosis of ovarian cancer plays a key role in providing effective treatment for the patients. Therefore, a significant endeavor has been placed on the development of novel approaches for the prediction and prognosis of ovarian cancer.

From the last few decades, machine learning approaches have been extensively used in various medical applications such as the classification and detection of tumors from the medical images (Cochran, 1997; Fielding et al., 1992) to the categorization of tumors from the complex proteomic and genomic measurements (Bach et al., 2003; Domchek et al.,

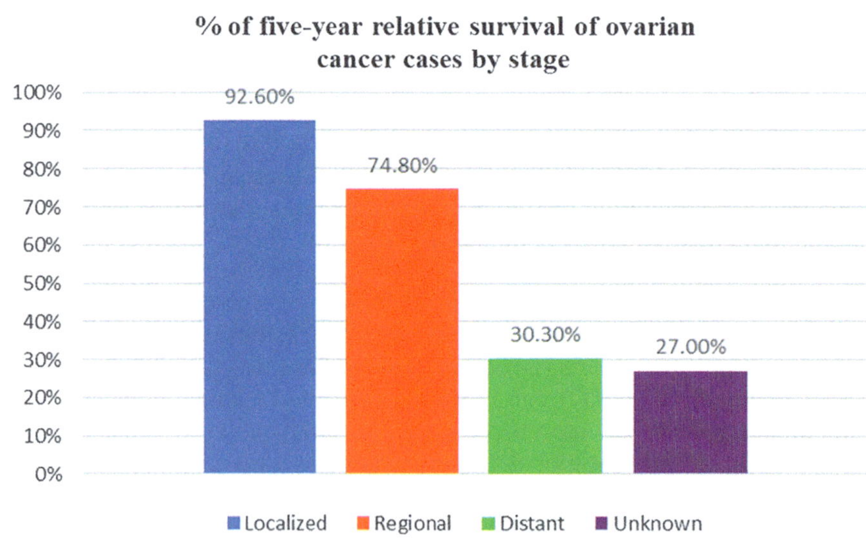

FIG. 3.1 Percent of five-year relative survival of ovarian cancer cases by stage (National Cancer Institute SEER, 2021).

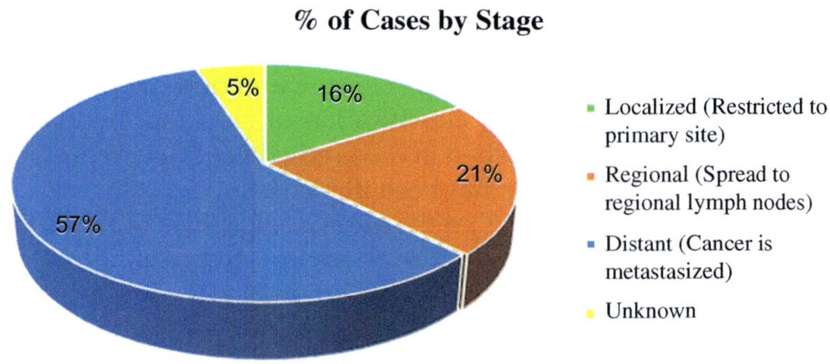

FIG. 3.2 Percent of cases by stage (National Cancer Institute SEER, 2021).

2003; Gascón et al., 2004). This is mainly due to the use of various probabilistic, statistical, and optimization approaches by machine learning to detect difficult patterns from large, noisy, and complex data.

In the past, physicians have used clinical, histological, and population-based data to conclude the decision on prognosis of cancer (Cochran, 1997; Fielding et al., 1992). Moreover, the features such as age, diet, family history, high-peril habits, vulnerability to surrounding carcinogens, and weight also present a vital part in the prediction of cancer (Bach et al., 2003; Domchek et al., 2003; Gascón et al., 2004). However, these features do not provide adequate information to conclude the decision on the prognosis of cancer. With the advancement of proteomic and genomic measures, a new type of molecular information can be acquired, which is useful in making robust decisions regarding the prognosis of cancer. The use of High-Throughput Technologies (HTTs) produces a large volume of cancer data for the medical research community. Even though a huge amount of cancer data is available, still the prediction of disease outcomes remains a challenging task to the physicians. From the last few decades, machine learning (ML) approaches have been extensively used in various medical applications such as the classification and detection of tumors from the medical radiographs (Bocchi et al., 2004; Petricoin & Liotta, 2004) to the categorization of tumors from the complex genomic and proteomic measurements (Dettling, 2004; Wang et al., 2005; Zhou et al., 2004). This is mainly due to the use of various probabilistic, statistical, and optimization approaches by machine learning to detect difficult patterns from large, noisy, and complex data. As machine learning tools have emerged as powerful tools in the prediction and prognosis of cancer, these approaches have also been used in the prediction of ovarian cancer.

The rapid advancement in the technology of artificial intelligence especially in deep learning (DL) and the significant enhancement of computational power in recent years has led to the development of more accurate and powerful models for the prediction of disease. In the medical field, the DL approaches have been successfully utilized in a variety of applications such as determination of pathological histological slides (Kather et al., 2019), finding the Aβ plague region in patients of Alzheimer's prediction of hospital readmission from the information of electronic health records (EHRs) (Rajkomar et al., 2018; Shameer et al., 2017), mortality rate (Elfiky et al., 2018), and so on. In addition, DL approaches also provide a reduction in the cost of large-scale next-generation sequencing and the accessibility of open-source databases such as TCGA and GEO databases to the medical research community. As DL approaches have been successfully utilized, these approaches have also been utilized in the prediction of various types of cancer prognosis such as identification of metastatic breast cancer (Wang et al., 2016; Zhou et al., 2020), categorization of lung cancer histopathology images (Coudray et al., 2018), prediction of lung cancer survival period (Doppalapudi et al., 2021), colorectal cancer (Wulczyn et al., 2021), classification of skin cancer (Manne et al., 2020), and so on. The number of articles published in the prediction of cancer prognosis using deep learning approaches for the last 5 years has been depicted in Fig. 3.3.

From the figure, it can be concluded that the number of publications on the prediction of cancer prognosis is increasing from 2016 to 2020. Therefore, in this study, a deep learning approach has been used in the prediction of ovarian carcinoma.

The main purpose of this study is to enhance the performance of the ovarian cancer prediction model by using DL approaches such as a convolutional neural network (CNN). Further, techniques such as data pre-processing and feature extraction have been applied to enhance the effectiveness of the model. Finally, to assess the performance of the models, the proposed models have been evaluated on 349 patient's raw datasets.

The remaining section of the paper is categorized as specified below. The research work carried on the prediction of ovarian cancer using ML and DL approaches has been described in "Related work" section. The description of the dataset and proposed methodology has been represented in "Methodology" section. "Result analysis" section describes the comparative analysis of results using DL approaches and machine learning approaches. Finally, the conclusion of the study has been represented in "Conclusion" section.

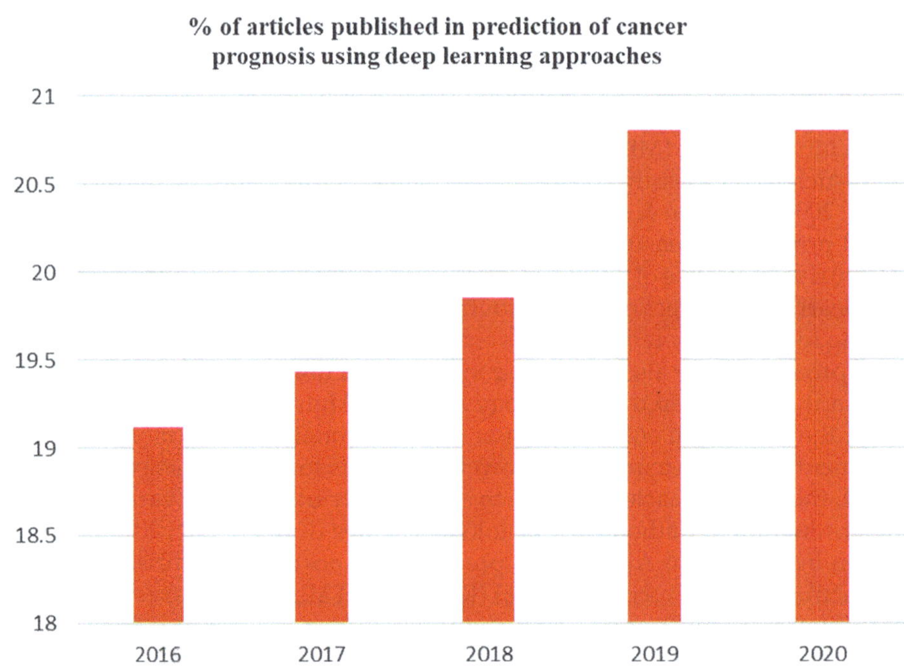

FIG. 3.3 Percentage of articles published in prediction of cancer prognosis using deep learning approaches.

Related work

Several research works have been accomplished on the design and development of automated systems for the prediction of ovarian carcinoma using ML and DL approaches in a precise way. To predict the pathological diagnosis of ovarian cancer from preoperative examinations, machine learning approaches have been suggested by Akazawa and Hashimoto (2020). The study considered 202 patients out of which 53 patients are having ovarian cancer, 23 patients are having borderline malignant tumors, and 126 patients are having benign ovarian tumors. The diagnostic results of 16 features available from blood test, background information of patients, and imaging tests were derived using machine learning approaches such as random forest, support vector machine, logistic regression, Naive Bayes, and XGBoost algorithms. Further, the significance of 16 features was analyzed for the prediction of the disease. From the experimental results, it was revealed that XGBoost attained the highest accuracy of 0.80 when compared with that of other conventional machine learning approaches.

For the early diagnosis of ovarian carcinoma, a systematic study on seventy-five RNA markers as promising biomarkers has been presented by Hulstaert et al. (2021). Most of the markers are microRNAs and are recognized through microarrays in blood-based fluids or RT-qPCR. Further, the biomarkers that were sufficient in ovarian carcinoma biofluids are correlated to controls in at least two independent studies. The biomarkers that are proved to be lower in at least two studies among the 75 RNA markers include miR-106b, miR-122, miR-126, miR-193a, and miR-223. Even though these markers are used as promising markers in the initial investigation of ovarian cancer, further confirmation is needed before enforcing routine clinical care. Moreover, the challenges associated with the validation of the biomarkers and their future perspectives to promote progress in the interpretation of ovarian carcinoma have also been discussed in the review.

For the early investigation of ovarian cancer and to enhance the efficiency of the CA125, the incorporation of additional biomarkers into multimarker longitudinal models has been suggested by Whitwell et al. (2020) (Whitwell et al., 2020). The study considered 490 serum specimens from 49 women identified with ovarian tumor and 31 women free from carcinoma. It is observed from the experimental results that the models including CA125, HE4, CHI3L1, PEBP4, and/or AGR2 produced 85.7% and 95.4% recall and specificity, respectively, up to 1 year. Furthermore, the models also provided a recall and specificity of 95.5% and 95.4%, respectively, for all type II cases. From the results, it is concluded that longitudinal multimarker models enhanced the performance of CA125 for the early interpretation of ovarian carcinoma.

For the diagnosis of ovarian carcinoma over distinct tissue types, sample specimens, and mass spectrometry systems, a biocompatible device connected to a mass spectrometer known as MasSpec Pen has been developed by Sans et al. (2019). For extraction of molecules and diagnosis of rapid tissue, the proposed system makes use of a discrete water droplet approach. The study was carried out on 192 samples of fallopian tube, ovarian, and peritoneum tissue. Utilizing orbitrap data, the proposed MasSpec pen has achieved high performance of 96.7% clinical sensitivity and 95.7% specificity for high-grade serous cancer and 94.0% clinical recall and 94.4% specificity for overall cancer. Moreover, the proposed device achieved clinical sensitivity and specificity of 100% for high-grade and normal ovarian carcinoma using ion trap data. From the outcomes, it is noted that the developed MasSpec pen along with the machine learning approach produces more robust molecular models for the prognostication of ovarian serous cancer.

To reduce the misdiagnosis rate of ovarian carcinoma, a comparative analysis on human epididymis secretory protein 4 (HE4), ROMA index, and cancer antigen 125 (CA125) has been carried out by Zhang et al. (2019) for the investigation of ovarian cancer. The study considered data of 373 ovarian cancer patients from Tianjin Medical University Cancer Institute and Hospital during the period July 2016 to July 2017. The efficiency of serum HE4, ROMA index, and serum CA125 in the interpretation of ovarian cancer has been analyzed using the ROC curve. From the results, it is concluded that serum HE4, ROMA index, and serum CA125 have attained better efficiency in the interpretation of post-menopausal ovarian cancer rather than pre-menopausal ovarian cancer. Even though ROMA and HE4 have attained enhanced efficiency than the CA125 in most cases, they are influenced by pathological types.

A systematic analysis on the usage of LncRNAs in the propagation, cell cycle, transfer, metastasis, infringement, apoptosis, and drug resistance of ovarian cancer has been performed by Wang et al. (2019). In addition, the functionality of LncRNAs in determining the prognosis of ovarian cancer has also been discussed in the review. Finally, a clear insight on the capability of LncRNAs for validating the interpretation and prognostication of ovarian cancer has also been provided.

To predict ovarian cancer, a 26-gene panel suggested by Yeganeh and Mostafavi (2019) is used as a candidate biomarker for training machine learning models. Further, selected features have been utilized for categorizing the integrated expression data of 530 ovarian tissues. From the analysis, it is observed that cancerous tissue can be predicted using highly sensitive and specific SVM (Support Vector Machine) and RF (Random Forest) pipelines. Moreover, the quality of the 26 selected mRNAs is evaluated by reverifying the pipeline on randomly chosen mRNA expression. The other systematic work carried out on the prediction of ovarian cancer using machine learning approaches has been represented in Table 3.1.

Methodology

This study aimed to predict the pathological diagnosis of ovarian cancer using patient information and data from preoperative examinations using a DL approach known as Convolutional Neural Network (CNN). Further, the proposed model has been compared with seven machine learning classifiers, namely, Random Forest (RF), Multilayer perceptron (MLP), DT, Linear Discriminant Analysis (LDA), AdaBoost, XGBoost, and Gradient Boosting Tree (GBT). As ovarian carcinoma is one of the most prevalent cancer in women, the accurate prediction of benign ovarian tumors (BOT) and ovarian cancer (OC) has importance for real-world usability. Information about a total of 349 patients with ovarian tumors is available in the dataset, including 146 of TYPE 0 to indicate OC and 203 of TYPE 1 to indicate BOT. We also observed the importance of the nine best features for the diagnosis obtained after applying an ensemble of two filter-based feature selection methods namely Chi-Square (CS) and Mutual Information (MI). The highest accuracy obtained is 99.16 using the CNN Approach when compared with the machine learning approaches. Thus, it is observed that DL techniques

TABLE 3.1 Other systematic work on the prediction of ovarian cancer using machine learning techniques.

Author and year	Approach	Dataset	Performance results
Hwangbo et al. (2021)	logistic regression (LR), RF, SVM, and deep neural network	1002 patients from the Seoul National University Hospital, Asan Medical Center, and Severance Hospital	AUC of LR model = 0.741
Chen et al. (2020)	L2-regularized logistic regression	273 patients with stage I and II ovarian cancer	AUC is 0.621, accuracy is 0.761, sensitivity is 0.130, positive predictive value is 0.659, and F1 score is 0.216
Paik et al. (2019)	Gradient Boosting (GB) model	Training cohort $n = 1{,}1{,}28$ Validation cohort $n = 229$ from Samsung medical center	For training cohort, AUC of GB = 0.830 For validation cohort, AUC of GB = 0.843
Kawakami et al. (2019)	Gradient Boosting Machine (GBM), SVM, Conditional RF (CRF), Naïve Bayes, Neural Network (NN), RF and Elastic Net	334 patients having Epithelial Ovarian Cancer and 101 patients having benign ovarian tumor from Jikei University School of Medicine	Accuracy for RF = 92.4% and AUC of RF = 0.968 for segregating EOC from benign ovarian tumor
Alqudah (2019)	Principal component analysis (PCA)-SVM	207 noncancers and 262 ovarian cancers cases	Accuracy = 99.0% Sensitivity = 99.45% Specificity = 100%
Zhou et al. (2018)	SVM Classifier	Three datasets, namely, GSE17260, GSE44104 and GSE51088 from the Gene Expression Omnibus	Prediction accuracies of GSE17260, GSE44104, GSE51088 and TCGA are 92.7, 93.3%, 96.6%, and 90.4%, respectively
Bhattacharjee et al. (2017)	NN, Decision Tree (DT), SVM, Discriminant Analysis (DA), Nearest Neighbor Classifiers (NNC), and Ensemble Classifiers (Enclassifier)	121 Cancer and 91 benign cases from FDA-NCI center	Accuracy of NN = 99.1% Sensitivity = 100% Specificity = 97.9%
Tseng et al. (2017)	Extreme learning machine (ELM), SVM, C5.0, RF, multivariate adaptive regression splines (MARS)	987 patients from Chung Shan Medical University Hospital Tumor Registry	Classification accuracy of proposed C5.0 is 90.0%

could perform an important role in the accurate prediction of pathological diagnosis of ovarian cancer from preoperative examinations when compared with the ML approaches.

Proposed methodology

In this section, the architecture of the proposed approach and the overall flow of the proposed methodology have been described.

CNN architecture

This study makes use of a very simple CNN architecture which consists of two convolutional layers consisting of 16, 32 filters accompanied by the activation function, max-pooling, and fully connected layer.

Input layer

The reading of the pre-processed dataset is carried out by the input layer.

Convolutional layer

As the conventional layer is liable for most of the computations, it is considered the most significant layer of the proposed model. The eliciting of features from the dataset is mainly carried out at the convolutional layer by using a set of filters. Moreover, the computation of the dot product also known as convolved features is carried out by the moving of filters over the input data.

Activation function

The neural network output is determined by the activation function which is linked to every neuron of the network. Various non-linearity functions such as sigmoidal, tangent, ReLU (rectified linear unit), and parametric ReLU have been developed by various researchers to learn from complex tasks. The most commonly used activation function among the available activation functions is the ReLU activation function. Element-wise operation is carried out by the ReLU activation function. The mathematical equation of the ReLU is shown in Eq. (3.1)

$$Relu(z) = max\,(0, 2) \tag{3.1}$$

Pooling

The dimensionality reduction is accomplished by pooling layer. Several researchers have proposed different types of pooling techniques such as average, max, and global pooling. The most frequently utilized pooling technique is the max-pooling technique. The mathematical equation of the max-pooling is shown in Eqs. (3.2) and (3.3)

$$f_{max}(x) = max_{1 \le i \le n}(x_i) \tag{3.2}$$

$$f_{avg} = \frac{1}{N} \sum_{i=1}^{N} x_i \tag{3.3}$$

Fully connected layer

Finally, the flattening and categorization of data are accomplished by the convolutional layer. The fully connected layer takes output acquired from the max-pooling or convolutional layer as the input. The precise weights of the neurons in the fully connected layer are optimized by using a back-propagation process.

The summary of the developed CNN model has been represented in Fig. 3.4.

```
Model: "sequential_2"
_________________________________________________________________
Layer (type)                 Output Shape              Param #
=================================================================
conv1d_4 (Conv1D)            (None, 48, 16)            48
_________________________________________________________________
batch_normalization_4 (Batch (None, 48, 16)            64
_________________________________________________________________
dropout_6 (Dropout)          (None, 48, 16)            0
_________________________________________________________________
conv1d_5 (Conv1D)            (None, 47, 32)            1056
_________________________________________________________________
batch_normalization_5 (Batch (None, 47, 32)            128
_________________________________________________________________
dropout_7 (Dropout)          (None, 47, 32)            0
_________________________________________________________________
flatten_2 (Flatten)          (None, 1504)              0
_________________________________________________________________
dense_4 (Dense)              (None, 32)                48160
_________________________________________________________________
dropout_8 (Dropout)          (None, 32)                0
_________________________________________________________________
dense_5 (Dense)              (None, 1)                 33
=================================================================
Total params: 49,489
Trainable params: 49,393
Non-trainable params: 96
_________________________________________________________________
```

FIG. 3.4 Summary of the proposed CNN architecture.

Overall process of the proposed framework

To effectively provide this high-dimensional data for learning techniques, a data dimension reduction strategy known as feature selection has been implemented. The data dimension reduction technique aids in obtaining a clean and the most relevant subset of data to efficiently determine the outcome. In this case, an ensemble of two filter methods has been proposed for feature selection, namely, CS and MI.

Among the ranking lists produced by the two individual filter methods, sets of the top 20 features have been considered and an intersection of the two sets has then been taken. This intersection produces nine common features, thus providing us the nine most crucial features and hence the reduced attribute set was obtained from all the attributes.

The results are evaluated on both training and testing datasets using the reduced subset of attributes. This evaluation is carried out by a deep learning algorithm that is by passing it through a CNN and observing its corresponding accuracy. For validating the model, the proposed CNN approach makes use of a learning rate of 0.0001, and the number of epochs is 200. At the same time, another set of observations are noted by validating the model using machine learning classifiers. The classifiers used are RF, DT, LDA, MLP, AdaBoost, GBT, and XGBoost. The flowchart of the same has been demonstrated in Fig. 3.5.

Although the CNN itself extracts features, for medical analysis, it is important to select only the necessary attributes and eliminate the noisy and redundant ones. This would be helpful for the doctors by reducing the diagnosis time so that the treatment can be started as early as possible. The lesser the attributes, the lesser the tests required and hence lesser would be the economic pressure on the patients. This is the reason why the ensemble of MI and CS is applied for feature selection to extract the best set of features.

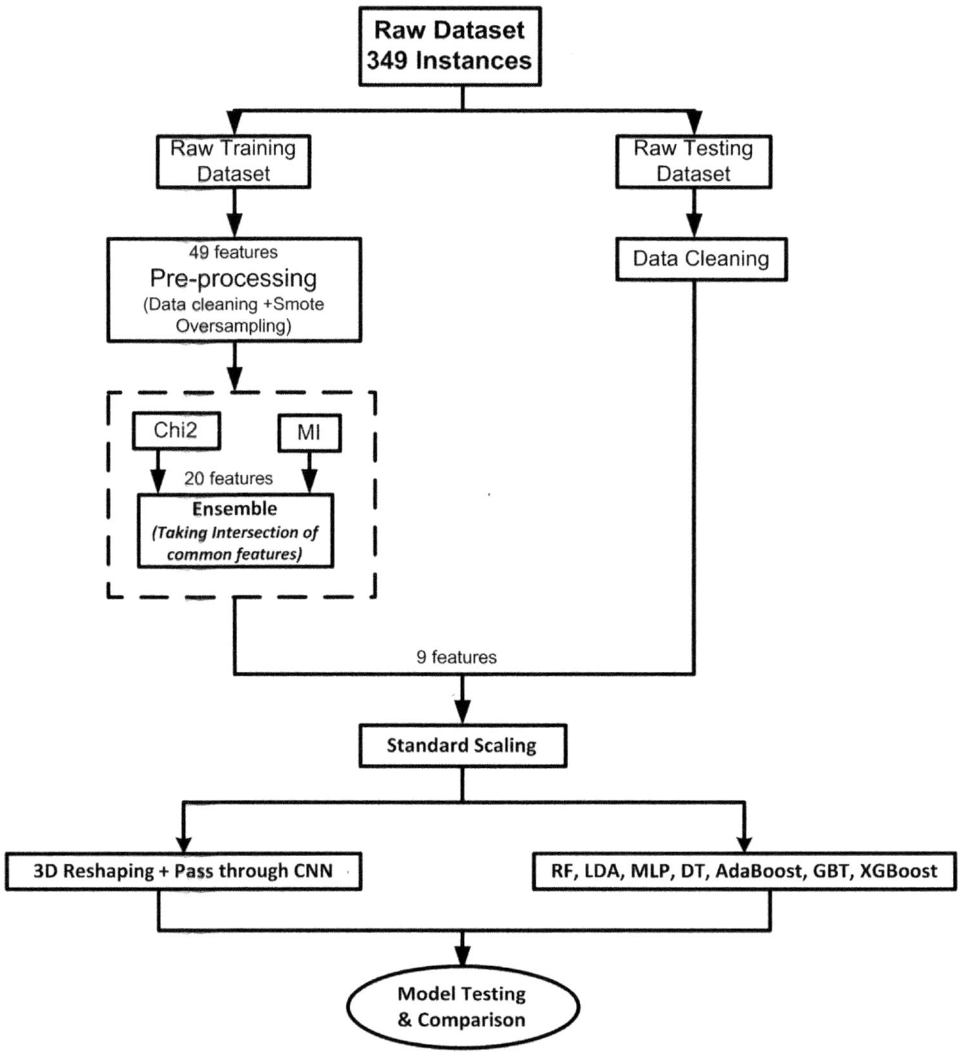

FIG. 3.5 A flowchart representing the modeling procedure of the proposed framework.

Result analysis

The dataset has been obtained from Kaggle (Predict Ovarian Cancer Dataset, 2021). In this proposed method, the model has been developed on the training dataset and validated on the testing dataset. A total of 349 patient's raw data are available here. Out of them, 235 instances have been separated out as training datasets and 114 instances as testing datasets. There are a total of 49 variables at this stage which include features like demographic, tumor markers, biodatasets of routine test, general chemistry, and histology. The biomarkers have been briefed in Table 3.2.

TABLE 3.2 A list of biomarkers, their abbreviations, and their descriptions used in the study.

Abbreviation	BioMarker name	Sample type	Instrument	Method	Lower limit	Upper limit	Unit
MPV	Mean platelet volume	Full blood	Sysmex XE-2100	Calculation method	7.4	12.5	fL
BASO#	Basophil cell count	Full blood	Sysmex XE-2100	FCM with semiconductor laser	0	0.06	10^9/L
PHOS	Phosphorus	Serum	Beckman Coulter AU5800	Phosphomolybdate method	0.7	1.62	mmol/L
GLU.	Glucose	Serum	Beckman Coulter AU5800	Glucose oxidase method	3.9	6.1	mmol/L
CA72-4	Carbohydrate antigen 72-4	Serum	Roche Cobas 8000	ECLIA	0	7	U/mL
K	Kalium	Serum	Beckman Coulter AU5800	Ion selective electrode method	3.5	5.3	mmol/L
AST	Aspartate aminotransferase	Serum	Beckman Coulter AU5800	MDH method	6	40	u/L
BASO%	Basophil cell ratio	Full blood	Sysmex XE-2100	Calculation method	0	1	%
Mg	Magnesium	Serum	Beckman Coulter AU5800	XB-I method	0.73	1.3	mmol/L
CL	Chlorine	Serum	Beckman Coulter AU5800	Ion selective electrode method	99	110	mmol/L
CEA	Carcinoembryonic antigen	Serum	Roche Cobas 8000	ECLIA	0	5	ng/mL

Continued

TABLE 3.2 A list of biomarkers, their abbreviations, and their descriptions used in the study—cont'd

Abbreviation	BioMarker name	Sample type	Instrument	Method	Lower limit	Upper limit	Unit
EO#	Eosinophil count	Full blood	Sysmex XE-2100	FCM with semiconductor laser	0.02	0.52	10^9/L
CA19-9	Carbohydrate antigen 19-9	Serum	Roche Cobas 8000	ECLIA	0	37	U/mL
ALB	Albumin	Serum	Beckman Coulter AU5800	Bromocresol green method	35	55	g/L
IBIL	Indirect bilirubin	Serum	Beckman Coulter AU5800	Calculation	2	15	μmol/L
GGT	Gama glutamyltransferase	Serum	Beckman Coulter AU5800	Rate method	3	73	u/L
MCH	Mean corpuscular hemoglobin	Full blood	Sysmex XE-2100	Calculation method	27	34	Pg
GLO	Globulin	Serum	Beckman Coulter AU5800	Calculation	20	40	g/L
ALT	Alanine aminotransferase	Serum	Beckman Coulter AU5800	Lactate dehydrogenase method	1	45	u/L
DBIL	Direct bilirubin	Serum	Beckman Coulter AU5800	Vanadate oxidation	1.5	7	umol/L
RDW	Red blood cell distribution width	Full blood	Sysmex XE-2100	Calculation method	10.6	15.5	%
PDW	Platelet distribution width	Full blood	Sysmex XE-2100	Calculation method	15.5	18.1	%
CREA	Creatinine	Serum	Beckman Coulter AU5800	Jaffe method	44	144	umol/L
AFP	Alpha-fetoprotein	Serum	Roche Cobas 8000	ECLIA	0	7	ng/mL
HGB	Hemoglobin	Full blood	Sysmex XE-2100	SLS-Hemoglobin method	110	150	g/L
Na	Natrium	Serum	Beckman Coulter AU5800	ion selective electrode method	137	147	mmol/L
HE4	Human epididymis protein 4	Serum	Roche Cobas 8000	ECLIA	0	140	pmol/L

TABLE 3.2 A list of biomarkers, their abbreviations, and their descriptions used in the study—cont'd

Abbreviation	BioMarker name	Sample type	Instrument	Method	Lower limit	Upper limit	Unit
LYM#	Lymphocyte count	Full blood	Sysmex XE-2100	FCM with semiconductor laser	1.1	3.2	10^9/L
CA125	Carbohydrate antigen 125	Serum	Roche Cobas 8000	ECLIA	0	35	U/mL
BUN	Blood urea nitrogen	Serum	Beckman Coulter AU5800	Urease-glutamic acid dehydrogenase	1.7	8.3	mmol/L
LYM%	Lymphocyte ratio	Full blood	Sysmex XE-2100	Calculation method	20	50	%
Ca	Calcium	Serum	Beckman Coulter AU5800	Arsenazo III method	1.12	1.32	mmol/L
AG	Anion gap	Serum	Beckman Coulter AU5800	Calculation	8	30	mmol/l
MONO#	Mononuclear cell count	Full blood	Sysmex XE-2100	FCM with semiconductor laser	0.1	0.6	10^9/L
PLT	Platelet count	Full blood	Sysmex XE-2100	Hydrodynamic Focusing DC method	125	350	10^9/L
NEU	Neutrophil ratio	Full blood	Sysmex XE-2100	Calculation method	40	75	%
EO%	Eosinophil ratio	Full blood	Sysmex XE-2100	Calculation method	0.02	0.52	10^9/L
TP	Total protein	Serum	Beckman Coulter AU5800	Biuret method	60	82	g/L
UA	Uric acid	Serum	Beckman Coulter AU5800	Urease method	90	450	μmol/L
RBC	Red blood cell count	Full blood	Sysmex XE-2100	Hydrodynamic Focusing DC method	3.5	5.5	10^{12}/L
PCT	Thrombocytocrit	Full blood	Sysmex XE-2100	Calculation method	0.114	0.282	L/L
CO2CP	Carbon dioxide-combining Power	Serum	Beckman Coulter AU5800	Enzymatic method	18	30	mmol/L

Continued

TABLE 3.2 A list of biomarkers, their abbreviations, and their descriptions used in the study—cont'd

Abbreviation	BioMarker name	Sample type	Instrument	Method	Lower limit	Upper limit	Unit
TBIL	Total bilirubin	Serum	Beckman Coulter AU5800	Vanadate oxidation	4	19	μmol/L
HCT	Hematocrit	Full blood	Sysmex XE-2100	Hydrodynamic Focusing DC method	0.35	0.45	L/L
MONO%	Monocyte ratio	Full blood	Sysmex XE-2100	Calculation method	3	10	%
MCV	Mean corpuscular volume	Full blood	Sysmex XE-2100	Calculation method	82	100	fL
ALP	Alkaline phosphatase	Serum	Beckman Coulter AU5800	NPP substrate-AMP buffer method	25	130	u/L

Data pre-processing is a very important step that obtains some information from datasets which helps in making accurate predictions. In this dataset, pre-processing involved data cleaning, standardizing, and normalizing approaches. Also, in the training dataset, ovarian cancer (OC) class has 146 records, whereas benign ovarian tumor (BOT) class (minority class) has 89 records. The Synthetic Minority Oversampling Technique (SMOTE) is used to remove this imbalance by creating samples of minority class using the present data.

Thus, after processing, the ensemble has been applied to obtain nine best features which are "ALB," "BUN," "CA125," "CA19-9," "CA72-4," "CEA," "CREA," "GLU," and "HE4." This implies that among all the 49 features which were there initially, these nine features form the best set for analysis. By applying this best set to the training data, the model is first trained by the training set. Using the best set of features, the accuracy of the testing set with respect to those predicted by the trained model was noted and compared. The parameter set up for the tested models has been tabulated in Table 3.3.

The efficiency of the developed CNN model and the ML approaches has been evaluated using various evaluation metrics such as Precision, Recall, F1score, Area under Receiver Operating Characteristics (ROC-AUC), and Accuracy (Nayak et al., 2021). The results obtained by the Deep Learning Algorithm CNN and that by the machine learning algorithms DT, LDA, MLP, RF, AdaBoost, GBT, and XGBoost on the testing dataset have been tabulated in Table 3.4.

It is clear from Table 3.3 when the dataset itself was split into 235 instances and 114 instances for training and testing, respectively, the CNN approach obtained better performance in terms of precision, recall, F1 score, ROC-AUC, and accuracy with values 0.9773, 1.0, 0.9914, 0.9921, and 99.16, respectively, when compared with the conventional machine learning approaches DT, LDA, MLP, RF, AdaBoost, GBT, and XGBoost.

TABLE 3.3 Parameters of all considered models.

Models	Parameter
DT	criterion='gini', splitter='best', min_samples_split=2, min_samples_leaf=1, min_weight_fraction_leaf=0.0, presort=False
LDA	solver='svd', store_covariance=False, tol=0.0001
MLP	hidden_layer_sizes=100, activation='relu', solver='adam', alpha=0.0001, batch_size='auto', learning_rate='constant', learning_rate_init=0.001, power_t=0.5, max_iter=200, shuffle=True, tol=0.0001, verbose=False, warm_start=False, momentum=0.9, nesterovs_momentum=True, early_stopping=False, validation_fraction=0.1, beta_1=0.9, beta_2=0.999, epsilon=1e-08, n_iter_no_change=10, max_fun=15000
RF	n_estimators=100, criterion='gini', min_samples_split=2, min_samples_leaf=1, min_weight_fraction_leaf=0.0, max_features='auto', min_impurity_decrease=0.0, bootstrap=True, oob_score=False, verbose=0, warm_start=False, ccp_alpha=0.0
AdaBoost	base_estimator=DecisionTreeClassifier,n_estimators=200, learning_rate=1.0,algorithm='SAMME.R'
GBT	loss='deviance', learning_rate=0.1, n_estimators=200, subsample=1.0, criterion='friedman_mse', min_samples_split=2, min_samples_leaf=1, min_weight_fraction_leaf=0.0, max_depth=3, min_impurity_decrease=0.0, verbose=0, warm_start=False, validation_fraction=0.1, tol=0.0001, ccp_alpha=0.0
XGBoost	n_estimators=200, max_depth=6, reg_lambda=0, reg_alpha=0
CNN	filters=16, kernel_size=2, activation='relu', optimizer=Adam (), loss=binary_crossentropy

TABLE 3.4 Comparative analysis of the evaluation metrics of the developed CNN model and machine learning approaches.

Prediction models	Precision	Recall	F1 score	ROC-AUC	Accuracy
DT	0.7508	0.7508	0.7508	0.7513	75.08
LDA	0.6456	0.6456	0.6456	0.6444	64.56
MLP	0.7228	0.7228	0.7228	0.7237	72.28
RF	0.8350	0.8350	0.8350	0.8349	83.50
AdaBoost	0.7649	0.7649	0.7649	0.7650	76.49
GBT	0.8385	0.8385	0.8385	0.8388	83.85
XGBoost	0.8315	0.8315	0.8315	0.8307	83.15
Proposed CNN model	0.9773	1.0	0.9914	0.9921	99.16

The ROC for the testing data as obtained in the case of each ML Classifier has been plotted in Figs. 3.6–3.13.

Furthermore, the performance of the CNN model has been analyzed through the training and validation accuracy and loss curves as shown in Figs. 3.14 and 3.15.

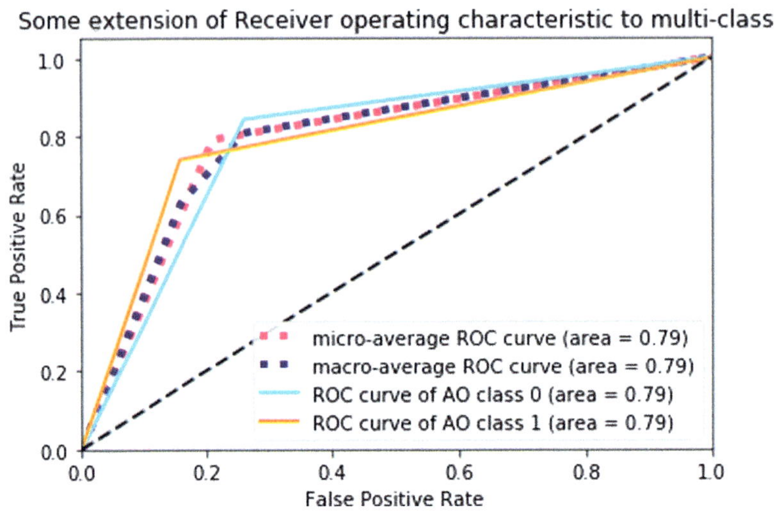

FIG. 3.6 ROC-AUC curve of DT.

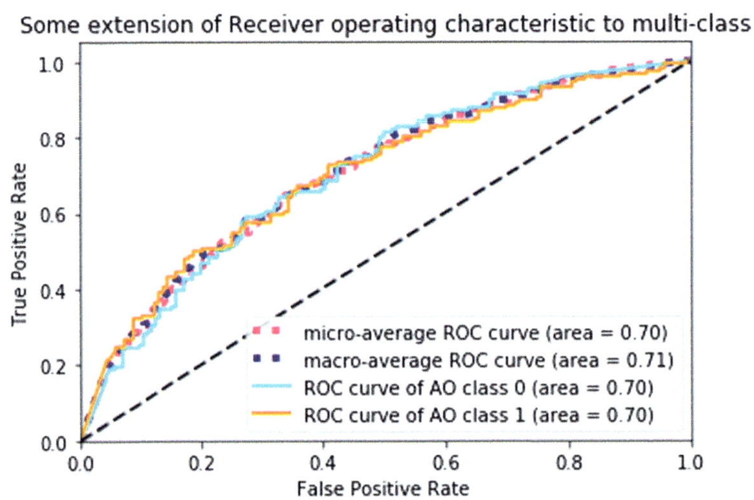

FIG. 3.7 ROC-AUC curve of LDA.

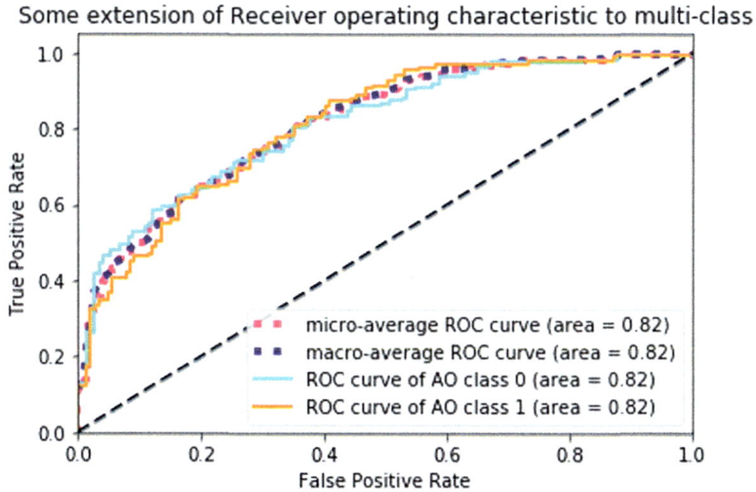

FIG. 3.8 ROC-AUC curve of MLP.

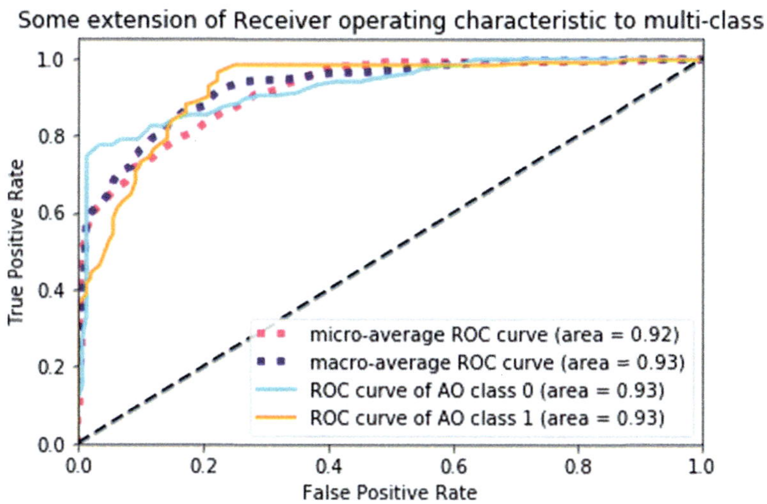

FIG. 3.9 ROC-AUC curve of RF.

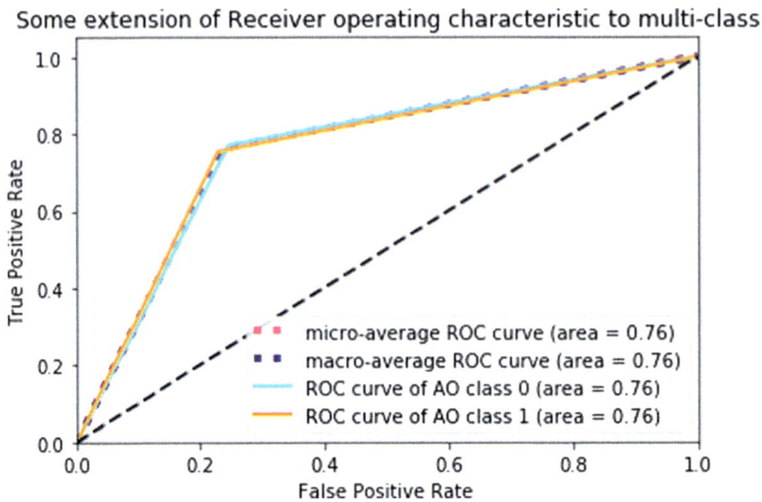

FIG. 3.10 ROC-AUC curve of AdaBoost.

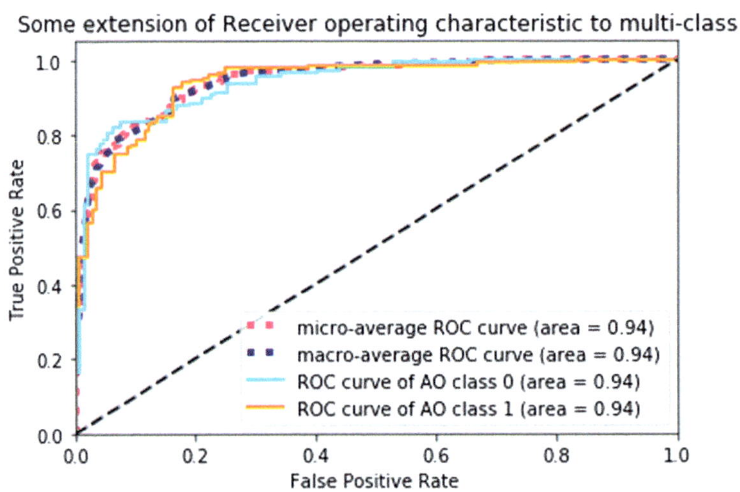

FIG. 3.11 ROC-AUC Curve of GBoost.

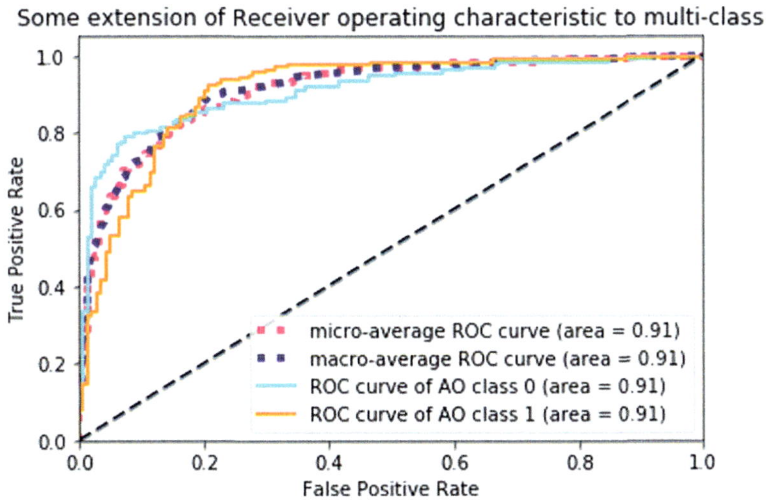

FIG. 3.12 ROC-AUC curve of XGBoost.

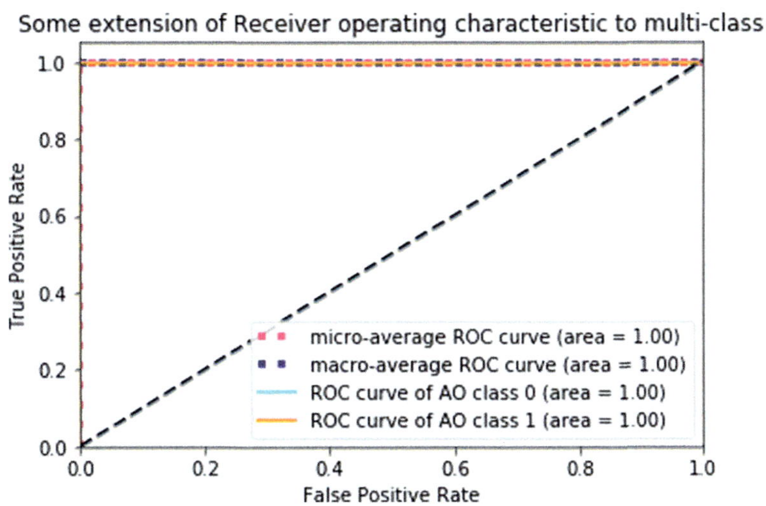

FIG. 3.13 ROC-AUC curve of CNN.

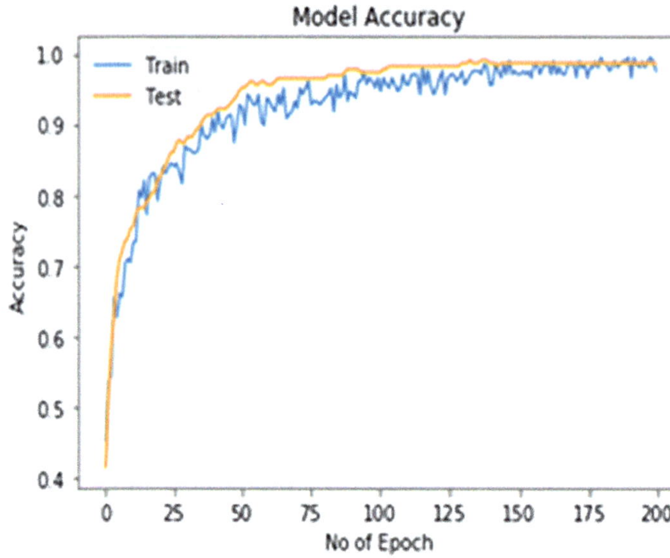

FIG. 3.14 Plot of the model accuracy with respect to the number of epochs.

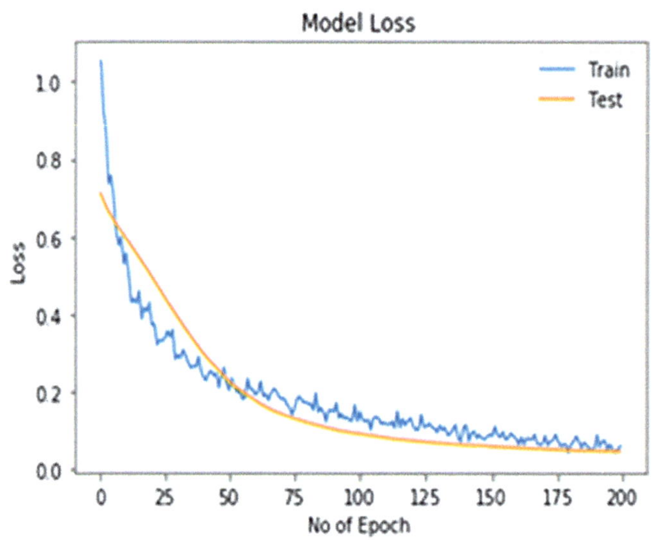

FIG. 3.15 Plot of the model loss with respect to the number of epochs.

Conclusion

Early identification of ovarian cancer plays a crucial step in reducing the rate of mortality of ovarian cancer in women. Further, accurate classification of ovarian cancer into benign and malignant ovarian carcinoma utilizing biomarkers has a profound impact on the lives of many people throughout the world. Therefore, the development of a more accurate prediction model is of great importance for the prediction and detection of ovarian cancer at an early stage. Hence, a deep learning approach known as the convolutional neural network has been developed for the interpretation of ovarian cancer from the dataset consisting of 349 instances of ovarian cancer patients. To make accurate prediction, data pre-processing approaches such as data cleaning have been applied on the dataset. As the dataset consists of 146 ovarian cancer and 89 benign ovarian cancer instances, the SMOTE approach has been applied to remove this imbalance. After pre-processing, the ensemble of feature extraction methods such as CS and MI has been applied to obtain nine best features out of the available 49 features from the dataset. Then, the model is trained and tested by using the best-selected features. The experimental outcomes confess that the developed CNN model obtains better performance in terms of precision, recall, F1 score, ROC-AUC, and accuracy with values 0.9773, 1.0, 0.9914, 0.9921, and 99.16, respectively, when compared with the conventional machine learning approaches DT, LDA, MLP, RF, AdaBoost, GBT, and XGBoost. Validating the model on large datasets to further enhance the accuracy of the developed CNN model can be considered the future scope of the work.

References

Akazawa, M., & Hashimoto, K. (2020). Artificial intelligence in ovarian cancer diagnosis. *Anticancer Research, 40*(8), 4795–4800. https://doi.org/10.21873/anticanres.14482.

Alqudah, A. M. (2019). Ovarian cancer classification using serum proteomic profiling and wavelet features A comparison of machine learning and features selection algorithms. *Journal of Clinical Engineering, 44*(4), 165–173. https://doi.org/10.1097/jce.0000000000000359.

Bach, P. B., Kattan, M. W., Thornquist, M. D., Kris, M. G., Tate, R. C., Barnett, M. J., Hsieh, L. J., & Begg, C. B. (2003). Variations in lung cancer risk among smokers. *Journal of the National Cancer Institute, 95*(6), 470–478. https://doi.org/10.1093/jnci/95.6.470.

Bhattacharjee, S., Singh, Y. J., & Ray, D. (2017). Comparative performance analysis of machine learning classifiers on ovarian cancer dataset. In *2017. Proceedings - 2017 3rd IEEE international conference on research in computational intelligence and communication networks, ICRCICN 2017* (pp. 213–218). Institute of Electrical and Electronics Engineers Inc. https://doi.org/10.1109/ICRCICN.2017.8234509.

Bocchi, L., Coppini, G., Nori, J., & Valli, G. (2004). Detection of single and clustered microcalcifications in mammograms using fractals models and neural networks. *Medical Engineering and Physics, 26*(4), 303–312. https://doi.org/10.1016/j.medengphy.2003.11.009.

Cancer Facts & Figures. (2021). https://www.cancer.org/research/cancer-facts-statistics/all-cancer-facts-figures/cancer-facts-figures-2021.html.

Cancer Facts for Women. (2021). https://www.cancer.org/healthy/cancer-facts/cancer-facts-for-women.html.

Chen, R. (2020). Machine learning for ovarian cancer: Lasso regression-based predictive model of early mortality in patients with stage I and stage II ovarian cancer. medRxiv, https://doi.org/10.1101/2020.05.01.20088294

Cochran, A. J. (1997). Prediction of outcome for patients with cutaneous melanoma. *Pigment Cell Research, 10*(3), 162–167. https://doi.org/10.1111/j.1600-0749.1997.tb00479.x.

Coudray, N., Ocampo, P. S., Sakellaropoulos, T., Narula, N., Snuderl, M., Fenyö, D., Moreira, A. L., Razavian, N., & Tsirigos, A. (2018). Classification and mutation prediction from non–small cell lung cancer histopathology images using deep learning. *Nature Medicine, 24*(10), 1559–1567. https://doi.org/10.1038/s41591-018-0177-5.

Dettling, M. (2004). BagBoosting for tumor classification with gene expression data. *Bioinformatics, 20*(18), 3583–3593. https://doi.org/10.1093/bioinformatics/bth447.

Domchek, S. M., Eisen, A., Calzone, K., Stopfer, J., Blackwood, A., & Weber, B. L. (2003). Application of breast cancer risk prediction models in clinical practice. *Journal of Clinical Oncology, 21*(4), 593–601. https://doi.org/10.1200/JCO.2003.07.007.

Doppalapudi, S., Qiu, R. G., & Badr, Y. (2021). Lung cancer survival period prediction and understanding: Deep learning approaches. *International Journal of Medical Informatics, 148*. https://doi.org/10.1016/j.ijmedinf.2020.104371.

Elfiky, A. A., Pany, M. J., Parikh, R. B., & Obermeyer, Z. (2018). Development and application of a machine learning approach to assess short-term mortality risk among patients with Cancer starting chemotherapy. *JAMA Network Open*, *1*(3), e180926. https://doi.org/10.1001/jamanetworkopen.2018.0926.

Fielding, L. P., Fenoglio-Preiser, C. M., & Freedman, L. S. (1992). The future of prognostic factors in outcome prediction for patients with cancer. *Cancer*, *70*(9), 2367–2377. https://doi.org/10.1002/1097-0142(19921101)70:9<2367::AID-CNCR2820700927>3.0.CO;2-B.

Fitzmaurice, C., Dicker, D., Pain, A., Hamavid, H., Moradi-Lakeh, M., & MacIntyre, M. F. (2013). Global burden of disease cancer collaboration. *JAMA Oncology*, *1*(4), 505–527.

Gascón, F., Valle, M., Martos, R., Zafra, M., Morales, R., & Castano, M. (2004). Childhood obesity and hormonal abnormalities associated with cancer risk. *European Journal of Cancer Prevention*, *13*(3), 193–197. https://doi.org/10.1097/01.cej.0000130021.16182.c3.

Hulstaert, E., Morlion, A., Levanon, K., Vandesompele, J., & Mestdagh, P. (2021). Candidate RNA biomarkers in biofluids for early diagnosis of ovarian cancer: A systematic review. *Gynecologic Oncology*, *160*(2), 633–642. https://doi.org/10.1016/j.ygyno.2020.11.018.

Hwangbo, S., Kim, S. I., Kim, J. H., Eoh, K. J., Lee, C., Kim, Y. T., Suh, D. S., Park, T., & Song, Y. S. (2021). Development of machine learning models to predict platinum sensitivity of high-grade serous ovarian carcinoma. *Cancers*, *13* (8). https://doi.org/10.3390/cancers13081875.

Jin, K., Li, T., van Dam, H., Zhou, F., & Zhang, L. (2017). Molecular insights into tumour metastasis: Tracing the dominant events. *Journal of Pathology*, *241*(5), 567–577. https://doi.org/10.1002/path.4871.

Kather, J. N., Krisam, J., Charoentong, P., Luedde, T., Herpel, E., Weis, C.-A., Gaiser, T., Marx, A., Valous, N. A., Ferber, D., Jansen, L., Reyes-Aldasoro, C. C., Zörnig, I., Jäger, D., Brenner, H., Chang-Claude, J., Hoffmeister, M., Halama, N., & Butte, A. J. (2019). Predicting survival from colorectal cancer histology slides using deep learning: A retrospective multicenter study. *PLoS Medicine*, *e1002730*. https://doi.org/10.1371/journal.pmed.1002730.

Kawakami, E., Tabata, J., Yanaihara, N., Ishikawa, T., Koseki, K., Iida, Y., Saito, M., Komazaki, H., Shapiro, J. S., Goto, C., Akiyama, Y., Saito, R., Saito, M., Takano, H., Yamada, K., & Okamoto, A. (2019). Application of artificial intelligence for preoperative diagnostic and prognostic prediction in epithelial ovarian cancer based on blood biomarkers. *Clinical Cancer Research*, *25*(10), 3006–3015. https://doi.org/10.1158/1078-0432.CCR-18-3378.

Manne, R., Kantheti, S., & Kantheti, S. (2020). Classification of skin cancer using deep learning, convolutional neural networks-opportunities and vulnerabilities—A systematic review. *International Journal for Modern Trends in Science and Technology*, 2455–3778.

National Cancer Institute SEER. (2021). https://seer.cancer.gov/statfacts/html/ovary.html.

Nayak, J., Naik, B., Dash, P. B., Souri, A., & Shanmuganathan, V. (2021). Hyper-parameter tuned light gradient boosting machine using memetic firefly algorithm for hand gesture recognition. *Applied Soft Computing*, *107*, 107478. https://doi.org/10.1016/j.asoc.2021.107478.

Ovarian Cancer Statistics. (2021). https://www.cancer.org/cancer/ovarian-cancer/about/key-statistics.html.

Paik, E. S., Lee, J. W., Park, J. Y., Kim, J. H., Kim, M., Kim, T. J., Choi, C. H., Kim, B. G., Bae, D. S., & Seo, S. W. (2019). Prediction of survival outcomes in patients with epithelial ovarian cancer using machine learning methods. *Journal of Gynecologic Oncology*, *30*(4). https://doi.org/10.3802/jgo.2019.30.e65.

Petricoin, E. F., & Liotta, L. A. (2004). SELDI-TOF-based serum proteomic pattern diagnostics for early detection of cancer. *Current Opinion in Biotechnology*, *15*(1), 24–30. https://doi.org/10.1016/j.copbio.2004.01.005.

Predict Ovarian Cancer Dataset. (2021). https://www.kaggle.com/Saurabhshahane/predict-ovarian-cancer.

Rajkomar, A., Oren, E., Chen, K., Dai, A. M., Hajaj, N., Hardt, M., Liu, P. J., Liu, X., Marcus, J., Sun, M., Sundberg, P., Yee, H., Zhang, K., Zhang, Y., Flores, G., Duggan, G. E., Irvine, J., Le, Q., Litsch, K., … Dean, J. (2018). Scalable and accurate deep learning with electronic health records. *Npj Digital Medicine*. https://doi.org/10.1038/s41746-018-0029-1.

Sans, M., Zhang, J., Lin, J. Q., Feider, C. L., Giese, N., Breen, M. T., Sebastian, K., Liu, J., Sood, A. K., & Eberlin, L. S. (2019). Performance of the MasSpec pen for rapid diagnosis of ovarian cancer. *Clinical Chemistry*, *65*(5), 674–683. https://doi.org/10.1373/clinchem.2018.299289.

Shameer, K., Johnson, K. W., Yahi, A., Miotto, R., Li, L. I., Ricks, D., Jebakaran, J., Kovatch, P., Sengupta, P. P., Gelijns, A., Moskovitz, A., Darrow, B., Reich, D. L., Kasarskis, A., Tatonetti, N. P., Pinney, S., & Dudley, J. T. (2017). Predictive modeling of hospital readmission rates using electronic medical record-wide machine learning: A case-study using Mount Sinai heart failure cohort. In *Pacific symposium on biocomputing* (pp. 276–287). World Scientific Publishing Co. Pte Ltd. https://doi.org/10.1142/9789813207813_0027. Vol. 0, Issue 212679.

Sung, H., Ferlay, J., Siegel, R. L., Laversanne, M., Soerjomataram, I., Jemal, A., & Bray, F. (2021). Global cancer statistics 2020: GLOBOCAN estimates of incidence and mortality worldwide for 36 cancers in 185 countries. *CA: A Cancer Journal for Clinicians, 71*(3), 209–249. https://doi.org/10.3322/caac.21660.

Tseng, C. J., Lu, C. J., Chang, C. C., Chen, G. D., & Cheewakriangkrai, C. (2017). Integration of data mining classification techniques and ensemble learning to identify risk factors and diagnose ovarian cancer recurrence. *Artificial Intelligence in Medicine, 78*, 47–54. https://doi.org/10.1016/j.artmed.2017.06.003.

Wang, D., Khosla, A., Gargeya, R., Irshad, H., & Beck, A. H. (2016). Deep learning for identifying metastatic breast cancer.

Wang, J.y., Lu, A.q., & Chen, L.j. (2019). LncRNAs in ovarian cancer. *Clinica Chimica Acta, 490*, 17–27. https://doi.org/10.1016/j.cca.2018.12.013.

Wang, J. X., Zhang, B., Yu, J. K., Liu, J., Yang, M. Q., & Zheng, S. (2005). Application of serum protein fingerprinting coupled with artificial neural network model in diagnosis of hepatocellular carcinoma. *Chinese Medical Journal, 118* (15), 1278–1286.

Whitwell, H. J., Worthington, J., Blyuss, O., Gentry-Maharaj, A., Ryan, A., Gunu, R., Kalsi, J., Menon, U., Jacobs, I., Zaikin, A., & Timms, J. F. (2020). Improved early detection of ovarian cancer using longitudinal multimarker models. *British Journal of Cancer, 122*(6), 847–856. https://doi.org/10.1038/s41416-019-0718-9.

WHO, The Global Health Estimates. (2021). https://www.who.int/data/gho/data/themes/mortality-and-global-health-estimates/ghe-leading-causes-of-death.

Wulczyn, E., Steiner, D. F., Moran, M., Plass, M., Reihs, R., Tan, F., Flament-Auvigne, I., Brown, T., Regitnig, P., Chen, P. H. C., Hegde, N., Sadhwani, A., MacDonald, R., Ayalew, B., Corrado, G. S., Peng, L. H., Tse, D., Müller, H., Xu, Z., … Mermel, C. H. (2021). Interpretable survival prediction for colorectal cancer using deep learning. *Npj Digital Medicine, 4*(1). https://doi.org/10.1038/s41746-021-00427-2.

Yeganeh, P. N., & Mostafavi, M. T. (2019). Use of machine learning for diagnosis of cancer in ovarian tissues with a selected mRNA panel. In *Proceedings - 2018 IEEE international conference on bioinformatics and biomedicine, BIBM 2018* (pp. 2429–2434). Institute of Electrical and Electronics Engineers Inc. https://doi.org/10.1109/BIBM.2018.8621371.

Zhang, L., Chen, Y., & Wang, K. (2019). Comparison of CA125, HE4, and ROMA index for ovarian cancer diagnosis. *Current Problems in Cancer, 43*(2), 135–144. https://doi.org/10.1016/j.currproblcancer.2018.06.001.

Zhou, J., Li, L., Wang, L., Li, X., Xing, H., & Cheng, L. (2018). Establishment of a SVM classifier to predict recurrence of ovarian cancer. *Molecular Medicine Reports, 18*(4), 3589–3598. https://doi.org/10.3892/mmr.2018.9362.

Zhou, X., Liu, K. Y., & Wong, S. T. C. (2004). Cancer classification and prediction using logistic regression with Bayesian gene selection. *Journal of Biomedical Informatics, 37*(4), 249–259. https://doi.org/10.1016/j.jbi.2004.07.009.

Zhou, L. Q., Wu, X. L., Huang, S. Y., Wu, G. G., Ye, H. R., Wei, Q., Bao, L. Y., Deng, Y. B., Li, X. R., Cui, X. W., & Dietrich, C. F. (2020). Lymph node metastasis prediction from primary breast cancer US images using deep learning. *Radiology, 294*(1), 19–28. https://doi.org/10.1148/radiol.2019190372.

CHAPTER

4

Learning from multiple modalities of imaging data for cancer diagnosis

Liping Song[a,b,c], *Shuai Liu*[a,b,c], *and Xiangbin Liu*[a,b,c]

[a]Hunan Provincial Key Laboratory of Intelligent Computing and Language Information Processing, Hunan Normal University, Changsha, People's Republic of China [b]College of Information Science and Engineering, Hunan Normal University, Changsha, People's Republic of China [c]Xiangjiang Institute of Artificial Intelligence, Changsha, People's Republic of China

Introduction

With the improvement of living standards today, people have more and more choices in diet. However, with the diversification of diet and the decrease of health awareness, the diseases faced by human beings not only increase dramatically but also have complex and diverse characteristics. When people mention the word "cancer," they cannot help but feel scared. Cancer is one of the leading causes of illness and death in the world. Judging from the "World Cancer Report 2020" released by the World Health Organization in February this year, the number of cancer cases worldwide may increase by 60%. In low - and middle - income countries, the enhancement may be as high as 81%. One in six people worldwide dies of cancer every year. In 2018, about 9.6 million people worldwide died of cancer. By 2040, the total number of patients worldwide will reach 27.5 million, and 16.3 million patients will die as a result. Lung cancer, breast cancer, colorectal cancer, prostate cancer, and stomach cancer are the top five cancer species in the world (Siegel et al., 2020).

At present, many countries have invested a great deal of human and financial resources to research the treatment of cancer. At the same time, hospitals have taken cancer screening, diagnosis, and treatment as the focus of research. When doctors or researchers perform quantitative analysis, real -time monitoring, and treatment planning of specific internal tissue and organ, in order to make correct treatment decisions, they usually need to know detailed information about this tissue and organ. The final medical image is obtained by selecting the appropriate medical imaging technology for the corresponding tissues and organs. And then

https://doi.org/10.1016/B978-0-323-85240-1.00005-5

the lesions in the medical image are analyzed. With the continuous development and advancement of computer technology, artificial intelligence has penetrated more and more into human life, including intelligent medical diagnoses. Using computer image processing technology, medical imaging improves the working efficiency of doctors and alleviates the problem of doctor resource shortages. Therefore, medical image processing has become an indispensable tool and technical means in medical research, clinical disease diagnosis, and treatment (Bhansali & Mehta, 2020).

With more advanced medical methods, various new types of medical imaging equipment are becoming more popular. The object of medical image processing is medical images with multiple modalities of image data. The types of medical imaging widely used clinically mainly include computer tomography (CT), X -ray, magnetic resonance imaging (MRI), positron emission tomography (PET), ultrasound imaging (UI), and pathological image. In addition, it also includes some common RGB images, such as microscopy and fundus retinal images. Doctors use CT and other medical images to judge the patient's condition, which has gradually become the mainstay for doctors' clinical diagnosis. Learning features for cancer diagnosis from multiple modalities of imaging data has become the focus of attention in the field of computer vision. Under this situation, the emergence of computer -aided diagnosis (CAD) and computer -aided detection (CAD) has finally become the need of development in this era. This chapter summarizes and analyzes the medical images with different characteristics produced by different imaging technologies, starting from the diseases where the CAD is commonly applied and the research methods used in medical images in recent years (Natarajan & Gokulachandran, 2021). The structure of this chapter is given in Fig. 4.1.

Diagnosis in CT

Roentgen, a German scientist, discovered X -ray in 1895, which opened the door to medical imaging. In the past, if doctors wanted to observe the inside of a patient's body, they could only rely on dissection of the patient's body. Obviously, the risks and resource consumption of this method are high. X -ray was used for clinical diagnosis shortly after its birth. Since then, scientists have invented other imaging diagnostic methods to further improve medical imaging diagnosis.

CT was invented in 1972. Different from X -ray imaging, CT is a reconstructed image obtained by scanning the human body with an X -ray beam and processing the obtained information through a computer. CT shows an anatomical cross -section, and its density resolution is much better than that of the X -ray image. Using it for cancer detection and diagnosis improves the detection rate of lesions and the accuracy of diagnosis (Yip et al., 2021).

Imaging characteristics of CT

CT is called computer tomography. It uses X -ray beams, γ ray, ultrasound, etc., combined with extremely sensitive detectors to scan a certain part of the human body one by one. According to the different rays used, it can be divided into: X -ray and γ -ray.

Since its discovery, X -ray has been used in medicine to detect human diseases. CT is X -ray scanning of a certain thickness of the human body inspection part. The detector receives X -ray that passes through human tissues in different directions on the plane and converts

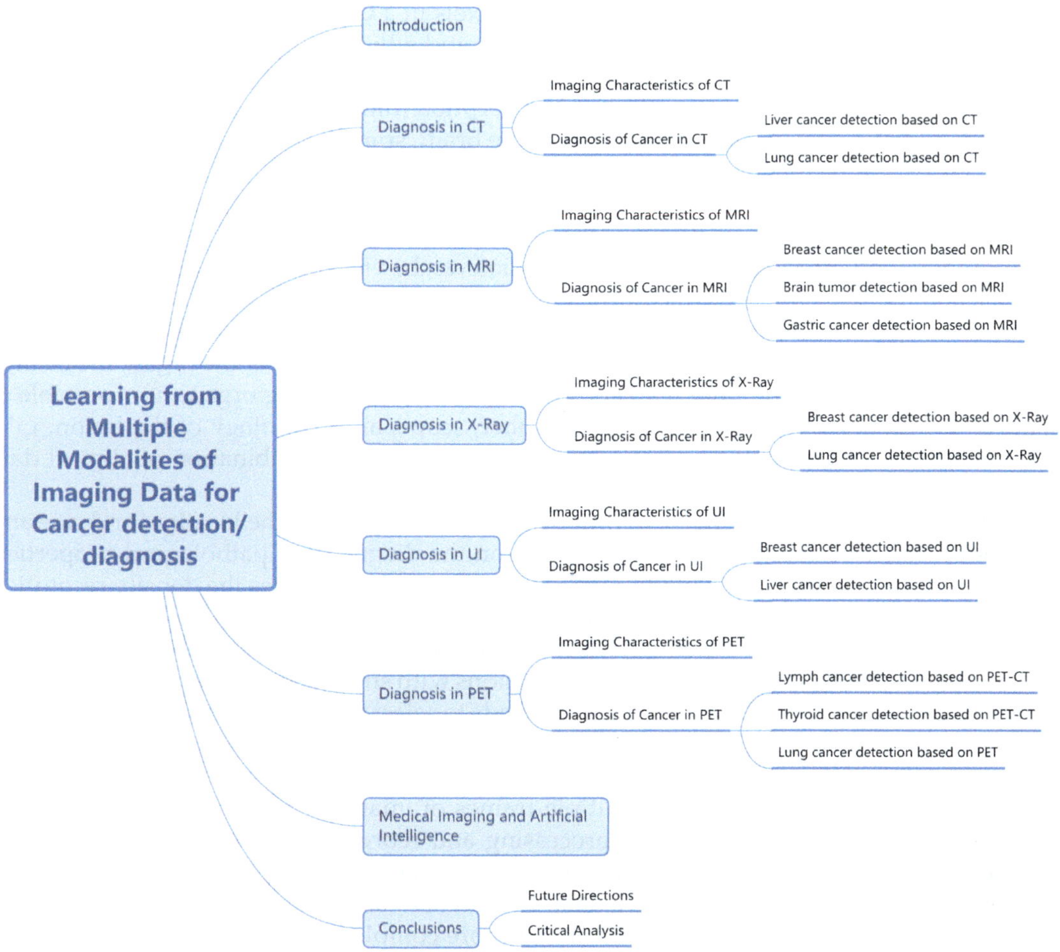

FIG. 4.1 Structure of the chapter.

photoelectric into electrical signals. The electrical signal obtained is converted into a digital signal by a digitizer and input into a computer for processing. If the selected plane is divided into several cuboids with the same volume, we call them voxels. The X -ray attenuation coefficient of each voxel is obtained after computer processing and then arranged into a matrix to obtain a digital matrix. Through the digitizer, each number in the digital matrix is displayed in black and white with different gray levels, that is, pixels.

CT is like slicing your body and the parts that need to be examined, and it is very thin. Compared with X -ray, the cross -sectional images obtained by CT have accurate layer thickness, clear images, high -density resolution, and no structural interference outside the layer. The pixels of different gray levels from black to white are arranged in a matrix to form the entire CT, which reflects the degree of X -ray absorption by organs and tissues. There are many types of CT devices, and different CT devices have different pixel sizes and numbers. Common CT pixel sizes are 0.5×0.5 mm, 1.0×1.0 mm. The number of pixels is generally

256 × 256, 512 × 512. The smaller the number of pixels in a CT, the more detailed the image, i.e., the higher the spatial resolution. Like the black and white images shown in an X -ray, the black shadows in the pictures indicate low -absorption areas (low -density areas), such as the lungs. White shadows indicate areas of high absorption (high -density areas), such as bones. Most organs composed of soft tissues, such as the brain, spinal cord, lung, liver, gallbladder, and pancreas, can be displayed with CT.

Multiple CT images can be obtained from a single CT scan. Combined with observation of a series of multi -frame images, the size, shape, and anatomical relationship between organs can be understood three -dimensionally. The lesion can be visualized when the lesion is large enough, and the adjacent tissue has sufficient density difference. The lesion can be better visualized on a good anatomical background. This is also the characteristic of CT images and the main basis for diagnosis. By comprehensively analyzing the size of the organ, the change in shape, the realization of the lesion, and the impact of neighboring organs, it is possible to make an evaluation on the location, size, number, scope, and pathology of the lesion. Like other imaging techniques, the analysis of CT also requires the combination of clinical data and comprehensive analysis with other modalities.

CT is more sensitive and reliable in finding lesions, determining the location, and size and number of lesions, but have certain limitations on the diagnosis of pathological properties, which are mainly related to the limitations of CT. CT greatly improves the density resolution of diagnostic images, but due to the influence of various factors, it also has its limitations. CT is a radiographic examination method, it is difficult to find the functional changes of the organ tissue structure, and it is difficult to find early lesions with small density changes or limited to the cell level. Therefore, pathological analysis needs to select pathological images. In addition, many artifacts in reconstructed images are also a major disadvantage (Popilock et al., 2008).

Tomography is not conducive to the overall display of organ image structure and lesions. It requires continuous observation of multiple frames of images, which can be integrated by human brain thinking or image post -processing and reorganization technology to form a complete concept. The emergence of the artificial intelligence medical industry has just solved the problem of the huge amount of CT data. Researchers use computers to perform various image processing and analysis on CT to obtain more complete information about the lesion (Wang et al., 2021) (Fig. 4.2).

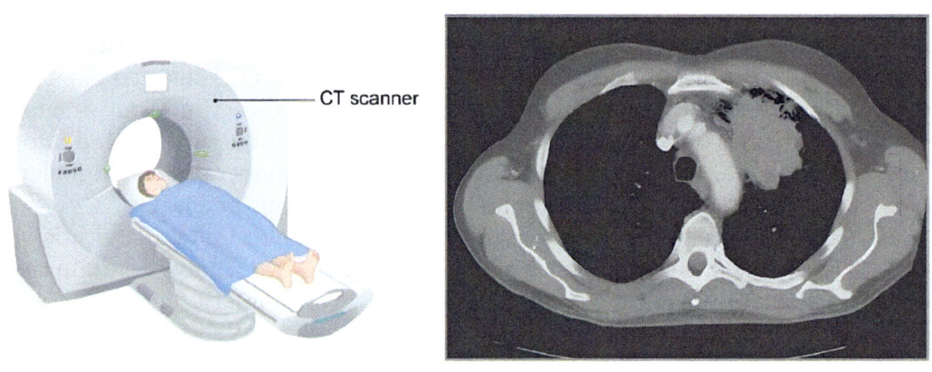

FIG. 4.2 CT scanning device and the image obtained after scanning.

Diagnosis of cancer in CT

CT images have many applications in this cancer, ranging from early screening, diagnosis, and treatment of cancer to follow -up after cancer treatment. Early -stage cancer tumors are small and are not easy to be detected if they have not spread. Medical imaging technology can be used for early detection or cancer screening. For example, early screening of lung cancer using low -dose chest CT often reveals small abnormalities in the lung, such as pulmonary nodules or masses. Compared with chest X -ray (CXR) screening, low -dose CT lung screening can reduce lung cancer mortality by 20%. This result is based on a large central clinical trial followed up for more than 6 years in the United States. There are three reasons to choose low - dose CT rather than other imaging techniques to screen for lung cancer. The first is because of the results of the NLST clinical trial, and another is because low -dose CT has less radiation than conventional CT. Radiation is also a risk factor for cancer. Third, there are certain conditions in the chest that cannot be detected by traditional CXR. For example, blood clots in the lungs cannot be shown on the chest radiograph. However, low -dose CT is superior in this regard (Gatsonis et al., 2011).

Liver cancer detection based on CT

The emergence of CT has made a qualitative leap in liver cancer and promoted the rapid development of liver surgery. The resolution of CT is quite high, the image is clear and stable, and it can reflect the characteristics of liver cancer more comprehensively and objectively. CT examination can clearly show the size, shape, location, boundary, tumor blood supply richness, and the relationship with the intrahepatic duct. It has important diagnostic value for portal vein, hepatic vein and inferior vena cava, lymph node metastasis of hepatic portal, and liver cancer invasion of adjacent organs. For example, the image segmentation algorithm using multi -scale gradient watershed transformation is used to segment the lesion area from the abdominal plain CT image of liver cancer patients, and then the gray level co -occurrence matrix is used to extract the texture features of the segmented image. Finally, the texture feature is used as the input of the genetic back propagation neural network for classification diagnosis. Experimental results show that the classification accuracy of the intelligent diagnosis method reaches 95%, which has a high clinical application value. Benefiting from big data technology, standardized management of accurate diagnosis and treatment of hepatobiliary tumors has been promoted to a new level. Based on deep learning technology, segmentation difficulties in CT images of liver tumors due to the small proportion of small targets in the background are studied. To achieve accurate segmentation of liver tumors, first, the larger target liver is segmented. Based on a U -Net that is most popular in medical image segmentation, author (Qing, 2019) used dilated convolution and common convolution crossover for downsampling in the encoding stage to increase the receptive field. Correspondingly, deconvolution is used to replace the usual bilinear interpolation method for upsampling in the decoding stage, which can automatically optimize the weights. In the second step, based on the liver segmentation in the first step, in order to minimize the unexpected noise interference in the target area, the area outside the liver was cut out and the processed image was uniformly sized. The same is based on U-Net, mixed dilated convolution is used to increase the image receptive field, and Tversky loss is further used to effectively improve the class imbalance problem in liver tumor segmentation.

Lung cancer detection based on CT

Lung cancer is the most common cancer. Early diagnosis of lung cancer and corresponding treatment measures are crucial factors in reducing mortality. Currently, CT can be used to help doctors detect lung cancer at an early stage. In many cases, the diagnosis of lung cancer depends on the doctor's experience, which can miss some instances of it in patients. In the field of medical imaging diagnosis, deep learning has proven to be a popular and effective method. Setio et al. (2016) detected pulmonary nodules in 1D CT of the chest. 2D patches centering on candidates were extracted from nine different directions, and then different combinations of convolutional neural networks (CNN) were used to classify each candidate. Tajbakhsh and Suzuki (2017) detected pulmonary nodules in CT images and distinguished benign from malignant pulmonary nodules and compared the performance of massive training artificial neural networks (MTANNs) and CNN, two kinds of artificial neural networks for end -to -end training. Anthimopoulos et al. (2019) collected 172 high -resolution CT images from multiple medical institutions and segmented lung parenchyma, healthy tissue, micronodules, and cellular structures in lung CT images using fully convolutional networks (FCN) with cavitary convolution structure and multi -scale feature fusion. Nasrullah et al. (2019) made significant modifications to the deep learning lung nodule detection model used in previous studies and used a deep learning model with multiple strategies to automatically detect and classify lung nodules. The CAD system uses two deep 1D customized mixed link network (CMixNet) architectures to detect and classify lung nodules. The fast region convolutional neural network is adopted to detect the CNN. Dey et al. (2018) studied the role of 1D CNN in the classification of benign and malignant lung cancer CT images. It can be seen that CAD can help doctors make discoveries and diagnoses in lung cancer examination by taking advantage of high -speed computing and automatic processing.

Diagnosis in MRI

The full name of MRI is magnetic resonance imaging. In 1946, Flelix Bloch of Stanford University and Edward Purcell of Harvard University independently discovered nuclear magnetic resonance. Nuclear magnetic resonance is a physical phenomenon, which is widely used in physics, chemical biology, and other fields as an analysis method. MRI technology is based on this physical phenomenon and was officially used in medical clinical testing in 1973. Doctors have discovered that hydrogen atoms in water molecules can produce nuclear magnetic resonance, which can be used to obtain information on the distribution of water molecules in the human body, thereby mapping the internal structure of the human body. In a series of heuristics and experiments in 1973, Paul Lauterball developed an imaging technology based on the phenomenon of nuclear magnetic resonance and used his equipment to successfully map the internal structure of a living clam. Since then, MRI technology has become increasingly mature and widely used. It has become a conventional means of medical detection.

Imaging characteristics of MRI

Although CT images have a wide range, not all organs, such as the base of the skull and spinal canal, are suitable for CT examination. In this case, MRI can provide very clear results.

It can not only reflect the anatomy of the human body but also reflect physiological function information such as human blood flow and cell metabolism (Vandenberghe & Marsden, 2015).

MRI is a measure of the size of the magnetic resonance signal generated by the hydrogen nuclei in human tissues, organs, and lesions under the action of a strong external magnetic field. Magnetic resonance imaging is a type of tomography that uses magnetic resonance to obtain electromagnetic signals from the human body and reconstruct human body information. The nuclear magnetic resonance method and computer processing technology are used to obtain an atomic nuclide of a certain cross -section inside organisms and objects. The nuclear magnetic resonance tomography currently used is hydrogen. This tomography has more advantages than the currently used X -ray. X -ray obtains the density distribution image of the imaging object, while MRI tomography is the distribution image of the nuclear density of the imaging object. Since hydrogen is the main chemical element that constitutes the human body and living organisms, it is currently limited to the density distribution imaging of hydrogen nuclei. From the perspective of scientific and technological development, it is predicted that more nuclides will be added. The hydrogen distribution image obtained from MRI has more information than the X -ray density distribution image. Using CT and X -ray on the head, the density of the outer skull is high, while the density of the inner brain tissue is low. Therefore, it is difficult to obtain a clear image of human brain tissue. But from the MRI of the head, a clear image of the distribution of hydrogen in the brain tissue can be obtained, so it can be seen whether the brain tissue is normal. At the time of cancer diagnosis, there is no significant difference between the normal tissue and tissue of patients with early tumor cancer. At this time, it is difficult to see the abnormality from the X -ray, but the abnormality can be seen from the MRI.

MRI is very similar to CT. The difference between it and CT is that it does not use destructive X -ray ionizing radiation and does little damage to the body. The main disadvantage of MRI is that it takes a long time to scan, so the examination of some patients who do not cooperate is often difficult. For a sporty organ like the gastrointestinal tract, the display is generally unclear due to the lack of a suitable contrast agent. For the lungs, the imaging effect is also unsatisfactory due to respiratory movement and low hydrogen density in the alveoli. MRI is not as accurate and sensitive as CT in displaying calcification and bone lesions. In addition, the spatial resolution of MRI needs to be further improved.

The grayscale of the magnetic resonance image ranges from white, gray, to black. The greater the brightness, the stronger the magnetic resonance signal. Normal adipose tissue and cancellous bone are white, while brain spinal cord and bone marrow are white gray. Internal organs and muscles are off -white. Liquid and normal blood flow are black. The bone cortex, gas, and air -bearing lungs are black (Fig. 4.3).

Diagnosis of cancer in MRI

Chest MRI can directly show the myocardium and the left and right ventricular chambers, thus understanding myocardial damage and determining cardiac function. Abdominal MRI can provide very valuable information for the diagnosis of the liver, kidney, pancreas, spleen, adrenal glands, and other substantial organ diseases. Pelvic MRI can show the lesions of the uterus, ovaries, bladder, prostate, seminal vesicles, and other organs. MRI has a wide range of applications for cancer detection, and MRI can easily obtain tumor information.

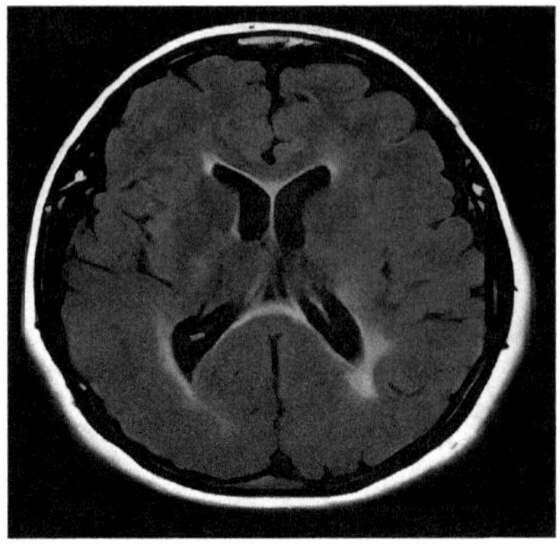

FIG. 4.3 MRI of the brain.

Breast cancer detection based on MRI

Breast cancer, the second most common cancer in the world, poses a serious threat to women's physical and mental health. Its incidence is gradually trending toward younger women. In the process of treatment, imaging techniques such as mammography, ultrasound, and MRI play an indispensable role. MRI has the advantages of the higher soft -tissue resolution, multiple angles, multiple planes, and multiple parameters. Moreover, the scan is less affected by the density of breast glands, making MRI the most sensitive examination technique for breast cancer screening and diagnosis. At the same time, MRI is safe, reliable, non -invasive, and non -radiative. In recent years, MRI has become an important method for breast cancer diagnosis and treatment evaluation. MRI examination technology has shown an irreplaceable role in the qualitative diagnosis of breast lesions, tumor staging, preoperative evaluation and treatment options, and postoperative follow -up. It provides a relevant morphological diagnosis basis for the detection of breast cancer and improves the detection rate of breast cancer. Routine MRI is mainly based on the morphological characteristics of breast lesions for direct diagnosis. Breast cancer lesions are mostly irregular, with unclear borders and rough edges, often showing crab foot or star -shaped changes. Signs such as thickening and edema of the skin and inverted nipples in peripheral structures are also common. The research of computer -aided detection and diagnosis system based on MRI is mainly in the tasks of lesion detection, segmentation, and classification. Agarap (2018) proposed a new algorithm that combines recurrent neural network (RNN), gated recurrent unit (GRU), and support vector machine (SVM) for breast cancer diagnosis. All the proposed machine learning (ML) algorithms used in breast cancer binary classification have shown high performance. Carneiro et al. (2015) used an ImageNet pre -trained CNN model to classify lumps based on multiple views of breast molybdenum target (including bilateral breast axes and bilateral breast oblique positions) as well as segmented microcalcifications and mass areas to assess the risk of breast cancer development in patients.

CADs can obtain more information through the analysis of multi -dimensional image features, so it is highly expected by scholars at home and abroad. However, most of them are focused on the detection of lesions, and there are few studies on the diagnosis of

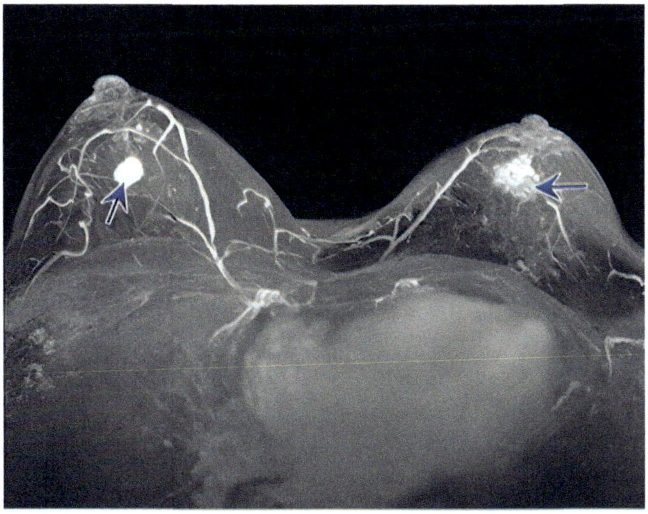

FIG. 4.4 MRI of bilateral breast cancer.

lesions. In the future, we hope that CAD can be used not only to detect lesions but also to evaluate the effect of neoadjuvant chemotherapy and predict the risk of metastasis in patients (Fig. 4.4).

Brain tumor detection based on MRI

MRI is the most used and most important examination method for brain lesions. Although CT and PET can also be used to study the brain, MRI is still the most popular. Quantitative analysis of brain MRI has been widely used to analyze the characteristics of many brain diseases. It can flexibly select axial, coronal, sagittal, and oblique scans according to needs to fully show the lesions. In a cranial examination, the most basic scanning sequences include T1 -weighted and T2 -weighted imaging, as well as some changes and derived T1 and T2 sequences on this basis. Generally speaking, T1 shows good anatomy, and T2 is more sensitive in showing lesions. To analyze the quantification of the corresponding brain structure changes, it is required to segment the MRI acquired at different time points. So far, much research has been invested in classic ML algorithms for the segmentation of normal (white matter and gray matter) and abnormal (brain tumor) MRI brain images. Brain tumor segmentation methods in MRI mainly include the region growing method, clustering segmentation method, segmentation method based on deformation model, and segmentation method based on abnormal detection. However, traditional ML algorithms are not easy to use. Automatic segmentation and detection of abnormal areas have been restricted. Moeskops et al. (2017) used adversarial training to improve the segmentation performance of brain MRI in the network structure of full convolution and dilated convolution. Myronenko (2018) proposed deep learning network 1D MRI brain tumor segmentation based on asymmetric FCN and combined with residual learning.

Gastric cancer detection based on MRI

Generally, when the liver lesions of gastric cancer patients cannot be confirmed by CT whether they are metastases or the patients are allergic to CT contrast agents, doctors recommend MRI to further confirm the diagnosis. For the detection of gastrointestinal tumors, MRI has a higher diagnostic value. It can show the depth of invasion of gastric and rectal malignant tumors and the invasion of peripheral organs and regional lymph node metastasis, which is of great significance to the preoperative staging of tumors. Enhanced MRI is the first choice for liver metastasis of gastric cancer. The accuracy of abdominal MRI for understanding the distant metastasis of gastric cancer is basically the same as that of enhanced CT. The accuracy of N staging of gastric cancer and the sensitivity of diagnosing lymph node invasion are constantly improving compared with CT.

Diagnosis in X-ray

X -ray is also called roentgen rays, which are invisible to the naked eye, but they can cause certain compounds to fluoresce or make photographic films sensitive. It does not deflect in an electric or magnetic field, but can reflect, refract, interfere, and diffract. It can penetrate matter, but substances have a different penetrating power. It has the effect of destroying cells. Different tissues of the human body have different sensitivity to X -rays, and the degree of damage is also different. Therefore, X -ray can make the human body form an image on a screen or film based on the difference in density and thickness of human tissue. After the German physicist Roentgen discovered X -ray in 1895, it was first used for medical diagnosis and then proposed for treatment in the second year.

Imaging characteristics of X -ray

The reason why X -ray can make the human body form images on the screen or film is on one hand based on the special penetration of X -ray, fluorescence effect, and photographic effect, and on the other hand, based on the difference of the density and thickness of human tissues. Because of this difference, X -ray is absorbed to different degrees as it penetrates the body's various tissues. So, there is a difference in the amount of X -ray that reach the film and you get black and white versus different images. Therefore, the formation of X -ray images should meet the following three basic conditions: (a) X -ray should have a certain penetrating power so that they can penetrate the irradiated tissue structure. (b) The tissue structure to be penetrated must have differences in density and thickness. (c) This different remaining X -ray is still invisible and must go through the process of imaging. For example, only X -ray film and screen display can obtain X -ray images with black and white contrast (Martz et al., 2016).

X -ray inspection is a traditional check method. It is the most common, and the price is relatively inexpensive. It is mainly used for the preliminary examination of some diseases to facilitate the discovery of more obvious tissues and structures. It is the first choice for the initial screening of diseases. X -ray is penetrating, but there are differences in density and thickness among human tissues. The dosage of X -ray absorbed by bones is more than that absorbed by

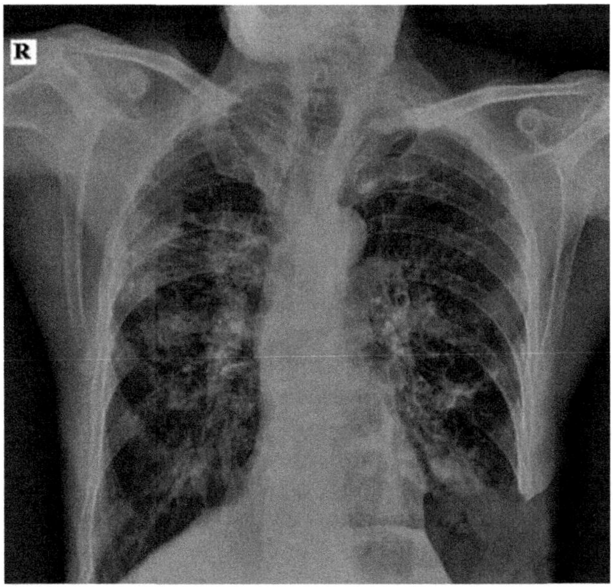

FIG. 4.5 Chest X -ray.

muscles. So, the dosage of X -rays that pass through the body is different, giving information about the density distribution of different parts of the body. According to the contrast of shades, combined with clinical manifestations, laboratory results, and pathological diagnosis, it can be judged whether a certain part of the human body is normal.

X -ray examinations are still used for some parts, such as the gastrointestinal tract, skeletal system, chest, and abdomen. The chest examination is mainly used for the diagnosis of pneumonia and lung cancer. Abdominal X -ray examination is mainly used for the diagnosis of tumors, liver cancer, intestinal tuberculosis, and other diseases. X -ray examination of bone and joint is mainly used for the diagnosis of fractures, joint disease, rheumatism, joint tumors, etc. The purpose of X -ray imaging is to allow the doctor to observe a certain diseased tissue and its condition in the examiner. Therefore, the quality of the medical image will directly affect the doctor's diagnosis. However, X -ray inspection can only provide flat images, and the quality of imaging is also easily affected by clothing and jewelry. And, when excessive X -rays irradiate biological organisms, biological cells may be destroyed. Therefore, the application of X -ray in medical imaging technology is still under constant research and progress. To obtain high -quality X -ray images, it is necessary to eliminate noise well while maintaining good image details (Fig. 4.5).

Diagnosis of cancer in X -ray

X -ray is a good tool for studying head, chest, abdomen, and spine problems. It can be used to determine the location and size of the tumor. With the development and application of artificial neural networks and CAD technology based on artificial intelligence, artificial intelligence has gradually been integrated into the daily workflow of X -ray technology (Polan et al., 2016).

Breast cancer detection based on X -ray

Mammography is a technique for examining breasts with low -dose X -ray. It has the characteristics of comprehensiveness, simple operation, safety, and low cost. It is currently the preferred and effective method for diagnosing early breast cancer (Heck & Herzen, 2020). It can detect some abnormalities in the breast before symptoms appear, so tumors can be found early. CADs are widely used in X -ray photography for breast cancer screening. It has some research on the use of computer image technology to detect or diagnose abnormal masses on breast imaging and to improve the accuracy of calcification and mass detection. X -ray photography has a high detection rate of microcalcifications, but the detection rate of masses is greatly affected by the density of the glands (Benjamin et al., 2016). Patel et al. (2017) developed natural language processing algorithms. The key features of breast X -rays of 543 breast cancer patients were extracted, and correlation analysis with breast cancer subtypes was performed. The accuracy rate of results obtained was as high as 80%, and the diagnosis speed was 30 times that of ordinary doctors. Dhungel et al. (2015) and others used a structured SVM to build a model that combines different types of latent functions, including positional priors, Gaussian mixture models, and deep confidence networks for mammogram quality segmentation. The accuracy rate can reach 96%, and the program runtime is short. Arevalo et al. proposed a feature learning framework for breast cancer diagnosis, using CNN to automatically learn distinguishing features and classify lesions in mammograms.

Lung cancer detection based on X -ray

A CXR examination can not only determine the location and size of the lesion but can also diagnose the extent and nature of the spread of cancer cells. Central lung cancer is manifested as a rounded shadow on one side of the hilum with burrs on the edge, and atelectasis and localized emphysema can be seen. In the early stage, peripheral lung cancer may show small patchy shadows with unclear edges and light density. When it reaches the late stage, the shadows can be found to be enlarged and round. Bronchiolar alveolar carcinoma is nodular and diffuse. Traditional CXR have a very low detection rate of early lung nodules. Researchers pointed out that CXR assisted CAD to detect lung nodules. Cha et al. (2019) designed a deep convolutional neural network based on ResNet. The author inserted a new FCN based on the deep residual network of ResNet -50 and ResNet -101. A total of 1483 cases of cancerous lung nodules diagnosed by surgery were selected from the LIDC -IDRI database (Armato et al., 2011) and local medical institutions as the test set. Comparing the detection rate of six radiologists and the deep learning model for nodules, the results show that the average area under curve detected manually is 82%, which has a good detection rate.

At present, the general trend of the development of X -ray imaging technology is faster imaging speed, clearer images, gradual reduction in dose, and more convenient operation. We have reason to believe that the accuracy of X -ray will increase steadily. X -ray imaging will become the leader in medical development.

Diagnosis in UI

As one of the most used imaging modalities, ultrasound has become an indispensable clinical scanning and diagnostic tool. Due to the advantages of relative safety, low cost, non -invasiveness, real -time imaging, and comfortable operation, UI is especially widely used in prenatal screening around the world (Reddy et al., 2008). Clinical practice has confirmed that UI has some obvious advantages compared with medical imaging modalities such as X-ray, MRI, and CT. Ultrasound has no ionizing radiation. It is convenient to carry, easy to access, and cost -effective. In current clinical practice, medical ultrasound has been applied in various specialized departments such as breast ultrasound, abdominal ultrasound, and prenatal diagnostic ultrasound.

Imaging characteristics of UI

UI is reflected in the difference of acoustic parameters in the medium, which can obtain information different from that of optics, X rays, etc. Unlike X -ray or CT examinations that use radiation, ultrasound exposes the body to be examined to high -frequency sound waves instead of ionizing radiation. Ultrasound has the ability to distinguish human soft tissues and can obtain useful signals with a dynamic range of up to 120dB or more, which is helpful for identifying small lesions in biological tissues.

The working principle of medical ultrasound is similar to that of sonar, that is, ultrasound is emitted into the human body. When it encounters interfaces in the body, it reflects and refracts and may be absorbed and attenuated in human tissues. Because the shape and structure of various tissues of the human body are different, the degree of reflection and absorption of ultrasonic waves is also different. Doctors distinguish them by the characteristics of the waveform, curve, or image reflected by the instrument. In addition, combined with anatomical knowledge, it can be diagnosed whether the examined organ is diseased.

UI is also divided into many types according to different inspection parts. Ultrasound has the capacity to distinguish soft tissues of the human body, which is conducive to identifying small lesions in biological tissues. There are many ways to display ultrasound information, which can be used according to different options, and various information can be obtained to achieve a wide range of applications. However, ultrasound also presents some unique challenges such as low imaging quality caused by noise and artifacts, highly dependent on the experience of the operator or diagnostician, and high differences between ultrasound systems of different institutions. For example, a study using UI for prenatal abnormality detection showed that the sensitivity range between different medical institutions is 27.5%–96% (Salomon et al., 2008). To meet these challenges, only the development of advanced automated UI analysis methods can make the ultrasound diagnosis more accurate and intelligent (Fig. 4.6).

Diagnosis of cancer in UI

The application of ultrasound in medicine is mainly to diagnose diseases, and it has become an indispensable diagnostic method in clinical medicine. Ultrasound diagnosis has

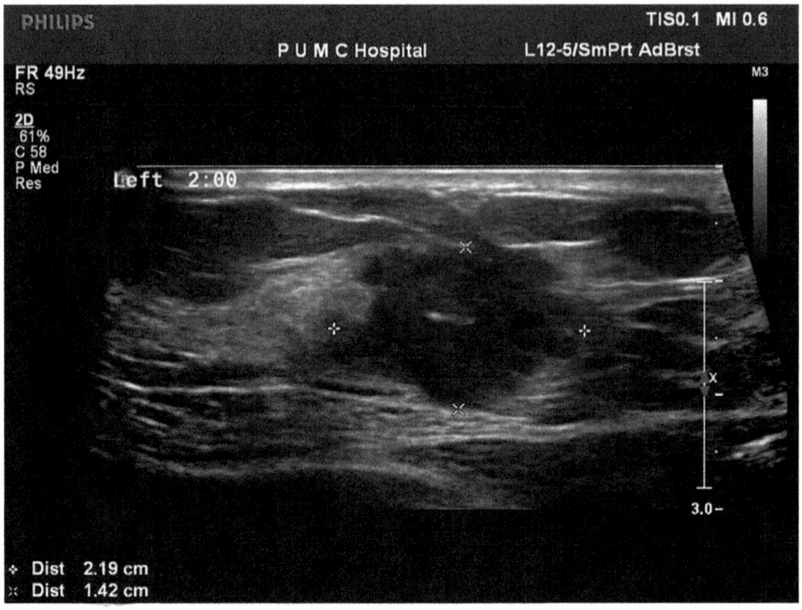

FIG. 4.6 Ultrasound image of breast cancer.

many advantages but also has great limitations. In the diagnosis of tumors, ultrasound needs to combine CT and MRI and then use other methods to diagnose tumors. Therefore, it can be said that ultrasound has never been an effective screening method for any cancer, whether it is breast cancer or ovarian cancer.

Ultrasound can help doctors find tumors that are not well -shown on X -ray in certain parts of the body. Ultrasound is usually one of the "first -line" inspection methods in cancer detection, but it is not a screening method for any cancer. It detects abnormal tissue and gives a diagnosis of a suspected cancer based on the appearance of these images. While it cannot diagnose cancer, it can help doctors detect abnormal tissue that may cause cancer.

Breast cancer detection based on UI

Color Doppler ultrasound is referred to as color ultrasound, which is a special application of ultrasound. It measures the direction and speed of blood as it passes through blood vessels. The movement of blood causes the tone change of reflected sound wave to be called the Doppler effect, which is used for detection. The blood flow in tumors is different from that in normal tissues, so Doppler ultrasound is also very helpful in tumor diagnosis. Color ultrasound makes it easier for doctors to find if cancer has spread to blood vessels, especially the liver and pancreas. Doppler ultrasound is usually the first choice in ultrasound breast examinations, which can show blood flow in breast lumps. The limitation of breast UI processing based on the traditional neural network method is that the features extracted by artificial design usually have good adaptability for a certain type of lesion with specific characteristics. The breast UI processing method based on deep learning can learn more generalized feature representation. To adapt to the detection of various breast pathological images in the actual diagnosis and treatment scene, Chiang et al. (2018) developed a CAD technology based on 1D CNN to detect breast cancer using automated whole breast UI. Yap et al. (2018) compared three different deep learning methods, Block -based LeNet, U -Net, and migration pre - training FCN -Alexnet, and carried out breast cancer detection on two ultrasonic image data sets obtained by two different ultrasonic systems. Qi et al. (2019) constructed a breast UI data set marked by breast surgeons and used the CNN to build a new model on this data set to identify malignant tumors in breast UI. Men et al. (2018) trained a deeply expanded ResNet for the segmentation of breast clinical target volume. Wang et al. (2016) showed that the deep learning model can not only accurately segment calcifications but also distinguish benign and malignant calcifications. Han et al. (2017) trained a GoogLeNet model to classify breast tumors into benign and malignant breast tumors based on a UI dataset composed of 4254 cases of benign tumors and 3154 cases of malignant tumors. Huang et al. (2017) constructed a CNN model to identify the area of breast lesions based on breast UI and further performed breast imaging reports and data system grading on the lesions. To sum up, artificial intelligence can realize automatic recognition of abnormal manifestations in breast images. It can help doctors to observe the suspected lesions and reduce the rate of missed diagnosis among radiologists.

Liver cancer detection based on UI

Another common tumor is liver cancer, which has become the sixth leading cause of cancer and the third leading cause of cancer death in the world. Biopsy is still the current gold standard for liver cancer diagnosis, and it relies heavily on traditional CAD methods. However, a biopsy is uncomfortable during the inspection and can easily cause other adverse effects. Therefore, ultrasound diagnostic technology has become the most important non -invasive method for liver cancer diagnosis. Wu et al. (2014) applied a 3 -layer dynamic Bayesian network to the time -intensity curve extracted from the contrast -enhanced ultrasound video sequence to achieve the classification of benign and malignant liver lesions. They had the highest accuracy rate of 86.36%, which is better than that of traditional ML methods. Guo et al. (2017) combined deep canonical correlation analysis with multi -kernel learning classifier to distinguish benign and malignant liver cancer. The experimental results show that making full use of these two methods can achieve high classification accuracy and low computational complexity.

Diagnosis in PET

Positron Emission Computed Tomography is a relatively advanced clinical examination imaging technology in the field of nuclear medicine. PET and PET -CT technologies are the world's highest -end medical imaging diagnostic equipment, which can be called "the crown of modern medical high –tech." It uses large -scale imaging equipment to observe the distribution of imaging drugs injected into the body in vivo. Compared with CT observation of tissue density and blood perfusion, MRI has observed the difference in water content and blood perfusion in tissues. Its mechanism is completely different. Most of the PET examination groups are tumor patients. Therefore, PET has been widely used in tumor diagnosis and treatment.

Imaging characteristics of PET

The general method of PET is to combine certain substances, generally essential substances in the metabolism of biological life, such as glucose, protein, and nucleic acid. After the marked short -lived radionuclide is injected into the human body, the accumulation of the substance in the metabolism reflects the life and metabolic activities so as to achieve the purpose of diagnosis. These short -lived substances release positrons during the decay process. A positron is annihilated after encountering an electron after traveling a few tenths of a millimeter to a few millimeters, resulting in a pair of photons with an energy of 511 KeV in opposite directions. The photons are captured by a highly sensitive camera and scattered by a computer. By performing the same analysis on different positions, we can obtain a 1D image of the aggregation in the organism.

CT shows obvious advantages in observing the morphology and density differences of tissues and lesions. MRI has no radiation and can provide more disease information through sequence changes in the observation of soft tissue, nervous system, and intraosseous diseases. Enhanced CT and MRI can also reflect the blood perfusion of the disease. The PET inspection process is more complicated and has more influencing factors. PET images reflect the functional status of various organs in the body and the metabolism of diseases. The hardware structure of PET instruments is more complicated, and the price is also the most expensive among large -scale medical equipment. So, PET also has its limitations.

A single PET image has a huge problem, that is, it cannot accurately display the anatomical structure. To solve this problem, the PET -CT machine was integrated in the early 21st century, and the PET -CT machine was born. Its appearance is another revolution in medical imaging, which has been recognized and widely concerned by the medical community (Boellaard, 2016) (Fig. 4.7).

Diagnosis of cancer in PET

Because PET is functional imaging, it can detect diseases earlier than morphological examinations such as CT and magnetic resonance. It is particularly prominent in the imaging of the nervous system, such as the diagnosis of epilepsy lesions, Parkinson's, and Alzheimer's disease. In many cases, when there is no morphological change in MRI, PET examination has

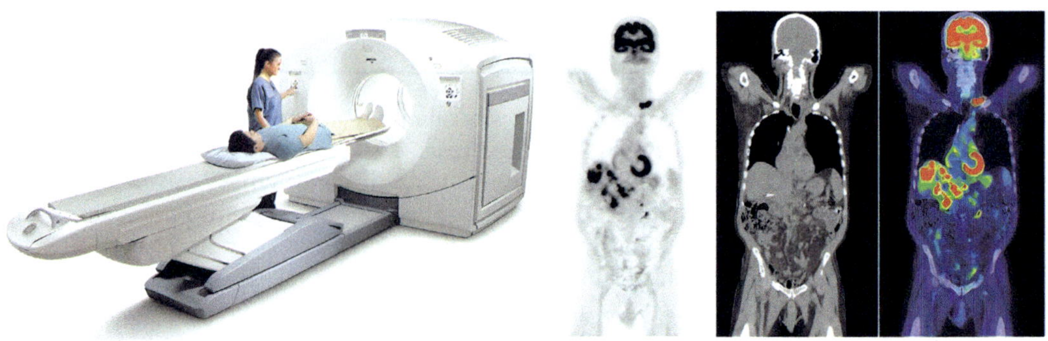

FIG. 4.7 PET scanning device and the image obtained after scanning.

outstanding performance in brain function, which can provide clinical information that is different from CT and MRI. In addition, it has unmatched advantages in discovering primary tumors, tumor patient staging, and efficacy evaluation. Therefore, the application in tumor imaging, neuroreceptor imaging, and brain function imaging has been covered.

PET examinations are used for 85% of tumor examinations. Because most malignant tumors have high glucose metabolism, fluorodeoxyglucose, as a compound with a similar structure to glucose, will accumulate in malignant tumor cells after intravenous injection. Thus, PET can distinguish malignant tumors from benign tumors and normal tissues. At the same time, the recurring tumor can be distinguished from surrounding necrosis and scar tissue. Now, it is mostly used for the examination of lung cancer, breast cancer, lymphoma, and other cancers, and its diagnosis accuracy rate is over 90%. For example, based on CNN, Ypsilantis et al. (2015) proposed a 3S -CNN model to detect PET images of 107 patients with esophageal cancer. The sensitivity and specificity of the dataset reached 80.7% and 81.6%, respectively. The effect is far superior to that of other models. This plays an important role in guiding the staging of tumor diagnosis, the scope of surgical resection, and whether surgery is required.

Lymph cancer detection based on PET -CT

For lymphoma, CT and ultrasound are often used to check whether the lymph nodes are enlarged. And, lymphadenitis can also make the lymph nodes enlarge, causing misdiagnosis. As an ultra -fine and accurate inspection, PET can reduce this occurrence. The principle of PET -CT to diagnose lymphoma is based on the principle that lymphoma has a higher metabolic capacity than that of normal tissue. During the PET -CT examination, an 18F -FDG imaging agent is injected into the subject. After PET -CT scan, it will show different phenomena. Where lymphomas gather, there will be luminescence.

Thyroid cancer detection based on PET -CT

The incidence of thyroid cancer has been on the rise in recent years. In the past, ultrasound and CT were usually used to check the thyroid gland, but the accuracy was not high. PET -CT can solve this problem. There are many kinds of thyroid diseases, all of which can manifest as lumps on the thyroid. As an advanced imaging examination, PET -CT not only has the positioning function of CT but also has the function of PET examination of metabolism. The PET -CT examination for thyroid cancer is characterized by the abnormally strong metabolic capacity of cancer tissues. The tracer 18F -FDG will accumulate in cancer tissues on a large scale, while there are few in normal tissues, so it is easy to distinguish cancer tissues from normal tissues. If it is a benign tumor, the intake is between cancer tissue and normal tissue. Therefore, PET -CT can detect thyroid cancer and benign tumors. Even if benign tumors are complicated by thyroid cancer, PET -CT is not difficult.

Lung cancer detection based on PET

PET plays an irreplaceable role in the diagnosis of lung cancer. With its precise positioning and higher accuracy in tumor diagnosis, it has become an indispensable examination method for tumor patients. Before the advent of PET, CT was generally considered to be a better imaging method to evaluate the morphological and density characteristics of lung cancer. At present, research on CT is more in -depth. The signs of lung cancer are complex and changeable, and qualitative diagnosis is sometimes difficult, especially the qualitative diagnosis of single lung nodules, and the diagnostic value of CT is limited. PET is used to distinguish benign and malignant pulmonary nodules by visually comparing the differences in the degree of radioactivity uptake between the nodules and the mediastinal blood pool to determine the metabolic activity of the nodules.

Medical imaging and artificial intelligence

Although the imaging techniques, imaging principles, and methods of various imaging examinations are different, the diagnostic value and limits are also different. However, the purpose of an examination is to make images of the internal structures and organs of the human body so as to understand anatomical and physiological functions of the human body and pathological changes in order to achieve the purpose of diagnosing cancer.

The combination of artificial intelligence and medical imaging technology can solve medical problems. Prof. Li, a tenured professor at Stanford University, has also expressed a similar view, letting artificial intelligence "people do their best" to play a huge role in the medical field. Currently, artificial intelligence can diagnose cancer through symptoms and medical records. After more than 4 years of training, IBM Watson has learned 200 textbooks in the field of oncology, 290 medical journals, and more than 15 million documents. IBM Watson has begun in clinical and can advise human physicians in the fields of lung, breast, rectal, colon, gastric, and cervical cancer. If artificial intelligence is applied to the early detection and diagnosis of cancer, countless lives can be saved.

High -incidence cancers such as lung and breast cancer are closely related to unhealthy lifestyle habits. Therefore, many cancers can be prevented. For example, liver cancer is caused by viral hepatitis and aflatoxin. Coastal cities are areas with a high incidence of liver cancer, which mainly meets two major conditions. One is that the coastal areas are humid, creating conditions for the breeding of aflatoxins. The other is the relatively high infectiousness of viral hepatitis in the coastal areas. Artificial intelligence can explore the connection between patient examination information, past medical records, and the social environment through comprehensive mining of multi -modal mass data such as text and images. Thus, a predictive analysis model is established to further explore the distribution and evolution rules of disease, determine risk factors, and predict the epidemic trend of the disease. There are many top hospitals in the world. Based on a large number of real clinical data of desensitization, the deep learning model of training after data preprocessing was used to analyze the correlation and contribution of multiple dimensional parameters to patient morbidity. They tap into potential correlations between patients' behavior, genes, and patients to create predictive models that can predict many types of cancer at an early stage.

Conclusion

Future directions

This article details the different imaging data and the auxiliary methods used in the diagnosis, and we provide a new understanding of the development of the medical industry.

Whether it is cancer or various diseases, the diagnosis is inseparable from the development of computers, and the favorable combination of the two can bring tremendous progress.

The research of the CAD system based on various medical images is still in the early stages. Establishing a robust and high -performance CAD system is the direction we need to work hard at present. There is still much space for improvement and further expansion in the application of system algorithms, complete databases, and standard performance evaluation indicators.

Critical analysis

With the innovation of the Internet of Things, medical data can be accessed through remote sensors on the Internet. The advances of these technologies are of great help to CAD systems, and the use of medical images has improved the accuracy of cancer detection and classification. At present, the most widely used medical image detection and classification technology are based on deep learning. Deep learning is used to detect and classify medical images by carrying out convolution operations at different levels and extracting features. However, the inexplicability of deep learning makes it impossible for researchers to explore the meaning behind the detection and classification of medical images, let alone combine the characteristics of medical images with deep learning, so as to make new breakthroughs and progress. During deep learning training, on one hand, weights and biases of different levels need to be randomly initialized, which means that the learning direction of deep learning is random at the beginning; on the other hand, the effect of convolution operation is related to the setting of two parameters in the convolution layer, steps and padding. The values of steps and padding in the deep learning algorithm are generally determined through multiple groups of comparative experiments, rather than theoretical values. Another reason for the shortcomings of deep learning in the detection and classification of medical images is that it requires a large data set and time for the model to fine -tune through repeated training. Deep learning requires huge amounts of data as support. Therefore, the uncertainty of deep learning parameters leads to its inexplicability. Although medical images have accumulated a certain amount of data in recent years, there are many types of diseases, and some hospitals are unable to share data due to privacy issues, so the number of data sets of a single category cannot meet the needs of deep learning since it requires data preprocessing, that is, manual data labeling and classification. This requires both professional knowledge and a number of people, which consumes manpower and material resources. In addition, although a high - performance GPU can accelerate deep learning calculations, it is still difficult to train an efficient model based on a new data set in a short period of time.

References

Agarap, A. F. M. (2018). On breast cancer detection: An application of machine learning algorithms on the Wisconsin diagnostic dataset. In *ACM international conference proceeding series* (pp. 5–9). Association for Computing Machinery. https://doi.org/10.1145/3184066.3184080.

Anthimopoulos, M., Christodoulidis, S., Ebner, L., Geiser, T., Christe, A., & Mougiakakou, S. (2019). Semantic segmentation of pathological lung tissue with dilated fully convolutional networks. *IEEE Journal of Biomedical and Health Informatics*, *23*(2), 714–722. https://doi.org/10.1109/JBHI.2018.2818620.

Armato, S. G., McLennan, G., Bidaut, L., McNitt-Gray, M. F., Meyer, C. R., Reeves, A. P., Zhao, B., Aberle, D. R., Henschke, C. I., Hoffman, E. A., Kazerooni, E. A., MacMahon, H., Van Beek, E. J. R., Yankelevitz, D., Biancardi, A. M., Bland, P. H., Brown, M. S., Engelmann, R. M., Laderach, G. E., & Clarke, L. P. (2011). The lung image database consortium (LIDC) and image database resource initiative (IDRI): A completed reference database of lung nodules on CT scans. *Medical Physics*, *38*(2), 915–931. https://doi.org/10.1118/1.3528204.

Benjamin, Q. H., Hui, L., & Giger, M. L. (2016). Digital mammographic tumor classification using transfer learning from deep convolutional neural networks. *Journal of Medical Imaging*, *3*(3).

Bhansali, P., & Mehta, S. (2020). *Medical image processing using soft computing techniques and mathematical morphology* (pp. 303–318). Springer Science and Business Media LLC. https://doi.org/10.1007/978-981-15-8820-4_29.

Boellaard, R. (2016). The engagement of FDG PET/CT image quality and harmonized quantification: From competitive to complementary. *European Journal of Nuclear Medicine and Molecular Imaging*, *43*(1), 1–4. https://doi.org/10.1007/s00259-015-3182-7.

Carneiro, G., Nascimento, J., & Bradley, A. P. (2015). Unregistered multiview mammogram analysis with pre-trained deep learning models. In *Vol. 9351*. *Lecture notes in computer science (including subseries lecture notes in artificial intelligence and lecture notes in bioinformatics)* (pp. 652–660). Springer Verlag. https://doi.org/10.1007/978-3-319-24574-4_78.

Cha, M. J., Chung, M. J., Lee, J. H., & Lee, K. S. (2019). Performance of deep learning model in detecting operable lung cancer with chest radiographs. *Journal of Thoracic Imaging*, *34*(2), 86–91. https://doi.org/10.1097/RTI.0000000000000388.

Chiang, T. C., Huang, Y. S., Chen, R. T., et al. (2018). Tumor detection in automated breast ultrasound using 3-D CNN and prioritized candidate aggregation. IEEE Transactions on Medical Imaging, 38(1), 240–249

Dey, R., Lu, Z., & Hong, Y. (2018). Diagnostic classification of lung nodules using 3D neural networks. In *2018. Proceedings—International symposium on biomedical imaging* (pp. 774–778). IEEE Computer Society. https://doi.org/10.1109/ISBI.2018.8363687.

Dhungel, N., Carneiro, G., & Bradley, A. P. (2015). Deep structured learning for mass segmentation from mammograms. In *2015. Proceedings—International conference on image processing, ICIP* (pp. 2950–2954). IEEE Computer Society. https://doi.org/10.1109/ICIP.2015.7351343.

Gatsonis, C. A., Aberle, D. R., Berg, C. D., Black, W. C., Church, T. R., Fagerstrom, R. M., Galen, B., Gareen, I. F., Goldin, J., Gohagan, J. K., Hillman, B., Jaffe, C., Kramer, B. S., Lynch, D., Marcus, P. M., Schnall, M., Sullivan, D. C., Sullivan, D., Zylak, C., … Dyer, S. (2011). The national lung screening trial: Overview and study design. *Radiology*, *258*(1), 243–253. https://doi.org/10.1148/radiol.10091808.

Guo, L., Wang, D., Xu, H., Qian, Y., Wang, C., Zheng, X., Zhang, Q., & Shi, J. (2017). CEUS-based classification of liver tumors with deep canonical correlation analysis and multi-kernel learning. In *Proceedings of the annual international conference of the IEEE engineering in medicine and biology society, EMBS* (pp. 1748–1751). Institute of Electrical and Electronics Engineers Inc. https://doi.org/10.1109/EMBC.2017.8037181.

Han, S., Kang, H. K., Jeong, J. Y., Park, M. H., Kim, W., Bang, W. C., & Seong, Y. K. (2017). A deep learning framework for supporting the classification of breast lesions in ultrasound images. *Physics in Medicine and Biology*, *62*(19), 7714–7728. https://doi.org/10.1088/1361-6560/aa82ec.

Heck, L., & Herzen, J. (2020). Recent advances in X-ray imaging of breast tissue: From two- to three-dimensional imaging. *Physica Medica*, *79*, 69–79. https://doi.org/10.1016/j.ejmp.2020.10.025.

Huang, Y., Han, L., Dou, H., et al. (2019). Two-stage CNNs for computerized BI-RADS categorization in breast ultrasound images. Biomedical Engineering Online, 18(1), 1–18.

Martz, H. E., Logan, C. M., Schneberk, D. J., & Shull, P. J. (2016). X-ray imaging: Fundamentals, industrial techniques and applications. In *X-ray imaging: Fundamentals, industrial techniques and applications* (pp. 1–571). CRC Press. https://doi.org/10.1201/9781315375199.

Men, K., Zhang, T., Chen, X., Chen, B., Tang, Y., Wang, S., Li, Y., & Dai, J. (2018). Fully automatic and robust segmentation of the clinical target volume for radiotherapy of breast cancer using big data and deep learning. *Physica Medica*, *50*, 13–19. https://doi.org/10.1016/j.ejmp.2018.05.006.

Moeskops, P., Veta, M., Lafarge, M. W., Eppenhof, K. A. J., & Pluim, J. P. W. (2017). Adversarial training and dilated convolutions for brain MRI segmentation. In *Vol. 10553*. *Lecture notes in computer science (including subseries lecture notes in artificial intelligence and lecture notes in bioinformatics)* (pp. 56–64). Springer Verlag. https://doi.org/10.1007/978-3-319-67558-9_7.

Myronenko, A. (2018). 3D MRI brain tumor segmentation using autoencoder regularization. In *International MICCAI brainlesion workshop* (pp. 311–320). Cham: Springer.

Nasrullah, N., Sang, J., Alam, M. S., Mateen, M., Cai, B., & Hu, H. (2019). Automated lung nodule detection and classification using deep learning combined with multiple strategies. *Sensors (Switzerland)*, *19*(17). https://doi.org/10.3390/s19173722.

Natarajan, K. K., & Gokulachandran, J. (2021). Application of artificial neural network techniques in computer aided process planning—A review. *International Journal of Process Management and Benchmarking*, *11*(1), 80–100. https://doi.org/10.1504/IJPMB.2021.112257.

Patel, T. A., Puppala, M., Ogunti, R. O., Ensor, J. E., He, T., Shewale, J. B., Ankerst, D. P., Kaklamani, V. G., Rodriguez, A. A., Wong, S. T. C., & Chang, J. C. (2017). Correlating mammographic and pathologic findings in clinical decision support using natural language processing and data mining methods. *Cancer*, *123*(1), 114–121. https://doi.org/10.1002/cncr.30245.

Polan, D. F., Brady, S. L., & Kaufman, R. A. (2016). Tissue segmentation of computed tomography images using a Random Forest algorithm: a feasibility study. *Physics in Medicine and Biology*, 6553–6569. https://doi.org/10.1088/0031-9155/61/17/6553.

Popilock, R., Sandrasagaren, K., Harris, L., & Kaser, K. A. (2008). CT artifact recognition for the nuclear technologist. *Journal of Nuclear Medicine Technology*, *36*(2), 79–81. https://doi.org/10.2967/jnmt.107.047431.

Qi, X., Zhang, L., Chen, Y., Pi, Y., Chen, Y., Lv, Q., & Yi, Z. (2019). Automated diagnosis of breast ultrasonography images using deep neural networks. *Medical Image Analysis*, *52*, 185–198. https://doi.org/10.1016/j.media.2018.12.006.

Qing, L. (2019). *Study on CT image segmentation of liver tumors based on deep learning*. Chang'an University.

Reddy, U. M., Filly, R. A., & Copel, J. A. (2008). Prenatal imaging: Ultrasonography and magnetic resonance imaging. *Obstetrics and Gynecology*, *112*(1), 145–157. https://doi.org/10.1097/01.AOG.0000318871.95090.d9.

Salomon, L. J., Winer, N., Bernard, J. P., & Ville, Y. (2008). A score-based method for quality control of fetal images at routine second-trimester ultrasound examination. *Prenatal Diagnosis*, *28*(9), 822–827. https://doi.org/10.1002/pd.2016.

Setio, A. A. A., Ciompi, F., Litjens, G., Gerke, P., Jacobs, C., Van Riel, S. J., Wille, M. M. W., Naqibullah, M., Sanchez, C. I., & Van Ginneken, B. (2016). Pulmonary nodule detection in CT images: False positive reduction using multiview convolutional networks. *IEEE Transactions on Medical Imaging*, *35*(5), 1160–1169. https://doi.org/10.1109/TMI.2016.2536809.

Siegel, R. L., Miller, K. D., Goding Sauer, A., Fedewa, S. A., Butterly, L. F., Anderson, J. C., Cercek, A., Smith, R. A., & Jemal, A. (2020). Colorectal cancer statistics, 2020. *CA: a Cancer Journal for Clinicians*, *70*(3), 145–164. https://doi.org/10.3322/caac.21601.

Tajbakhsh, N., & Suzuki, K. (2017). Comparing two classes of end-to-end machine-learning models in lung nodule detection and classification: MTANNs vs. CNNs. *Pattern Recognition*, *63*, 476–486. https://doi.org/10.1016/j.patcog.2016.09.029.

Vandenberghe, S., & Marsden, P. K. (2015). PET-MRI: A review of challenges and solutions in the development of integrated multimodality imaging. *Physics in Medicine and Biology*, *60*(4), R115–R154. https://doi.org/10.1088/0031-9155/60/4/R115.

Wang, B., Jin, S., Yan, Q., Xu, H., Luo, C., Wei, L., Zhao, W., Hou, X., Ma, W., Xu, Z., Zheng, Z., Sun, W., Lan, L., Zhang, W., Mu, X., Shi, C., Wang, Z., Lee, J., Jin, Z., … Dong, J. (2021). AI-assisted CT imaging analysis for COVID-19 screening: Building and deploying a medical AI system. *Applied Soft Computing*, *98*, 106897. https://doi.org/10.1016/j.asoc.2020.106897.

Wang, J., Yang, X., Cai, H., Tan, W., Jin, C., & Li, L. (2016). Discrimination of breast cancer with microcalcifications on mammography by deep learning. *Scientific Reports*, *6*. https://doi.org/10.1038/srep27327.

Wu, K., Chen, X., & Ding, M. (2014). Deep learning based classification of focal liver lesions with contrast-enhanced ultrasound. *Optik*, *125*(15), 4057–4063. https://doi.org/10.1016/j.ijleo.2014.01.114.

Yap, M. H., Pons, G., Martí, J., Ganau, S., Sentís, M., Zwiggelaar, R., Davison, A. K., & Martí, R. (2018). Automated breast ultrasound lesions detection using convolutional neural networks. *IEEE Journal of Biomedical and Health Informatics*, *22*(4), 1218–1226. https://doi.org/10.1109/JBHI.2017.2731873.

Yip, R., Jirapatnakul, A., & Hu, M. (2021). Added benefits of early detection of other diseases on low-dose CT screening. *Translational Lung Cancer Research*, *10*(2), 1141–1153.

Ypsilantis, P. P., Siddique, M., Sohn, H. M., Davies, A., Cook, G., Goh, V., & Montana, G. (2015). Predicting response to neoadjuvant chemotherapy with PET imaging using convolutional neural networks. *PLoS One*, *10*(9). https://doi.org/10.1371/journal.pone.0137036.

CHAPTER

5

Neural network for lung cancer diagnosis

Shuai Liu[a,b,c], *Liping Song*[a,b,c], *and Xiangbin Liu*[a,b,c]

[a]Hunan Provincial Key Laboratory of Intelligent Computing and Language Information Processing, Hunan Normal University, Changsha, People's Republic of China [b]College of Information Science and Engineering, Hunan Normal University, Changsha, People's Republic of China [c]Xiangjiang Institute of Artificial Intelligence, Changsha, People's Republic of China

Introduction

In recent years, lung cancer has had the highest mortality rate in the world, which threatens people's lives. According to the official journal report of the American Cancer Society, Cancer Statistics of 2020, the number of deaths from lung cancer has declined (Siegel et al., 2020). There are many reasons for the decline in regards to lung cancer mortality. The steady decline in smoking is the primary reason for the decline of lung cancer incidence and mortality. Secondly, with the rise and the wide application of artificial intelligence (AI), the treatment of lung cancer has also been rapidly improved. These include better diagnostic imaging technology, more precise diagnostic methods, and a new generation of anti-cancer therapies, which will further increase the rate of decline with lung cancer mortality (Fig. 5.1).

According to the staging standard of lung cancer, lung cancer is clinically divided into five stages, which include lung cancer in situ, early lung cancer, middle lung cancer, middle and late lung cancer, and advanced lung cancer. The cancer cells of lung cancer spread as the disease gets worse, which greatly increases the difficulty of treatment. Therefore, it is very important to diagnose the prevalence of lung cancer quickly and accurately in patients as well as gain more treatment time for lung cancer patients. In the clinical preliminary screening of lung cancer, the first examination is performed using chest Computed Tomography (CT). For lung cancer patients, it usually appears as a solitary lung nodule. If chest CT finds a large and irregularly shaped nodule, in order to clarify the nature of the nodule, enhanced CT can be used to judge. At the same time, a needle biopsy can be performed for the related

https://doi.org/10.1016/B978-0-323-85240-1.00004-3

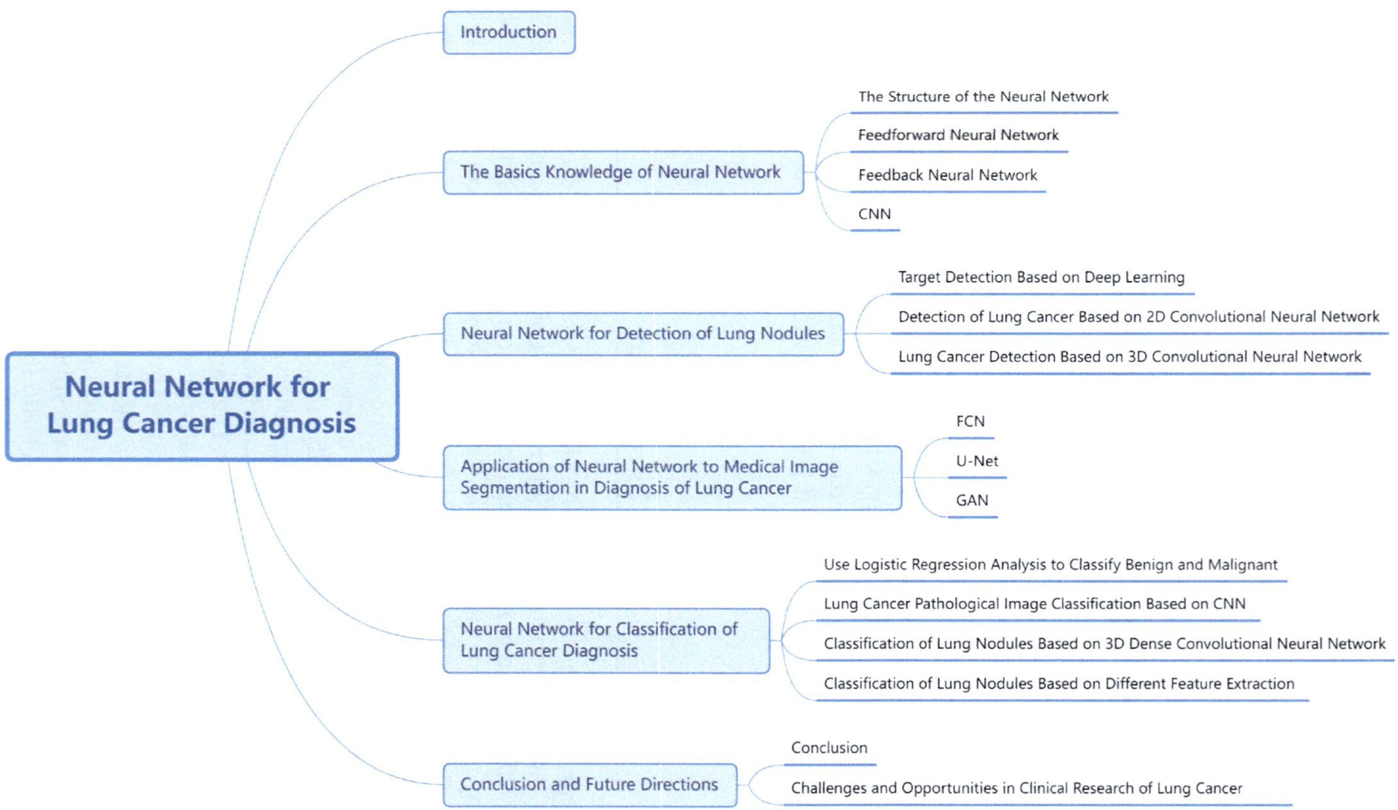

FIG. 5.1 The structure of the chapter.

pathological examinations in order to determine the type and severity of the tumor. Once the lung cancer is diagnosed, CT or color ultrasound of other body parts should be performed in order to observe whether there are metastases in other parts. In addition, it can also be checked through a bronchoscopy, which can visually observe the condition of the mass as well as take a part of the tissue for pathological and other related examinations.

With the rapid development of deep learning, medical image processing that is based on deep convolutional neural networks (CNN) has become a research hotspot. Using deep learning to diagnose lung cancer alleviates a shortage of the doctors' resources and buys treatment time for the patients' treatment. The diagnosis of lung cancer depends mostly on the diagnosis of the lung nodules. Lung nodules appear as highlights and shadows on the CT. Generally, it refers to a spot with a diameter of less than 3 cm, which is also known as a bulbous lesion in the lung (Jiang et al., 2021). According to the spatial location of lung nodules, lung nodules can be divided into Solitary Pulmonary Nodules (SPNs), Juxta-Pleural Nodules (JPNs), and Juxta-Vascular Nodules (JVNs) (Krishnamurthy et al., 2016). According to the density classification, SPNs are divided into solid lung nodules and subsolid lung nodules. The latter is divided into pure ground glass nodules (non-solid) and mixed lung nodules (partial solid nodules) (Wyker & Henderson, 2021), which is shown in Fig. 5.2.

In order to detect the entire nodule, we need to complete multiple tasks, which include lung segmentation, nodule detection, and nodule segmentation. After that, a carry out follow-up and a retrieval of the nodules as well as quantitative and qualitative analysis of nodules are conducted, such as judging benign and malignant nodules. These tasks have their own training data, but they share the characteristics of the deep learning network layer between them and perform different tasks at the end. According to the different tasks, the training labeling mode is different during training.

The basics knowledge of neural network

With the development and research of neuroscience and cognitive science, we have a deeper understanding of human intelligent behavior. The human brain is an organ that can produce consciousness, thoughts, and feelings. Human intelligent behaviors are all related to brain activity. Early neural network scientists were inspired by this, so they simulated the brain nervous system and constructed a mathematical model, which is called an artificial neural network (ANN) or a neural network in short. When we talk about neural networks in machine learning, we mean *neural network learning*, which is a network structure model composed of many artificial neurons. The connection strength between these artificial neurons is a learnable parameter (Okwu & Tartibu, 2020).

In the field of machine learning, the computational model of an ANN is inspired by the central nervous system of animals. A neural network is composed of many neurons. Each neuron can be regarded as a small information processing unit. The neurons are connected to each other in a certain way in order to form an entire deep neural network. A general neural network consists of the following parts: an input layer, a hidden layer, and an output layer. A classic neural network is shown in Fig. 5.3. This is a neural network with three levels. The yellow level is the input layer, the blue level is the middle/hidden layer, and the black level is the output layer. The input layer has three input units, the hidden layer has four units, and the output layer has two units.

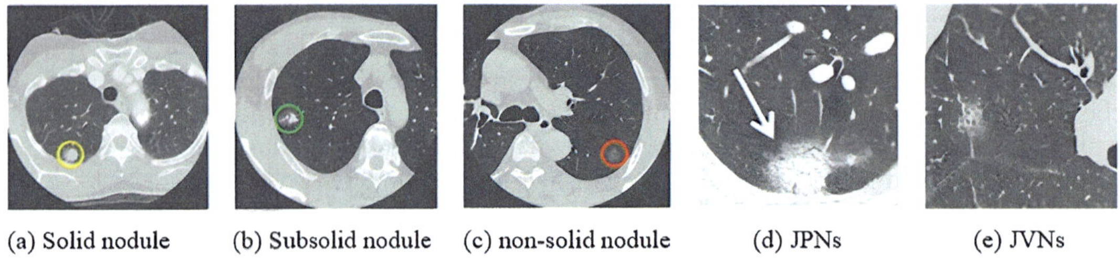

FIG. 5.2 Five different types of lung nodules.

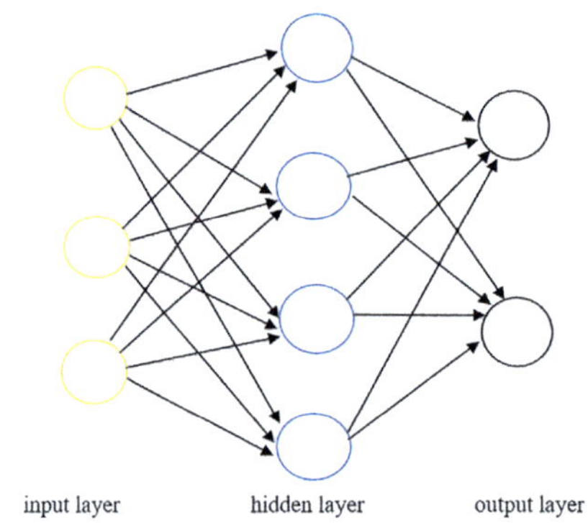

FIG. 5.3 Classic neural network structure.

The structure of the neural network

An artificial neuron, which is referred to as a neuron, is the basic unit of a neural network. It mainly simulates the structure and characteristics of biological neurons, receives a set of input signals, and produces output. As early as the early 20th century, biologists discovered the structure of biological neurons. A biological neuron usually has multiple dendrites and one axon. Dendrites are used to receive information, and axons are used to send information. When the accumulation of the input signals that are obtained by a neuron exceeds a certain threshold, it is in an excited state and generates electrical impulses. There are many ends of an axon tail that can connect to the dendrites of other neurons and transmit electrical impulse signals to other neurons. The neuron model includes an input, an output, and calculation functions. The input can be analogous to the dendrite of a neuron, the output can be analogous to the axon of a neuron, and the calculation can be analogous to the nucleus. Fig. 5.4 is a typical neuron model.

In 1943, psychologist McCulloch and mathematician Pitts referred to the structure of biological neurons and published an abstract neuron model, which is called McCulloch-Pitts (MP). The MP proposes a formal mathematical description of the neurons and a network structure method, which proves that a single neuron can perform logical functions, and opened the era of neural network research. After more than 70 years of tortuous development, its related theories and methods have become a cross-discipline between physics, mathematics, computer

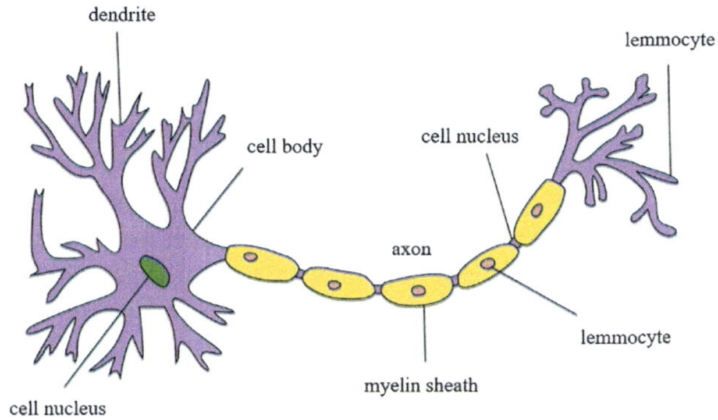

FIG. 5.4 Typical neuron model.

science, and neurobiology, and it has become a major realization technology for AI. It has been widely used in the fields of pattern recognition, image processing, intelligent robots, automatic control, financial forecasting, optimization combination, man-machine game, data communication, industrial control, and expert systems. It solves many practical problems that are difficult to solve using traditional computer programs, and it shows good intelligence characteristics and learning evolution characteristics. It has become the main foundation of deep learning technology.

The neuron model is a model that includes input, output, and calculation functions. The input can be analogous to the dendrites of the biological neurons, the output can be analogous to the axons of the biological neurons, and the calculations can be analogous to the cell bodies of the biological neurons. A typical artificial neuron model is shown in Fig. 5.5. Among them, X_1, X_2, …, X_m represent the initial features. W_{11}, W_{22}, …, W_{mk} represent the weights, which are the scaling factors of the features. b_k represents the bias of the neuron. f(·) represents the activation function. After the features are scaled and offset, they are all added up. Also, after an activation operation, they are the output. The arrow lines are called connections, and each connection has a weight.

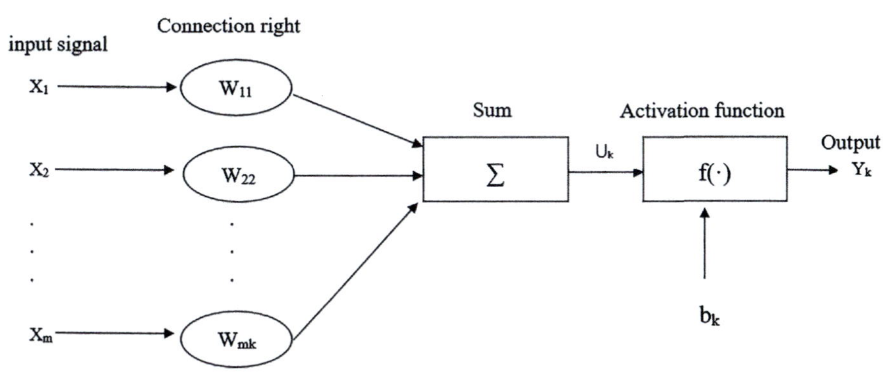

FIG. 5.5 Artificial neuron model.

The ANN is equivalent to a multi-input single-output nonlinear threshold device. If the weighted sum of the input signal exceeds the threshold, the ANN will then be activated. The process of a neuron reading the features, performing calculations, and producing output is called the activation. The activation function is a nonlinear function, which is used to add nonlinear transformation features to the neural network model so that it can handle complex nonlinear classification tasks. The commonly used activation functions are the Sigmoid function, which are in addition to the common hyperbolic Tanh and Rectified Linear Units.

The input layer is the first layer of the neural network, which accepts signals or data from the external world and corresponds to the characteristic input of the sample. Each node represents a characteristic variable in the characteristic vector X of the sample. It does not perform any operations on the input signal, and there is no corresponding weight and offset value. The output layer node corresponds to the predicted output of the sample. Each node represents the predicted probability of the samples in different categories and realizes the output of the system that processes the results. The hidden layer has nodes that apply different transformations to the input data, which correspond to the intermediate activation calculations. These nodes are called hidden units. In a neural network, the role of the hidden unit can be understood as transforming the features of the input layer and passing it layer by layer to the output layer for result prediction.

Feedforward neural network

The network acquires knowledge through learning. The direction of the information flow in a neural network is divided into forward and backward. Therefore, the neural network can be divided into two basic types: feedforward and feedback. The feedforward network is a unidirectional multilayer structure network where the neurons are divided into different layers according to the order of receiving information. There is no interconnection between the neurons in the same layer, the flow of information between the layers only goes in one direction, and there is no feedback in the entire network. Each neuron starts from the input layer, receives the input of the previous layer, and outputs to the next layer until the output layer.

The feedforward network has complex nonlinear mapping capabilities. However, the output of the feedforward network is only determined by the current input and the weight matrix and has nothing to do with the previous output state of the network. The feedforward neural network has a simple structure and a wide range of applications. It can approximate any continuous function and square-integrable function with arbitrary precision, and it can accurately realize any limited training sample set. From the system point of view, the feedforward network is a static non-linear mapping. Through the composite mapping of simple non-linear processing units, complex non-linear processing capabilities can be obtained. Most feedforward networks are supervised learning networks. Its classification ability and pattern recognition ability are generally stronger than those of the feedback network.

Common feedforward neural networks include perceptron networks, Backpropagation (BP) networks, and Radial Basis Function (RBF) networks. The perceptron is the simplest feedforward network. It is mainly used for pattern classification, and it can also be used in learning control and multi-modal control that is based on pattern classification. BP refers to a feedforward network that uses a backpropagation learning algorithm to adjust the connection weight. The difference from the perceptron is that the neuron transformation function of the BP network uses the Sigmoid function. Therefore, the output quantity is a continuous quantity between 0 and 1, which can realize any non-linear mapping from the input to the output (Werbos, 1974). RBF refers to a feedforward network that is composed of hidden layer neurons and RBF neurons.

Feedback neural network

A feedback neural network is a neural network with feedback connections from the output to the input, and its structure is more complex than that of the feedforward network. Typical feedback networks are the Elman network and the Hopfield network. The Hopfield network is composed of a group of interconnected neurons and belongs to the recurrent neural network model. Its characteristic is that each neuron is both an input unit and an output unit, and there is no hidden unit. The Hopfield network is also called the associative memory network. It often stores one or more stable target vectors. When similar vectors are input at the input of the network, these stable target vectors will wake up the network memory mode and present it through the output. The Elman network is a two-layer backpropagation network with neurons connected to the hidden layer and the input vector. The output is used as the input of the output layer, and it is also connected to some other neurons in the hidden layer. Which are fed back to the input of the hidden layer. Since its input represents the spatial information of the signal, which the feedback branch is a delay unit that reflects the timing information of the signal, the Elman network can perform pattern recognition in the time domain and the spatial domain.

CNN

CNN is a classic supervised learning deep model that is produced by the combination of deep learning and image processing technology. It is one of the most representative neural networks in the field of deep learning technology, and it has made many breakthroughs in the field of image analysis and processing (Khan et al., 2020). On ImageNet, which is a standard image annotation set that is commonly used in academia, many achievements have been made based on a CNN, which includes image feature extraction and classification and pattern recognition. The basic idea is to share the weights of the feature maps in different positions of the previous layer network and use spatial relative relationships to reduce the number of parameters in order to improve the training performance.

From the proposal of a CNN to the current widespread application, it has roughly experienced the theoretical embryonic stage, the experimental development stage, large-scale applications, and the in-depth research stage. The proposal of receptive fields and neurocognitive machines in human visual information is an important theory in the embryonic stage of theory. Hubel and Wiesel (1962) showed through biological research that the transmission of visual information in the brain from the retina is accomplished through multi-level receptive stimulation, and they first proposed the concept of a receptive field. In 1980, Fukushima (Fukushima & Miyake, 1982) proposed a neurocognitive machine that is based on the concept of receptive fields. The neurocognitive machine can be understood as the first version of a CNN. The core point is to model the visual system, which is not affected by the position and size of the vision. LeCun et al. (1998) proposed LeNet5 using a gradient-based backpropagation algorithm for the supervised training of the network, which entered the experimental development stage. It was not until the introduction of the AlexNet network in 2012 that the CNN status in deep learning applications was established. The AlexNet, which was proposed by Krizhevsky et al. (2017), won the image classification championship on the ImageNet training set, which the CNN become a key research object in computer vision that continues to deepen.

In the image recognition task, the general structure of a CNN works as follows: input layer -> convolutional layer -> pooling layer -> fully connected layer -> output result. In medical image processing, the picture size of the input layer is generally an integer multiple of 2, such as 224×224, 448×448, and 896×896. The convolution layer uses smaller convolutions, such as 3×3 and 5×5. The convolution layer is used to extract the features of an image, and the pooling layer is used to reduce the dimensionality of the convolution result. The classic CNN takes LeNet-5 as an example. LeNet5 uses convolution, parameter sharing, pooling, and other operations to extract features through clever design, which avoids a lot of computational costs, and finally uses a fully connected neural network for classification and recognition. This network is also the starting point for many neural network architectures. The LeNet-5 model has a total of seven layers, which include two convolutional layers, two pooling layers, and three fully connected layers. Each layer contains a different number of trainable parameters, each layer has multiple feature maps, and each feature map has multiple neurons. The input image size of LeNet-5 is 32×32. The convolution layer uses a 5×5 convolution kernel. Each time the convolution slides by one pixel, a feature map uses the same convolution kernel. The pooling layer uses a 2×2 input field, which means that four nodes in the upper layer are used as the input of one node in the next layer, which slides 2 pixels at a time so that the input fields do not overlap.

2D CNN

A CNN consists of an input layer, an output layer, and several hidden layers. Each layer in the hidden layer performs a specific operation, such as convolution, pooling, and activation. The input layer is connected to the input image, and the number of neurons in this layer is the pixel of the input image. The middle convolutional layer performs the feature extraction on the input data through a convolution operation in order to obtain a feature map. The result of the convolution operation depends on the setting of the parameters in the convolution kernel. The pooling layer is behind the convolutional layer filters and selects the feature maps, which simplifies the computational complexity of the entire network. Through the fully connected layer, all the neurons in the previous layer are then fully connected, and the obtained output value is sent to the classifier, which gives the classification result. A general CNN is a 2D CNN, its input image is 2D, and the convolution kernel is a 2D convolution kernel, such as ResNet (He et al., 2016) and VGG (Simonyan & Zisserman, 2014). Suppose the input image size is $H \times W$ with three channels, RGB. The convolution kernel of size (c, h, and w) slides on the spatial dimension of the input image, which is where c, h, and w denote the number of channels and the height and width of the convolution kernel, respectively. You need to enter

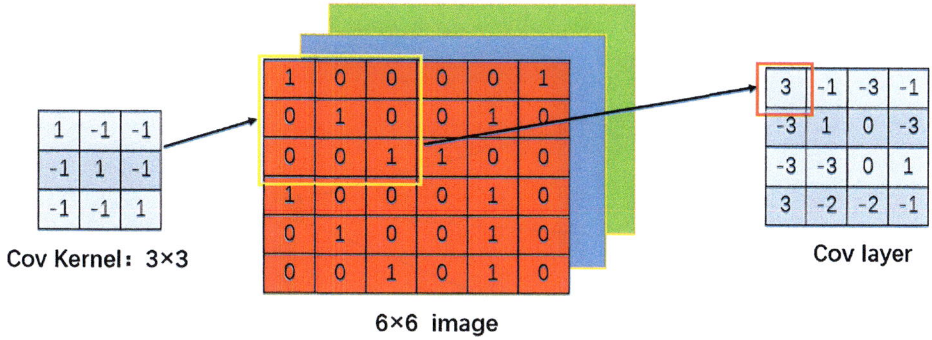

FIG. 5.6 2D CNN convolution.

the value of the image and the value of (h and w) on each channel to perform a convolution operation in order to obtain a value. Fig. 5.6 (Liu et al., 2021).

3D CNN

Most medical images are usually 3D, such as CT and MRI. Although the CT usually shows a 2D picture, it is just a slice. Therefore, if you want to segment some diseased tissues, you must use a 3D convolution kernel, for example, the convolution kernel that is used by the segmentation network 3D U-Net is 3D. It changed the 2D convolution kernel in the U-Net network to a 3D convolution kernel, which is suitable for 3D medical image segmentation (Minaee et al., 2021). A 3D CNN can extract a more powerful volume representation on the three axes of X, Y, and Z. The use of three-dimensional information in segmentation makes full use of the advantages of spatial information. The 3D convolution kernel has one more depth than the 2D convolution kernel, which means the number of 2D slices of the medical images. In a 3D picture, $C \times N \times H \times W$, where C, N, H, and W represent the number of channels, the number of slice layers, and the height and width of the convolution kernel, respectively. A value is obtained by sliding the window on the height, width, and the number of layers on each channel, which is similar to the 2D convolution operation. Fig. 5.7 (Liu et al., 2021).

Neural network for detection of lung nodules

The first step in the diagnosis of lung cancer is abnormality detection. Many lung cancers of computer-aided diagnosis (CAD) systems divide lung nodule detection into candidate nodule detection and false-positive reduction. First, the lung CT images are extracted using the traditional image segmentation methods in order to extract the candidate nodules, and these candidate nodules are then sent to the second stage according to the combination of the shape, location, gradient feature, texture feature, and context information of the nodule in order to reduce the false positives. When performing the preliminary detection of the lung nodules in the image, you need to try to extract as many true nodules as candidate nodules in order to reduce the missed detection rate. Most of the lung nodules in the candidate nodules are

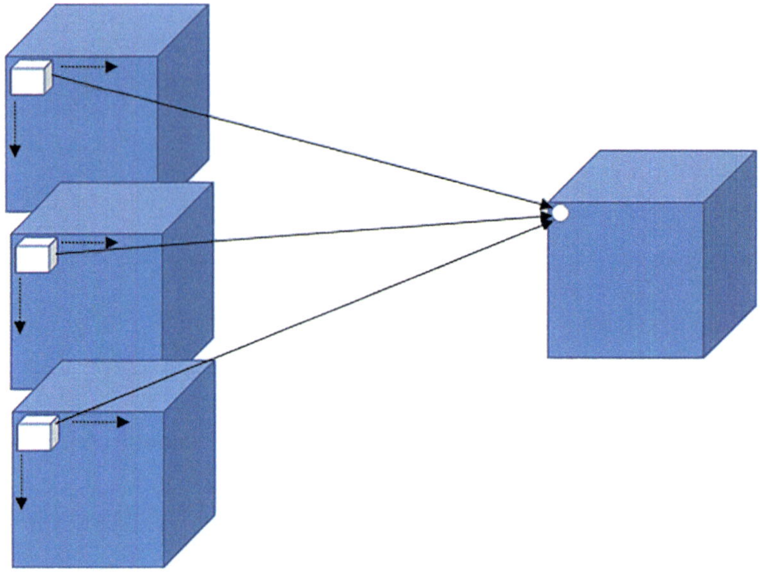

FIG. 5.7 3D CNN convolution.

extracted in the first step, which means that the false positives are too high. Therefore, the second step is needed in order to reduce the false positives.

The traditional algorithms for candidate nodule extraction include threshold, region growth, morphology, and clustering. These detection methods have complex processes and low efficiency rates. The feature value design for the detection with different types of diseases is mainly realized through the feature function of texture analysis, edge detection, and object detection, such as SIFT and HoG. The choice of these features is entirely dependent on the choice of the experts. The experts need to evaluate many cases and sum up the experiences from a large amount of data. Some important information is often ignored during the feature selection, which leads to a lower diagnosis rate of the disease. At the same time, in the face of different diseases, it is necessary to design a set of completely different feature vectors, which is the reason why traditional CAD technology cannot meet the processing of large amounts of medical image data. With the development of deep learning, it realizes the automatic learning of the most effective features from a large amount of data. Deep learning simulates the human visual system and some implementations in the recognition system. Using the principle of gradient backpropagation to automatically extract feature vectors instead of artificially selecting features through image reviewing is the biggest difference.

Target detection based on deep learning

In lung CT images, the lung nodules are the target of the detection. The target detection task is different from image classification. It not only needs to distinguish the target category, but it also needs to determine the target's position and accurately locate it. Therefore, it needs to pay attention to the high-level semantic information. Target detection that is based on a

CNN uses itself as a feature extractor to judge the location attributes of the image in order to generate a limited frame that can frame the target. The target detection that is based on a CNN is two-stage at the beginning, which includes the region proposal and detection. This means that it first generates some candidate boxes in a certain way, and it then classifies the content in these candidate boxes and corrects the position of the candidate boxes. The R-CNN proposed by Girshick et al. (2014) is the earliest method that used a CNN to achieve target detection tasks. It filters some candidate region proposals through a selective search. These regional proposals are scaled to a specified size and used to input a CNN for the classification of the positive and negative samples. The SVM is then trained with the feature vectors that are extracted from each proposal. The obtained classification results use the non-maximum suppression to screen the final region proposals. Finally, the region proposals are regressed, and the coordinate values in of the region proposals are corrected to obtain more accuracy. The R-CNN takes more time in the Selective Search. A CNN must be passed through for each proposal, so the efficiency is low. He et al. (2015) proposed the SPP-net to solve the shortcomings that each region proposal in an R-CNN must go through the CNN once, which thereby improves the efficiency. Spatial pyramid pooling in SPP-net avoids the problem of target shape distortion that is caused by image scaling in order to adapt to the input size of the CNN. From the improvement analysis of SPP-net, we know that what restricts the speed of an R-CNN is the repeated convolution of the multiple ROIs of an image. Therefore, the author of a Fast R-CNN (Girshick, 2015) improved along this line of thinking and proposed a Fast R-CNN. The Fast R-CNN proposes an ROI pooling structure, which is equivalent to a special SSP with a pyramid level of 1 and calculates the pooling only once. This is an end-to-end network structure, and the final SVM classification is removed. The Faster R-CNN (Ren et al., 2015) has improved based on the Fast R-CNN. The main improvement is to use a region proposal network instead of a Selective Search to generate the region proposals. In addition, the concept of the anchor is introduced, which has been used in the subsequent models. The idea of one-stage is to directly predict and classify the region proposals at various positions on the image, which is different from two-stage, without pre-generating some candidate frames. The one-stage method is represented by YOLO and SSD. The YOLO model divides the image into multiple blocks and directly outputs a tensor with the same number of blocks. The prediction vector of each block contains whether it has a target, what category it is, and the correction of the specific location. The YOLO model has now been developed for the fifth generation. SSD is based on the basic idea of YOLO and considers the multi-scale situation. GT Bounding boxes of different scales are allocated to different scales for the prediction. The prediction of different scales is carried out through the feature maps of different layers, which is unlike most computer vision tasks that involve dozens or even hundreds of categories. There are only two categories with the diagnosis of lung cancer: benign and malignant. Using deep learning to detect the lung CT images belongs to the detection of small targets in the image. Small target detection has always been a difficult practical and common problem. Some solutions have also emerged in order to improve the performance of small target detection with its development in recent years. For example, the image pyramids with different resolutions can be built from the original image for targets of different scales. A classifier with a fixed input resolution is then used to slide in this layer in order to detect small targets at the bottom of the pyramid for each layer of the pyramid. Lung nodule detection that is based on commonly used deep learning includes 2D CNN, 3D CNN, and faster regional convolutional neural networks.

Detection of lung cancer based on 2D convolutional neural network

The 2D CNN uses a 2D convolution kernel to slide on a 2D image, but this framework can also retain the 3D information of the nodule by using the adjacent lung slices or different axial cuts. Setio et al. (2016) proposed multi-view 2D ConvNets for CAD systems in order to reduce the false positives. The composition of this network structure corresponds to the two important steps in the previous lung nodule detection, which include (1) extracting the candidate nodules and (2) reducing the false positives. The author first trained three candidate detection models for a solid nodule, a subsolid nodule, and a large nodule. Finally, these detection models are combined in order to improve the sensitivity of the detection of various nodules. The sensitivity reaches 94.4% when the three algorithms are combined. The detection performance has been improved, but it also brings the problem of false positives. Therefore, the second part is to reduce the false positives. This stage is composed of different ConvNets streams, which is called a multi-view structure. The input of each stream is the image blocks of different views captured at the candidate positions. This method uses nine views, and the result is significantly better than that of the 3-view or 1-view method. The two-dimensional convolutional network that is used is composed of three consecutive convolutional layers and a maximum pooling layer. The last layer of each channel is a fully connected layer. These fully connected layers will be cascaded in different ways in order to get the final classification result. The method was evaluated on the ANODE09 and DLCST datasets. The CPM reached 0.637 on the ANODE09 dataset. With the DLCST data set, a sensitivity of 76.5% is achieved when there are six false positives in each scan. The experiments show that the method has better robustness.

Xie et al. (2019) proposed an improved nodule detection network model that is based on a 2D Faster R-CNN. Compared with the original network, this method uses a two-region proposal network and a deconvolution structure. The lung nodule detection network is divided into three parts. First, the feature extraction network extracts features based on the VGG16. The region proposal network then generates the detected area. Finally, the generated region of interest frame-interest region classification network is classified. Second, an enhanced structure that is based on a 2D CNN is designed in order to reduce the false positives that are generated in the previous step. At this stage, in order to solve the imbalance between the false-positive candidate nodes and the real nodes, the methods of translation, the horizontal reflection data expansion, the pre-screening of the negative samples, and the down-sampling of the negative samples are used. In order to introduce the 3D information of the nodes, three models were trained on the three slices. Finally, these three models are fused in order to obtain a candidate model. In order to enhance the robustness of the model and improve the sensitivity of the nodule detection task, the representative and difficult to distinguish nodules in the dataset that were pre-judged incorrectly were screened out in order to train the model again during the training process. The results show that the detection sensitivity is 86% when evaluated on the LUNNA16 dataset.

As early as 2017, in the lung cancer detection competition with the lung CT held on the Kaggle, deep learning was used for lung cancer detection. The National Cancer Institute provided high-resolution lung CT for the competition. In the competition, the contestants needed to predict the risk of cancer based on 3D CT images of a patients' lung. Wit and his collaborators used 17-layer 3D convolutional neural networks to construct a nodule detector in order

TABLE 5.1 3D Convnet network structure.

Layer	Params	Activation
Input		
Avg pool	2*1*1	
3D conv	3*3*3	ReLU
Max pool	1*2*2	
3D conv	3*3*3	ReLU
Max pool	2*2*2	
3D conv (2×)	3*3*3	ReLU
Max pool	2*2*2	
3D conv (2×)	3*3*3	ReLU
Max pool	2*2*2	
3D conv	2*2*2	ReLU
3D conv	2*2*2	Sigmoid
3D conv	2*2*2	none

to predict the possibility of cancer. In the end, the second place in the competition (Wit & Hammack, 2018) candidate nodule detection process is used to extract the sub-images of the original image for the diagnosis. Therefore, each CT image needs to be processed hundreds of times regardless of whether there is an abnormal area, which causes huge computational waste. In order to avoid this problem, a model is built to identify whether the processing area is abnormal at first. On this basis, the region of interest is detected. The input data of the neural network model are $64\,mm^3$ CT images. The model consists of five convolutional blocks, a maximum pooling layer, and an activation function. The size of the 3D convolution kernel is $3 \times 3 \times 3$, and the size of the pooling layer is $2 \times 2 \times 2$. The model training uses progressive learning to accelerate the training. The model is first trained on 32×32, which is where the batch size is 128. Two thousand parameters are approximately updated. 64×64 data is then used for the training, which is where the batch size is 64. Table 5.1 shows the network structure of the model.

Lung cancer detection based on 3D convolutional neural network

The 3D convolutional neural network uses 3D data for the 3D convolution. Using the 3D convolution kernel to learn the 3D features of the nodules helps improve the performance of the nodule detection and classification. A combination of 2D and 3D methods can also be used for different stages. Golan et al. (2016) used a 3D deep convolutional neural network in order to detect the lung nodules, and the author did not perform the second stage of the false-positive reduction. A 20-layer deep convolutional neural network is trained through the backpropagation algorithm to extract valuable 3D features from 3D input data in order to detect lung nodules in CT image sample blocks. The size of the 3D CT image is [65,764] × 512 × 512. In order to apply the sliding window method on the CNN, the output of the CNN needs to be averaged at different positions, and a 3D voting network with the same size as the CT scan is calculated. This algorithm has been explored in the detection of the 3D lung nodules. However, there is no false-positive reduction, so the problems of low sensitivity and more false positives exist. Hamidian et al. (2017) proposed an automatic lung nodule detection algorithm that is based on a 3D convolutional neural network. Using the FCN network structure, the 3D convolutional neural network with a fixed input size is converted into a 3D FCN that can be input at any size. An FCN can efficiently generate a score map of the entire image in one scan. The model first uses a 3D CNN to classify the region of interest sample blocks extracted from the LIDC-IDRI dataset into nodules and no nodules. You need to covert this CNN into an FCN network and use the FCN to screen out the negative samples that are difficult to classify. The CNN for classification tasks needs to be trained this way, and it then needs to be converted to a new FCN network in order to generate the candidate nodes. In order to reduce the false positives, the false-positive results that are obtained are used to train the third CNN. Compared with the sliding window method of applying a CNN to the entire input image mentioned in the previous method, the speed of using an FCN is increased by 800 times. The author successfully extended the 2D CNN to 3D. It is also the first time that a 3DFCN has been used for the detection of lung nodule pairs. The downside is that after the positive is reduced, the sensitivity still needs to be improved.

Ding et al. (2017) proposed a new type of lung nodule detection method that is based on the improvement of a deep convolutional neural network. The method of combining 2D and 3D that introduces a deconvolution structure to improve the Faster R-CNN network greatly improves the detection performance of the nodules. We used a 3D DCNN for false-positive reduction. The candidate node detection network is composed of an RPN and an ROI classifier. The RPN is used to extract the suspected nodule region, and the classifier recognizes whether the ROIs are nodes. Adding a deconvolution layer structure to the convolution layer of the VGG-16 solves the problem that the lung nodule size is too small to extract the features. At the same time, it can save the computational cost of training two deep convolutional neural networks as a shared feature layer. This algorithm has demonstrated its superior performance in the detection of nodules on the LUNA16 dataset and achieved the first result.

At the LUNA16 challenge last year, there were also many algorithms that performed well with the direction of nodule detection. Among them, the team of Setio et al. (2019) proposed a 3D region proposal U-Net model based on the outstanding network framework U-Net in the field of medical image processing research. This network also shows the superiority of DenseNet and ResNet. Unlike the two-stage mode in the target detection model based on deep learning, this network is a separate end-to-end detection network. Detection can be achieved in one stage, and no false-positive reduction stage is added. The proposed network structure draws on the RPN with the Faster R-CNN. It has been improved in order to adapt to the input of a 3D patch. In addition, the model draws on the intensive learning in DenseNet, the residual learning in ResNet, and the experience of U-Net in small target detection for the small target detection of the lung nodules. This network structure improves the detection performance while reducing model complexity and reasoning time. Except for the Zhongliu_Xie team, the LUNA16FONOVACAD (Setio et al., 2019) also put forward a good plan. They were inspired by the work of the first team in the Kaggle2017 Data Science Bowl competition (Xie et al., 2019), which was mentioned earlier. They also made an improvement to U-Net and proposed a nodule detection architecture based on a 3D U-Net network. They added a stage to reduce the false positives and integrated the three 3D CNN structures in order to reduce the false positives. Using 10-fold cross-validation for evaluation on the LUNA16 dataset, the sensitivity of the nodule detection network is 99.1%, and the sensitivity of the nodule detection network is 99.1%. The PAtech (Setio et al., 2019) team developed a Feature Pyramid Network (FPN) based on the network model of the first place in the Data Science Bowl 2017 competition and used two 3D CNN classifiers in the false-positive reduction part. The network also consists of two parts: nodule detection and false-positive reduction. The nodule detection task is based on the feature pyramid network in order to independently predict each output layer. In the false-positive reduction stage, two 3D DCNNs are used for the classification. The first network uses multi-scale input, and the second network is based on a 3D U-Net architecture. The result is the fusion of the two classification models.

Application of neural network to medical image segmentation in diagnosis of lung cancer

After detecting the CT images of lung cancer with lesions, which are the abnormal areas, the next step is to study the process of segmentation of the abnormal areas, which is to segment the abnormal areas from the medical image based on the similarities and differences between the areas. Accurately separating the lung nodules from the lung CT images for assessment and diagnosis of lung cancer can effectively improve the diagnosis and treatment of lung cancer. Generally speaking, medical image processing has the characteristics of higher complexity compared with ordinary images. This is because the medical images have the characteristics of a large grayscale and unclear boundaries. The precision and accuracy of segmentation are higher than those of ordinary images. Although the traditional segmentation method can segment the contour boundary of the nodule, there is a serious problem of over-segmentation, which leads to unsatisfactory segmentation accuracy.

When performing image segmentation operations, a CNN has excellent feature extraction capabilities and good feature expression capabilities, so there is no need to manually extract the image features or perform excessive pre-processing on the image. Therefore, the CNN has achieved great success in the field of medical image segmentation and auxiliary diagnosis in recent years. Common CNNs for segmenting pathological regions in lung cancer diagnosis mainly include Fully Convolutional Networks (FCN), U-Net, and Generative Adversarial Networks (GAN) (Liu et al., 2021).

FCN

For general CNNs, such as VGG and ResNet, some fully connected layers are added at the end of the network. The category probability information can be obtained after softmax.

However, this probability information is one-dimensional, which means that it can only identify the category of the entire picture and not the category of each pixel, so this fully connected method is not suitable for image segmentation. Long et al. (2015) proposed the

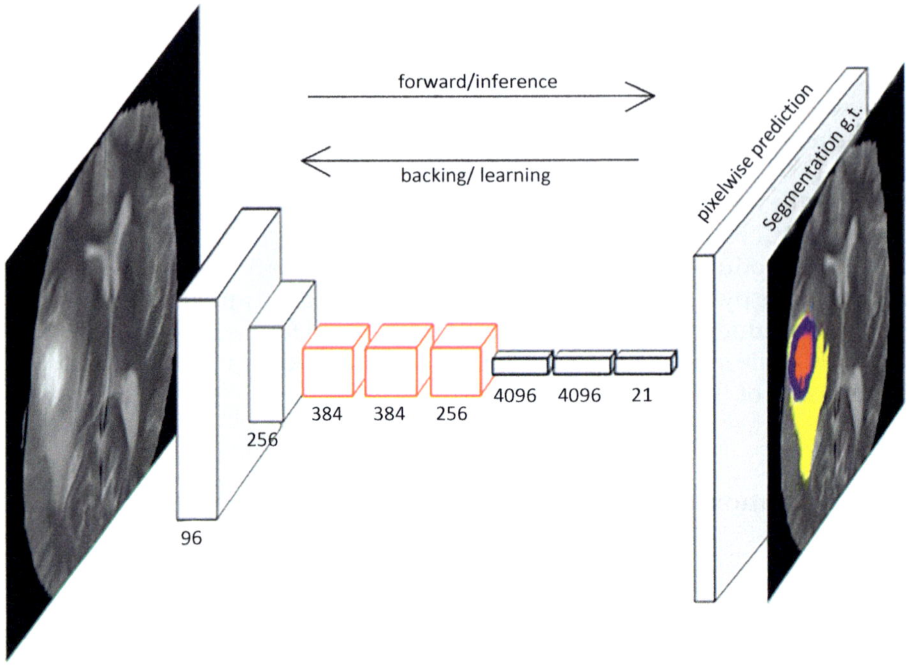

FIG. 5.8 The structure of FCN.

Fully Convolutional Network (FCN) in response to the above problems. An FCN replaces the following fully connected layers with convolutional layers, so a 2D feature map can be obtained. Softmax is then followed in order to obtain the classification information of each pixel, which solves the segmentation problem. An FCN is the pioneering work of the most successful and advanced deep learning technology for semantic segmentation. The network structure of an FCN is shown in Fig. 5.8 (Liu et al., 2021). The deconvolution layer is used to up-sample the feature map of the last convolution layer in order to restore it to the same size of the input image. Thus, a prediction can be generated for each pixel, while retaining the spatial information in the original input image. Finally, we performed pixel-by-pixel classification on the up-sampled feature map in order to complete the final image segmentation. According to the magnification of up-sampling, it is divided into FCN-32s, FCN-16s, and FCN-8s. However, the shortcomings of an FCN are also very prominent. First, the results of its upsampling are relatively fuzzy and insensitive to the details of the image, which results in the segmentation results not being fine enough. Second, the idea of segmentation is essential in order to classify each pixel. As a result, the relationship between the pixels is not fully considered and spatial consistency is lacking. Many researchers have proposed improved schemes for the problems of a FCN, such as SegNet.

The segmentation of lung parenchymal tissue and the corresponding nodules and blood vessels is the prerequisite and initial step for a subsequent quantitative lung CT image analysis. An FCN also has a place in the challenge hosted by the medical image computing and computer-assisted intervention conference (MICCAI). In the 2018 challenge hosted by

MICCAI, Astono et al. (2018) used an automatic feature extraction network like an FCN to segment the liver. In the up-sampling process, the reserved spatial information was used to realize the prediction output of the pixels and won the championship of the competition. Regarding the lung structure segmentation, Anthimopoulos et al. (2019) collected 172 high-resolution CT images from multiple medical institutions. They used an FCN with dilated convolution and multi-scale feature fusion in order to segment the lung parenchyma, healthy tissue, micro-nodules, and honeycomb structures in the lung CT images. The network test result reached a segmentation accuracy of close to 82%.

Because chest X-rays are quick and easy, it is the most common medical image that is used in medicine. Chest X-rays use a small dose of radiation in order to produce images of the chest. In the chest X-ray examination, we can achieve segmentation of the lung area. It can be used to help diagnose and monitor various lung diseases, such as pneumonia and lung cancer. Jue et al. (2019) used U-Net and FCN to improve a learning method derived from cross-modality, which used MR information obtained from a CT to reconstruct an MRI in order to improve the CT segmentation.

U-Net

Ronneberger et al. (2015) designed a U-Net for biomedical images, which was based on an FCN. It has been widely used in medical image segmentation since it was proposed. Due to its excellent performance, U-Net and its variants have now been widely used in various subfields of computer vision. This method was presented at the 2015 MICCAI conference and has been cited more than 4000 times. U-Net has had many variants thus far. Currently, there are many new design methods of convolutional neural networks, but many of them continue the core idea of U-Net by adding new modules or other design concepts.

U-Net is composed of a U channel and a skip connection. A U channel is like the code structure of SegNet. The encoder has four sub-modules. Each sub-module contains two convolutional layers, and after each sub-module, there is a maximum pool in order to achieve downsampling. The decoder contains four sub-modules, and the resolution is sequentially increased through upsampling. Finally, the value of each pixel is predicted. The network structure is shown in Fig. 5.9 (Liu et al., 2021). The input is 572×572, and the output is 388×388. As can be seen from the figure, there are no fully connected layers in the network. There are only convolution and subsampling. The network also uses skip connections to connect the upsampling results to the output of the sub-module. The resolution of the output of this sub-module in the encoder is the same as the input of the next sub-module in the decoder.

The reason why U-Net is suitable for medical image segmentation is that its structure can combine low-level and high-level information at the same time. Low-level information helps improve the accuracy, whereas high-level information helps extract the complex features.

The improvement of U-Net has become the research hotspot of medical image segmentation, and many variants have been developed on this basis. For example, Çiçek et al. (2016) proposed a 3D U-Net. This model aims to make the U-Net structure have richer spatial information. The network structure is shown in Fig. 5.10 (Liu et al., 2021). This structure realizes 3D image segmentation by inputting a continuous 2D slice sequence of the 3D images. This network can be trained on a sparsely labeled data set and predict other unlabeled places on this

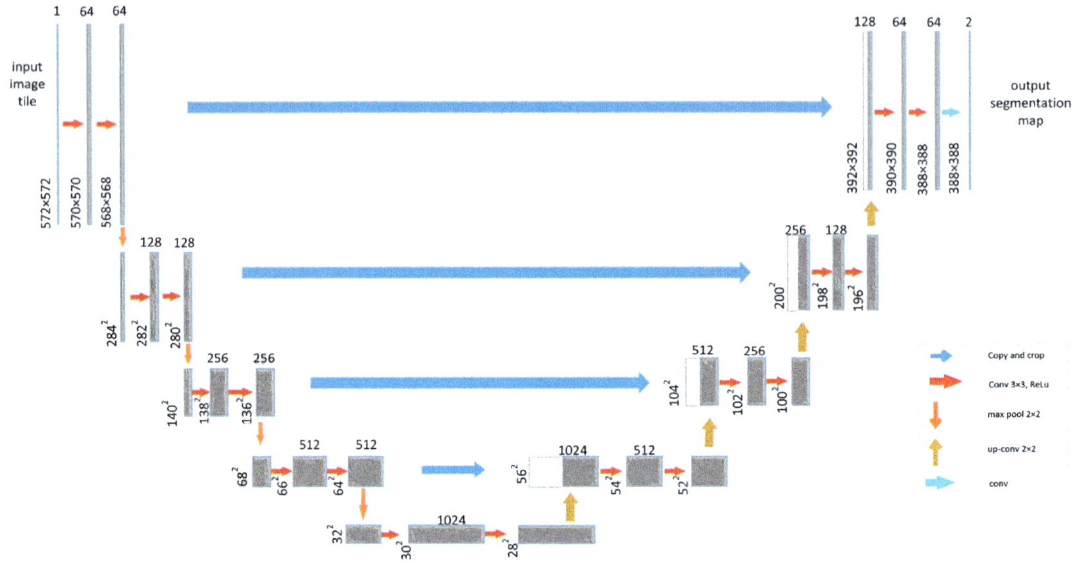

FIG. 5.9 The network structure of U-Net.

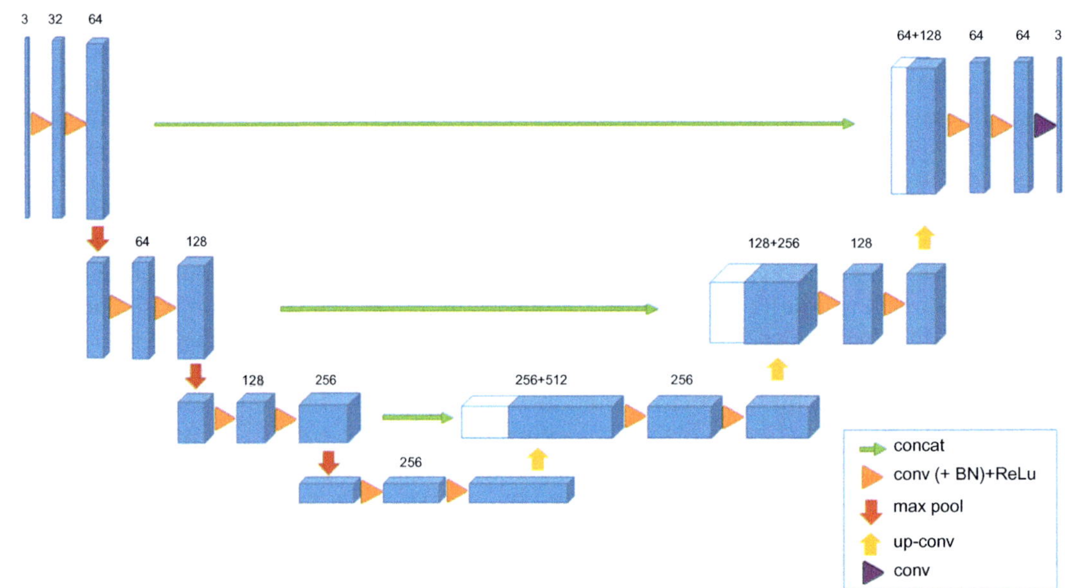

FIG. 5.10 The network structure of 3D U-Net.

data set. It can train on multiple sparsely labeled datasets and then predict new data. Compared with the U-Net input, the input here is a stereo image ($132 \times 132 \times 116$) and three channels. The size of the output picture is $44 \times 44 \times 28$. The birth of 3D U-Net in medical image segmentation is of great help especially for volumetric images, which retain the excellent features originally possessed by an FCN and U-Net. Milletari et al. (2016) proposed a 3D deformation structure V-net of the U-Net. The V-net uses the Dice coefficient loss function instead of the traditional cross-entropy loss. V-net uses a 3D convolution kernel to convolve the image, and it reduces the channel dimension through a $1 \times 1 \times 1$ convolution kernel.

GAN

A generative adversarial network, which is a new method of training generative models, has recently been introduced. Goodfellow et al. (2014) proposed an adversarial method in order to learn deep generative models in 2014, such as a GAN. The model consists of two parts, and its structure is shown in Fig. 5.11. The first part is the generation network, which receives a random noise z, which is a random number and generates an image through this noise. The second part is an adversarial network, which determines whether an image is real. Its input parameter is X, which is a picture, and the output D (x) represents the probability that X is a real picture. If it is 1, it represents 100% of the real picture. It represents an impossible real picture when the output is 0, which is to make two networks compete through training. Generative networks generate fake data, and the adversarial networks use a discriminator to determine the authenticity. Finally, it is hoped that the data generated by the generator can be fake.

A GAN has been cited tens of thousands of times since it was proposed. It has also made great contributions in the field of medical image processing. The research on a GAN is mainly divided into two directions, which include a GAN itself in order to adapt the particularity of medical images and expand the application level of a GAN in the field of medical image processing. In the field of medical image segmentation, the improvement of a GAN has attracted many researchers. Xue et al. (2018) proposed U-Net as the generator of a GAN network, which is called a Segmentation Adversarial Network. In medical image segmentation, U-Net cannot effectively solve the problem of unbalanced pixel categories in the image. Based on the above problems, the author designed a new segmentation network based on the ideas of a GAN and proposed Multi-scale L1 Loss to optimize the segmentation network. In addition, a GAN can not only improve the performance of medical image segmentation but can also contribute to the data processing of medical image segmentation. In the training of the medical image segmentation network model, the model is over-fitted due to an insufficient data set. This problem is very common in medical image analysis. One solution to solve the shortage of a training data set is data enhancement. Data enhancement techniques based on a GAN used in segmentation tasks are widely used in different medical images. CycleGANs

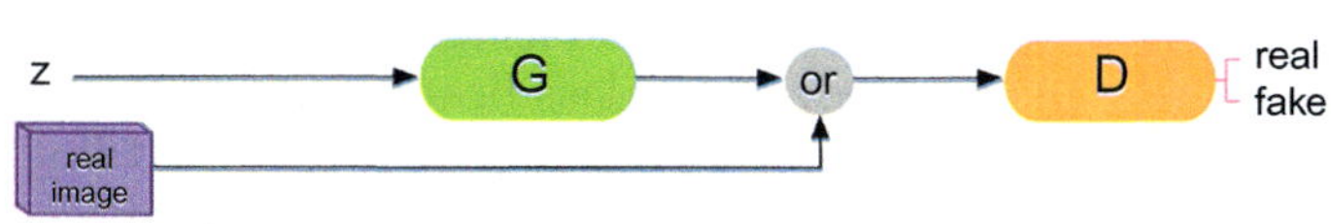

FIG. 5.11 The model of GAN.

and CGANs have been used in various ways to synthesize certain types of medical images. On the 2020 CVPR, the distributed and asynchronous discriminator GAN (AsynDGAN), which is published by Rutgers University and others, is a distributed GAN learning framework that does not need to share data, protect data security, and realize efficient communication in order to realize the training of a central generator with a distributed discriminator (Chang et al., 2020).

Dai et al. (2018) proposed a Structure Correction Adversarial Network (SCAN) to segment the lung field and heart in CXR images. The network adopted the idea that Luc2016 first used a GAN for image segmentation. The difference is that both the segmentation network and the discriminant network use a fully convolutional network. For the first time, the fully convolutional network is used for the segmentation and discrimination network. The discriminant network imposes structural regularity from human physiology on the convolutional segmentation network. During the training process, the discriminant network learned to distinguish the *ground truth organ annotation* from the mask synthesized by the segmentation network. Through this confrontation process, the discriminant network learns higher-order structures and guides the segmentation model to achieve realistic segmentation results. In addition, the SCAN simplified the down-sampling module based on the particularity of the Chest X-rays images.

Neural network for classification of lung cancer diagnosis

Traditional lung cancer pathological image classification methods often use classic image processing techniques, such as nucleus-based segmentation methods. Also, many commonly used image analysis algorithms, such as the watershed algorithm, the active contour, the level set algorithm, the K-means clustering algorithm, and regional growth. At the same time, traditional classifiers, such as Support Vector Machine (SVM) or classifiers designed based on specific problems, are used to classify the images.

The limitation of the lung cancer classification effect based on the traditional neural network method is that the features extracted by the artificial design usually have good adaptability for a certain type of lesion with specific characteristics. For the pathological image classification method based on deep learning, more generalized feature representations can be learned in order to detect the pathological image of lung cancer with diverse changes in the actual diagnosis and treatment scenes.

The early manifestation of lung cancer is the pulmonary nodule, so the improvement of the classification method of benign and malignant pulmonary nodules is the key and difficult point in clinical practice. The traditional classification of benign and malignant pulmonary nodules mainly uses the underlying features of the CT images, which are generally artificially designed and have subjective differences. The deep learning method is used to extract the features, and a multi-level learning model is established to automatically extract the image features in order to improve the accuracy of the classification. Common deep learning models include convolutional neural networks, self-encoding, and restricted Boltzmann machines. Among them, a CNN is not sensitive to the extraction of abstract features, such as the direction, size, and location of the objects, which helps improve the accuracy of the classification.

Use logistic regression analysis to classify benign and malignant

At present, the domestic researchers have proposed many classification methods. These methods are all performed by extracting the image features of the lung nodules. In the process of feature extraction, most of the early classification methods are based on artificially designed image bottom-level features plus classifiers. These bottom-level features include the size, shape, texture, and edges of the lung nodules. First, the CT image of the segmented lung nodule is convolved in order to obtain a feature map based on a CNN through a sliding window. After multiple convolutions, the dimensionality of the obtained features is reduced through the pooling layer. Finally, the image is transformed into a low-dimensional feature vector through the fully connected layer. The features extracted by a CNN are classified into benign and malignant using logistic regression analysis. Logistic regression is a common probability model for dealing with dichotomous problems, which is suitable for the automatic diagnosis of diseases and economic predictions. The value of the variables in the results of logistic regression must be dichotomous or multinomial. In the analysis of practical problems using regression models, the variables studied are often not all interval variables but sequential variables or attribute variables, such as the binomial distribution problem. By analyzing age, sex, body mass index, mean blood pressure, disease index, and other indicators, we can determine whether pulmonary nodules are benign or malignant, $Y=0$ means benign, $Y=1$ means malignant, and the response variable here is the variable distributed in 0–1. A total of 750 images of lung nodules were selected from the LIDC database with at least two radiologists who judged the same degree of malignancy. A CNN was used for feature extraction. The number of convolution layers and pooling layers is 2. The size of the convolution kernel is 5×5. The pooling layer uses a 2×2 input field, so the down sampling reduction is $1/4$. Each pulmonary tuberosity image was converted into a 192-dimensional feature map, which is then classified by the logistic classifier. The classification accuracy of the pulmonary nodules was the highest with a value of 84.4%.

Lung cancer pathological image classification based on a CNN

The diagnosis of lung cancer is unreliable only through a CT of the patient's chest, so it can only provide a preliminary examination and evaluation. Further diagnosis of lung cancer and the subtypes of lung cancer require a professional pathologist to observe the biopsies under a microscope. This process takes a long time, and it is difficult for the doctors to reach a consensus diagnosis. The equipment for full-slice pathological images can now reach billions of pixels. With the continuous improvement of computer processing capabilities, the problem of slow neural network training parameters has also been solved, which makes neural networks achieve better applications in processing medical image classification. A convolutional neural network is an improved neural network, and it is also the main network in deep learning. Small blocks were extracted from pathological images through a CNN network, and the network parameters were adjusted according to the verification results to train the classification network in order to obtain better classification accuracy. The size of the convolution kernel is 3×3, and the number of convolution kernels increases from 32 to 64 as the number of network layers deepens. In order to avoid the gradient saturation of the Sigmoid activation function and the problem of non-zero output, the ReLU activation function is used here. After the convolutional layer, the dimensionless feature map is obtained through the maximum pooling of size 2×2. The pixels of the slice pathological image is much higher than the pixels of the network input image. The input image of the neural network model is generally 256×256. Therefore, it is cut before being inputted into the network model. The convolutional neural network classification model is trained based on small blocks. In the test, the voting method is used to aggregate the block-based classification results to obtain the final classification results of the big picture. The dataset used includes 95 lung gland slices, small cell carcinoma slices, and 14 squamous cell carcinoma slices. Good results are obtained by randomly fetching blocks for the training.

Classification of lung nodules based on a 3D dense convolutional neural network

In the diagnosis of benign and malignant pulmonary nodules, deep learning technology has been used thoroughly, so researchers continue to focus on the diagnosis of the lung nodules using deep learning. In the realization of the classification of lung nodules, the problem involved is not only the classification, so we also still need to do a lot of pre-processing before the classification. The lung structure is complex, and the CT obtained by CT scanning is very large. If the original lung CT is directly sent to the convolutional neural network for calculation, a large number of computing resources will be consumed, and the calculation cost is very high. In addition, the CT of the lungs contains pulmonary vessels, which include the bronchi and other tissues, which may interfere with the diagnosis of the pulmonary nodules. Therefore, it is necessary to preprocess the original lung CT to segment the lung parenchyma from the lung CT image. On one hand, it can reduce the cost of computations, and it can improve the efficiency of pulmonary nodule detection and the accuracy of the classification. Traditional segmentation methods, such as the threshold method and the morphological method were used to segment the lung parenchyma. The pulmonary nodules were then detected and diagnosed as benign or malignant on the CT images of the segmented lung parenchyma. A common problem when training deep learning models is the disappearance of the gradients. The training data were enhanced by randomly flipping the CT images of the pulmonary nodules, adding noise, and randomly clipping. Moreover, the positive and negative sample ratio in the training set should be balanced.

In the diagnosis of the benign and malignant lung nodules, most of them are based on 2D CT of the lungs, so the convolutional neural network is also 2D. However, this method ignores the 3D spatial features of the pulmonary nodules and loses some features, which makes the accuracy of the diagnosis decrease. The diagnosis problem of the benign and malignant lung nodules corresponds to the classification problem in image processing. In the image classification network, the improvements in recent years are mainly from two aspects, which include (a) a depth from the network, such as ResNet. After deepening the network depth, the problem of the gradient disappearance is solved, and the classification accuracy is also significantly improved. (b) Another aspect is from the width of the network, such as GoogLeNet and Inception. By deepening the width of the network to control the parameters of the calculation, the accuracy can be improved under the premise of a small number of parameters and the calculation amount. In the classification task, the method with a good classification effect is a 3D dense convolutional neural network. We use this network to diagnose benign and malignant lung nodules. For the three-dimensional image of the lung nodules, the intermediate dense projection is used to restore the 3D spatial information of the lung nodules. 3D information was input into the 3D DenseNet for the training, and the loss function replaced the traditional cross-entropy loss with the focal loss, which made the network pay more attention to learning samples that are easy to be misclassified. It reduces the difficulty of the network training but improves the accuracy of the pulmonary nodule benign and malignant diagnosis. Starting with the features, the 3D DenseNet achieves a better classification effect and uses fewer parameters through the extreme use of the features. The dense connection mechanism it uses connects each layer in the network with all the previous layers as the input of the next layer, which strengthens the transferability of the features and effectively utilizes the features. Compared to ResNet, this is a dense connection. 3D DenseNet directly connects the feature maps from the different layers in order to achieve the purpose of feature reuse. Fig. 5.12 shows the network structure of 3D DenseNet. The depth of the model is 52 layers. The size of the input picture is $50 \times 50 \times 25$. The network contains three DenseBlocks. The size of the 3D convolution kernel is $7 \times 7 \times 7$, which is followed by a $3 \times 3 \times 3$ max pooling layer with a stride of 2.

3D DenseNet uses the 3D features of the lung nodules and calculates the projection of the 3D lung nodules in the x, y, and z directions through the intermediate dense projection. The three 2D MIP images that are obtained are synthesized into a 3D tensor, and the tensor is then sent to 3D DenseNet for training.

Classification of lung nodules based on different feature extraction

Research, which is based on CT images, shows that deep learning requires a sufficiently large training data set to better reflect its advantages in feature extraction in order to achieve an accuracy that meets the actual applications. In the task of medical image processing, there is no type of large-scale medical image annotation dataset with ImageNet. There is not enough discrimination for the features extracted by the not too deep network, and the training of the too deep network is costly and difficult. For the high-dimensional feature redundancy that exists after feature extraction, Gao (2019) proposed an adaptive feature selection algorithm in order to reduce the redundant or irrelevant features by selecting a subset of the features that is related to the benign and malignant lung nodules. After the model is trained, the adaptive feature selection algorithm extracts the 4096-dimension activation features of the F2 layer in the forward propagation process for the four small nodule blocks that are clipped from each node in the training set. We used 2-norm feature pooling to calculate the features of the four nodule blocks that are generated by each node and finally generate the final single-feature vector for each node. The formula for 2-norm feature pooling is:

$$f(V) = \left(\frac{1}{M} \sum_{i=1}^{M} V_i^P \right)^{\frac{1}{P}} \tag{5.1}$$

Among them, P is 2, M is the number of cropped nodule images 4 and V_i is the 4096-dimensional feature vector of the ith nodule block. With the average activation value of the n-dimensional feature, what we selected accounts for 20% of the average activation value

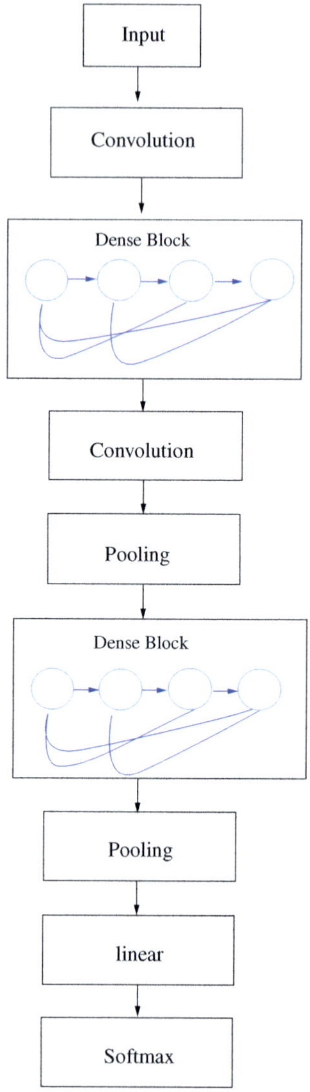

FIG. 5.12 3D DenseNet.

of the original 4096-dimensional feature, which means that the adaptive value is greater than or equal to 0.2 in order to stop the feature selection. After the feature selection, the article uses the SVM to classify the benign and malignant nodules. In the process of the SVM training, the author used the features and labels that were obtained after the adaptive feature selection as the input. Hussein et al. (2017) proposed a multi-view deep convolutional neural network, which is based on a multi-view deep CNN model of 3D CT images in order to determine the features of the lung nodules. The median intensity projection is used to generate three 2D patches that correspond to each dimension, and the patches are then concatenated into a three-dimensional tensor. After data enhancement, Gaussian regression is used to judge whether the tumor is benign or malignant. The accuracy rate can reach 82.47%. Dou et al. (2016) introduced a 3D CNN that learns directly from the 3D CT images, which can reduce this feature from having false positives, and the sensitivity of detecting lung cancer is 94.4%.

Conclusion and future directions

Conclusion

Medical images are combined with AI and image analysis to automatically distinguish the medical images. Machines are used to replace humans because this process saves a lot of labor and sometimes is even more accurate than manual judgments. In the diagnosis of lung cancer, AI can help doctors in finding the nodule area in order to assist them with the diagnosis of lung cancer through CT images. It can also segment the lung nodules through medical image segmentation technology in order to further identify the probability of the sample that is suffering from cancer. This result can also be used by the doctors to assist in a diagnosis. From the perspective of the entire diagnosis process of lung cancer, CAD can indeed greatly increase the diagnosis rate of the doctors, increase the diagnosis speed, and prevent missed detections and misdiagnoses.

Challenges and opportunities in the clinical research of lung cancer

Some research on the diagnosis of lung cancer has been conducted in the medical community, and some practical results have been achieved. However, due to the various types of lung cancer and the various related factors, the existing diagnosis has considerable limitations in terms of accuracy and practicability. There are also many challenges involved in the process of using neural networks to diagnose lung cancer, which are described below.

In the training data set, many nodules are marked by the doctors and experts, which can easily lead to many misjudgments and missed detections. The data set used in a lung cancer diagnosis is LIDC-IDIR, but this data set does not have a unified standard for the definition of benign and malignant nodules. Different researchers use different standards for the benign and malignant nodules in the dataset, so the experiment contrast is uneven. The standardization of the datasets still has a long way to go. LUNA16 is currently a relatively large-scale lung CT pathology image dataset, but its data capacity is still small compared to that of ImageNet. We need to build a larger dataset in order to design a more stable pathological image classification algorithm for the lung nodules.

The lung CT images are in 3D, which makes any detection algorithm expensive in terms of computational costs. In future research work, the multi-scale 3D convolutional neural network model will be further optimized in order to speed up the network training. Among the differences in the number of balanced categories, data preprocessing that only relies on the traditional data augmentation is a bit monotonous. In the future, 3D generative adversarial networks can be used to generate more nodule samples in order to further improve the detection performance.

In medical images, due to the resolution of pathological slices tens of billions of pixels and global pathological features can be obtained at low resolutions. At high resolution, more detailed pathological features can be identified. How to measure the size of the pathological slices directly affects the detection and diagnosis of the lung nodules. The multiscale fusion method can greatly enhance the robustness of the model and improve classification accuracy.

In the diagnosis of lung cancer using neural networks, what we need to do is not to replace the doctors but better assist them. Improving the work efficiency of the doctors allows them to concentrate on determining the patient's diagnosis and treatment plan.

References

Anthimopoulos, M., Christodoulidis, S., Ebner, L., Geiser, T., Christe, A., & Mougiakakou, S. (2019). Semantic segmentation of pathological lung tissue with dilated fully convolutional networks. *IEEE Journal of Biomedical and Health Informatics*, *23*(2), 714–722. https://doi.org/10.1109/JBHI.2018.2818620.

Astono, I., Welsh, J. S., & Chalup, S. (2018). Adjacent network for semantic segmentation of liver CT scans. In *Proceedings - 2018 IEEE 18th international conference on bioinformatics and bioengineering, BIBE 2018* (pp. 35–40). Institute of Electrical and Electronics Engineers Inc. https://doi.org/10.1109/BIBE.2018.00015.

Chang, Q., Qu, H., Zhang, Y., Sabuncu, M., Chen, C., Zhang, T., & Metaxas, D. N. (2020). Synthetic learning: Learn from distributed asynchronized discriminator gan without sharing medical image data. In *Proceedings of the IEEE/CVF conference on computer vision and pattern recognition* (pp. 13856–13866).

Çiçek, Ö., Abdulkadir, A., Lienkamp, S. S., Brox, T., & Ronneberger, O. (2016). 3D U-Net: learning dense volumetric segmentation from sparse annotation. In *International conference on medical image computing and computer-assisted intervention* (pp. 424–432). Cham: Springer.

Dai, W., Dong, N., Wang, Z., Liang, X., Zhang, H., & Xing, E. P. (2018). Scan: Structure correcting adversarial network for organ segmentation in chest x-rays. In *Vol. 11045. Lecture notes in computer science (including subseries lecture notes in artificial intelligence and lecture notes in bioinformatics)* (pp. 263–273). Springer Verlag. https://doi.org/10.1007/978-3-030-00889-5_30.

Ding, J., Li, A., Hu, Z., & Wang, L. (2017). Accurate pulmonary nodule detection in computed tomography images using deep convolutional neural networks. In *International conference on medical image computing and computer-assisted intervention* (pp. 559–567). Cham: Springer.

Dou, Q., Chen, H., Yu, L., Qin, J., & Heng, P. A. (2016). Multilevel contextual 3-D CNNs for false positive reduction in pulmonary nodule detection. *IEEE Transactions on Biomedical Engineering*, *64*(7), 1558–1567.

Fukushima, K., & Miyake, S. (1982). Neocognitron: A self-organizing neural network model for a mechanism of visual pattern recognition. In *Competition and cooperation in neural nets* (pp. 267–285). Berlin, Heidelberg: Springer.

Gao, H. (2019). *Research on detection and benign and malignant classification method of pulmonary nodules based on convolutional neural network.* Master's thesis, Taiyuan University of Technology.

Girshick, R. (2015). Fast R-CNN. In *Vol. 2015. Proceedings of the IEEE international conference on computer vision* (pp. 1440–1448). Institute of Electrical and Electronics Engineers Inc. https://doi.org/10.1109/ICCV.2015.169.

Girshick, R., Donahue, J., Darrell, T., & Malik, J. (2014). Rich feature hierarchies for accurate object detection and semantic segmentation. In *Proceedings of the IEEE computer society conference on computer vision and pattern recognition* (pp. 580–587). IEEE Computer Society. https://doi.org/10.1109/CVPR.2014.81.

Golan, R., Jacob, C., & Denzinger, J. (2016). Lung nodule detection in CT images using deep convolutional neural networks. In *2016 International joint conference on neural networks (IJCNN)* (pp. 243–250). IEEE.

Goodfellow, I., Pouget-Abadie, J., Mirza, M., et al. (2014). Generative adversarial nets. Neural information processing systems. MIT Press.

Hamidian, S., Sahiner, B., Petrick, N., et al. (2017). 3D convolutional neural network for automatic detection of lung nodules in chest. In *Proceedings of SPIE-Int Soc Opt Eng.*

He, K., Zhang, X., Ren, S., & Sun, J. (2015). Spatial pyramid pooling in deep convolutional networks for visual recognition. *IEEE Transactions on Pattern Analysis and Machine Intelligence, 37*(9), 1904–1916. https://doi.org/10.1109/TPAMI.2015.2389824.

He, K., Zhang, X., Ren, S., & Sun, J. (2016). Deep residual learning for image recognition. In (Vols. 2016). *Proceedings of the IEEE computer society conference on computer vision and pattern recognition* (pp. 770–778). IEEE Computer Society. https://doi.org/10.1109/CVPR.2016.90.

Hubel, D. H., & Wiesel, T. N. (1962). Receptive fields, binocular interaction and functional architecture in the cat's visual cortex. *The Journal of Physiology, 160*(1), 106–154. https://doi.org/10.1113/jphysiol.1962.sp006837.

Hussein, S., Gillies, R., Cao, K., Song, Q., & Bagci, U. (2017). TumorNet: Lung nodule characterization using multi-view convolutional neural network with Gaussian process. In *Proceedings - International symposium on biomedical imaging* (pp. 1007–1010). IEEE Computer Society. https://doi.org/10.1109/ISBI.2017.7950686.

Jiang, H., Shen, F., Gao, F., & Han, W. (2021). Learning efficient, explainable and discriminative representations for pulmonary nodules classification. *Pattern Recognition, 113*, 107825. https://doi.org/10.1016/j.patcog.2021.107825.

Jue, J., Jason, H., Neelam, T., Andreas, R., Sean, B. L., Joseph, D. O., & Harini, V. (2019). Integrating cross-modality hallucinated MRI with CT to aid mediastinal lung tumor segmentation. In *Vol. 11769. Lecture notes in computer science (including subseries lecture notes in artificial intelligence and lecture notes in bioinformatics)* (pp. 221–229). Springer. https://doi.org/10.1007/978-3-030-32226-7_25.

Khan, A., Sohail, A., Zahoora, U., & Qureshi, A. S. (2020). A survey of the recent architectures of deep convolutional neural networks. *Artificial Intelligence Review, 53*(8), 5455–5516. https://doi.org/10.1007/s10462-020-09825-6.

Krishnamurthy, S., Narasimhan, G., & Rengasamy, U. (2016). Three-dimensional lung nodule segmentation and shape variance analysis to detect lung cancer with reduced false positives. *Proceedings of the Institution of Mechanical Engineers, Part H: Journal of Engineering in Medicine, 230*(1), 58–70. https://doi.org/10.1177/0954411915619951.

Krizhevsky, A., Sutskever, I., & Hinton, G. E. (2017). ImageNet classification with deep convolutional neural networks. *Communications of the ACM, 60*(6), 84–90. https://doi.org/10.1145/3065386.

LeCun, Y., et al. (1998). Gradient-based learning applied to document recognition. *Proceedings of the IEEE, 86*(11), 2278–2324.

Liu, X., Song, L., Liu, S., & Zhang, Y. (2021). A review of deep-learning-based medical image segmentation methods. *Sustainability, 13*(3), 1224. https://doi.org/10.3390/su13031224.

Long, J., Shelhamer, E., & Darrell, T. (2015). Fully convolutional networks for semantic segmentation. In *Proceedings of the IEEE conference on computer vision and pattern recognition* (pp. 3431–3440).

Milletari, F., Navab, N., & Ahmadi, S. A. (2016). V-Net: Fully convolutional neural networks for volumetric medical image segmentation. In *Proceedings - 2016 4th international conference on 3D vision, 3DV 2016* (pp. 565–571). Institute of Electrical and Electronics Engineers Inc. https://doi.org/10.1109/3DV.2016.79.

Minaee, S., Boykov, Y. Y., Porikli, F., Plaza, A. J., Kehtarnavaz, N., & Terzopoulos, D. (2021). Image segmentation using deep learning: A survey. *IEEE Transactions on Pattern Analysis and Machine Intelligence*. https://doi.org/10.1109/TPAMI.2021.3059968.

Okwu, M. O., & Tartibu, L. K. (2020). *Metaheuristic optimization: Nature-inspired algorithms swarm and computational intelligence, theory and applications. Vol. 927*. Springer Nature.

Ren, S., He, K., Girshick, R., & Sun, J. (2015). Faster R-CNN: Towards real-time object detection with region proposal networks. In (Vols. 2015). *Advances in neural information processing systems* (pp. 91–99). Neural Information Processing Systems Foundation.

Ronneberger, O., Fischer, P., & Brox, T. (2015). U-Net: Convolutional networks for biomedical image segmentation. In *Vol. 9351. Lecture notes in computer science (including subseries lecture notes in artificial intelligence and lecture notes in bioinformatics)* (pp. 234–241). Springer Verlag. https://doi.org/10.1007/978-3-319-24574-4_28.

Setio, A. A. A., Ciompi, F., Litjens, G., Gerke, P., Jacobs, C., Van Riel, S. J., … Van Ginneken, B. (2016). Pulmonary nodule detection in CT images: False positive reduction using multi-view convolutional networks. *IEEE Transactions on Medical Imaging, 35*(5), 1160–1169.

Setio, A. A. A., Traverso, A., & De Bel, T. (2019). *LUNA16-Results*. https://luna16.grand-challenge.org/Results/.

Siegel, R. L., Miller, K. D., & Jemal, A. (2020). Cancer statistics, 2020. *CA: A Cancer Journal for Clinicians, 70*(1), 7–30. https://doi.org/10.3322/caac.21590.

Simonyan, K., & Zisserman, A. (2014). *Very deep convolutional networks for large-scale image recognition*. Computer Science.

Werbos, P. (1974). *Beyond regression: New tools for prediction and analysis in the behavioral sciences*. Harvard Univ. Diss.

Wit, & Hammack, D. (2018). *2nd place solution for the 2017 national datasicence bowl.* http://juliandewit.github.io/kaggle-ndsb2017/.

Wyker, A., & Henderson, W. W. (2021). *Solitary pulmonary nodule.* StatPearls.

Xie, H., Yang, D., Sun, N., Chen, Z., & Zhang, Y. (2019). Automated pulmonary nodule detection in CT images using deep convolutional neural networks. *Pattern Recognition, 85*, 109–119.

Xue, Y., Xu, T., Zhang, H., Long, L. R., & Huang, X. (2018). SegAN: Adversarial network with multi-scale L 1 loss for medical image segmentation. *Neuroinformatics, 16*(3–4), 383–392. https://doi.org/10.1007/s12021-018-9377-x.

CHAPTER

6

Machine learning for thyroid cancer diagnosis

Feyzullah Temurtas[a], *Kutlucan Gorur*[a], *Onursal Cetin*[a], *and Ilyas Ozer*[b]

[a]Electrical and Electronics Engineering, Bandirma Onyedi Eylul University, Balikesir, Turkey

[b]Computer Engineering, Bandirma Onyedi Eylul University, Balikesir, Turkey

Introduction

Thyroid cancer is considered a disease in which the first symptom is a nodule in the thyroid area of the neck. It is recorded as one of the most common cancers in the literature, and it poses a great risk to human health, having the highest increase rate of about 2.4-fold in the last 30 years (Wang, Zhang, et al., 2020). With worldwide rise in thyroid cancer being observed recently, it has turned it into the sixth most common type of cancer, especially among women, according to the latest cancer statistics for 2019 (Bhalla et al., 2020). Moreover, the thyroid malignancy incidence rate has been still increasing in the world health system. Furthermore, this cancer has been observed to have increased at a higher rate compared to the other cancers. In the United States alone, thyroid cancer has been reported to have increased by more than 300% in the past 30 years (Wang, Song, et al., 2020).

Exposure to radiation and environmental carcinogens can be the most important of the possible factors that play a role in its rise. There are four types of thyroid cancer that are histopathologically stated as papillary, follicular, medullary, and anaplastic (Bhalla et al., 2020). Papillary thyroid carcinoma (PTC) and follicular thyroid carcinoma (FTC) have been accepted as Differentiated Thyroid Cancer (DTC). These types of cancers have created the majority kind of thyroid malignancies (Bhalla et al., 2020). PTC is mentioned in the literature as the most common thyroid cancer, accounting for about 80% of DTC. Although PTC is known as an affirmative prognosis with a 1.2% mortality rate within 20 years, approximately 10%–15% of patients with PTC have recurrent disease after the first surgery. And then, 35% of these patients die after this recurrent thyroid cancer (Wang, Song, et al., 2020). According to the

https://doi.org/10.1016/B978-0-323-85240-1.00011-0

American Cancer Society, in the early stage (stage I and stage II), the survival rate of PTC patients has been reported to be almost 100%, while the survival rate at stage 4 has been reported to reduce to 55% (Bhalla et al., 2020). However, the aggressiveness of PTC is often described by the clinician with a pathological evaluation of thyroid tissue samples, although the pre-operative study may help introduce some aggressive imaging (Wang, Song, et al., 2020). These statistical results and pre-operative work show the requirements for new methods or biomarkers for early detection of thyroid cancer (Thomas et al., 2020).

Ultrasound (US) is seen as the most important and favorable imaging method for thyroid nodules (Wang, Song, et al., 2020). US imaging provides one of the most sensitive and widely used technique to detect thyroid nodules (Wang, Zhang, et al., 2020). Besides, computed tomography (CT) and magnetic resonance imaging (MRI) are other screening/imaging techniques for detection of thyroid nodules (Wang, Song, et al., 2020). Cytopathology/histopathology images and molecular markers are other features to provide information to the physicians in labeling a thyroid nodule as benign or malignant (Thomas et al., 2020). However, all these current methods have been observed as subjective in thyroid nodule risk stratification. At this point, machine learning algorithms have been implemented to overcome this deficiency (Thomas et al., 2020).

It takes a great deal of effort to classify complex histopathological patterns of cancerous cells and tissues. Hence, pattern recognition in thyroid cancer and thyroid gland diseases has turned into one of the important research areas of machine learning/deep learning (Pathpedia, n.d.; Temurtas, 2009). In the current literature, machine learning algorithms have emerged as an essential technique for detecting, predicting, and diagnosing different diseases in classification processes (such as heart disease, breast cancer, and lung cancer). Especially, these artificially based algorithms have also been used to detect various diseases, including thyroid disorders (Abbad Ur Rehman et al., 2021; Temurtas, 2009). It is also shown on a statistical figure that ML plays a vital role in the classification tasks presented in Fig. 6.1 (Abbad Ur Rehman et al., 2021).

Computer-aided diagnosis (CAD) systems have the potential to improve the diagnostic accuracy of thyroid cancer by using ML algorithms (Park et al., 2019). In the literature, supervised machine learning algorithms were generally split into two methods. The first one/old is known as traditional/shallow algorithms. The second one/newest is defined as deep learning algorithms. The success of the shallow ML algorithms has been investigated in years on thyroid cancer/diseases. One of these algorithms is described as K-Nearest Neighbor (KNN), Probabilistic Neural Network (PNN), Decision Tree (DTC), Support Vector Machine (SVM), Linear Discriminant Analysis (LDA), Multilayer Neural Network (MLNN), and Random Forest (RF). The main common process of these shallow algorithms needs handcrafted feature extraction. Forming handcrafted features often include challenges in striking the right balance between accuracy and computational efficiency. And, handcrafted features are pre-designed by human experts to achieve a specific set of selected features (Nanni et al., 2017). The popularity of deep learning algorithms has increased in recent years because outperformed performances can be achieved compared to the traditional (i.e., shallow) machine learning methods, and automatic feature extraction is made from raw data with almost no pre-processing (Chauhan et al., 2019). Among the deep learning models, particularly Convolutional Neural Network (CNN) has so many applications and very suited for medical image processing (Chauhan et al., 2019; Sun et al., 2020). Moreover, pre-trained models of

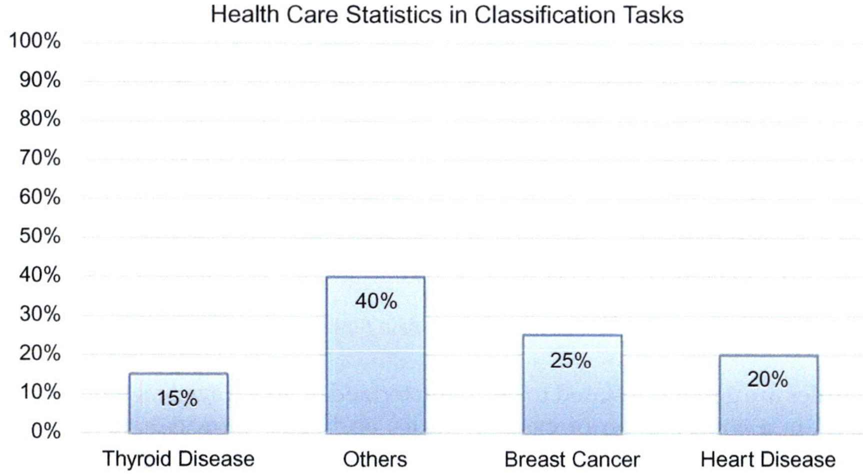

FIG. 6.1 Machine learning usage in classification tasks of diseases presented via health care statistics (Abbad Ur Rehman et al., 2021).

CNN (i.e., GoogleNet, VGG-19) have so many layers that enable superior feature extraction from the dataset. Transfer learning and fine-tuning skills of CNN models can provide feature maps from the pre-trained networks to the shallow MLs or small-sized deep models (Chi et al., 2017; Yang et al., 2021; Yu et al., 2020; Zhu et al., 2021). For example, the network (VGG-19) architecture includes 47 layers, containing 16 convolutional and 3 fully connected learnable weight layers. The filter size (3×3) of each convolution layer was formed with a different number of kernels. Approximately 144 million weight parameters of the model have high feature extraction potential. These convolutional layers extract local features comprising lines, shapes, edges, and textures. Then, these feature maps can be transferred to other ML models for similar pattern recognition tasks (i.e., cancer recognition from the ultrasound images) (Zhu et al., 2021). Furthermore, Auto-Encoder-based deep learning models are also employed in thyroid nodule classification for benign and malignant. The number of layers (i.e., hidden layers) is structured in both parts named encoder and decoder (Turk et al., 2021). R-CNN (Region-CNN) is a state-of-the-art object detection network and can also be employed on ultrasound imaging for thyroid cancer nodules. Training and testing procedures can be achieved by using dataset images/molecular markers in all ML models to provide an accurate state of the thyroid nodule (Turk et al., 2021). Fig. 6.2 presents the illustration of the thyroid nodule and main flow for signal processing.

Literature survey

The thyroid secretes thyroid hormone to regulate the body's metabolism of the human and is known as a significant endocrine organ. This thyroid gland organ is located just below the thyroid cartilage in the front of the neck (Chen, You, & Li, 2020). The edges of thyroid nodules are generally in well-defined and regular shape, while others have irregular forms. The

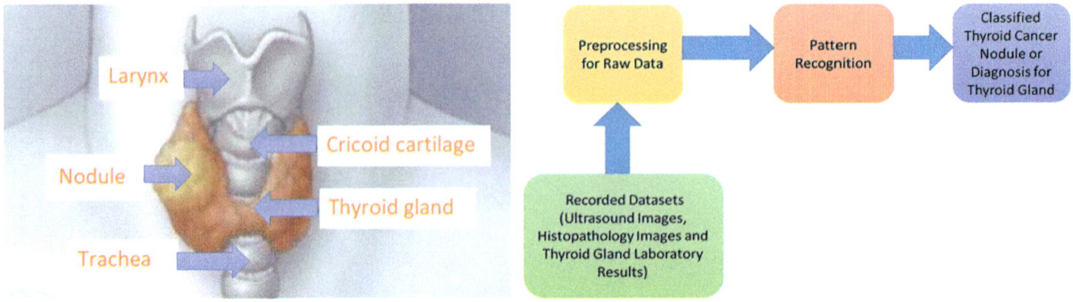

FIG. 6.2 Thyroid nodule presentation on the neck and main flow for automatic recognition.

thyroid nodules are generally stated to be characterized as solid, cystic, while some others are characterized in a combination of cystic nodules and solid form (Chen, You, & Li, 2020). According to findings including epidemiological area, thyroid nodules (the type of palpable) were observed among the population in the range of 4%–7%. The rates of nodules found incidentally on ultrasound examination were observed to be between 19% and 67% (Chen, You, & Li, 2020). Different imaging techniques are available to diagnose thyroid disease, including computed tomography, magnetic resonance imaging, ultrasound imaging, and radionuclide imaging (Chen, You, & Li, 2020; Turk et al., 2021; Wang, Song, et al., 2020). It has been observed that most of the studies using imaging methods have focused on ultrasound imaging technology. The reason for this is that the US is a real-time, inexpensive, non-invasive, and non-radioactive method, and it is stated in the literature that it is the most commonly used technique in the diagnosis of thyroid diseases (Chen, You, & Li, 2020). Therefore, in our literature survey, ultrasound imaging and deep learning methods have been stated in most of the surveys. The other modalities for thyroid cancer detection are observed as whole slide cytopathology/histopathology images, and genetical datasets (Dov et al., 2021; Tahmasebi et al., 2020). Furthermore, thyroid gland disorders may cause death, according to the existing literature (Temurtas, 2009). Therefore, this disorder and machine learning applications were investigated under another subtitle within the survey. According to the Scopus Index, thyroid and machine learning/deep learning search results are shown in Figs. 6.3 and 6.4, respectively. Totally, 371 (machine learning) and 195 (deep learning) research papers associated with thyroid disease were listed chronologically (Scopus, n.d.).

Year	Count	Year	Count	Year	Count
☐ 2021	(14) >	☐ 2011	(6) >	☐ 1994	(1) >
☐ 2020	(99) >	☐ 2010	(4) >	☐ 1993	(2) >
☐ 2019	(79) >	☐ 2009	(5) >	☐ 1992	(1) >
☐ 2018	(52) >	☐ 2008	(4) >	☐ 1990	(1) >
☐ 2017	(29) >	☐ 2007	(3) >	☐ 1977	(1) >
☐ 2016	(27) >	☐ 2006	(2) >		
☐ 2015	(9) >	☐ 2005	(2) >		
☐ 2014	(11) >	☐ 2004	(2) >		
☐ 2013	(6) >	☐ 1997	(3) >		
☐ 2012	(6) >	☐ 1995	(2) >		

FIG. 6.3 The list of the chronologic search results shows the paper number for thyroid and machine learning keywords (Scopus, n.d.).

☐ 2021	(15) >	☐ 2011	(2) >
☐ 2020	(91) >	☐ 2009	(1) >
☐ 2019	(41) >	☐ 2003	(1) >
☐ 2018	(18) >	☐ 2001	(1) >
☐ 2017	(14) >		
☐ 2016	(4) >		
☐ 2015	(4) >		
☐ 2014	(1) >		
☐ 2013	(1) >		
☐ 2012	(1) >		

FIG. 6.4 The list of the chronologic search results shows the paper number for thyroid and deep learning keywords (Scopus, n.d.).

Accordingly, it is observed that most of the papers have been abruptly revealed since 2016 and 2017 because of the advancement of deep learning algorithms on thyroid disease applications.

Thyroid cancer detection by ultrasound imaging and machine learning/deep learning

The detection of thyroid nodules (TN) has been reported to have increased significantly in the last two decades. Thus, more nodules detected by chance could be determined. Most of these nodules are benign or passive (Chen, You, & Li, 2020). However, it is important to determine whether thyroid nodules are benign or malignant, and this may reduce the risk of patients. Thus, it can reduce significant medical healthcare costs within the healthcare system for Fine Needle Analysis (FNA) biopsy and/or surgical operations (Chen, You, & Li, 2020).

However, examinations over the ultrasound imaging in clinical diagnoses are a common process. Interpretation of the thyroid nodule is dependent on the experience of the clinicians. Therefore, in order to be sophisticated and specialized in ultrasound examinations, clinicians need long learning processes, and it is difficult to eliminate subjective factors during diagnostic processes (Chen, You, & Li, 2020). Important sonographic features to accurately identify thyroid nodules are characterized in five categories in the thyroid imaging reporting and data system (TI-RADS) published by the American College of Radiology. These features are listed as echogenicity, margin, composition, shape, and echogenic foci. Shapes and borders are presented as key features to distinguish and identify the type of thyroid nodules that is benign or malignant. In order to determine the thyroid nodules related to the malignant, spiculated margin and the shape with taller-than-wide contour have been seen as the primary characteristics (Chen, You, & Li, 2020). According to the reported study, determining the impact of the taller-than-wide contour and the features of spiculated margin on the outcomes has been noted as 91.4% (specificity) and 40.0% (sensitivity) and 91.8% (specificity), and 48.3% (sensitivity), respectively. It has been stated that erroneous segmentation results may lead to misdiagnoses based on boundary features. Hence, correct thyroid nodule segmentation is an important part of making the correct distinction between benign and malignant nodules in terms of clinical practice (Chen, You, & Li, 2020).

Besides, accurate segmentation over the ultrasound images in order to analyze thyroid volume is essential to detect abnormalities in the secretion of thyroid hormone related to the thyroid gland. Because of these requirements, ultrasound segmentation has been used in clinical diagnosis since the 1980s (Chen, You, & Li, 2020). The size, shape, and regularity of the boundary gathering from the echogenicity are obtained in different textures for thyroid glands and nodules. If there is a texture difference or change between muscle and the thyroid, the blurred effect is found on ultrasound images (Chen, You, & Li, 2020).

The thyroid gland area is seen on the ultrasound image, and it is much larger than the thyroid nodule area. It has been reported that the thyroid gland area on the ultrasound image is more sensitive to noise, artifacts, spots, and signal attenuation. However, there are structures such as blood vessels, trachea, and muscle tissues around the thyroid gland. Thus, poorly defined boundaries and fuzzy border regions are structured by the thyroid gland. This is because surrounding tissues are affected by artifacts and spots (Chen, You, & Li, 2020).

However, machine learning/deep learning methods have a great promising approach to accurately detect and segment thyroid gland and thyroid nodules through ultrasound images in recent years. These artificial-based methods automatically provide classification and interpretation of medical image data related to thyroid cancer in rapid progress (Li et al., 2018; Wang, Zhang, et al., 2020).

In the research study, papillary thyroid carcinoma (PTC), a kind of cancer, was detected to find normal or benign tissues over the ultrasound images using the Region-CNN method. In the reported paper, ultrasound images are usually obtained as monochrome and low-resolution (Li et al., 2018). Therefore, cancer regions are formed as irregular shapes, and they are stated generally to have a blurred and vague margin. Besides, the texture of the cancer regions resembles normal tissue or benign tissue. Hence, researchers stated that CNN is not enough to discriminate the region of thyroid cancer. Hence, R-CNN was implemented as a detector of suspicious PTC region (Li et al., 2018).

In CNN or R-CNN deep learning algorithms, the deeper layers need more images, and the more data images mean greater accurate performances (Gorur et al., 2019; Zhang et al., 2019). In this reported study, the ultrasound images from 300 cases were collected to be processed by state-of-the-art object detection network R-CNN. The result was stated 93.5% as True Positive Rate (TPR), which means PTC detection (Li et al., 2018).

In another research study by Chi et al., thyroid nodule classification was carried out using deep CNN and fine-tuning operation with probabilistic Random Forest (Chi et al., 2017; Reis et al., 2018). First, the last layers (named as Loss1, Loss2, and Loss3) have been modified for class output from 1000 to 2. After training with ultrasound images, the pre-trained GoogLeNet architecture was fine-tuned for a high-level feature extraction process over the ultrasound images. Then, feature maps have been implemented in Random Forest classifier to predict the nodule type, which is benign or malignant. Classification performances achieved by GoogLeNet have been stated as 98.29% of accuracy, 99.10% sensitivity, and 93.90% specificity. The algorithm of the prediction process is shown in Fig. 6.9. A total of 428 ultrasound images (357 positives and 71 negative cases in database 1) with the size of 560×360 were processed with the hybrid system. Moreover, a total of 164 ultrasound images in database-2 were used to classify due to the not enough dataset images. The second technique to enrich the input dataset was chosen as the uncommon augmentation method (commons are re-scaling, rotation, or contrast-adjusting). The focus of this method has been chosen as the region where the cancer nodule tissue is located. Hence, it is possible to learn features of cancer textures from different ultrasound images.

In addition to the diagnostic thyroid nodules over the ultrasound imaging, according to the literature survey, the machine learning-based study has been carried out using magnetic resonance imaging (Wang, Song, et al., 2020). In this article, the aim is to evaluate the aggressiveness of PTC via machine learning methods. The best performance achievement has been succeeded by probabilistic prediction-based Gradient Boosting Classifier (AUC=0.92) in the testing phase. Gradient Boosting can provide prediction via a probabilistic model (Duan et al., 2019). It is noted that there are some advantages to using MRI-based thyroid nodule diagnostics in clinical applications. For example, although US-guided Fine-Needle Aspiration biopsy can produce an accurate and cost-effective solution to have pathological confirmation of thyroid nodules, its ability to supply information on tumor aggressiveness has been limited. Moreover, the US method has been existed in a lack of diagnostic accuracy due to the deep structure of neck anatomy that can be blocked by air and bone (Wang, Song, et al., 2020). However, MRI-based thyroid diagnostics can provide superior contrast in soft tissues, thus allowing anatomical details to be evaluated (Wang, Song, et al., 2020). Hence, this technique may produce high-throughput extraction for machine learning algorithms (Wang, Song, et al., 2020).

Moreover, ultrasound-based machine learning approaches aim to overcome inter-observer and intra-observer variability of clinicians in accurately diagnosing thyroid nodules (Ha & Baek, 2021). The other ultrasound imaging-based machine learning/deep learning thyroid cancer nodule detection are listed in Table 6.1.

TABLE 6.1 Selected papers using machine learning/deep learning models and statistical methods for thyroid nodule classification/detection over the ultrasound images in recent years.

Reference/ year	Method	Dataset	Evaluation metric	Performance/explanation
Yang et al. (2021)/2021	Multi-Task Cascade Deep Learning Model/Transfer Learning/SVM	1030 ultrasound images	Accuracy, AUC	Accuracy=90.01% AUC=91.07%
Zhu et al. (2021)/2021	Deep Convolutional Neural Networks	719 Thyroid and 672 Breast ultrasound images	Accuracy, AUC, TPR, TNR	Accuracy=86.5%, AUC=0.861, TPR=84.4%, TNR=87.7% for thyroid cancer nodules
Yu et al. (2020)/2020	Transfer Learning Radiomics (TLR), Clinical Statistical Mode (SM), Traditional Radiomics Model (RM), Nontransfer Learning Model (NTLR), Statistical Analysis was also made by Chi-squared test	533 PTC cases images	AUC, accuracy, sensitivity, specificity, F1 score	TLR model: ACC=0.89 AUC=0.96 NTLR model: ACC=0.76 AUC=0.81 RM model: ACC=0.64 AUC=0.67 SM model: ACC=0.70 AUC=0.77

Continued

TABLE 6.1 Selected papers using machine learning/deep learning models and statistical methods for thyroid nodule classification/detection over the ultrasound images in recent years—cont'd

Reference/ year	Method	Dataset	Evaluation metric	Performance/explanation
Chen, Hu, et al. (2020)/ 2020	Random Forest (RF), Extreme Learning Machine (ELM), SVM, LDA, KNN, Naive Bayesian (NB), Logistic Regression (LR), Neural Network (NET), Logistic Regression (LOG), Adaptive Boosting (ADAB)	1558 Thyroid nodule images from 1480 patients	AUC, sensitivity, F1, specificity, positive predictive value (PPV), negative predictive value (NPV)	RF: AUC = 0.96 F1 = 0.701 PPV = 0.96 NPV = 0.612 ELM: AUC = 0.952 F1 = 0.657 PPV = 0.971 NPV = 0.524 The other results are listed in the related paper
Sun et al. (2020)/2020	Pre-trained ImageNET (CNN model) and SVM	1037 ultrasound images	Statistical analysis and classification metrics: Sensitivity, specificity, PPV, NPV, accuracy, and ROC curve	The sensitivity, specificity, PPV, NPV, and accuracy of the system were obtained as 96.4%, 83.1%, 94.0%, 89.5%, and 92.5%, respectively
Li et al. (2020)/2020	k-means and Region-CNN	30,000–40,000 ultrasound images from 2000 patients	Accuracy	Accuracy = 89.8% PTC detection rate

Transfer learning is a technique that improves the learning process in new tasks by transferring information from related tasks that have already been learned. In recent years, transfer learning has many applications in many medical image analysis fields such as image segmentation, lesion/nodule localization, and lesion/nodule pattern recognition (Chi et al., 2017; Yu et al., 2020). Deep transfer learning features based on CNN are also implemented to transfer high-level extracted features to the shallow machine learning methods (such as SVM) in pattern recognition of medical images (Nanni et al., 2017). Fig. 6.5 shows the deep transfer learning feature-based classification process (Nanni et al., 2017).

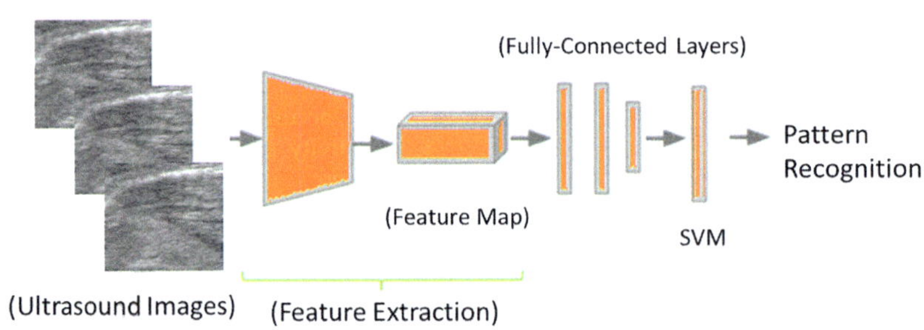

FIG. 6.5 Schematic of the deep transfer learning (Nanni et al., 2017).

Machine learning/deep learning based thyroid cancer prediction using cytopathology whole-slide images and histopathology images

The most important preoperative diagnosis test is known as the Fine Needle Aspiration Biopsy (FNAB) for thyroid malignancy (Dov et al., 2021). This test is performed by staining and then spreading onto a glass slide. The FNAB sample is analyzed under an optical microscope by a cytopathologist to determine the risk of malignant regions based on various characteristics of the follicular thyroid cells, such as their size, color, and cell group architecture. Each sample contains scanned whole slide images (WSI) and recorded at an image resolution (Dov et al., 2021).

Multiple sample learning (MIL) approaches, generally implemented in WSI analysis, separate the image (bag) into patches (samples) for estimating the tag at a single bag level. These approaches perform poorly on cytopathology slides because of the unique bag structure. They are infrequently placed for rare conditions with different abnormality characteristics. At this point, machine learning/deep learning algorithms with statistical approaches can offer a framework of maximum likelihood estimation (MLE) and deep learning to identify informative instances. Therefore, these local instances take scores of malignancy prediction. After that, the total prediction score for global malignancy is estimated by this technique (Dov et al., 2021).

Buddhavarapu et al. have implemented the research study on histopathology image analysis with machine learning algorithms (Buddhavarapu & Angel Arul Jothi, 2020). This article proposes an automated thyroid histopathology image classification system with deep neural networks using transfer learning theory and pre-trained Convolutional Neural Networks (CNNs). The state-of-the-art CNN architectures were applied to the thyroid histopathology images. This pre-trained network was stated as VGGNet, ResNet, InceptionNet, and DenseNet. The performance metrics were noted as accuracy, precision, sensitivity, specificity, the area under the receiver operating characteristic (AUROC) analysis, and F1-score.

According to the related article, 70:30, 80:20 split criteria, and 10-fold cross-validation results had been obtained by these pre-trained networks (as feature extractor) and CNN. A total of 221 images (156 normal and 65 PTC) were used to classify the type of thyroid diagnostic prediction. It is reported that DenseNet had achieved 100% accuracy with the value of sensitivity (100%) and specificity (100%) (Buddhavarapu & Angel Arul Jothi, 2020). Fig. 6.6 shows the histopathology image examples for the type of thyroid texture. In another research, pre-trained VGG11-based CNN architecture was employed to predict and evaluate thyroid FNAB over the whole slide images. It was noted that this deep learning-based model had achieved high sensitivity (92%) and specificity (90.5%) scores (Elliott Range et al., 2020). Moreover, there are a few literature reviews for artificial intelligence in thyroid FNAB (Kezlarian & Lin, 2020).

Thyroid gland disorders and machine learning approaches

Thyroid disorder is also seen as an increasing problem today. There are two types of thyroid disorders in the literature. Hyperthyroidism is caused by a lack of thyroid hormone in the blood, and hypothyroidism is caused by an excess of thyroid hormone in the blood

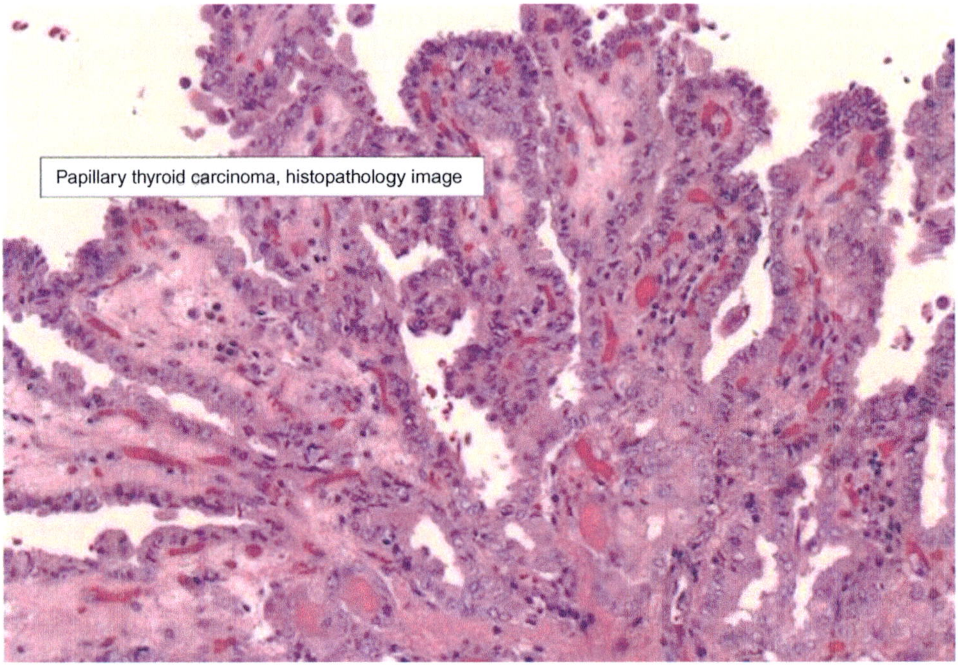

FIG. 6.6 Histopathology images of the Papillary Thyroid Carcinoma (PTC) (Pathpedia, n.d.).

(Mishra et al., 2021). Thyroid hormone is an important hormone that is secreted from the thyroid gland and regulates metabolism. Among its tasks, it has many functions such as rate control of metabolism, burning calories, and protein. It also has the function of regulating excessive hormone secretion from other hormonal glands (Mishra et al., 2021). If mentioned disorders are underestimated, they may cause death in a significant number of cases (Temurtas, 2009). The sample features in thyroid disorder prediction were listed in an example study (Temurtas, 2009):

- T3-resin uptake test
- Total serum thyroxin
- Total serum triiodothyronine
- Basal thyroid-stimulating hormone (TSH)
- Maximal absolute difference of TSH value after injection of 200 mg of thyrotropin

Correct diagnosis of the disease is important and requires adequate experience and knowledge. This situation makes it difficult to diagnose thyroid disorder. The machine learning approach makes this problem simple in a more accurate manner (Mishra et al., 2021). There are numerous machine learning approaches to predict these thyroid disorders.

Temurtas implemented Levenberg Marquardt (LM) with the back-propagation algorithm in Multilayer Perceptron (MLNN), Probabilistic Neural Network (PNN), and Learning Vector Quantization (LVQ) to diagnose thyroid disorder. The results were provided as 93.19% (MLNN), 94.81% (PNN) and 90.05% (LVQ) accuracies in 10-fold cross-validation

(Temurtas, 2009). It was reported that the dataset consists of 215 samples (150 normal, 35 hyperthyroidism, and 30 hypothyroidism).

Reddy SD et al. have aimed to predict thyroid gland disorders using SVM, Decision Tree, KNN, Random Forest, XGBoost, and Logistic Regression ML algorithms. XGBoost has provided the highest classification score with 99.57% accuracy. It is noted that 20 features were processed in the related dataset (Reddy et al., 2020). Hemalatha D. et al. have also carried out a prediction of thyroid disorders over the 21 features using Naive Bayes, SVM, Decision Tree, and KNN. The best result has been obtained by KNN (99.72% accuracy) (Hemalatha & Poorani, 2020).

Methods

Linear Discriminant Analysis

Linear Discriminant Analysis (LDA), derived from Kernel Fisher Discriminant Analysis, is a type of projection technique classifier that reduces the size of the data (Muller et al., 2001). LDA aims at maximizing the ratio of between-class variance and minimizing the ratio of within-class variance. This method effectively addresses situations where in-class frequencies are unequal and accuracy on random-generated test data is examined (Alpaydın, 2010). LDA finds the projection direction to distinguish the spatial model in the direction of maximum separability as possible (Gorur et al., 2020). Eqs. (6.1)–(6.3) form the mathematical foundation of LDA (Alpaydın, 2010):

$$z = w^T x \tag{6.1}$$

where x samples are projected onto w. The LDA projection method is shown in Fig. 6.7.

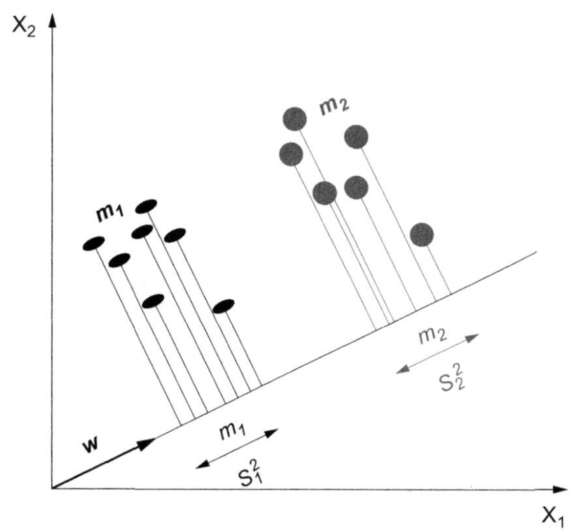

FIG. 6.7 The projection method of LDA.

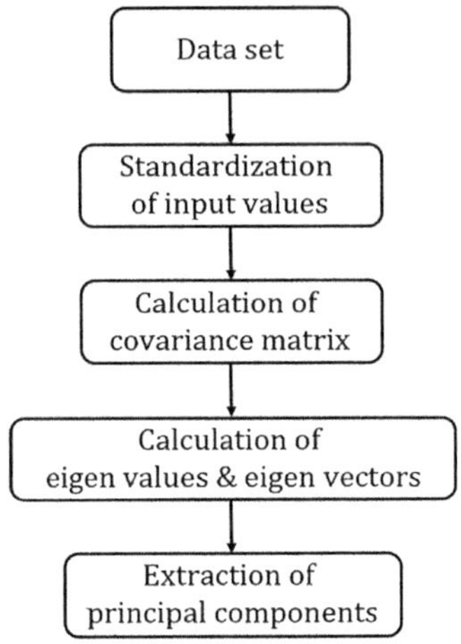

FIG. 6.8 Flow diagram of PCA.

where the bold terms $\boldsymbol{m_1}$ and $\boldsymbol{m_2}$ refer to the samples in classes of C1 and C2 before and after projection, respectively. The terms are m_1 and m_2 which define the scatter of samples from C1 and C2.

Principal Component Analysis

Principal Component Analysis (PCA) is a useful statistical technique used in recognition, classification, and dimensionality reduction. The motivation behind PCA is to bring together highly correlated variables in the data, thus showing a multidimensional data with fewer variables called "principal components" (Alpaydın, 2010). The stages of extracting the principal components are shown in Fig. 6.8.

While finding the principal components, the centering process is performed by subtracting the average value of each variable from the data. In the next step, the covariance matrix is calculated as a measure of the strength of the correlation between two or more sets of random variables. Then, the eigenvectors and eigenvalues of the covariance matrix are calculated (Gorur et al., 2020). The eigenvalues are ordered in ascending order, and the eigenvectors corresponding to these eigenvalues are selected. The large eigenvalues indicate that the relevant eigenvector contains more information about the dataset (Mohanty & Palo, 2020). Thus, a new dataset with a reduced size is obtained.

K-Nearest Neighbor

The Euclidean-based K-Nearest Neighbor (KNN) algorithm assigns the data classified as input to the class having the most examples in the training set (Wang et al., 2009). The KNN algorithm is shown in Fig. 6.9. All neighbors have equal votes, and the class with the highest number of voters among its neighbors is selected (Gorur et al., 2021). The algorithm's hyperparameter "k" refers to the number of nearest neighbors used to classify or predict results in a dataset (Alpaydın, 2010). There are no pre-defined mathematical techniques to find

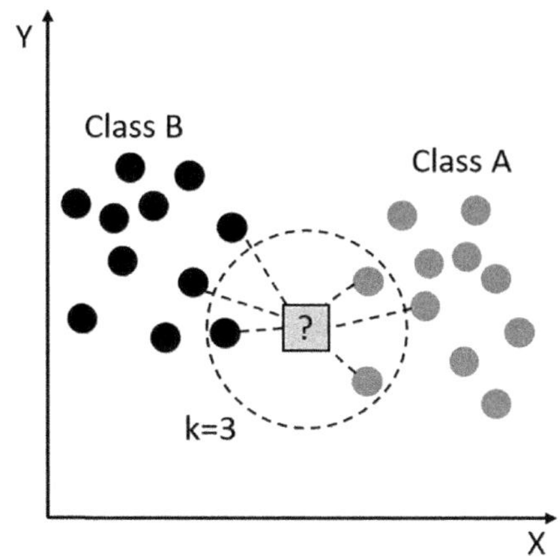

FIG. 6.9 KNN algorithm.

the optimal value of "k". If the "k" number is chosen small, the prediction region is limited, and the general distribution of the classifier becomes less precise.

Decision Tree

A Decision Tree (DTC), which is a rule-based algorithm that implements the divide and conquer strategy, is used for both regression and classification (Gorur et al., 2021). For classification, the number of rows in the input matrix and the number of class labels must be equal. From the top-down, it consists of the root, the decision nodes, and the leaves that give the result. It is a nonparametric method that uses a dataset by dividing it into smaller sets with decision rules (Alpaydın, 2010; Wang et al., 2009). The number of segments (the number of branches in the tree) that the problem space will be divided into is important for the solution. However, the main problem is determining the number of branches required for the best classification while creating the tree. Also, over-fitting is observed when a decision tree is constructed to perfectly match all samples in the training dataset. The decision tree structure is indicated in Fig. 6.10.

Random Forest

Random Forest (RF) is a supervised learning algorithm that consists of many individual decision trees working as a community (Mishra et al., 2021). The fact that the model gives good results without hyperparameter estimation and can be applied to both classification and regression problems makes it the popular model among the machine learning models. The Random Forest structure is shown in Fig. 6.11.

Random Forest randomly selects and trains different subsets from both the dataset and the feature set to overcome over-fitting. This is one of the biggest problems of decision trees. Each of the created decision trees makes individual predictions. For the classification problem, the highest number of votes is chosen among these predictions, while the predictions are averaged for regression (Mishra et al., 2021).

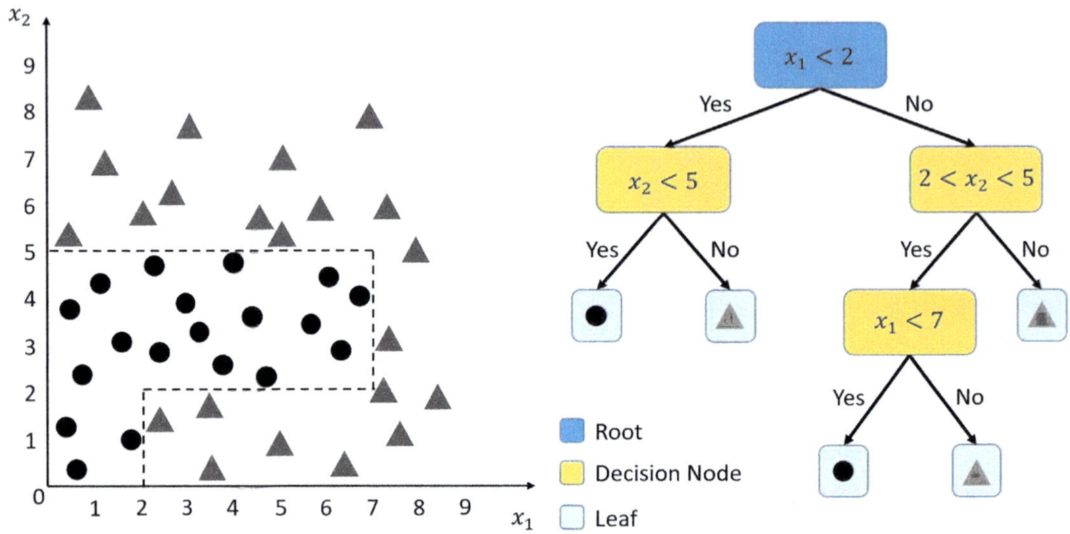

FIG. 6.10 Decision Tree.

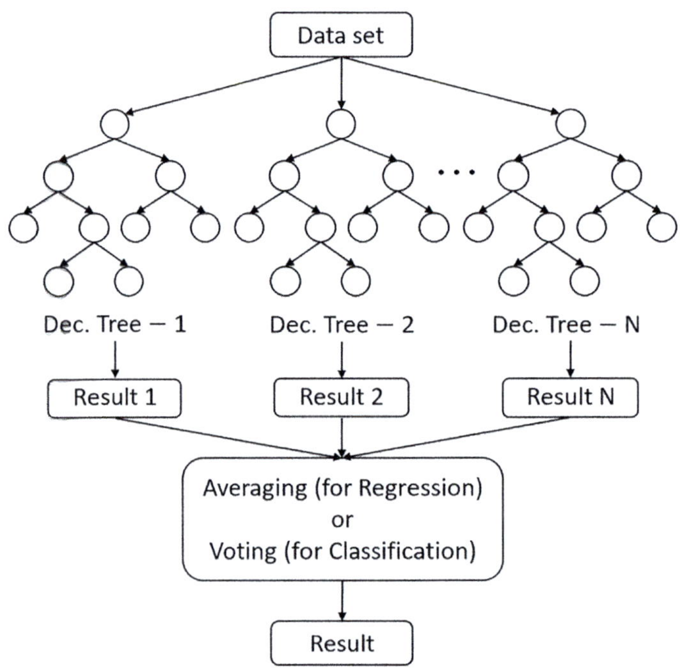

FIG. 6.11 Random Forest structure.

Support Vector Machine

Corinna Cortes and Vladimir Vapnik proposed the Support Vector Machine (SVM) as a machine learning algorithm (Gorur et al., 2018a). The Support Vector Machine has taken its place in the literature as an effective machine learning algorithm focusing on the statistical learning principle used for classification and regression analysis (Alpaydın, 2010). The SVM is a very common and useful tool for classifying bio-signal patterns and is a kernel-based method of classification algorithms. In the SVM, the weight vector is calculated after the training process and called support vectors. These are situations that are close to boundaries (hyperplane) to obtain information in which ambiguous and erroneous states occur (Liu et al., 2019). Thus, the hyperplane allocates input space for decision in the classification process. The distance from the hyperplane to the closest support vectors on both sides is referred to as the margin. The optimal hyperplanes are determined by maximizing the margins (dashed lines) of class boundaries, as shown in Fig. 6.12 (Alpaydın, 2010).

According to Fig. 6.12, circled instances stand for support vectors for a two-class problem, in which a square-pattern and a dot-pattern represent instances of classes. The SVM only deals with near-border instances for generalization while ignoring instances that lie in the interior (Gorur et al., 2019).

The values of *+1/−1* are used as labels for two classes, $wx+b$ defines the localized hyperplane. SVM does not use heuristic parameters such as learning rate, initialization, and convergence control. This does not imply that there is no need for hyperparameters. The main hyperparameter of the SVM is the kernel, which significantly affects the efficiency of the algorithm. For this function, one of the regular kernels is used, e.g., the linear kernel, the radial kernel, and the polynomial kernel (Alpaydın, 2010).

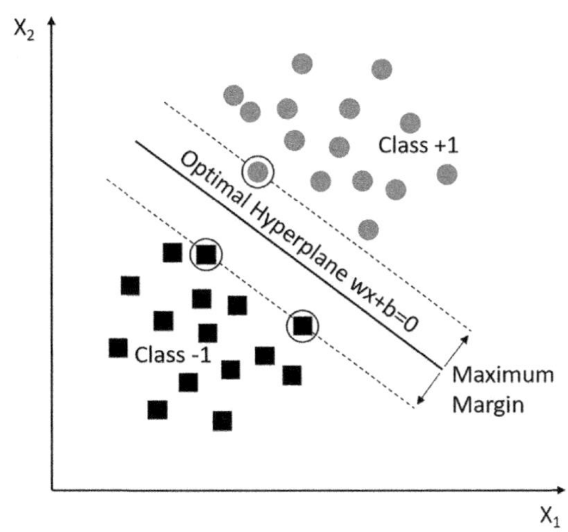

FIG. 6.12 The fundamental notions of SVM structure.

Multilayer Neural Network

In artificial neural networks, which are inspired by the information processing capability of the brain, the key structure used to solve a problem is the interconnected elements called neurons (Bascil et al., 2015). These algorithms, which have a trainable structure, can adapt their mathematical model to new problems. The artificial neural network consisting of many artificial neurons with linear or nonlinear transfer (activation) functions mimics the neural structure of the human brain (Gorur et al., 2018b; Shafi et al., 2006).

Multilayer Neural Network (MLNN) is a type of artificial neural network widely used in biomedical signal processing. The motivation behind using this algorithm is based on capturing the nonlinear behavior of a system or data samples and predicting them (Cetin & Temurtas, 2021). The MLNN is capable of learning and generalizing the pattern. Training in MLNN is defined as the act of updating the weights in order to obtain better convergence to desired results. The structure of the MLNN consists of the input layer, the output layer, and one or more hidden layers located between these two layers (Ilah et al., 2020). An example of MLNN structure consisting of two hidden layers is shown in Fig. 6.13.

The output of a neuron is expressed in terms of the sum of the weighted inputs and the bias value. Then, an activation function is applied to this weighted sum value to produce an output. An example of a commonly used nonlinear activation function is sigmoid. As seen in Fig. 6.13, the logarithmic sigmoid function is used in the hidden layers and the tangent sigmoid function in the output layer. The formulas for logarithmic sigmoid and tangent sigmoid activation functions are given in Eqs. (6.2) and (6.3), respectively (Cetin & Temurtas, 2021).

$$\frac{1}{1+e^{-x}} \tag{6.2}$$

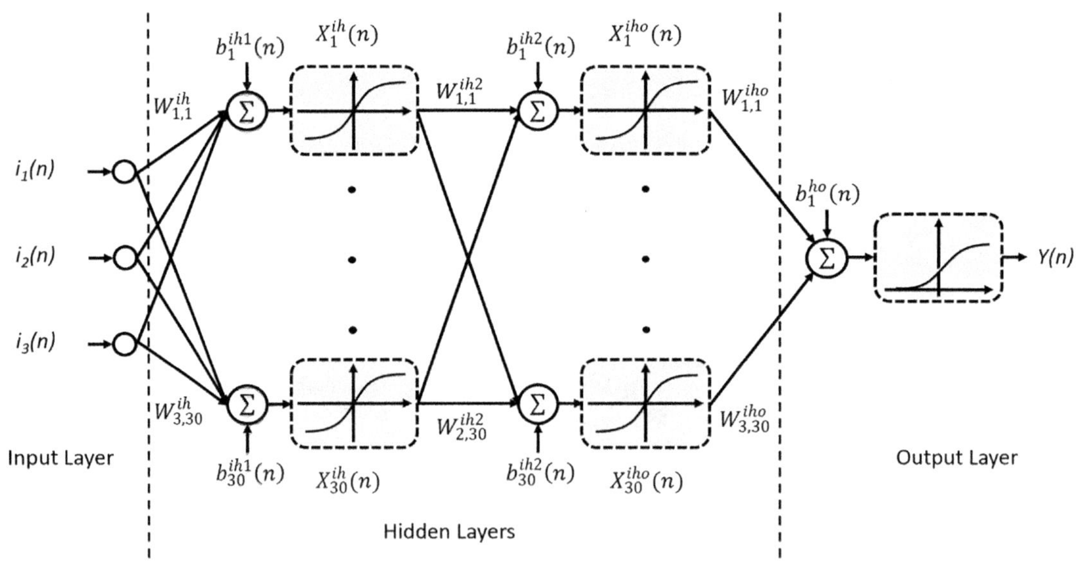

FIG. 6.13 MLNN structure.

$$\frac{e^x - e^{-x}}{e^x + e^{-x}} \tag{6.3}$$

Performance varies depending on network structure as well as the number of hidden layers and activation functions used (Çetin et al., 2020).

One of the most important problems in the MLNN structure is determining the number of hidden layers and neurons. There is no absolute rule for this procedure. The optimum network structure should have the lowest number of hidden layers capable of performing the given task and maximizing generalization ability (Gorur et al., 2018b).

Probabilistic Neural Network

Probabilistic Neural Networks (PNNs) are a type of feedforward network structure with two middle layers called radial base and competition layers (Er et al., 2016). PNN, developed by Donald Specht, estimates the probability density function using Gaussian kernels and performs the Bayes decision discriminant function (Bascil et al., 2015; Gorur et al., 2018b). It is known as a distance-based artificial neural network that uses a bell-shaped activation function. This method implements nonlinear decision boundaries and uses Bayes optimal approximation. PNN is suitable for biomedical use due to the Bayes strategy, a process familiar to medical decision makers. Furthermore, since PNN does not use a back-propagation algorithm, it is not necessary to train the network on the entire dataset. Thus, it can perform the classification process faster than MLNN. However, the network structure is larger (Cetin & Temurtas, 2021). A typical PNN sample is shown in Fig. 6.14.

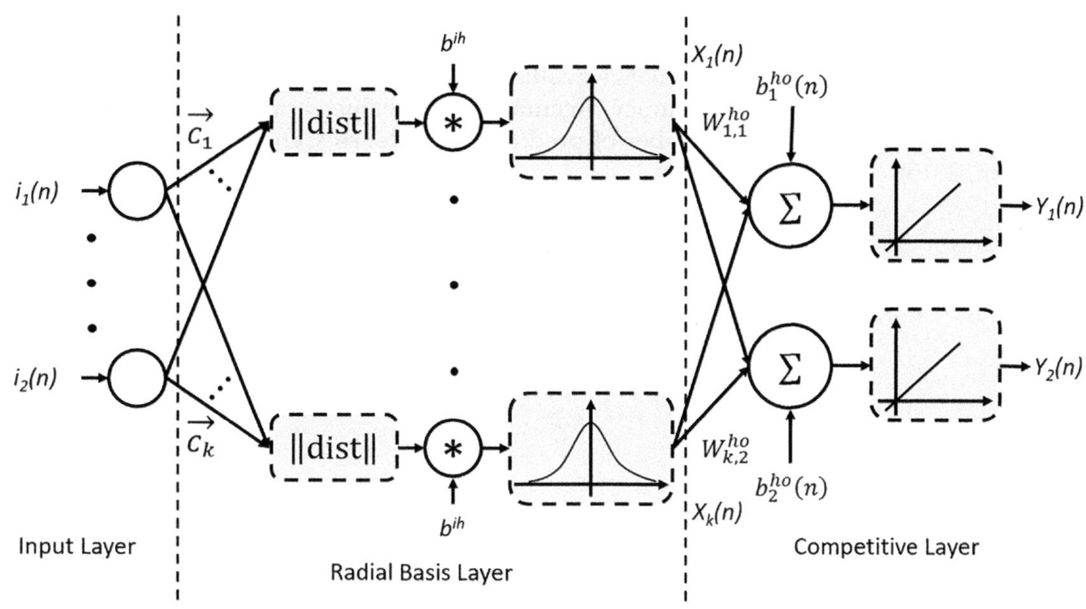

FIG. 6.14 A typical PNN structure.

PNN calculates the distances between the input vector and the weight matrix in the radial basis layer. The second middle layer (competitive layer) determines the classes with maximum accuracy. All hidden units receive the input vector carried to the hidden layer nodes via connection weights. The hidden layer consists of a series of radial-based functions (Gorur et al., 2018b). The center of the parameter vector associated with the jth hidden unit is named c_k. Hidden layer nodes determine the Euclidean distance between the center and the network input layer. The results are then moved to radial-based Gaussian functions in the network. w represents to the weight between the output node and hidden node (Cetin & Temurtas, 2021).

Performance evaluation metrics

The selection of appropriate evaluation metrics and their interpretation play a critical role in evaluating the performance of a classifier. Actual and predicted values in a classification problem can be shown with the Confusion Matrix presented in Fig. 6.15. The data in the matrix are widely used to visualize the performance of machine learning algorithms (Polat et al., 2008).

A confusion matrix has rows representing instances in an actual class and columns representing instances in a predicted class (Hossin & Sulaiman, 2015). According to this matrix;

- TP (True Positive) represents the number of correct predictions for a positive instance,
- FP (False Positive) represents the number of incorrect predictions for a negative instance,
- TN (True Negative) represents the number of correct predictions for a negative instance, and
- FN (False Negative) represents the number of incorrect predictions for a positive instance.

The most frequently used performance evaluation criterion to obtain the optimum classifier is accepted as classification accuracy. Accuracy (ACC) value is calculated by the ratio of correct predictions to the total number of predictions made, as seen in Eq. (6.4) (Hossin & Sulaiman, 2015).

$$ACC = \frac{TP + TN}{TP + TN + FP + FN} \tag{6.4}$$

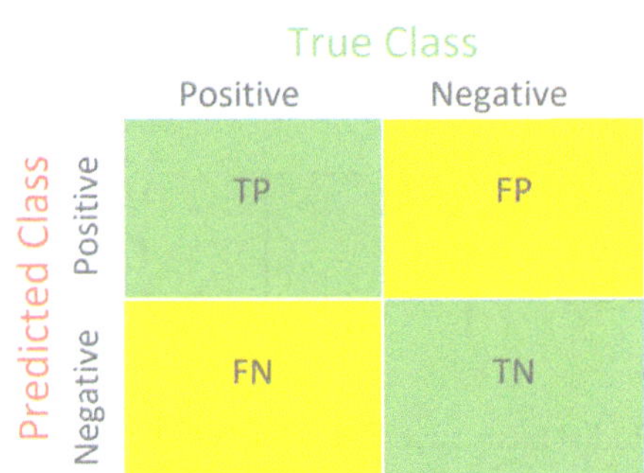

FIG. 6.15 Confusion matrix.

In datasets where the distribution between classes is not uniform (imbalanced datasets), accuracy alone is insufficient as an evaluation criterion (Hossin & Sulaiman, 2015). It would be, therefore, appropriate to evaluate the classification accuracy in conjunction with other performance criteria to optimize a classifier. Sensitivity (also known as recall) and specificity are statistical measures that indicate the proportions of correctly identified positives and negatives, respectively. Mathematical expressions of sensitivity (*SENS*) and specificity (*SPEC*) are given in Eqs. (6.5)–(6.6) (Cetin & Temurtas, 2021).

$$SENS = \frac{TP}{TP + FN} \tag{6.5}$$

$$SPEC = \frac{TN}{TN + FP} \tag{6.6}$$

The precision (*PREC*) given in Eq. (6.7) is used to measure the proportion of relevant instances among the retrieved instances (Abdar et al., 2018).

$$PREC = \frac{TP}{TP + FP} \tag{6.7}$$

The *F1 score* value, which is an important evaluation metric for datasets that are not uniformly distributed, is obtained by the harmonic average of sensitivity and precision values. The largest value it can take is 1.0, and its mathematical formulation is given in Eq. (6.8) (Abdar et al., 2018).

$$F_1\ score = \frac{2TP}{2TP + FP + FN} \tag{6.8}$$

It is also possible to use the receiver operating characteristic (ROC) and area under the curve (AUC) evaluation metrics to visualize classifiers' performance. The ROC curve is a two-dimensional probability graph that is plotted with the FP rate (X axis) and the TP rate (Y axis) (Fawcett, 2006; Metz, 1978). The area under the entire ROC curve is used to calculate AUC, which corresponds to the degree of differentiation between classes. High AUC values, measured in the range of [0,1], indicate better classifier performances (Fawcett, 2006).

Autoencoders

Autoencoders are often used to obtain a lower-dimensional representation of the data. For this process, neural networks are used to produce the output as the same input data. With this feature, they are a good example of representation learning. The training of autoencoders takes place through unsupervised learning. They consist of two parts as the encoder and decoder (Tschannen et al., 2018). The encoder enables obtaining the low-dimensional representation of the data. The decoder comes after the encoder. Here, the low-dimensional representation obtained in the encoder is taken as input, and the reconstruction of the same image as the original is produced as output. In summary, the basic principle of autoencoders is to try to reconstruct input data at model output. Fig. 6.16 shows an example autoencoder structure.

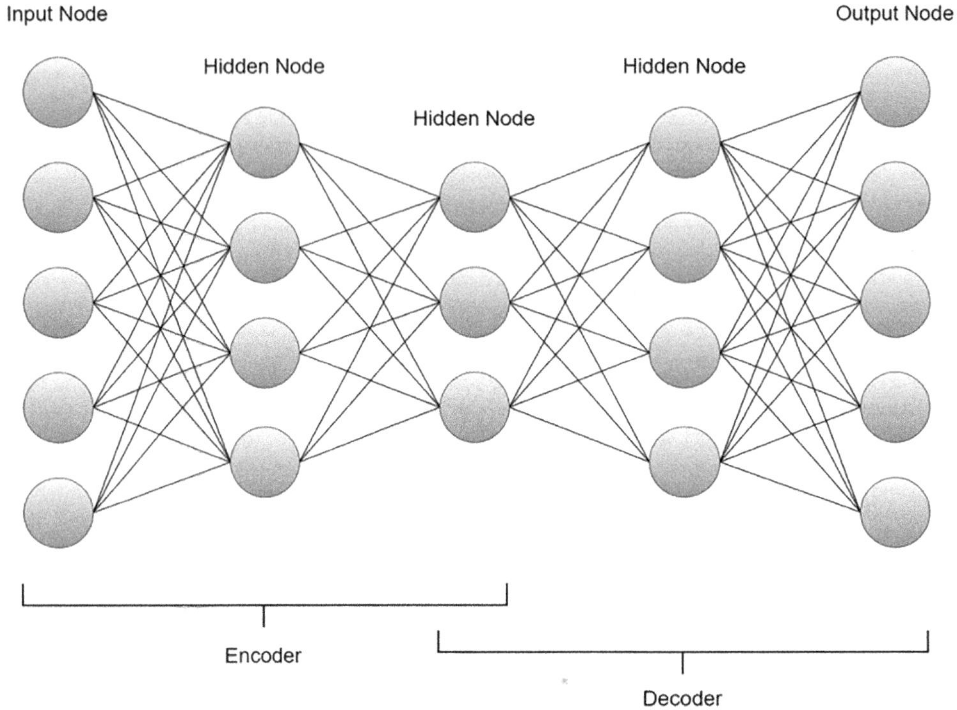

FIG. 6.16 An example autoencoder structure.

Convolutional Neural Networks

CNN is a well-known and widely used deep learning algorithm. It was developed by taking inspiration from the vision mechanisms of living things (Hubel & Wiesel, 1968). It is possible to consider the operation of CNN architectures in two stages. In the first step, feature extraction is performed over the given input. In the second stage, these features are used for tasks such as classification. Fully connected layers are frequently used for the classification process. However, instead of these layers, different algorithms such as support vector machines can be used. CNN architectures are commonly created by combining convolutional, fully connected, and pooling layers in various ways. Information on these three layers is provided in the sections below. Fig. 6.17 shows the process steps of a CNN architecture.

Convolutional layer

The evolution layer is the most basic and most important structure that makes up a CNN (Ozer et al., 2018). This layer consists of various convolution cores (kernels) used to determine feature maps (Ölmez et al., 2020). Neurons in a feature map make connections with only a certain region of the previous layer. These regions that interact with a CNN filter are called receptive fields. In this way, local characteristics of the input are extracted by using filters.

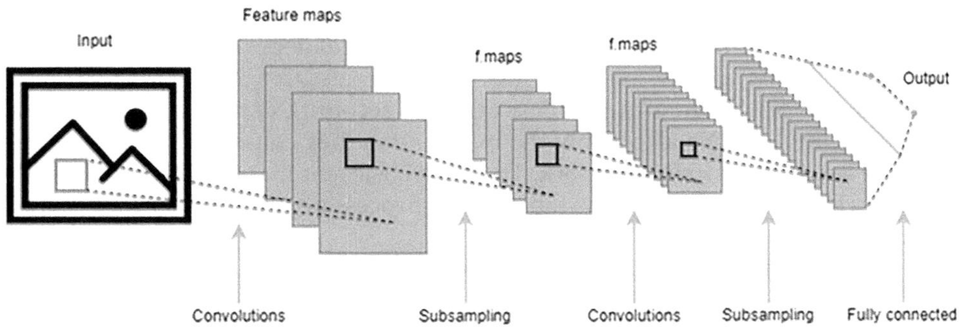

FIG. 6.17 Standard CNN architecture.

Filters with different weight vectors sliding along the inlet's width and height are used for this process. Thanks to the activation function, nonlinearity is provided to the CNN. This process is desirable in multilayer networks to detect nonlinear features.

Pooling layer

The purpose of this layer is to filter out unnecessary information from small image segments. Thus, the resolution of the feature maps is reduced. Accordingly, the spatial dimension of the inputs decreases. In addition, the number of parameters and computational costs in the CNN architecture decreases. Additionally, it is used to deal with the pooling layer overfitting problem. Each feature map produced by this layer is linked to the feature map of the previous layer.

There are different pooling methods used in CNN architectures. Maximum and average pooling are the two most frequently preferred methods. In both methods, the input is split into non-overlapping rectangles. In the maximum pooling method, the highest value in each rectangle (Zare & Ayati, 2021) is taken. In average pooling, the average value of each rectangle is taken.

Fully-connected layer

The neurons in this layer make connections with all the neurons in the layers before and after it. Softmax function is widely used to obtain normalized class probabilities in the last layer of CNNs (Alqahtani et al., 2021). This function is defined as follows:

$$f_t(z) = \frac{e^{zt}}{\sum_k e^{zk}} \tag{6.9}$$

Long short-term memory network

The use of deep learning models has become widespread in recent years. Depending on the different qualities of the input data, the deep learning model used varies. For example, CNN is very good on inputs with spatial dependencies. On the other hand, recurrent neural networks (RNN) are specialized in detecting temporal dependencies. Accordingly, they are widely used on data in the form of sequences with temporal dependencies such as video, audio, and text.

Traditional feedforward neural networks take into account existing samples to which they are exposed as input. RNNs also apply the information they calculate overtime, along with existing samples, as input to the network. The most basic feature of RNN models is that they have a cyclical structure that performs the learning process depending on past situations and current inputs.

The *k* value is a variable that differs depending on the input length. A hidden state information in the form of $[h_1, h_2,.. h_k]$ is generated in each processing step of the RNN model and $[x_1,x_2, \ldots x_k]$ are input sequences. To show *t* the current time, x_t the current input and h_{t-1} the previous hidden state, it is possible to express the hidden state at time *t* as

$$h_t = f(x_t, h_{t-1}) \tag{6.10}$$

It can be said that RNN models have a memory that stores the information calculated so far. This information is reapplied to the model with existing inputs. Fig. 6.18 shows the structure of a standard RNN.

As mentioned above, RNNs take into account the past information thanks to their repeating structure. However, the larger the time interval between the input data, the less effective the historical data. Long Short-Term Memory (LSTM) networks, which were proposed in the mid-90s, were developed to solve the problems experienced in the transfer of long-term dependencies (Hochreiter & Schmidhuber, 1997). Internal structures of LSTM modules differ from standard RNN modules.

According to Fig. 6.19, the LSTM module consists of three separate gates. These are the forgetting gate, the input gate, and the output gate. Forget gates decide how much of the information will be forgotten and how much will be passed on to the next level. For this operation, it uses the sigmoid function that produces a value between 0 and 1. 0 means the information will never be transmitted, and 1 means that all of them must be transmitted. The next step is to decide what information should be stored. This is done with the sigmoid function at the input gate. The tanh function then generates a vector of

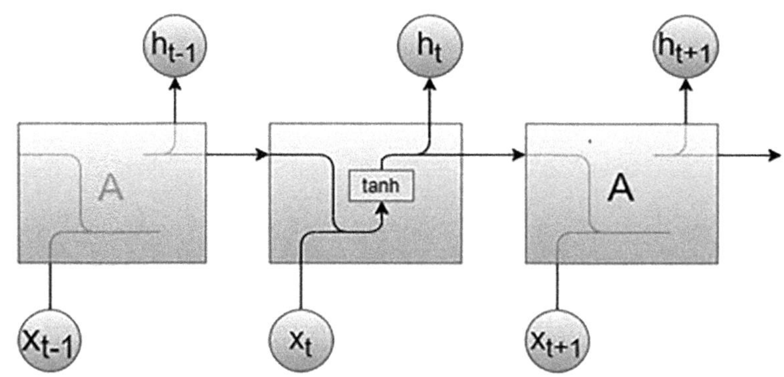

FIG. 6.18 Standard RNN structure.

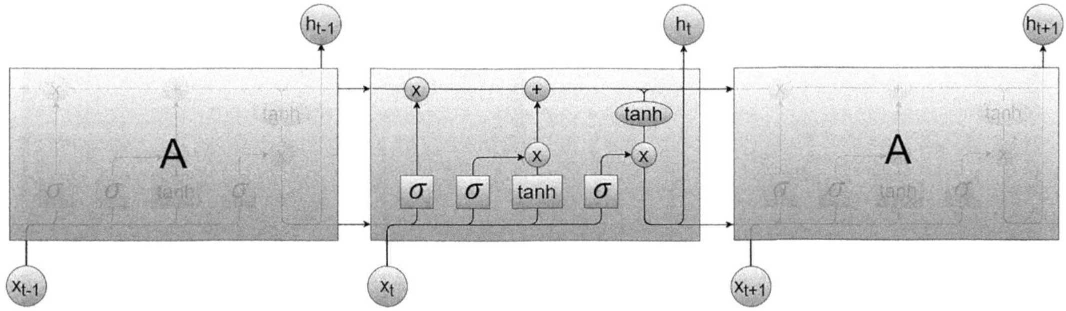

FIG. 6.19 LSTM module structure.

candidate values expressed as $\widetilde{C}_t$. These two processes are then combined. After this process, the new state information of the memory cell is calculated. Finally, the output of the system is calculated.

Results

In this section, the performance of six different machine learning algorithms consisting of deep and shallow models is compared on a publicly available thyroid nodule dataset (Pedraza et al., 2015). This dataset is widely used in studies on thyroid (Chi et al., 2017; Nguyen et al., 2019; Sai Sundar et al., 2019). The dataset consists of $560 \times 360 \times 3$ size ultrasound images in RGB format. It also includes thyroid imaging reporting and data system (TI-RADS) scores as well as images. TI RADS scores are widely used for risk assessment of thyroid lesions. In the evaluation here, if TI-RADS scores are 2 and 3, they are grouped as benign cases. On the other hand, if TI RADS scores were 4a, 4b, 4c, and 5, it was considered malignant case. A total of 52 patients in the dataset were grouped as benign cases and 246 patients as malign cases. The dataset includes one or more images of 298 patients. The samples of the benign and malign pictures from the dataset are presented in Fig. 6.20. The main focusing aims on these two processes (determining and segmentation for the type of the Thyroid nodule) over ultrasound images.

In cases where the distribution between classes is extremely unbalanced, it will not be sufficient to compare ACC values alone. For this reason, SENS, SPEC, and AUC values were compared as well as ACC values for comparison. The benign case is considered a negative class and malignant cases a positive class. 5-fold cross-validation results are given for all machine learning models.

Before the models were trained, all images were sized to $140 \times 90 \times 3$. Two separate convolution layers are used in the CNN model. There are 32 filters of 3×3 sizes in each layer. After each convolution layer, the 2×2 maxpooling process was carried out. These are followed by a fully connected layer consisting of 100 and 2 nodes. Another model used is LSTM. In the first layer of this model, there is an LSTM layer consisting of 100 modules. In the last layer, there is a fully connected layer consisting of two nodes for the classification process. After the images were adjusted to 140×270 dimensions, they were applied to the

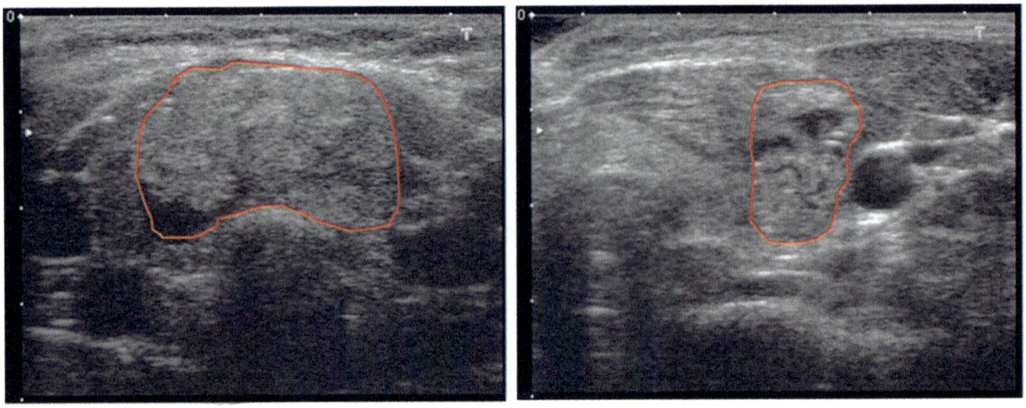

FIG. 6.20 Benign Thyroid nodule image (*left*) and Malignant Thyroid nodule image (*right*) (Pedraza et al., 2015).

TABLE 6.2 Performance comparison of various machine learning models.

	ACC (%)	SPEC (%)	SENS (%)	AUC
LSTM	87.36	65.57	91.99	0.79
CNN	86.78	68.85	90.59	0.80
MLNN	85.34	62.30	90.24	0.76
SVM	80.75	57.38	85.71	0.72
KNN	81.90	45.90	89.55	0.68
DTC	77.31	42.86	85.71	0.64

LSTM model. In models other than CNN and LSTM, the features are flattened and applied to the models. MLNN model has three fully connected layers. Their neuron numbers are 100,100, and 2, respectively. The Gini impurity algorithm was used in the decision tree. Neighborhood value is set to 5 in the KNN method.

Table 6.2 shows the results obtained from the classification process. It is seen that the performance of ANN-based models is better than that of SVM, KNN and DTC models. On the other hand, deep learning models are more successful than other models. It is considered that this is due to the ability of deep learning models to detect more abstract structures. ACC of the LSTM model was 87.36%. CNN's ACC value is 86.78%. The performance of both models is very similar to each other.

In Table 6.2 and Fig. 6.21, SENS and SPEC comparisons are also shown besides the performance values. The SPEC value of the CNN model is much better than that of other models. The SPEC value for SVM, KNN, and DTC was below 60%. In the MLNN model, the SPEC value was 62.30%. In the LSTM and CNN models, it was 65.57% and 68.57%, respectively. Image data contain large spatial dependencies. However, LSTM models are generally much more successful in temporally dependent data (Ozer et al., 2021). CNN models are quite

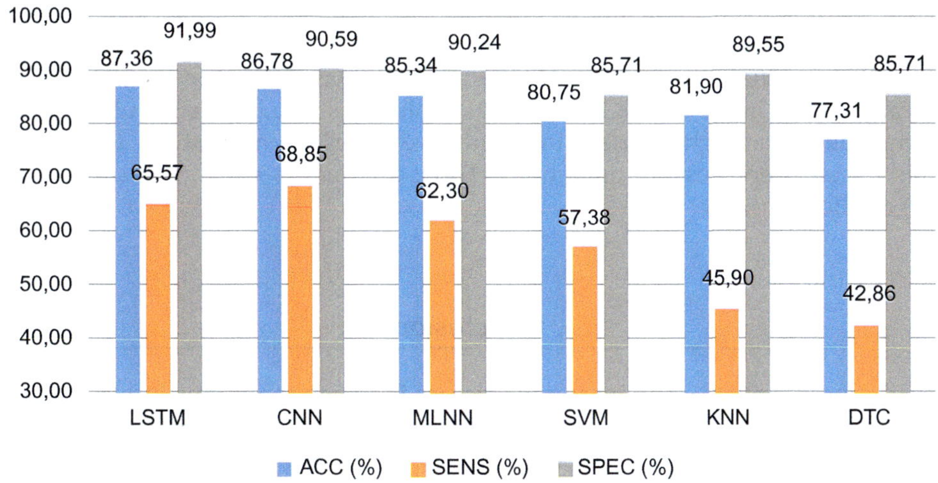

FIG. 6.21 Performance evaluations of models.

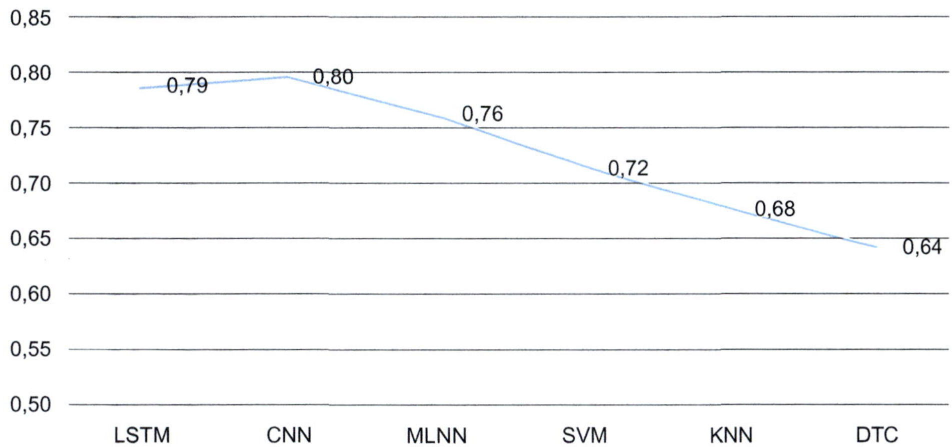

FIG. 6.22 AUC score evaluations of models.

successful in detecting spatial dependencies (Ozer, 2021). In addition, structures such as thyroid lesions have poor absolute locality. So, they can be found in any region on the image. CNN models are also very good at detecting weak absolute localities. For these reasons, it is evaluated that CNN has a better performance than LSTM. In addition, as seen in Fig. 6.22, the AUC value of the CNN model is better than that of the other models.

The lowest SENS value belongs to SVM and DTC with 85.71%. The SENS value for KNN was 89.55%. It was 90.24%, 90.59%, and 91.99% in MLNN, CNN, and LSTM models, respectively. In all models, SENS value is much higher than SPEC values. As mentioned above, the number of samples in the positive class is much higher than in the negative class. Machine learning models often tend to the majority class. Accordingly, SENS values are considered to be higher.

Overall, the ultrasound imaging techniques, histopathology images, and thyroid gland datasets can be implemented with machine learning algorithms in recent articles (Ciompi, 2021; Dov, 2021; Li, 2021; Liu, 2021; Olatunji, 2021; Temurtas, 2009), and our machine learning models can be used in the future studies. Data volume is a challenge for ultrasound image processing due to the high elapsed time with deep learning algorithms. Raw data image pixels also need preprocessing operations. Thus, this means extra time consumption for real-time applications.

Conclusion

Thyroid cancer is noted as an important disease in which symptoms form a nodule in the thyroid area of the neck. It is one of the most common types of cancer and has been reported to have the highest increase rate in the last 30 years. In recent years, machine learning/deep learning capabilities and statistical probabilistic methods (PNN or Random Forest) can provide high prediction results in diagnostic thyroid cancer nodules and regions. In particular, CNN, the deep learning model, came to the fore in image processing, but it achieved better results by LSTM in classification task in our image processing results with the same dataset (Table 6.2). LSTM models can provide generally much more successful results in temporally dependent data. In this chapter, especially it is aimed to present machine learning/deep learning approaches and probabilistic methods in helping clinicians for thyroid cancer prediction over the ultrasound images and Whole-Slide Images of histopathology. In our findings, the raw data images were processed for ultrasound images; however, the preprocessing operations were able to increase the performances of machine learning models. Moreover, thyroid gland disorder and machine learning models have also been handled in a concise manner.

References

Abbad Ur Rehman, H., Lin, C.-Y., Mushtaq, Z., & Su, S.-F. (2021). Performance analysis of machine learning algorithms for thyroid disease. *Arabian Journal for Science and Engineering*. https://doi.org/10.1007/s13369-020-05206-x.

Abdar, M., Yen, N. Y., & Hung, J. C. S. (2018). Improving the diagnosis of liver disease using multilayer perceptron neural network and boosted decision trees. *Journal of Medical and Biological Engineering*, *38*, 953–965. https://doi.org/10.1007/s40846-017-0360-z.

Alpaydın, E. (2010). *Introduction to machine learning* (2nd ed.). The MIT Press.

Alqahtani, H., Bharadwaj, S., & Ray, A. (2021). Classification of fatigue crack damage in polycrystalline alloy structures using convolutional neural networks. *Engineering Failure Analysis*, *119*, 104908. https://doi.org/10.1016/j.engfailanal.2020.104908.

Bascil, M. S., Tesneli, A. Y., & Temurtas, F. (2015). Multi-channel EEG signal feature extraction and pattern recognition on horizontal mental imagination task of 1-D cursor movement for brain computer interface. *Australasian Physical & Engineering Sciences in Medicine*, *38*, 229–239. https://doi.org/10.1007/s13246-015-0345-6.

Bhalla, S., Kaur, H., Kaur, R., et al. (2020). Expression based biomarkers and models to classify early and late-stage samples of papillary thyroid carcinoma. *PLoS One*, *15*, e0231629. https://doi.org/10.1371/journal.pone.0231629.

Buddhavarapu, V. G., & Angel Arul Jothi, J. (2020). An experimental study on classification of thyroid histopathology images using transfer learning. *Pattern Recognition Letters*, *140*, 1–9. https://doi.org/10.1016/j.patrec.2020.09.020.

Çetin, O., Dalcalı, A., & Temurtaş, F. (2020). A comparative study on parameters estimation of squirrel cage induction motors using neural networks with unmemorized training. *Engineering Science and Technology, An International Journal, 23*, 1126–1133. https://doi.org/10.1016/j.jestch.2020.03.011.

Cetin, O., & Temurtas, F. (2021). A comparative study on classification of magnetoencephalography signals using probabilistic neural network and multilayer neural network. *Soft Computing, 25*, 2267–2275. https://doi.org/10.1007/s00500-020-05296-7.

Chauhan, S., Vig, L., De Filippo De Grazia, M., et al. (2019). A comparison of shallow and deep learning methods for predicting cognitive performance of stroke patients from MRI lesion images. *Frontiers in Neuroinformatics, 13*, 1–12. https://doi.org/10.3389/fninf.2019.00053.

Chen, D., Hu, J., Zhu, M., et al. (2020). Diagnosis of thyroid nodules for ultrasonographic characteristics indicative of malignancy using random forest. *BioData Mining, 13*, 14. https://doi.org/10.1186/s13040-020-00223-w.

Chen, J., You, H., & Li, K. (2020). A review of thyroid gland segmentation and thyroid nodule segmentation methods for medical ultrasound images. *Computer Methods and Programs in Biomedicine, 185*, 105329. https://doi.org/10.1016/j.cmpb.2020.105329.

Chi, J., Walia, E., Babyn, P., et al. (2017). Thyroid nodule classification in ultrasound images by fine-tuning deep convolutional neural network. *Journal of Digital Imaging, 30*, 477–486. https://doi.org/10.1007/s10278-017-9997-y.

Ciompi, F. (2021). Editorial computational pathology. *IEEE Journal of Biomedical and Health Informatics*. https://doi.org/10.1109/JBHI.2021.3052029.

Dov, D. (2021). Weakly supervised instance learning for thyroid malignancy prediction from whole slide cytopathology images. *Medical Image Analysis*. https://doi.org/10.1016/j.media.2020.101814.

Dov, D., Kovalsky, S. Z., Assaad, S., et al. (2021). Weakly supervised instance learning for thyroid malignancy prediction from whole slide cytopathology images. *Medical Image Analysis, 67*, 101814. https://doi.org/10.1016/j.media.2020.101814.

Duan, T., Avati, A., Ding, D. Y., et al. (2019). *NGBoost: Natural gradient boosting for probabilistic prediction*. arXiv. https://doi.org/10.48550/arXiv.1910.03225.

Elliott Range, D. D., Dov, D., Kovalsky, S. Z., et al. (2020). Application of a machine learning algorithm to predict malignancy in thyroid cytopathology. *Cancer Cytopathology, 128*, 287–295. https://doi.org/10.1002/cncy.22238.

Er, O., Cetin, O., Bascil, M. S., & Temurtas, F. (2016). A comparative study on Parkinson's disease diagnosis using neural networks and artificial immune system. *Journal of Medical Imaging and Health Informatics, 6*, 264–268. https://doi.org/10.1166/jmihi.2016.1606.

Fawcett, T. (2006). An introduction to ROC analysis. *Pattern Recognition Letters, 27*, 861–874. https://doi.org/10.1016/j.patrec.2005.10.010.

Gorur, K., Bozkurt, M. R., Bascil, M. S., & Temurtas, F. (2018a). Glossokinetic potential based tongue–machine interface for 1-D extraction. *Australasian Physical & Engineering Sciences in Medicine, 41*, 379–391. https://doi.org/10.1007/s13246-018-0635-x.

Gorur, K., Bozkurt, M. R., Bascil, M. S., & Temurtas, F. (2018b). Glossokinetic potential based tongue–machine interface for 1-D extraction using neural networks. *Biocybernetics and Biomedical Engineering, 38*, 745–759. https://doi.org/10.1016/j.bbe.2018.06.004.

Gorur, K., Bozkurt, M., Bascil, M., & Temurtas, F. (2019). GKP signal processing using deep CNN and SVM for tongue-machine interface. *Trait du Signal, 36*, 319–329. https://doi.org/10.18280/ts.360404.

Gorur, K., Bozkurt, M. R., Bascil, M. S., & Temurtas, F. (2020). Comparative evaluation for PCA and ICA on tongue-machine interface using glossokinetic potential responses. *Celal Bayar University Journal of Science, 16*, 35–46. https://doi.org/10.18466/cbayarfbe.571994.

Gorur, K., Bozkurt, M. R., Bascil, M. S., & Temurtas, F. (2021). Tongue-operated biosignal over EEG and processing with decision tree and kNN. *Academic Platform Journal of Engineering and Science, 9*, 112–125. https://doi.org/10.21541/apjes.583049.

Ha, E. J., & Baek, J. H. (2021). Applications of machine learning and deep learning to thyroid imaging: Where do we stand? *Ultrasonography, 40*, 23–29. https://doi.org/10.14366/usg.20068.

Hemalatha, D., & Poorani, S. (2020). Supervised machine learning models for classification of thyroid data. *International Journal of Scientific and Technology Research, 9*, 1683–1685.

Hochreiter, S., & Schmidhuber, J. (1997). Long short-term memory. *Neural Computation, 9*, 1735–1780. https://doi.org/10.1162/neco.1997.9.8.1735.

Hossin, M., & Sulaiman, M. N. (2015). A review on evaluation metrics for data classification evaluations. *International Journal of Data Mining & Knowledge Management Process*, *5*, 01–11. https://doi.org/10.5121/ijdkp.2015.5201.

Hubel, D. H., & Wiesel, T. N. (1968). Receptive fields and functional architecture of monkey striate cortex. *The Journal of Physiology*, *195*, 215–243. https://doi.org/10.1113/jphysiol.1968.sp008455.

Ilah, S., Jae Kim, Y., et al. (2020). Application of machine learning to ultrasound images to differentiate follicular neoplasms of the thyroid gland. *Ultrasonography*, *39*, 257–265. https://doi.org/10.14366/usg.19069.

Kezlarian, B., & Lin, O. (2020). Artificial intelligence in thyroid fine needle aspiration biopsies. *Acta Cytologica*, *10065*. https://doi.org/10.1159/000512097.

Li, L. R. (2021). Artificial intelligence for personalized medicine in thyroid cancer: Current status and future perspectives. *Frontiers in Oncology*. https://doi.org/10.3389/fonc.2020.604051.

Li, H., Weng, J., Shi, Y., et al. (2018). An improved deep learning approach for detection of thyroid papillary cancer in ultrasound images. *Scientific Reports*, *8*, 6600. https://doi.org/10.1038/s41598-018-25005-7.

Li, H., Zhang, B., Zhang, Y., et al. (2020). A semi-automated annotation algorithm based on weakly supervised learning for medical images. *Biocybernetics and Biomedical Engineering*, *40*, 787–802. https://doi.org/10.1016/j.bbe.2020.03.005.

Liu, W. C. (2021). Machine learning for the prediction of bone metastasis in patients with newly diagnosed thyroid cancer. *Cancer Medicine*. https://doi.org/10.1002/cam4.3776.

Liu, J., Song, S., Sun, G., & Fu, Y. (2019). Classification of ECG arrhythmia using CNN, SVM and LDA. In *Lecture notes in computer science (including Subser Lect Notes Artif Intell Lect Notes bioinformatics)* (pp. 191–201). https://doi.org/10.1007/978-3-030-24265-7_17. 11633 LNCS.

Metz, C. E. (1978). Basic principles of ROC analysis. *Seminars in Nuclear Medicine*, *8*, 283–298. https://doi.org/10.1016/S0001-2998(78)80014-2.

Mishra, S., Tadesse, Y., Dash, A., et al., & Mishra, D. (2021). Thyroid disorder analysis using random forest classifier. In R. Buyya, P. Mohapatra, & S. Patnaik (Eds.), *Intelligent and cloud computing* (pp. 385–390). Singapore: Springer Singapore.

Mohanty, M. N., & Palo, H. K. (2020). Child emotion recognition using probabilistic neural network with effective features. *Measurement*, *152*, 107369. https://doi.org/10.1016/j.measurement.2019.107369.

Muller, K.-R., Mika, S., Ratsch, G., et al. (2001). An introduction to kernel-based learning algorithms. *IEEE Transactions on Neural Networks*, *12*, 181–201. https://doi.org/10.1109/72.914517.

Nanni, L., Ghidoni, S., & Brahnam, S. (2017). Handcrafted vs. non-handcrafted features for computer vision classification. *Pattern Recognition*, *71*, 158–172. https://doi.org/10.1016/j.patcog.2017.05.025.

Nguyen, D. T., Pham, T. D., Batchuluun, G., et al. (2019). Artificial intelligence-based thyroid nodule classification using information from spatial and frequency domains. *Journal of Clinical Medicine*, *8*, 1976. https://doi.org/10.3390/jcm8111976.

Olatunji, S. O. (2021). Early diagnosis of thyroid cancer diseases using computational intelligence techniques: A case study of a Saudi Arabian dataset. *Computers in Biology and Medicine*. https://doi.org/10.1016/j.compbiomed.2021.104267.

Ölmez, E., Akdoğan, V., Korkmaz, M., & Er, O. (2020). Automatic segmentation of meniscus in multispectral MRI using regions with convolutional neural network (R-CNN). *Journal of Digital Imaging*, *33*, 916–929. https://doi.org/10.1007/s10278-020-00329-x.

Ozer, I. (2021). Pseudo-colored rate map representation for speech emotion recognition. *Biomedical Signal Processing and Control*, *66*, 102502. https://doi.org/10.1016/j.bspc.2021.102502.

Ozer, I., Efe, S. B., & Ozbay, H. (2021). A combined deep learning application for short term load forecasting. *Alexandria Engineering Journal*, *60*, 3807–3818. https://doi.org/10.1016/j.aej.2021.02.050.

Ozer, I., Ozer, Z., & Findik, O. (2018). Noise robust sound event classification with convolutional neural network. *Neurocomputing*, *272*, 505–512. https://doi.org/10.1016/j.neucom.2017.07.021.

Park, V. Y., Han, K., Seong, Y. K., et al. (2019). Diagnosis of thyroid nodules: Performance of a deep learning convolutional neural network model vs. radiologists. *Scientific Reports*, *9*, 17843. https://doi.org/10.1038/s41598-019-54434-1.

Pathpedia, n.d. https://www.pathpedia.com.

Pedraza, L., Vargas, C., Narváez, F., et al. (2015). An open access thyroid ultrasound image database. In E. Romero, & N. Lepore (Eds.), *10th International symposium on medical information processing and analysis* (p. 92870W).

Polat, K., Güneş, S., & Arslan, A. (2008). A cascade learning system for classification of diabetes disease: Generalized discriminant analysis and Least Square support vector machine. *Expert Systems with Applications, 34*, 482–487. https://doi.org/10.1016/j.eswa.2006.09.012.

Reddy, D. S., Vaishnavi, O. S., Vidya, J., et al. (2020). Diagnosis of thyroid gland disorder using machine learning techniques. *International Journal of Advanced Science and Technology, 29*, 4752–4761. https://doi.org/10.1007/978-3-319-52048-3_2.

Reis, I., Baron, D., & Shahaf, S. (2018). Probabilistic random forest: A machine learning algorithm for noisy datasets. *The Astronomical Journal*. https://doi.org/10.3847/1538-3881/aaf101.

Sai Sundar, K. V., Rajamani, K. T., & Siva Sankara Sai, S. (2019). Exploring image classification of thyroid ultrasound images using deep learning. In *Lecture notes in computational vision and biomechanics* (pp. 1635–1641). Springer International Publishing.

Scopus, n.d., https://www.scopus.com.

Shafi, I., Ahmad, J., Shah, S. I., & Kashif, F. M. (2006). Impact of varying neurons and hidden layers in neural network architecture for a time frequency application. In *2006 IEEE international multitopic conference* (pp. 188–193). IEEE.

Sun, C., Zhang, Y., Chang, Q., et al. (2020). Evaluation of a deep learning-based computer-aided diagnosis system for distinguishing benign from malignant thyroid nodules in ultrasound images. *Medical Physics, 47*, 3952–3960. https://doi.org/10.1002/mp.14301.

Tahmasebi, A., Wang, S., Daniels, K., et al. (2020). Ultrasonographic risk stratification of indeterminate thyroid nodules; a comparison of an artificial intelligence algorithm with radiologist performance. In *2020 IEEE international ultrasonics symposium (IUS)* (pp. 1–4). IEEE.

Temurtas, F. (2009). A comparative study on thyroid disease diagnosis using neural networks. *Expert Systems with Applications, 36*, 944–949. https://doi.org/10.1016/j.eswa.2007.10.010.

Thomas, J., Ledger, G. A., & Mamillapalli, C. K. (2020). Use of artificial intelligence and machine learning for estimating malignancy risk of thyroid nodules. *Current Opinion in Endocrinology, Diabetes, and Obesity, 27*, 345–350. https://doi.org/10.1097/MED.0000000000000557.

Tschannen, M., Bachem, O., & Lucic, M. (2018). Recent advances in autoencoder-based representation learning. *arXiv*, 1–25.

Turk, G., Ozdemir, M., Zeydan, R., et al. (2021). On the identification of thyroid nodules using semi-supervised deep learning. *International Journal for Numerical Methods in Biomedical Engineering, 37*, 1–18. https://doi.org/10.1002/cnm.3433.

Wang, H., Song, B., Ye, N., et al. (2020). Machine learning-based multiparametric MRI radiomics for predicting the aggressiveness of papillary thyroid carcinoma. *European Journal of Radiology, 122*, 108755. https://doi.org/10.1016/j.ejrad.2019.108755.

Wang, B., Wong, C. M., Wan, F., et al. (2009). Comparison of different classification methods for EEG-based brain computer interfaces: A case study. In *2009. 2009 IEEE Int Conf Inf Autom ICIA* (pp. 1416–1421). https://doi.org/10.1109/ICINFA.2009.5205138.

Wang, L., Zhang, L., Zhu, M., et al. (2020). Automatic diagnosis for thyroid nodules in ultrasound images by deep neural networks. *Medical Image Analysis, 61*, 101665. https://doi.org/10.1016/j.media.2020.101665.

Yang, W., Dong, Y., Du, Q., et al. (2021). Integrate domain knowledge in training multi-task cascade deep learning model for benign–malignant thyroid nodule classification on ultrasound images. *Engineering Applications of Artificial Intelligence, 98*, 104064. https://doi.org/10.1016/j.engappai.2020.104064.

Yu, J., Deng, Y., Liu, T., et al. (2020). Lymph node metastasis prediction of papillary thyroid carcinoma based on transfer learning radiomics. *Nature Communications, 11*, 4807. https://doi.org/10.1038/s41467-020-18497-3.

Zare, S., & Ayati, M. (2021). Simultaneous fault diagnosis of wind turbine using multichannel convolutional neural networks. *ISA Transactions, 108*, 230–239. https://doi.org/10.1016/j.isatra.2020.08.021.

Zhang, Z., Zhang, X., Lin, X., et al. (2019). Ultrasonic diagnosis of breast nodules using modified faster R-CNN. *Ultrasonic Imaging, 41*, 353–367. https://doi.org/10.1177/0161734619882683.

Zhu, Y., AlZoubi, A., Jassim, S., et al. (2021). A generic deep learning framework to classify thyroid and breast lesions in ultrasound images. *Ultrasonics, 110*, 106300. https://doi.org/10.1016/j.ultras.2020.106300.

PART 2

Prediction of cancer susceptibility

CHAPTER

7

Machine learning-based detection and classification of lung cancer

Shubham Dodia and B. Annappa

Department of Computer Science and Engineering, National Institute of Technology, Mangaluru, Karnataka, India

Introduction

Lung cancer holds the highest mortality and morbidity rate among all the cancers in the world. According to the study conducted by the American Cancer Society on lung cancer, the number of deaths occurring from lung cancer alone is approximately 1,42,670, consisting of 76,650 men and 66,020 women for the year 2019 (DeSantis et al., 2019). Lung cancer is the formation of irregular tissue in the thoracic/lung region named tumors/nodules. The irregular tissue formation does not necessarily mean the presence of lung cancer. The tumors can be of two types, noncancerous (benign) or cancerous (malignant) nodules. A patient is diagnosed with lung cancer if he/she has malignant nodules in the lung region. The primary reason for lung cancer having the highest mortality rate is that it is difficult to diagnose in the early stages of the disease. The symptoms are visible in the later stages of the disease. The lung cancer patient's survival rate is up to 5years if an early diagnosis is performed (Baldwin, 2015). Therefore, it is critical to diagnose lung cancer in the early stages and improve the patients' survival rate.

Image processing is a technique adapted to extract vital information from an image, where firstly, the image is transformed into digital form, and then several operations are carried out on that image to get an enhanced image (Bhalerao et al., 2019). Improved image processing methods, including artificial intelligence (AI) algorithms, are the current cutting edge tools for enhancing prediction for prognosis and treatment, particularly in cancer tumors in various medical applications. Liu et al. suggest that image recognition methods can be used to detect lung cancer, as there are apparent differences between balanced healthy pulmonary images and unhealthy pulmonary images (Y. Liu et al., 2019). Based on histopathology, lung cancer can be classified into two types, small cell lung cancer (SCLC) and non-small-cell lung

https://doi.org/10.1016/B978-0-323-85240-1.00002-X

TABLE 7.1 Stages of lung cancer.

Lung cancer stages	Parts of the body affected
Stage I	Restricted to the lung
Stages II and III	Restricted to the chest
Stage IV	Extensive metastases outside the chest/thorax or pleural involvement

cancer (NSCLC). Approximately 80% of lung cancer deaths are due to NSCLC type of cancer (Potghan et al., 2018). The presence of NSCLC is commonly observed in non-small-cells. The preliminary classification of NSCLC can be of 2 types, namely squamous carcinoma of the cell and nonsquamous carcinoma of the cell. Further, nonsquamous is classified as large-cell carcinoma and adenocarcinoma (Antony, 2012).

The staging of lung cancer can be done in four levels. These levels are illustrated in Table 7.1. Based on the order of grade, the stages from I to IV are stated (Baldwin, 2015).

Many imaging techniques are used for diagnosing tumors, such as chest radiographs (popularly known as X-ray), positron emission tomography (PET), and computed tomography (CT). One of the popular and best imaging techniques widely used for lung cancer nodule detection is CT imaging. This technique is more reliable and capable of detecting cancerous and noncancerous nodules (Gindi & Sami, 2014). However, the CT scan images consist of the variance of intensity. Also, there can be a possibility of radiologists or doctors misjudging an anatomical structure as a cancerous cell. To address these issues, assistance or a second opinion to the doctors and radiologists can be provided by computer-aided diagnosis (CAD) systems in accurately identifying cancerous and noncancerous nodules. The research on a developing reliable CAD system is going on for almost three decades. The process of lung cancer detection and classification is divided into two tasks: First, the potential lung cancer nodules/lesions must be detected from the CT images and this is performed by CADe systems; second is the task of further classifying these detected nodules into malignant and benign nodules and this is performed using CADx systems (Suzuki, 2012).

In this work, the lung cancer nodule detection system's evaluation is performed on the LUNA16 dataset (Setio et al., 2016). Initial steps include the preprocessing of CT scans, extraction of features from the images, selection of prominent features, and the final step is the classification. The preprocessing of the images is performed using various filters. This step is performed to improve the visibility of the CT as the image is of low quality. Applying filter also removes noise in the images. In order to get better classification performance, selection of features plays a crucial role as appropriate features provide good results. In this work, multiple image processing features are extracted from the preprocessed CT images. The features obtained from the preprocessed CT images may contain redundant and unnecessary information that does not add to the classification task's performance. Therefore, feature selection techniques are applied to the features to retain only the useful and necessary classification task information. The use of feature selection techniques reduces time complexity, as there are reduced feature sets for classification tasks. The final step is the detection of nodules and nonnodules from the CT images. To perform this task, various sets of classifiers are used.

After the detection of nodules from the CT scan images, in addition, the nodules found are further categorized into cancerous and noncancerous nodules. To perform this classification, a majority voting scheme is adapted from all the classifiers. Among all the classifiers used in this work, neural network classifiers provided better performance than other conventional machine learning algorithms for lung cancer nodule detection and classification task.

Related work

One of the popular techniques for lung cancer nodule detection is the combination of statistical methods, image processing features, and machine learning algorithms. To perform the clustering task, a K-mean unsupervised learning algorithm was adopted (Gopi & Selvakumar, 2017). A classification model was proposed by Aggarwal et al. (2015) to differentiate between lung cancer nodules and normal lung anatomy structure. The features extracted for performing the classification task are statistical, gray level, and geometric features. The classifier used is a linear discriminant analysis (LDA) classifier, and the results obtained from the method used in this work are considerably low, making it unreliable for real-time usage.

Many CAD systems have used segmentation techniques to extract region of interest (ROI) from the CT images, and then these extracted regions are used for further classification steps. Potghan et al. (2018) have used the local thresholding method to extract lung volumes and performed classification using K-nearest neighbor (KNN) and multilayer perceptron (MLP) classifiers. This method provided good results; however, a data imbalance issue was observed in their work. This may result in a nongeneralizable model. Morphological operations were used for segmentation by Kuruvilla and Gunavathi (2014) for lung cancer detection. The features extracted on the segmented regions were statistical features such as mean, standard deviation, skewness, kurtosis, and fifth and sixth central moments. These features were fed to feed-forward backpropagation neural network for further classification. But, the method resulted in relatively less sensitivity value which is one of the essential performance metrics in medical system evaluation. Therefore, a low sensitivity value is unacceptable.

Manikandan and Bharathi (2011) did a region-based segmentation for lung cancer nodule detection. A wiener filter was used for preprocessing the images. Spatial features such as diameter, intensity, and shape were extracted from the preprocessed images. The classifier used in their work is the fuzzy inference system (FIS), and this approach has a sensitivity of 92.3%. However, the only drawback of the system is that the model was tested on very few samples. The model may not work in a similar fashion if a huge dataset is provided. An ensemble and random forest (RF) classifiers were used by Y. Liu et al. (2019), and a two-dimensional OTSU's curve thresholding method was used for extracting ROIs from the input images. The model performed well for both texture and spatial features extracted from the segmented regions of the input images. Even though a lot of work has been proposed for lung cancer nodule detection, the CAD systems' performance is still not reliable for hospital deployment. An analysis of various preprocessing algorithms, feature extraction techniques, feature selection methods, and classifiers is provided in this work.

Problem definition

In this work, detection and classification of lung cancer is performed using machine learning algorithms. A pipeline of preprocessing techniques, feature extraction techniques, feature selection techniques, and classification algorithms is explored in order to achieve the best suitable framework to detect and classify lung cancer nodules from a given raw CT scan input.

Proposed methodology

In this work, in order to detect and classify lung cancer nodules, a comparison is carried out between various preprocessing, feature extraction, feature selection, and classification techniques. The dataset used in this work and the detailed description of the techniques are described below.

Overview

The proposed work explores a wide variety of techniques for all the steps involved in the detection of nodules and the classification of lung cancer nodules into benign and malignant nodules. The preprocessing step is crucial as noise from the images can be discarded. By varying the contrast, intensity, etc., preprocessing can provide better visibility of images than the original CT scan image. The preprocessing filters used in this work to improve the visibility of the image are Gabor, Gaussian, Hessian, Laplace, Median blur, and Sobel. From these preprocessed images, multiple features are extracted. These features consist of essential information in an image, which provides discriminative information to classify the lung nodules from other anatomical lung organs. The features considered in this work are local binary patterns (LBPs), histogram of oriented gradients (HOGs), scale-invariant feature transformation (SIFT), and morphological and statistical features. The system's performance can be improved by the combination of various features. This, in turn, results in a large feature vector. However, to avoid redundancy in the features, feature selection algorithms are applied. This is helpful to not only reduce the number of features but also reduce the time complexity of the system. The classification of lung nodules and nonnodules based on the features is the last step of the detection task. The lung nodules obtained from the nodule detection system are further classified into cancerous and noncancerous nodules. The nodules are annotated with the above-mentioned two classes based on the malignancy score provided by the expert radiologists for each nodule. A majority voting scheme is adopted in this work to classify detected lung nodules. The architecture of the proposed methodology is illustrated in Fig. 7.1.

Dataset

The dataset used for evaluating lung nodule detection is the LUng Nodule Analysis challenge (LUNA-16) dataset released in 2016 (Setio et al., 2016). The dataset is a curated version of the LIDC-IDRI dataset. This dataset is intended to encourage researchers to develop an

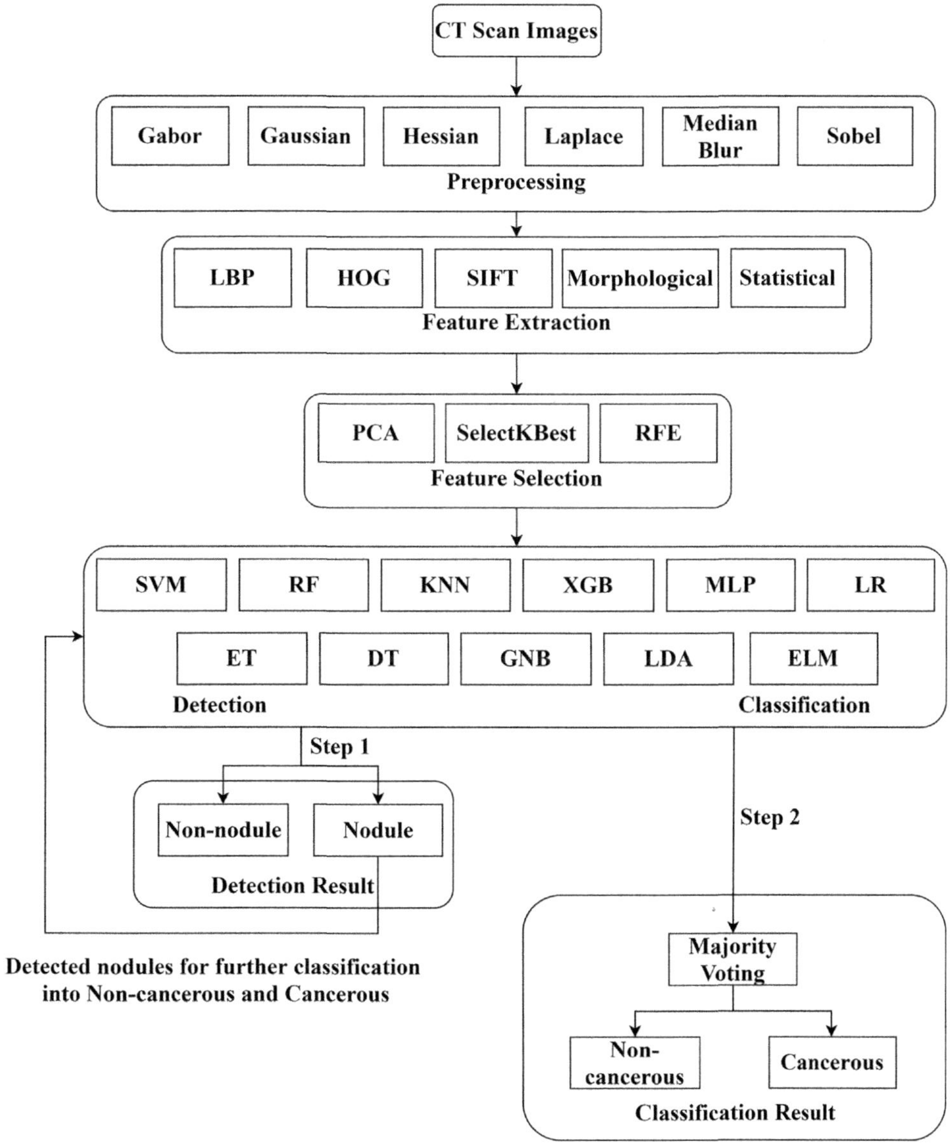

FIG. 7.1 Schematic diagram of the proposed methodology for lung cancer detection and classification. *No permission required.*

automated lung cancer nodule detection system. The dataset consists of CT images collected from 888 patients with 1186 nodule annotations. The candidates' labels in the CT scan images are provided based on the agreement of 3 out of 4 expert radiologist annotators. The dataset consists of a total of 551,065 nonnodules and 1186 nodules. Each CT scan consists of varying nodule sizes, and the candidate locations are provided in X, Y, and Z coordinates. The location of these coordinates is positioned at the center when it is viewed as a two-dimensional image.

Preprocessing

The preliminary step of the lung nodule detection task is to bifurcate the lung nodules and nonnodules into two separate classes for training the model. The candidate information in the CT scan is provided by X, Y, and Z axes coordinates. Based on these coordinates, the CT candidate is located, and a 32×32 region is cropped from the CT scan. The cropping of the region is performed to get the exact structure of a candidate and remove unnecessary background information as this additional information is merely background noise. After the candidate is generated, the candidate is stored in its corresponding class based on the label annotated, which can be a nodule or a nonnodule. The candidate stored is a cube. In this work, a center slice of the cube retaining maximum information of the candidate structure is used for performing the next steps. For the obtained CT slice, various filters are applied. The description of each filter is mentioned below.

Gabor filter

The Gabor filter is known as a linear filter whose impulse ratio is determined by the harmonic function multiplied by the Gaussian function (Zainudin et al., 2011). The functionality of the Gabor filter can be said as a multiplication-convolution property. The Gabor filter is given in Eq. (7.1):

$$\lambda, \Theta, \Phi, \sigma, \gamma(x,y) = \exp\left[\frac{-x^2 - \gamma^2 y^2}{2\sigma^2}\right] \cos\left[2x\frac{x'}{\lambda} + \phi\right] \tag{7.1}$$

where $x' = x\cos\theta + y\sin\theta$ and $y' = -x\sin\theta + y\cos\theta$. x and y in the above equation denote the impulse ratio in the image, and the parameters are λ the wavelength, θ the orientation, Φ the phase offset, σ the aspect ratio, and γ the bandwidth. Given all the parameters, the output of the Gabor filter is an intensity map image whose dimension is the same as the input image. The output image obtained from the Gabor filter is an enhanced image.

Gaussian filter

Gaussian filter is applied on the CT images to minimize the noise in the image. The noise is removed based on an averaging technique. The noisy pixel in the image is replaced by the average value of the neighboring pixels, which is performed based on Gaussian distribution. The Gaussian smoothing filter for image processing is given by Eq. (7.2):

$$g[x,y] = e^{\frac{x^2+y^2}{2\sigma^2}} \tag{7.2}$$

where $g[x,y]$ is a pixel point in an image, and σ is width spread parameter in the Gaussian (N. Kumar & Nachamai, 2017).

Hessian filter

Hessian matrix filter for a 2D image is applied to enhance the image. For a 2D image, the Hessian matrix $H(x)$ for an intensity value at location (x,y) in an image is given by Eq. (7.3):

$$H(x) = \begin{bmatrix} \frac{\partial^2 f}{\partial x^2} & \frac{\partial^2 f}{\partial x \partial y} \\ \frac{\partial^2 f}{\partial y \partial x} & \frac{\partial^2 f}{\partial y^2} \end{bmatrix} \triangleq \begin{bmatrix} f_{xx} & f_{xy} \\ f_{yx} & f_{yy} \end{bmatrix} \tag{7.3}$$

Let $\lambda = (\lambda 1, \lambda 2)$ be the eigenvalues of the Hessian matrix $H(x)$. The Hessian-based enhancement filter's response function is given by Eq. (7.4):

$$\phi(x) = \begin{cases} \phi\{\Lambda\} \text{ if condition } \phi_a \text{ is satisfied;} \\ 0, \text{otherwise} \end{cases} \tag{7.4}$$

where $\varphi(x)$ is the response function, and Λ is the condition used to the enhancement of the image (Yang & Cheng, 2014). Whether an element x in an image belongs to the structure or not is determined by checking if Hessian eigenvalues satisfy the condition φ Λ, if not the response of the filter is zero.

Laplace filter

Edge detection techniques are also one of the critical preprocessing techniques used to locate candidates in the CT image. Laplacian filter is a popular edge detection technique in image processing. The Laplacian of an image $L(x, y)$ is given in Eq. (7.5). Application of this filter on the input image highlights the image (Islam & Mondal, 2019).

$$L(x, y) = \frac{d^2 I}{dx^2} + \frac{d^2 I}{dy^2} \tag{7.5}$$

Median filter

This filtering technique is also known as the order-statistic filter. It is a nonlinear digital filter used for impulse noise removal from the images. The pixel value of the image is replaced by the median of the neighborhood pixels in this filtering technique. The size of the window is fixed for computing the neighborhood pixels. However, the filters tend to change both noisy and nonnoisy pixels in the image which may lead to blurring of the input image. There is a chance of nonnoisy pixels getting replaced causing the image to get more corrupt. But, still, a median filter is considered as one of the popular image denoising techniques for its improved mathematical accuracy (Ali, 2018). The median filter is given by Eq. (7.6):

$$m(x, y) = \text{median}(g\{s, t\}) \tag{7.6}$$

where (s, t) denotes the window from the image g in the coordinates (x, y).

Sobel operator

Sobel operator is also one of the popular edge detection techniques. This operator is used to find the edges in the x and y axes of the image using two operators, namely horizontal direction detector and vertical direction director as shown in Fig. 7.2. The operator computes the gradients along the x and y axes and then combines the output of these two operators.

X-direction Kernel

-1	0	1
-2	0	2
-1	0	1

Y-direction Kernel

-1	-2	-1
0	0	0
1	2	1

FIG. 7.2 Kernels of Sobel filter. *No permission required.*

Feature extraction

In this work, various feature extraction techniques are applied to preprocessed images obtained from the above-mentioned preprocessing filters. The features used in this work are LBP, HOG, SIFT, morphological gradient, and statistical features. The description of these features is provided below.

LBP

LBP features are the most commonly used image processing features for applications such as facial recognition, and texture description. With the use of local neighborhoods on the gray scale, the images' invariant texture is measured using LBP features. One of the reasons that LBP features are successful is the easier computation of features. Labels replace the pixel values in an image by performing binarization of a 3×3 neighborhood surrounding each pixel with the threshold value attained from the center pixel. This is the working of an LBP operator. Later, these pixel labels are converted into decimal numbers (Camlica et al., 2015). The acceptance or rejection of a pixel is completely dependent on the threshold value. If the value of the pixel is more than the threshold value, then the pixel is accepted. Else, the value of the pixel is rejected. LBP features are extracted by building histograms from the input images.

Advantages:

- Robust to local illumination changes in the image
- Computationally faster and simple and takes less time for training

Disadvantages:

- Results in high false-positive rate
- Accuracy is less

HOG

HOG is also another popular feature extraction method. HOG extracts correct information of the images using the attributes obtained from the local shape and the distribution in intensity gradients. The invariance property in the illumination of the images is normalized. The image is further subdivided into smaller cells for computing HOG features (Alhindi et al., 2018).

Advantages:

- Captures better object information in the image
- Easier and quicker to compute

Disadvantages:

- Information is captured globally, therefore resulting in loss of information

SIFT

SIFT is an image processing approach used to extract discriminative invariant features such as edges, and blobs. Extreme points in the entire scale space are detected from a SIFT descriptor. The key point detectors obtained from SIFT must be repeatable. From these key point detectors, around each key point, a neighborhood region is selected, and discriminative feature descriptors are computed from these selected regions. Computation of SIFT features is a four-step procedure. The first step is the scale-space extrema detection. This step is used to find the position of key points and the scales. Difference-of-Gaussian (DoG) functions are used to do this. The second step is the localization of key points. In this step, by removing the key points with low contrast points, the key points are localized and refined. The orientation assignment is the third step. Based on the gradient of a local image, the key point's orientation is obtained. The invariance property in the orientation is achieved by rotating the coordinates of the descriptor and the orientation of the gradients is relative to the orientation of the key points. The fourth step is the key point descriptor. By creating a histogram over a sample region of 4×4, a shift in gradient positions can be attained (Meng & Lv, 2016).

Advantages:

- Local information of the image is captured better
- SIFT features are both shift invariant and rotation invariant

Disadvantages:

- It is computationally complex and heavy
- Does not work well on low-powered devices
- Very sensitive to light changes and blurring of images

Morphological gradient

The morphological operations consist of two primary operations, namely erosion and dilation. These operations are based on the shape properties of the image. Erosion means removing the additional pixels in the image, whereas dilation means adding extra pixels to the image. The morphological gradient is the difference between the erosion and dilation operations in an image.

Advantages:

- Simpler implementation
- Provides low error rate
- Noise suppression is better

Disadvantages:

- Does not perform well if the edges are disconnected in the image

Statistical features

Statistical features considered in this work are mean, median, variance, standard deviation, range, skewness, and kurtosis. The mathematical formulas are as follows:

Mean

Mean is the arithmetic average of the data x where N is the total number of samples. The mean is shown in Eq. (7.7).

$$\text{mean}(x) = \frac{\sum x_i}{N} \tag{7.7}$$

Median

The middle number in the dataset when the values are arranged in a particular order. The median is shown in Eq. (7.8).

$$\text{median}(x) = \left(\frac{n+1}{2}\right) \tag{7.8}$$

Variance

The dispersion or how spread the data X is given by variance. The variance is shown in Eq. (7.9).

$$\text{Variance}(X) = \frac{\sum_{i=1}^{N}\left(X - \widetilde{X}\right)^2}{n} \tag{7.9}$$

Standard deviation

It is defined as the measure of spread or dispersion of the mean of the data X. The standard deviation is shown in Eq. (7.10).

$$\text{Std}(X) = \sqrt{\frac{\sum_{i=1}^{N}\left(X - \widetilde{X}\right)^2}{n}} \tag{7.10}$$

Range

Range is value obtained from the difference between the maximum and minimum value in the data.

Skewness

The asymmetry measure of the distribution of a real-valued random variable X and σ is the standard deviation. The skewness is shown in Eq. (7.11).

$$\text{Skew}(X) = \frac{1}{N}\sum_{i=1}^{N}\left[\frac{\left(X_i - \widetilde{X}\right)}{\sigma}\right]^3 \tag{7.11}$$

Kurtosis

The measure of flatness or peakness of a data X is given by kurtosis. The kurtosis is shown in Eq. (7.12).

$$\text{Kurtosis}(X) = \frac{1}{N}\sum_{i=1}^{N}\left[\frac{\left(X_i - \widetilde{X}\right)}{\sigma}\right]^4 \tag{7.12}$$

Advantages:

- Computationally simple and less time-consuming
- More accurate results

Disadvantages:

- Captures only global information of the image which results in loss of information in the local regions

Feature selection

The reduction of input variables by eliminating unnecessary information in the data is known as feature selection. In this work, multiple combinations of feature representations have been explored. A drawback that can be observed when multiple features are combined is that the feature's dimension increases. There can be redundant features in the feature matrix as well. The feature selection algorithms used in this work are described below.

Principal component analysis (PCA)

PCA dimension reduction algorithm is known to be one of the popular feature selection algorithms. Consider data D with dimensionality $b \times a$, where b denotes the number of rows and a denotes the number of features or columns in the data. After applying PCA on D, the dimensionality of data is reduced from $b \times a \geq b \times d$ where $d \ll m$. In the PCA technique, the data are projected into another subspace, and feature reduction is made while retaining the data's critical information. The first step performed in the PCA technique is calculating the means of the columns in the data. Followed by this step, the values of the means of each column are subtracted with the mean column value. Once the centered mean matrix is obtained, the covariance matrix is computed. From the covariance matrix, eigendecomposition is applied, which returns a list of eigenvectors and eigenvalues. Eigenvectors represent the direction of the reduced subspace of the data, and eigenvalues represent the directions' magnitude. The eigenvectors are sorted in descending order by the eigenvalues. If the eigenvalues are close to zero value, then they can be discarded, and the remaining values provide information. If not, then the data are already compressed enough, and further decomposition of data may lose information. The remaining eigenvalues and eigenvectors are considered as principal components: the projected data from the original data (Pearson, 1900).

Advantages:

- Returns smaller database representation
- Low complexity of the features
- Redundant features are discarded
- Low noise sensitivity is obtained
- Decrease in both time complexity and memory

Disadvantages:

- Slight invariance is also not captured if not trained explicitly
- Computation of covariance matrix is difficult to evaluate in an accurate manner

SelectKBest

SelectKBest method is also known as the univariate selection method. In this feature selection method, multiple statistical tests are applied to the features to check for the features' strongest relationship with the target variable. Different statistical tests available are ANOVA F-value, mutual information, and chi-square test.

Advantages:

- The feature selection method is scalable
- Faster computation and is independent of any classifier

Disadvantages:

- Feature dependencies are overseen
- No classifier interaction

Recursive feature elimination (RFE)

RFE algorithm is based on the recursive removal of attributes in the model and building the model with the remaining attributes. The evaluation metric used to identify the model's attributes is accuracy; i.e., it returns the attributes that provide better accuracy in the prediction.

Advantages:

- High-level features are selected from the input features
- More accurate features

Disadvantages:

- High computational complexity

Classification

In this work, various machine learning classifiers and neural-network-based classifiers are used to perform lung nodule detection and classification. The classifiers used in this work are given below.

Support vector machine (SVM)

A supervised learning-based classification used for binary and multiclass classifications is the SVM classifier. The data points in this algorithm are plotted in an n-dimensional space where "n" represents the data's number of characteristics. A specific coordinate value is associated with each characteristic. By changing the hyperplane, the separation of the two groups is carried out until it separates all the data points present in the data. The hyperplane is placed based on the support vectors. These support vectors are maximum marginal hyperplanes, i.e., the hyperplanes which separate the two classes with the nearest-neighboring data point. The optimal hyperplane is decided based on the support vectors, in which the optimal hyperplane is placed exactly in the center of the two support vectors. This provides a maximum margin for the hyperplane (Cortes & Vapnik, 1995). The technical steps involved in SVM model are given as follows:

Algorithm

Step 1: Prepare and format dataset
Step 2: Normalize the dataset
Step 3: Select suitable activation function
Step 4: Optimize parameters c and g using search algorithm after cross-validation
Step 5: Train SVM network
Step 6: Test SVM network
Step 7: Evaluate model performance

Random forest:

This supervised learning algorithm used for both classification and regression problems can be called an ensemble method. Random forest algorithm consists of multiple trees; the more the tree, the better the robustness. The data are provided as input to all the trees, based on the results obtained from all the trees using a voting mechanism; the final classification result is obtained. Random forest algorithm works better than a single-tree classification as the overfitting does not occur in this method (Ho, 1995). The steps involved in random forest algorithm are given as follows:

Algorithm

Step 1: Let 'K' be the number of trees to be generated
Step 2: At step 'k' ($1 < k < K$), a vector θ_k
Step 3: Construct tree using decision tree classifier
Step 4: Each tree casts a vote choosing the most popular class
Step 5: The class is predicted on the basis of maximum votes

K-nearest neighbor (KNN)

KNN algorithm is one of the pattern recognition algorithms. The K in the name represents the number of nearest neighbors from the data point. This algorithm uses a nonparametric estimation technique for both tasks. The algorithm's working is based on similarity measures computed using distance measures such as Euclidean, Manhattan, and Minkowski. The centers are chosen randomly. The distance between the centers and the data point is calculated. The distances of all the centers from the data points are sorted in ascending order. The top rows from the sorted distances array are chosen. The final predicted class is returned (Altman, 1992). The algorithm for KNN is as follows:

Algorithm

Input: Given 'X' is the training data, 'Y' is the class labels of X, and 'x' is the unknown sample.

Step 1: Given a query 'x' to be classified, let K = {x1, x2, …, xk} be the k instances from training that are near to x

Step 2: For all $i = 1, \ldots, l$, $vote_i = \{x \text{ E } X \mid class(x) = v_i\}$

Step 3: Return v_i such that $vote_i = largest$

XGBoost

XGBoost algorithm is a gradient boost algorithm combined with decision trees. The primary purpose of designing this machine learning algorithm is faster computation and improved performance (Chen & Guestrin, 2016). The algorithm working is similar to the working of the gradient boosting algorithm, in which the errors or residuals obtained from the previous models are corrected in the next model. Lastly, the error of the complete model is reduced. The XGBoost model handles even sparse data and supports parallelization as well.

Algorithm

Input: Dataset and hyperparameters

Step 1: Calculate first-order and second-order derivatives for all the attributes

Step 2: Determine the structure by choosing splits with maximized gain

Step 3: Determine the leaf weights

Step 4: Determine the base learner

Step 5: Add trees

Step 6: Combine the trees of all the attributes

Multilayer perceptron (MLP)

MLP algorithm can be said as an upgraded version of the perceptron. MLP is a supervised learning algorithm, also popularly known as artificial neural network (ANN) (Rumelhart et al., 1985). The difference between perceptron and MLP is that when the error is obtained in MLP, the weights are updated to learn misclassified data samples. The weight updation is shown in Eq. (7.13).

$$w = w + lr^{*}(\text{true\_label} - \text{pred\_label})^{*}x \tag{7.13}$$

where w is the weight of the neuron in the layer, lr is the learning rate, true label is the actual label of the data, and pred label is the predicted label from the MLP model, and x is the data.

MLP basically consists of three layers: the input layer, hidden layer, and output layer. Any number of hidden layers can be added to the model based on the task that is being performed. Each hidden layer consists of a nonlinear activation function that plays a major role in the network's weight updation. The steps involved in MLP classifier is stated as follows:

Algorithm

Step 1: Select the synaptic weights and the threshold value without any assumptions
Step 2: Compute the node by multiplying the weights and the input by moving layer by layer
Step 3: Determine the local gradients of the network
Step 4: Adjust the weights iteratively till the error is significantly reduced

Logistic regression

The learning of the logistic regression algorithm is in a supervised manner (Tolles & Meurer, 2016). This algorithm is used for the classification task, in which the probability of the output variable is predicted. The output or target variable is binary. The logistic regression model's output is either a success (1 value) or a failure (0 value). In terms of mathematics, a probability value $P(Y=1)$ is predicted for data X in the logistic regression model. The steps involved in logistic regression are as follows:

Algorithm

Step 1: Consider 'X' as the training samples and 'Y' as the class labels
Step 2: Multiply X with weights
Step 3: Calculate gradients and update weights
Step 4: Compute cost
Step 5: Evaluate the performance using metric

ExtraTrees

ExtraTrees is the abbreviation of extremely randomized trees. This classifier is one of the ensemble machine learning algorithms (Geurts et al., 2006). To be specific, the ExtraTrees algorithm is more related to the decision trees algorithm. The working of the ExtraTrees algorithm is mainly based on the unpruned decision trees constructed using the data. From the average predictions generated from the ExtraTrees algorithm for the testing data, a majority voting scheme is applied to obtain the predicted value.

Decision tree

The decision tree algorithm also belongs to the supervised learning-based machine learning algorithm family (Kamiński et al., 2018). The algorithm is rule-based, whose working is dependent on the decision rules applied to the training data samples. These rules, in turn, return a predicted value, which is the output or target variable. The algorithm works hierarchically, in which the label prediction is performed at the top level, i.e., at the root for the data attributes, and then is headed toward the branches of the tree. The algorithm for decision tree classifier is given as follows:

Algorithm

Step 1: All the training samples are placed at the root of the tree

Step 2: Attributes are usually categorical, and the training samples are partitioned recursively based on the attributes

Step 3: Test attributes are selected based on statistical or heuristic measure like information gain

Step 4: The recursive partitioning of the samples is stopped when all the samples in a given node belong to same class

Step 5: When there are no attributes left for partitioning, majority voting scheme is applied for the classification of the leaf

Gaussian Naive Bayes

The Naive Bayes algorithm is one of the predictive modeling algorithms, which surprisingly works well for classification problems. The algorithm can perform two-class and multiclass classification. In this algorithm, each attribute's values assume that the probability values are independent of each other conditionally when their target or output value is specified (Hand & Yu, 2007). But this assumption can seriously fail considering the real-time data. But, somehow, this algorithm works well for classification problems. For the Gaussian Naive Bayes algorithm, the Naive Bayes algorithm is extended for real values by assuming that the data are of Gaussian distribution. The steps involved in Gaussian Naive Bayes classifier are as follows:

Algorithm

Step 1: Calculate prior probabilities for given class labels

Step 2: Calculate conditional probability with each attribute for each class

Step 3: Multiply same class conditional probability

Step 4: Multiply prior probability with Step 3 probability

Step 5: Check for the highest probability. Higher probability belongs to the given input set

Linear discriminant analysis (LDA)

LDA algorithm is developed to overcome the issues of the logistic regression model. The logistic regression model works only for the two-class problems. For performing multiclass classification, LDA is used. In this algorithm, the data are assumed to be Gaussian, and for multiple attributes, multivariate Gaussian is applied to attain the mean and covariance matrix from the data. The estimated statistics are used in the LDA algorithm to get predicted outputs. The predictions are made using probability estimates achieved from Bayes' algorithm (Ramos & Liow, 2012). The steps involved for the LDA algorithm are listed as follows:

Algorithm

Input: Data matrix and its corresponding label vectors
Step 1: Compute mean vector for each class
Step 2: Compute total mean vector
Step 3: Compute within-class and between-class scatter matrices
Step 4: From Steps 2 and 3, compute eigen decomposition
Step 5: Projection matrix P is generated which consists of top eigenvectors to the largest eigenvalues

Extreme learning machine (ELM)

ELM is a type of feed-forward neural network that works better than other networks in terms of generalizability, and fast computation, hence the term "Extreme" (Huang et al., 2006). The key distinction between other neural network algorithms and ELM is that there is only one hidden layer in the network. The weight updation in the network is based on Moore-Penrose generalized inverse and not on gradient-based backpropagation. Instead of backpropagation, matrix inversion is performed in the ELM network. Also, due to a single hidden layer, the computation is not repeated and is performed only once. The steps involved in the ELM neural network are given as follows:

Algorithm

Step 1: Assign random input of biases and weights
Step 2: Calculate the hidden layer input
Step 3: Calculate the output weight
Step 4: The final weights are used to predict the unseen data

Result analysis and discussion

Nodule detection

Table 7.2 provides results obtained from all the preprocessing filters, features, and various classifiers. A total of 6 filters, 5 feature representations, and 11 classifiers are explored. The performance metric considered for the evaluation of the system is accuracy. The model best worked for statistical and morphological features. The neural networks performed well in the nodule classification task. However, the performance of the neural network classifier improves with larger datasets. Therefore, improvement of the neural network classifiers' performance can be done by providing more data samples or augmented data samples. The classifiers' names in the table are provided in abbreviations. The feature representations such as LBP and HOG did not perform that well in the nodule detection as compared to other feature representations. The reason for the low performance of LBP features is that the LBPs capture limited structural information. From the input image, only pixel differences are

TABLE 7.2 Accuracy values (in %) obtained from various preprocessing filters, features, and classifiers for lung nodule detection task.

Preprocessing	Feature	SVM	RF	KNN	XGB	MLP	LR	ET	DT	GNB	LDA	ELM
Gabor	LBP	50.2	50.7	50.61	50.8	49.68	50.63	50.66	50.67	50.48	50.48	50.56
	HOG	57.97	58.83	56.9	59.14	55.35	53.9	60.17	58.68	53.08	54.02	53.88
	SIFT	58.9	52.14	48.88	53.29	51.29	49.68	49.14	51.35	50.17	49.57	49.5
	Morphological	67.65	66.5	64.95	63.97	65.85	58.46	66.77	69.72	57.26	58.42	57.58
	Statistical	70.5	69.4	68.1	67.6	68.21	67.01	63.29	67.15	65.41	67.26	71.4
Gaussian	LBP	50.57	52.51	52.21	52.59	50.78	50.72	52.57	52.56	50.44	50.57	51.59
	HOG	58.46	63.07	60.8	62.54	62.2	53.52	63.06	56.16	51.27	53.54	52.86
	SIFT	60.34	59.94	55.31	57.67	56.03	54.75	59.91	55.64	53.84	54.9	52.96
	Morphological	66.09	68.69	61.16	63.06	60.88	55.47	65.13	62.26	55.07	55.63	55.03
	Statistical	59.55	70.98	57.32	62.91	67.61	67.71	68.28	65.67	64.74	67.71	71.34
Hessian	LBP	51.45	64.84	62.61	64.92	64.95	42.06	64.87	64.86	48.53	51.45	64.98
	HOG	56.98	55.3	55.41	55.59	55.58	53.28	54.61	53.68	51.23	53.16	53.22
	SIFT	49.72	48.87	50	48.97	50	48.63	48.97	48.97	49.14	49.15	48.97
	Morphological	57.61	58.16	57.24	58.26	57.56	55.6	58.6	57.75	54.26	55.25	54.95
	Statistical	64.77	68.65	60.79	67.61	59.84	61.93	69.5	59.65	57.19	61.42	61.74
Laplace	LBP	49.98	59.84	59.14	59.9	59.16	46.2	59.84	59.84	49.89	49.87	58.09
	HOG	54.98	53.54	52.24	53.15	52.38	52.5	52.93	51.48	51.52	52.24	51.78
	SIFT	58.03	59.41	52.57	55.58	53.18	50.81	59.5	54.64	50.24	51.37	50.59
	Morphological	59.77	64.95	57.59	61.54	51.41	52.51	67.7	62.1	51.4	52.61	52.2
	Statistical	60.79	69.59	63.63	68.18	52.08	67.42	70	58.52	52.65	66.28	63.82

MedianBlur	LBP	49.46	52.04	50.43	52.3	49.52	49.52	51.83	51.82	50.26	49.46	51.2
	HOG	55.87	58.3	57.6	57.76	57.69	54.75	57.95	52.93	53.18	54.68	54.15
	SIFT	61.06	59.22	56.96	58.97	57.55	54.84	59.24	56.91	54.44	55	53.61
	Morphological	67.09	65.74	61.35	63.14	60.65	55.33	67.59	62.92	54.98	55.4	54.98
	Statistical	56.7	68.69	52.03	69.91	52.84	65.24	70.93	60.16	60.77	62.19	68.99
Sobel	LBP	51.13	61.35	59.03	61.48	60.12	52.73	61.13	61.34	50	51.13	59.9
	HOG	59.76	58.51	56.58	57.33	57.7	54.33	58.05	53.13	53.46	54.68	53.92
	SIFT	60.91	60.39	54.55	57.64	55.89	54.58	60.64	55.57	53.76	54.58	53.83
	Morphological	65.52	70.22	62.28	67.71	60.1	54.48	65.15	70.67	54.69	54.71	54.47
	Statistical	63.82	69.75	60.48	70.12	62.89	65.12	69.38	59.92	64.56	64.19	65.12

considered, and the magnitude information is ignored. Whereas HOG features capture gradient information of the image, the features are extracted for the entire image as a whole. The HOG features do not capture the local information.

Table 7.3 provides different combinations of feature representations and filters. The best-performing combinations of feature representations are presented in the table. The combinations explored are HOG+ Morphological, HOG+Morphological+SIFT, HOG+SIFT, and SIFT+Morph. We could not consider few combinations due to an issue in dimensionality for concatenating the features. The best recognition accuracies are achieved from the combination of HOG + SIFT + Morphological features. This feature combination provided the discriminative information of all the features, resulting in the highest accuracy.

From the combinations, best-performing combinations are chosen, and feature selection techniques are applied. The feature selection techniques explored in this work are PCA, SelectKBest, and RFE. The results obtained for different feature combinations and feature selection methods are presented in Table 7.4.

Nodule classification

After lung nodules are detected from the CT scan, the nodules are further classified into cancerous and noncancerous nodules. The nodule's annotation to be either of two classes is decided based upon the average malignancy score provided by the expert radiologists in the LIDC-IDRI dataset annotation file. The malignancy score varies from 1 to 5, where 5 being the most probable cancer nodule and 1 being the least probable cancer nodule. Therefore, a threshold value is set to label a nodule as a cancerous and noncancerous nodule. If the nodule's average malignancy score is below the threshold, the nodule is considered a noncancerous nodule. If the average malignancy score is above the threshold, the nodule is considered a cancerous nodule. If average malignancy score is equal to the threshold, the nodule is discarded due to uncertainty.

The nodules obtained from the lung nodule detection system are given to a majority voting scheme, where the nodule is assigned to its particular class. The results obtained from the majority voting scheme from all the classifiers are presented in Table 7.5. The best feature combinations with the PCA feature selection technique from the nodule detection system are considered to classify lung cancer nodules in the table.

Confusion matrices for nodule detection and classification

The most common approach to depict the performance of a system is the confusion matrix. The confusion matrices resulted from the best methods for both the tasks are shown in Fig. 7.3. The highest recognition accuracy obtained for the nodule detection task is 94.31% from the combination of Gabor, HOG, Morphological, and SIFT feature representations and ELM classifier. The highest recognition accuracy obtained for the nodule classification task is 93.83% from the majority voting scheme.

TABLE 7.3 Accuracy values (in %) obtained for various feature combinations chosen from the best performing features for lung nodule detection task.

Preprocessing + feature or features	SVM	RF	KNN	XGB	MLP	LR	ET	DT	GNB	LDA	ELM
Gabor + HOG + Morph	74.02	76.24	75.89	68.27	76.09	73.89	71.21	74.8	73.42	69.31	78.02
Hessian + HOG + Morph	66.71	68.29	65.41	68.77	71.88	75.93	73.22	67.81	66.85	71.33	75.88
Laplace + HOG + Morph	67.22	67.18	70.3	68.13	72.18	65.13	71.11	72.39	69.99	68.14	75.93
Sobel + HOG + Morph	66.92	63.34	68.73	68.56	70.98	62.93	70.06	70.03	66.81	67.73	73.82
MedianBlur + HOG + Morph	66.71	66.94	69.71	65.18	69.43	65.04	69.01	70.94	69.01	67.03	72.93
Gabor + HOG + Morph + SIFT	76.3	73.3	72.73	74.51	79.88	76.41	78.82	80.31	77.78	81.32	84.37
Hessian + HOG + Morph + SIFT	74.26	71.83	71.54	74.07	78.71	75.39	79.9	79.74	78.5	80	83.88
Laplace + HOG + Morph + SIFT	67.81	69.96	69.48	71.73	80.97	65.72	66.81	79.74	75.91	79.51	80.59
Sobel + HOG + Morph + SIFT	74.11	70.67	71.94	72.89	76.2	75.23	78.82	78.48	76.02	80.56	83.88
MedianBlur + HOG + Morph + SIFT	72.88	74.21	70.66	77.2	80.01	75.32	77.9	81.7	76.45	82.54	84.09
Gabor + HOG + SIFT	66.56	62.44	69.08	68.13	79.29	65.22	68.51	74.28	68.32	70.01	75.29
Hessian + HOG + SIFT	64.92	59.32	62.4	65.66	70.26	61.47	62.76	70.93	62.54	69.25	73.71
Laplace + HOG + SIFT	63.02	66.27	71.63	73.28	68.01	69.51	70.25	76.44	70.31	74.87	78.64
Sobel + HOG + SIFT	66.2	70.31	69.48	68.36	70.36	71.22	73.76	74.49	72.51	75.29	76.94
MedianBlur + HOG + SIFT	68.41	71.2	72.5	70.61	72.51	70.6	77.31	75.21	77.41	79.41	78.75
Gabor + SIFT + Morph	77.91	71.2	75.93	80.72	82.94	74.02	78.03	79.4	80.51	81.63	86.82
Hessian + SIFT + Morph	78.42	73.94	78.5	79.64	81.83	78.7	79.93	81.04	79.99	83.29	85.86
Laplace + SIFT + Morph	80.77	76.87	78.61	82.74	83.1	77.94	80.89	83.21	82.59	82.8	89.31
Sobel + SIFT + Morph	79.41	77.02	75.41	79.84	79.99	78.72	78.69	81.54	80.64	81.63	85.21
MedianBlur + SIFT + Morph	77.91	74.73	73.9	78.43	75.83	76.31	77.42	79.96	78.99	80.16	83.91

TABLE 7.4 Accuracy values (in %) obtained from various feature combinations along with different feature selection techniques for lung nodule detection task.

Preprocessing + feature or features	Feature selection	SVM	RF	KNN	XGB	MLP	LR	ET	DT	GNB	LDA	ELM
Gabor + HOG + Morph + SIFT	PCA	80.51	76.51	76.31	77.21	82.21	79.31	80.31	82.7	80.04	88.96	94.31
Hessian + HOG + Morph + SIFT		77.31	73.39	72.4	78.51	79.21	77.4	81.21	80.9	80.89	83.72	85.21
Laplace + HOG + Morph + SIFT		70.46	72.88	75.6	75.21	84.7	69.48	72.31	81.35	77.94	81.56	85.77
Gabor + SIFT + Morph		80	73.8	78.3	85.01	84.32	77.41	81.96	81.21	83.33	84.87	89.31
Laplace + SIFT + Morph		84.37	79.37	83.46	85.29	87.42	80.65	81.36	85.48	84.33	83.31	91.51
Gabor + HOG + Morph + SIFT	SelectKBest	76.3	73.3	72.73	74.51	79.88	76.41	78.82	80.31	77.78	81.32	84.37
Hessian + HOG + Morph + SIFT		76.38	73.83	72.59	76.21	77.41	74.31	78.57	78.66	80.04	81.89	82.75
Laplace + HOG + Morph + SIFT		69.92	71.41	74.8	75.03	82.84	68.49	70.42	81	75.2	79.95	84.48
Gabor + SIFT + Morph		79.05	71.39	75.27	82.36	81.95	75.32	79.88	80.11	81.21	81.58	88.38
Laplace + SIFT + Morph		82.38	80.35	82.9	84.32	85.7	78.81	80.22	82.85	82.54	84.32	92.11
Gabor + HOG + Morph + SIFT	RFE	76.3	73.3	72.73	74.51	79.88	76.41	78.82	80.31	77.78	81.32	84.37
Hessian + HOG + Morph + SIFT		75.26	72.8	71.8	77.29	77.36	76.83	79.8	79.51	81.74	82.63	86.22
Laplace + HOG + Morph + SIFT		70.01	70.24	74.9	73.88	82.24	68.22	70.32	80.07	75.32	80.99	86.83
Gabor + SIFT + Morph		78.31	72.37	76.32	84.22	82.89	76.32	78.79	80.01	81.24	83.92	87.44
Laplace + SIFT + Morph		83.39	80.48	82.43	83.7	85.31	80.01	70.97	84.38	83.59	81.45	93.28

TABLE 7.5 Nodule classification results obtained from majority voting scheme.

Feature combinations (with PCA)	Accuracy (in %)
Gabor + HOG + Morph + SIFT	93.83
Hessian + HOG + Morph + SIFT	89.43
Laplace + HOG + Morph + SIFT	91.06
Gabor + SIFT + Morph	90.21
Laplace + SIFT + Morph	88.21

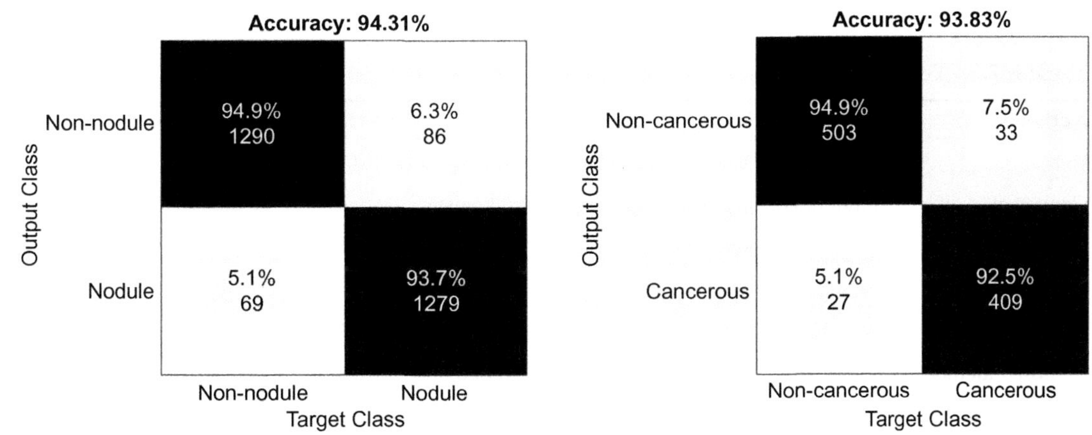

FIG. 7.3 Confusion matrices obtained for lung nodule detection and classification tasks using proposed methodology for highest recognition accuracies. *No permission required.*

Performance of the nodule detection and classification task using various evaluation metrics

Multiple performance metrics such as precision, specificity, sensitivity, accuracy, false-positive rate (FPR), and F-score are used to evaluate the proposed method. These metrics are computed from the confusion matrices. The results obtained for the best methods in detecting and classifying lung nodules are presented in Table 7.6.

Performance comparison of the proposed method with other CAD systems

Research on lung cancer CAD systems is being performed for decades. Table 7.7 provides a performance comparison of previously proposed CAD systems for LUNA16 and LIDC-IDRI datasets. The methods chosen for comparison include both machine learning classifiers and recent deep learning architectures. The proposed method performed well for conventional machine learning classifiers and also neural networks. D. Kumar et al. (2015) performed a nodule detection task using deep features and fed it to an autoencoder. The method resulted

TABLE 7.6 Performance comparison (in %) of the proposed method with previously proposed CAD systems.

Performance metrics	Nodule detection	Nodule classification
Accuracy	94.31	93.83
Specificity	94.88	93.80
Sensitivity	93.75	93.84
Precision	94.92	94.90
FPR	0.0512	0.0619
F-score	0.9433	0.9437

TABLE 7.7 Performance comparison of the proposed method with previously proposed CAD systems.

Authors	Methods used	Accuracy	Sensitivity	Specificity
Kumar et al.	Deep features from autoencoder	75.01%	–	83.35%
Gong et al.	Random Forest, J48 Decision Tree, SVM, Logistic	–	84.62%	–
Anum et al.	Deep Fully Connected Network (DFCNet)	84.58%	–	–
Hua et al.	Deep Belief Network, CNN	–	73.4%	82.2%
Liao et al.	Deep Neural Network	81.42%	–	–
Proposed method	Gabor, HOG, Morphological, SIFT, PCA, ELM	93.83%	93.84%	93.80%

in a good performance. However, the accuracy obtained from their method is relatively low and not reliable for critical applications such as CAD systems. Various machine learning classifiers were used by Gong et al. (2018) for performing lung nodule detection tasks. The method achieved good specificity but resulted in less sensitivity. In medical applications, sensitivity is one of the important evaluation metrics used to measure the CAD system's reliability and credibility. Therefore, a low sensitivity value is not acceptable. Deep neural networks such as fully connected networks, convolution neural network, and deep belief network are used for lung nodule detection. Surprisingly, the results achieved from the machine learning algorithms are better in this case as compared to the deep architectures. However, the scenario may be different when larger datasets are fed to machine learning classifiers. The deep learning models work better with larger datasets.

Conclusion

In this work, lung cancer nodule detection and classification are performed. An in-depth analysis is provided for different machine learning algorithms. Various preprocessing techniques are explored for improving visibility and removing noise from the CT images.

Different feature extraction and selection methods are explored. The selected prominent features are tested with multiple classifiers. The highest performance was obtained from the ELM classifier for Gabor, HOG, Morphological, and SIFT features. The proposed system provided a good performance of 94.31% accuracy for the detection tasks. The detected nodules are further classified into cancerous and noncancerous nodules by a majority voting schemes adapted from all the classifiers. The accuracy achieved for the classification task is 93.83%. The performance of neural networks was found to be better than the traditional machine learning algorithms. In future work, deep learning models can be used for developing lung cancer CAD systems.

References

Aggarwal, T., Furqan, A., & Kalra, K. (2015). Feature extraction and LDA based classification of lung nodules in chest CT scan images. In *2015 international conference on advances in computing, communications and informatics, ICACCI 2015* (pp. 1189–1193). Institute of Electrical and Electronics Engineers Inc. https://doi.org/10.1109/ICACCI.2015.7275773.

Alhindi, T. J., Kalra, S., Ng, K. H., Afrin, A., & Tizhoosh, H. R. (2018). Comparing LBP, HOG and deep features for classification of histopathology images. *2018 International joint conference on neural networks (IJCNN)* (pp. 1–7), https://doi.org/10.1109/IJCNN.2018.8489329.

Ali, H. M. (2018). MRI medical image denoising by fundamental filters. *High-Resolution Neuroimaging—Basic Physical Principles and Clinical Applications*, 14, 111–124.

Altman, N. S. (1992). An introduction to kernel and nearest-neighbor nonparametric regression. *American Statistician, 46*(3), 175–185. https://doi.org/10.1080/00031305.1992.10475879.

Antony, A. (2012). Categorizing the stages of lung cancer using multi SVM classifier. *International Journal of Research in Pharmaceutical Sciences, 10*, 2323–2328.

Baldwin, D. R. (2015). Prediction of risk of lung cancer in populations and in pulmonary nodules: Significant progress to drive changes in paradigms. *Lung Cancer, 89*(1), 1–3. https://doi.org/10.1016/j.lungcan.2015.05.004.

Bhalerao, R. Y., Jani, H. P., Gaitonde, R. K., & Raut, V. (2019). A novel approach for detection of lung cancer using digital image processing and convolution neural networks. In *2019 5th international conference on advanced computing and communication systems, ICACCS 2019* (pp. 577–583). Institute of Electrical and Electronics Engineers Inc. https://doi.org/10.1109/ICACCS.2019.8728348.

Camlica, Z., Tizhoosh, H. R., & Khalvati, F. (2015, December). Medical image classification via SVM using LBP features from saliency-based folded data. In *2015 IEEE 14th international conference on machine learning and applications (ICMLA)* (pp. 128–132). IEEE.

Chen, T., & Guestrin, C. (2016). XGBoost: A scalable tree boosting system. In (Vols. 13–17). *Proceedings of the ACM SIGKDD international conference on knowledge discovery and data mining* (pp. 785–794). Association for Computing Machinery. https://doi.org/10.1145/2939672.2939785.

Cortes, C., & Vapnik, V. (1995). Support-vector networks. *Machine Learning, 20*(3), 273–297. https://doi.org/10.1023/A:1022627411411.

DeSantis, C. E., Miller, K. D., Goding Sauer, A., Jemal, A., & Siegel, R. L. (2019). Cancer statistics for African Americans, 2019. *CA: A Cancer Journal for Clinicians, 69*(3), 211–233. https://doi.org/10.3322/caac.21555.

Geurts, P., Ernst, D., & Wehenkel, L. (2006). Extremely randomized trees. *Machine Learning, 63*(1), 3–42. https://doi.org/10.1007/s10994-006-6226-1.

Gindi, A., & Sami, A. T. (2014). A comparative study for comparing two feature extraction methods and two classifiers in classification of early stage lung cancer diagnosis of chest x-ray images. *Journal of American Science, 10*, 13–22.

Gong, J., Liu, J. Y., Wang, L. J., Sun, X. W., Zheng, B., & Nie, S. D. (2018). Automatic detection of pulmonary nodules in CT images by incorporating 3D tensor filtering with local image feature analysis. *Physica Medica, 46*, 124–133. https://doi.org/10.1016/j.ejmp.2018.01.019.

Gopi, K., & Selvakumar, J. (2017). Lung tumor area recognition and classification using EK-mean clustering and SVM. In *2017 international conference on Nextgen electronic technologies: Silicon to software, ICNETS2 2017* (pp. 97–100). Institute of Electrical and Electronics Engineers Inc. https://doi.org/10.1109/ICNETS2.2017.8067906.

Hand, D., & Yu, K. (2007). Idiot's Bayes: Not so stupid after all? *International Statistical Review, 69*, 385–398.

Ho, T. K. (1995, August). Random decision forests. In *Proceedings of 3rd international conference on document analysis and recognition* (Vol. 1, pp. 278–282). IEEE.

Huang, G. B., Zhu, Q. Y., & Siew, C. K. (2006). Extreme learning machine: Theory and applications. *Neurocomputing, 70* (1–3), 489–501. https://doi.org/10.1016/j.neucom.2005.12.126.

Islam, S. M., & Mondal, H. S. (2019). Image enhancement based medical image analysis. In *2019 10th international conference on computing, communication and networking technologies, ICCCNT 2019*Institute of Electrical and Electronics Engineers Inc. https://doi.org/10.1109/ICCCNT45670.2019.8944910.

Kamiński, B., Jakubczyk, M., & Szufel, P. (2018). A framework for sensitivity analysis of decision trees. *Central European Journal of Operations Research, 26*(1), 135–159. https://doi.org/10.1007/s10100-017-0479-6.

Kumar, N., & Nachamai, M. (2017). Noise removal and filtering techniques used in medical images. *Oriental Journal of Computer Science and Technology*, 103–113. https://doi.org/10.13005/ojcst/10.01.14.

Kumar, D., Wong, A., & Clausi, D. A. (2015). Lung nodule classification using deep features in CT images. In *Proceedings—2015 12th conference on computer and robot vision, CRV 2015* (pp. 133–138). Institute of Electrical and Electronics Engineers Inc. https://doi.org/10.1109/CRV.2015.25.

Kuruvilla, J., & Gunavathi, K. (2014). Lung cancer classification using neural networks for CT images. *Computer Methods and Programs in Biomedicine, 113*(1), 202–209.

Liu, Y., Wang, H., Gu, Y., & Lv, X. (2019). Image classification toward lung cancer recognition by learning deep quality model. *Journal of Visual Communication and Image Representation, 63*. https://doi.org/10.1016/j.jvcir.2019.06.012.

Manikandan, T., & Bharathi, N. (2011). Lung cancer diagnosis from CT images using fuzzy inference system. In *Vol. 250. Communications in computer and information science* (pp. 642–647). https://doi.org/10.1007/978-3-642-25734-6_110.

Meng, Q., & Lv, Z. (2016). An image registration method based on improved SIFT algorithm. In *Proceedings—2015 7th international conference on advanced communication and networking, ACN 2015* (pp. 7–10). Institute of Electrical and Electronics Engineers Inc. https://doi.org/10.1109/ACN.2015.18.

Pearson, K. (1900). On lines and planes of closest fit to points in space. *Philosophical Magazine, 2*, 559–572.

Potghan, S., Rajamenakshi, R., & Bhise, A. (2018). Multi-layer perceptron based lung tumor classification. In *Proceedings of the 2nd international conference on electronics, communication and aerospace technology, ICECA 2018* (pp. 499–502). Institute of Electrical and Electronics Engineers Inc. https://doi.org/10.1109/ICECA.2018.8474864.

Ramos, S. D. S., & Liow, S. R. (2012). Discriminant function analysis. *The Encyclopedia of Applied Linguistics, 1*.

Rumelhart, D. E., Hinton, G. E., & Williams, R. J. (1985). *Learning internal representations by error propagation*. California Univ. San Diego La Jolla Inst. for Cognitive Science.

Setio, A., Traverso, A., Bel, T., Berens, M., Bogaard, C., Cerello, P., Chen, H., Dou, Q., Fantacci, M., Geurts, B., Gugten, R., Heng, P., Jansen, B., Kaste, M., Kotov, V., Lin, J., Manders, J., Mengana, A., Naranjo, J. C., & Papavasileiou, E. (2016). Validation, comparison, and combination of algorithms for automatic detection of pulmonary nodules in computed tomography images: The luna16 challenge. *Medical Image Analysis, 42*.

Suzuki, K. (2012). A review of computer-aided diagnosis in thoracic and colonic imaging. *Quantitative Imaging in Medicine and Surgery, 2*, 163–176.

Tolles, J., & Meurer, W. J. (2016). Logistic regression: Relating patient characteristics to outcomes. *JAMA—Journal of the American Medical Association, 316*(5), 533–534. https://doi.org/10.1001/jama.2016.7653.

Yang, S. F., & Cheng, C. H. (2014). Fast computation of Hessian-based enhancement filters for medical images. *Computer Methods and Programs in Biomedicine, 116*(3), 215–225. https://doi.org/10.1016/j.cmpb.2014.05.002.

Zainudin, M. N. S., Mohd Said, M., & Ismail, M. M. (2011). Feature extraction on medical image using 2D Gabor filter. *Applied Mechanics and Materials, 52–54*, 2128–2132. https://doi.org/10.4028/www.scientific.net/AMM.52-54.2128.

CHAPTER

8

Deep learning techniques for oral cancer diagnosis

Ijaz Ul Haq[a], *Fath U Min Ullah*[b], *Khan Muhammad*[c], *and Sung Wook Baik*[d]

[a]Department of Computer Science & Engineering, School of Engineering and Computer Science, Oakland University, Rochester, MI, United States [b]Department of Electronic and Electrical Engineering, University of Sheffield, Sheffield, South Yorkshire, United Kingdom [c]Department of Applied Artificial Intelligence, Sungkyunkwan University, Seoul, South Korea [d]Sejong University, Seoul, South Korea

Introduction

Cancer is one of the most dangerous and life-threatening diseases if not diagnosed and treated in its early stages. Oral cancer, which may involve any part of the oral cavity, is a subcategory of a larger group of cancers known as head and neck cancers. According to the World Health Organization (WHO), carcinoma-based oral cancer is the sixth most common type of cancer in males and tenth in females in developing countries and is mostly caused by smoking, smokeless tobacco, excessive use of alcohol, excessive sun exposure, and genetic predisposition. Oral cancer may arise as a primary lesion originating in the oral cavity tissues; it is commonly associated with regional metastasis and appears as a growth or sore in the mouth. Oral cancer can be further categorized into six groups based on the organs of the mouth, i.e., cancer of the hard palate cancer, inner cheek, tongue, lip, gums, or floor of the mouth. Fig. 8.1 shows these categories in more detail.

There are several methodologies for the diagnosis of cancer, of which the pathological diagnosis of resected specimens is the gold standard. However, in most cases, the primary step involves biomedical image analysis, which requires image acquisition with special capturing technologies and an analysis of these images by medical specialists to avoid the resection of specimens. There are six imaging technologies that are available to clinicians for the diagnosis and treatment of human cancers: X-ray (plain film and computed tomography (CT)),

https://doi.org/10.1016/B978-0-323-85240-1.00015-8

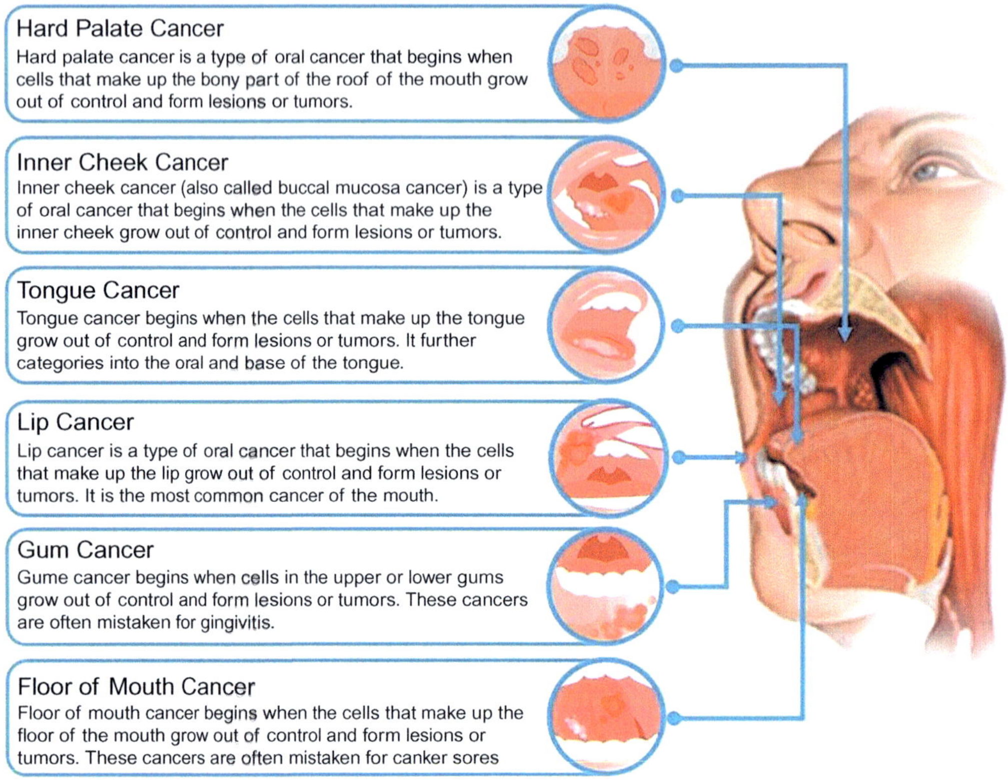

FIG. 8.1 Categorization of oral cancer based on regional appearance in the oral cavity.

ultrasound (US), magnetic resonance imaging (MRI), single-photon emission CT (SPECT), positron emission tomography (PET), and optical imaging. Four of these (CT, MRI, SPECT, and PET) are capable of 3D detection of cancer anywhere in the human body (Frangioni, 2008). However, the manual analysis of biomedical images faces four major challenges: (i) manual analysis is limited by individual experience, meaning that its effectiveness differs between specialists; (ii) training a qualified specialist involves a great deal of expense and years of effort; (iii) the time-consuming and tedious nature of this work puts enormous pressure on specialists; and (iv) it is difficult for the human eye to distinguish between subtle changes in tissues, especially if the specialists are fatigued, which may lead to misdiagnosis and hence pose a great risk to patients. To a certain extent, these challenges constitute a barrier between patients and resources, especially in underdeveloped countries. Researchers in the domain of computer vision are, therefore, actively working on the computer-aided analysis of biomedical images to enhance this process.

Deep learning is a recent development that takes advantage of current computing power and the availability of large-scale labeled data such as the ImageNet (Deng et al., 2009) and COCO (Lin et al., 2014) datasets and has been deployed in a diverse range of fields, including science, business, and government. In particular, convolutional neural network (CNN)

(Lecun et al., 2015) models have achieved human levels of accuracy, unlike traditional shallow methods of image classification, object detection and semantic segmentation. Similarly, the rapid increase in the use of biomedical imaging technologies in clinical practice has resulted in the development of big medical datasets (Siuly & Zhang, 2016). The literature now contains many deep learning-based techniques that have been applied to the computer-aided analysis of biomedical images, such as the detection of diabetic retinopathy (Gulshan et al., 2016), the diagnosis of skin cancer (Esteva et al., 2017), the detection of pulmonary disease based on X-ray radiography (Rajpurkar et al., 2018), the prediction of heart disease risk from fundus retinal images (Poplin et al., 2018), and the classification of brain tumors (Sajjad et al., 2019). Although biomedical images are captured in large numbers, the requirements for confidentiality of personal information and the lack of perfectly annotated data by specialists mean that small sample learning (SSL) issues arise for a particular category of disease diagnosis. This chapter, therefore, discusses the technical details of recently proposed deep learning-based techniques, particularly for the diagnosis of oral cancer; the results that have been obtained; and the challenges that are currently faced in this field. Finally, we suggest some future research directions. The overall structure of the chapter is shown in Fig. 8.2.

Deep neural networks (DNNs)

Deep neural networks (DNNs) are machine learning algorithms that were inspired by the biology of the brain (using artificial neurons). A computer program that could find itself representing a model from the set of defined data is not a new concept. The perceptron was one of the earliest approaches to conceptualize data directly from the environment (Rossenblatt, 1957). The principle of operation of DNNs involves simulating the transfer of

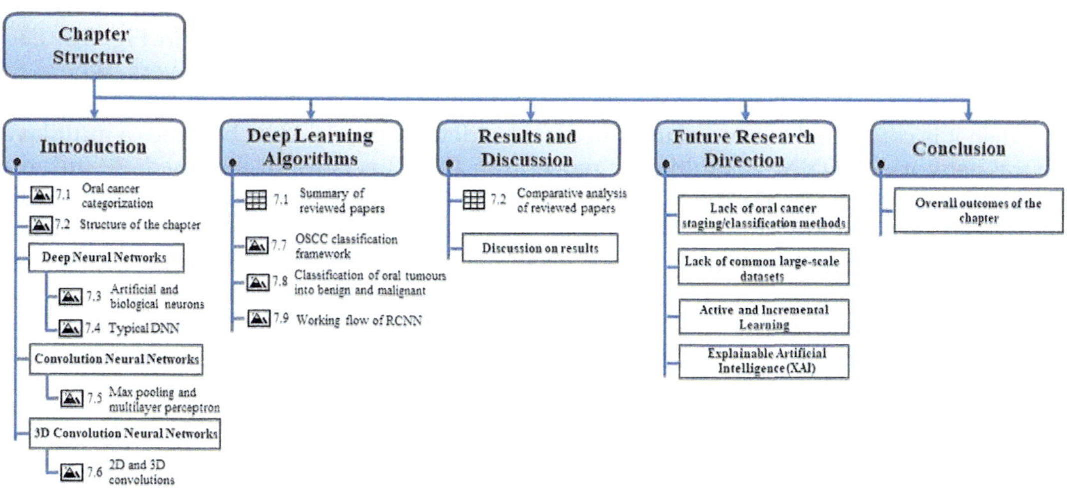

FIG. 8.2 The overall structure of the chapter.

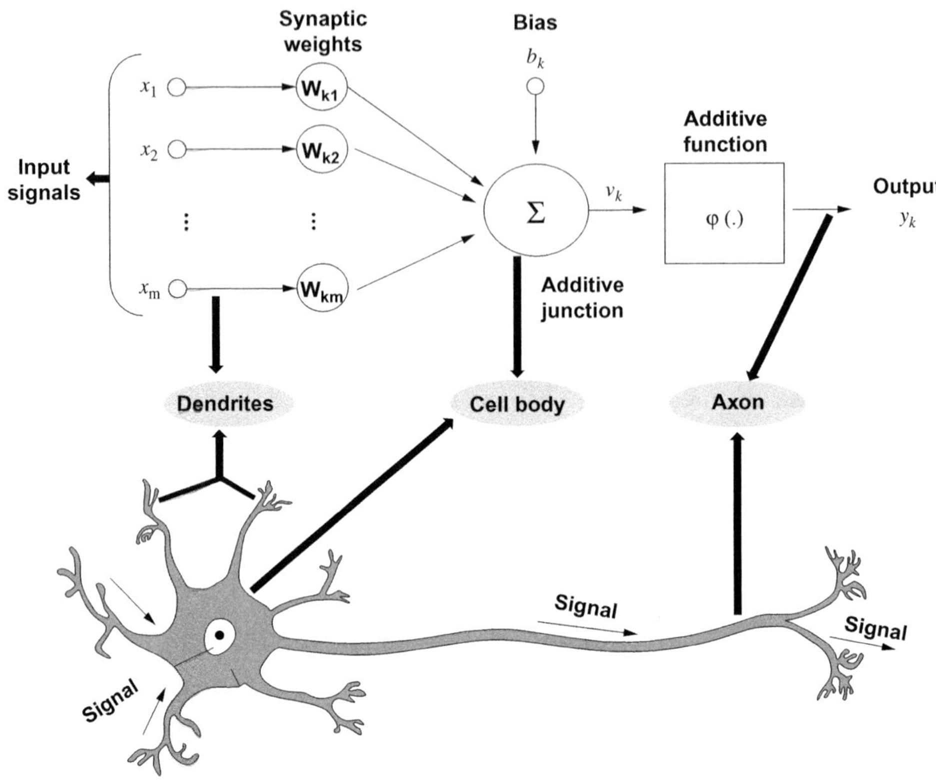

FIG. 8.3 Similarity between an artificial and a biological neuron.

information through a neuron, where a weighted node (synapses) receives the inputs, sums them to produce an activation (axon), and passes this activation to a nonlinear function called an activation or transfer function to generate the output signal, as shown in Fig. 8.3. Each neuron acts as an elementary processing unit in the form of organized layers, where the output signal of one unit is fed into the other units forming an artificial neural network (ANN).

When it is composed of several intermediate hidden layers, an ANN becomes a multilayer neural network or DNN. A multilayer perceptron (MLP) acts as an intermediate link that transforms a simple perceptron into a DNN. In this way, DNNs conceptually expand the MLP by adding a significant number of layers. There are three basic layers in an MLP, which are the input, hidden, and output layers, as shown in Fig. 8.4. The first is the input layer, which receives the dataset, and the last is the output layer, which delivers the result. The layers in between that transform the signal are called hidden layers. DNNs have multiple hidden layers consisting of nonlinear processing units for feature extraction and transformation, where each successive layer receives output from the previous layer as input. Different DNN architectures use different numbers, sizes, and compositions of these successive layers to extract a variety of high-level features. However, the most challenging aspect is to find an appropriate level of granularity, given the size and dimension of the available training dataset and the complexity of the task.

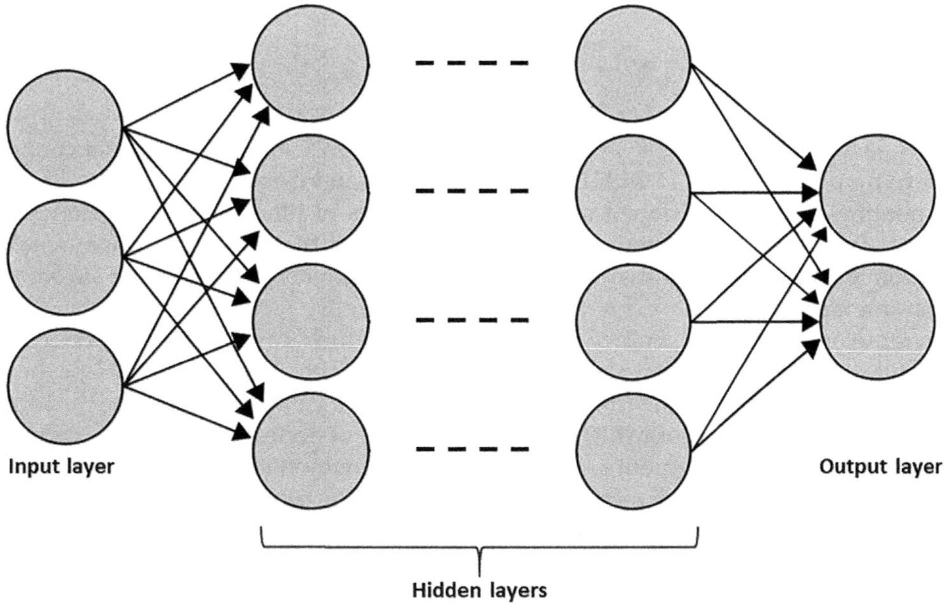

FIG. 8.4 Graphical representation of a typical deep neural network.

Convolution neural networks (CNNs)

Convolutional neural networks (CNNs) are the most commonly used DL-based networks for image analysis. The popularity of CNNs increased after they achieved stunning results in the ImageNet competition (Krizhevsky et al., 2017). A typical CNN consists of several successive data processing layers that extract representative features from an input image. The output from each neuron in a given layer of a fully connected neural network is passed to the input node of each neuron in the succeeding layer, while neurons in a given layer are completely independent. Medical images are typically spatially structured, meaning that neighboring pixels share similar intensity characteristics; we can, therefore, specify a parameterized patch of neurons to detect pixels from the same structure. There are also variations in the sample data, such as differences in the location of the structure between patients and in the image acquisition conditions, which create considerable differences between corresponding images. The patch of neurons mentioned above needs to have a certain level of invariance to the spatial location in order to act in the same way for the entire image. CNNs are, therefore, specially designed to take advantage of this spatially structured information and have advantages over low-level or hand-crafted features for medical images (Antipov et al., 2015). These networks are composed of three basic layers called the convolution, pooling, and fully connected layers, which carry out different tasks. The main task of a convolution layer is to detect local conjunctions in the features from the previous layer and to map their appearance to a feature map. The number of feature maps produced by a single layer is equivalent to the number of filters in that layer. The process of convolution over a given image can be represented as in Eq. (8.1):

$$Y_i^{(l)} = B_i^{(l)} + \sum_{j=1}^{m_1(l-1)} K_{i,j}^{(1)} \times Y_j^{(l-1)} \tag{8.1}$$

where $B_i^{(l)}$ is a bias matrix and $K_{i,j}^{(l)}$ is a the filter of size $2h_i^{(l)} + 1 \times 2h_2^{(l)} + 1$ that connects the jth feature map in layer $(l-1)$ with the ith feature map in layer l.

The pooling layer is positioned after multiple stages of other layers (i.e., convolutional layers) and is responsible for reducing the spatial size of the activation maps, resulting in a reduction in the computational complexity of the network. The pooling layer has two hyperparameters: the size and the stride. The most common settings for these hyperparameters are the size of 2×2 and a stride of 2, which down-samples each depth slice in the input by a factor of 2 along both the width and height and discards 75% of the activations, as shown in Fig. 8.5A. The fully connected layers in a CNN basically take the form of an MLP (i.e., a two- or three-layer MLP) in which each output node of each layer is connected to each input node of the subsequent layer and takes an activation volume instead of a vector, as shown in Fig. 8.5B. This layer is responsible for mapping the activation volume from previous different layers into a class probability distribution, by tuning the weight parameters to create a stochastic probability representation of each class based on the activation map.

3D convolution neural networks (3D CNNs)

One-dimensional (1D) CNNs can be used to extract spectral features from the input data, while 2D CNNs can extract spatial features. However, 3D CNNs have the advantages of both 1D and 2D CNNs and can extract both the spectral and spatial features simultaneously from the input volume. The mathematical formulation and operation of 3D CNNs are similar to those of 2D CNNs, except for the convolution layer, which has an extra dimension. Unlike 2D convolution, the 3D convolution convolves in three dimensions, as shown in Fig. 8.6.

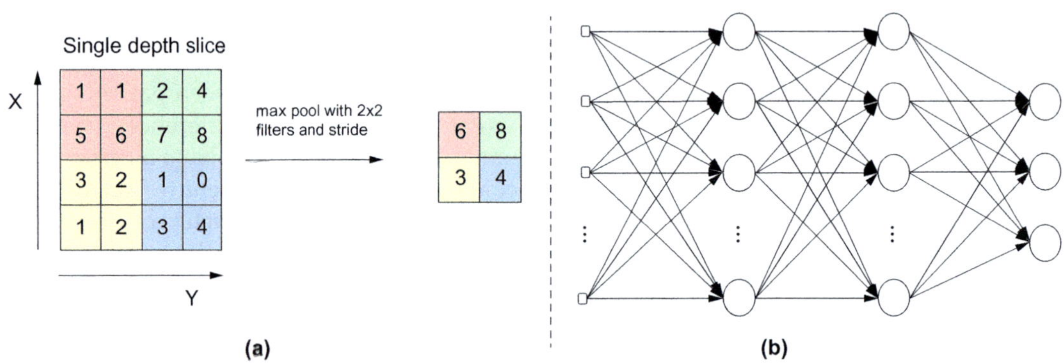

FIG. 8.5 Illustration of the pooling layer and MLP: (A) a non-overlapping maxpooling layer with 4×4 input, resulting in a 75% reduction in the data; (B) a three-layer fully connected MLP structure, which is identical to a fully connected layer in a CNN.

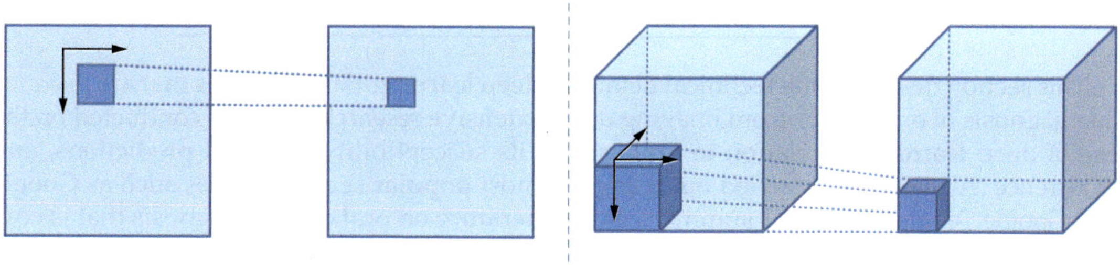

FIG. 8.6 Comparison between (A) 2D convolution and (B) 3D convolution.

These features of a 3D CNN are very useful in terms of analyzing volumetric data from medical imaging, such as CT scans or MRIs, and hence can be used for detection, diagnosis, and segmentation tasks. Recently, advanced medical imaging has been achieved by capturing slices across the depth of the tissue, meaning that these images must be examined in a 3D context. Combining these static images in volumetric or spatial context can be initially processed by a 3D CNN, thereby resulting in a reduction in the human effort and time required for the evaluation and ensuring faster patient care.

Spatial 3D data can take several forms, such as voxel images, red green blue depth (RGB-D), and 3D point clouds. That organize the depth information into 3D space based on planar image information. A voxel image represents the pose and shape of the 3D object. It is then an easy task to carry out the transplant analysis and processing of an ordinary 2D image into a voxel image. The 3D cloud data occupy limited spatial resources, which makes it easy to represent the complex features. An RGB image represents the color of the surface and the texture details of the object, while a depth image contains spatial information about the object. Several variants of 3D CNNs have been used for the diagnosis of oral cancer with different parameter settings. These 3D CNNs are made up of a number of layers N and contain N modules that carry out automatic feature learning for classification. For instance, the feature extraction module in the network in Xu et al. (2019) is composed of eight layers that are divided into four pairs, followed by a downsampling layer. The 3D convolution operation is shown in Eq. (8.2).

$$u_{ki}^{1}(x,y,z) = \sum\nolimits_{m,n,t} h_k^{l-1}(x-m, y-n, z-t)^{*}W_{ki}^{l}(m,n,t) \tag{8.2}$$

Here, W_{ki}^l represents the kernel where 3D space h_k^{l-1} is the convolution operation and W_{ki}^l (m, n, t) represents the voxel weights in the convoluted kernel. The corresponding output feature space is given in Eq. (8.3).

$$h_i^l = \sigma\left(\sum_k u_{ki}^l + b_i^l\right) \tag{8.3}$$

Since an oral CT scan can acquire the 3D structure of a tumor, a 3D CNN can then be applied to the CT image to obtain 3D spatial information on the tumor. A single CT image is processed sequentially to extract the features of the tumor from multiple angles.

CNN-based methods for oral cancer diagnosis

This section discusses the technical details of deep learning-based models that are used for the diagnosis of oral cancer from imaging data. Extensive research has been conducted on the use of deep learning in relation to oral cancer, its susceptibility, survival predictions, and recurrence. A search was carried out using the most popular search engines such as Google and Google Scholar. Table 8.1 summarizes the literature on oral cancer diagnosis that is covered in this chapter, in terms of the architecture of the model, the datasets used, the accuracy of the results, and specific applications. The data usage is different for each method; some practitioners have used public datasets, while a few have created their own private datasets. The practitioners collected the data from several different search engines to perform these experiments.

Oral cancer has remained the most reported disease out of 4.5 million in 2017 according to the social group of WHO and caused death of 85% of cases over several decades. Diagnosis at an early stage can help the researchers reduce the death rate by 70% (Erickson et al., 2018). According to another report from the WHO, early treatment can prevent 80% of cervical cancer (WHO, 2020). Oral cancer is similar to cervical cancer in several ways, and its incidence is increasing worldwide. Several techniques have been proposed for oral cancer diagnosis so far but still, there are some challenges where cancer before its diagnosis causes high mortality and morbidity rate. However, the recent advancement in deep learning obtained overwhelming growth and exploded in distinct areas of computer vision such as activity analysis (Ullah, Ullah, et al., 2019), forecasting problems (Ullah, Muhammad, et al., 2019), and video summarization (Hussain et al., 2020). Researchers in these domains have often applied CNNs, which form the core concept of deep learning networks and have been shown to achieve high levels of accuracy on image recognition tasks (Russakovsky et al., 2015).

In a similar way, deep learning approaches such as CNNs have been applied to the diagnosis of oral cancer for some time. For instance, Das et al. (2020) recently attempted the classification of OSCC into four types based on a histological grading system. Their OSCC framework is illustrated in Fig. 8.7 and was applied to images of oral biopsies using transfer learning via pre-trained deep CNNs including VGG-19, VGG-16, ResNet-50, and AlexNet. Deep learning algorithms that use binary classifiers such as loss and sigmoid functions have achieved significant results but invariably cause a reduction in accuracy, and this is one of the major challenges that is yet to be tackled. In view of this, the method proposed in Bhandari et al. (2020) aimed to increase the accuracy of the detection and classification of oral tumors within a short processing time. This method was based on a CNN with a modified loss function that minimized the prediction error and classified oral tumors by reducing data overfitting to support multi-classification.

Another method (Yan et al., 2020) presented an assembled CNN framework based on deep learning that distinguished between tongue squamous cell carcinoma and non-tumor tissue. The data were collected using a fiber-optic Raman system, and two convolutional kernels were used to extract non-linear feature representations from the spectral region. Next, a feature vector was created by fusing the layers, which were then propagated to fully connected layers to classify images of squamous cell carcinoma of the tongue in order to identify it and offer proper treatment and care.

TABLE 8.1 Summary of the reviewed papers on oral cancer diagnosis.

Ref	Methodology		Application			Dataset
	Preprocessing	Architecture	Classification	Detection	Region	
Xu et al. (2019)	✖	2D CNNs 3D CNNs	Binary classification (benign; malignant)	–	Overall oral cavity	Private data
Das et al. (2020)	✖	AlexNet VGG-16 VGG-19 Resnet-50	Classification of four classes (benign; well-differentiated; moderately differentiated; poorly differentiated)	–	Oral squamous cell carcinoma (OSCC)	Private data
Bhandari et al. (2020)	Bagging, boosting	CNN	Binary classification (benign; malignant)	✓	Squamous carcinoma, lymphoma, Osteosarcoma, andfibrosarcoma	Private data
Panigrahi and Swarnkar (2019)	Data augmentation	CNN	Binary classification (benign; malignant)	✖	Oral squamous cell	Genomic data (Genomic Data Commons Portal, n.d.)
Welikala et al. (2020)	Data annotation	DNNs ResNet-101 Faster R-CNN	Classification into five classes (no lesion; no referral needed; refer for other reason; refer low-risk oral potentially malignant disorders (OPMDs); refer cancer/ high risk OPMD; macro-averaged)	✓	Oral lesions	Private data
Anantharaman et al. (2018)	✖	Mask-R-CNN	✖	✓	Oral pathology	Private data
Jeyaraj et al. (2019)	Data normalization, subsampling	ResNet	Binary classification(benign; malignant)	✓	Overall oral cavity	Private data, tongue dataset (Tongue Dataset, n.d.)
Rajan et al. (2020)	Vesselness, filtering, deblurring	DCNN	Classification into three classes: (cancer; non-cancerous; pre-cancerous)	✖	Overall oral cavity	Private data, MIDAS (MIDAS Dataset, n.d.)
Shamim et al. (2019)	✖	DCNN VGG19 ResNet50	Classification into five classes (hairy tongue; fissured tongue; geographic tongue; strawberry tongue; oral hairy leukoplakia)	✓	Tongue	Private data

Continued

TABLE 8.1 Summary of the reviewed papers on oral cancer diagnosis—cont'd

Ref	Methodology		Application			Dataset
	Preprocessing	Architecture	Classification	Detection	Region	
Guo et al. (2021)	Image translation, scaling, rotation, flipping, enhancement	Residual Network	Binary classification (normal; ulcerative)	✖	Oral mucosa	Private data
Aubreville et al. (2017)	Data augmentation	CNN	Binary classification (clinically normal; carcinogen)	✖	Hard palate, alveolar ridge, and inner labium	Private data
Ariji et al. (2020)	Cropping	AlexNet	✓	✖	Cervical lymph nodes	Private data
Folmsbee et al. (2018)	Cropping	CNN	Classification into seven classes (stroma; lymphocytes; tumor; mucosa; keratin pearls; blood; background/adipose)	✖	Whole oral cavity	Private data
Chan et al. (2019)	Semantic segmentation	DCNN	✖	✓	Overall oral cavity	Buccal mucosa image dataset (Chan et al., 2019)
Fu et al. (2020)	✖	CNN	✖	✓	OSCC	Private data
Lee et al. (2018)	Data augmentation	DCNN	Binary classification (dental caries; non-dental caries)	✓	Dental caries	Private data

These networks were found to be suitable for classification. Using a four-patch CNN as a feature extraction method, Panigrahi and Swarnkar (2019), classified oral cancer images by applying augmentation techniques to prevent overfitting, including rotation, inversion, and flip functions. They classified oral cancers into benign and malignant, as shown in Fig. 8.8. There are several methods that can be used to classify oral lesions, and these represent an important step in oral cancer diagnosis as the region of interest can be annotated for detailed investigation and further processing. Hence, DNNs have been used to build an annotation system for oral lesions (Welikala et al., 2020) whereas the ResNet-101 models are widely used for classification of oral cancers. For region detection, Faster R-CNNs are used.

However, the detection of oral cancer remains a challenging step and is an important aspect of its diagnosis. Several techniques have used existing trained models for detection; for

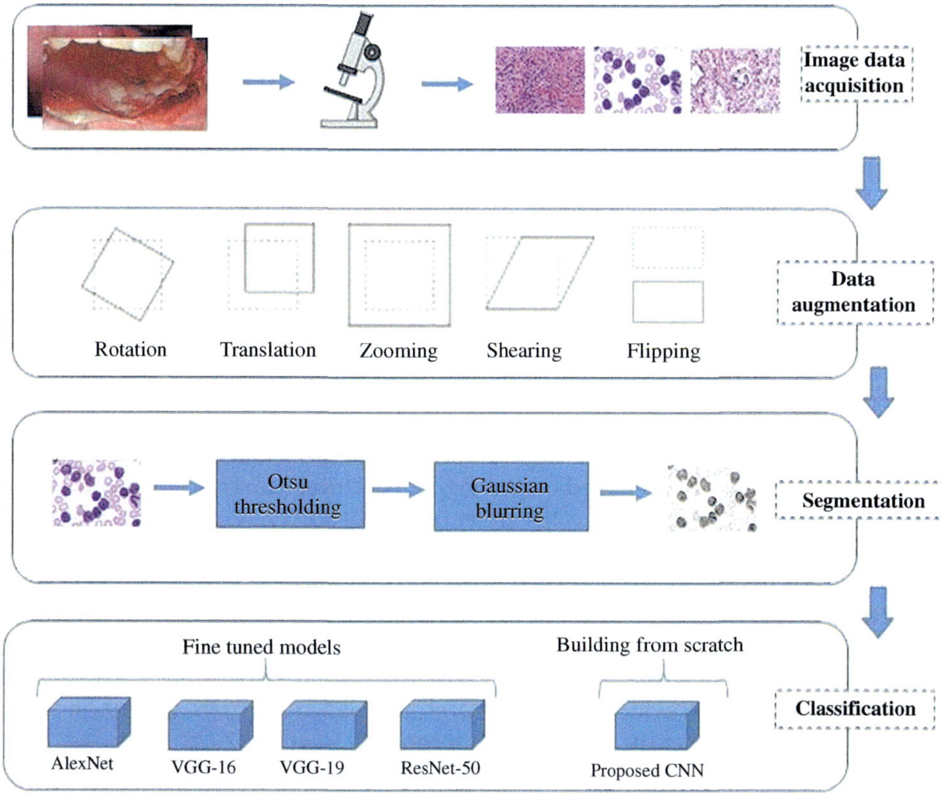

FIG. 8.7 OSCC classification framework proposed in Das et al. (2020).

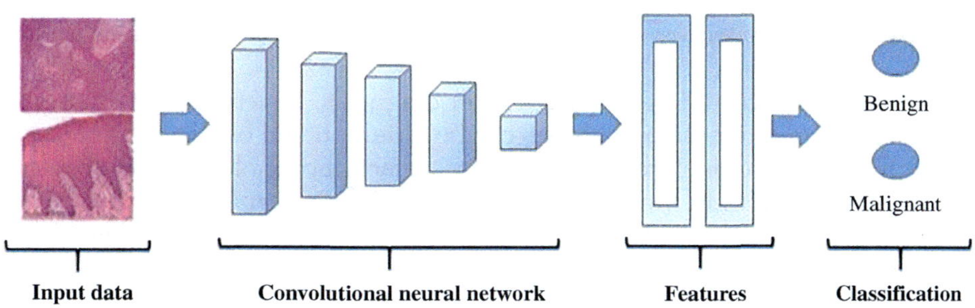

FIG. 8.8 Classification of cancers of the oral cavity into benign and malignant (Panigrahi and Swarnkar, 2019).

instance, Anantharaman et al. (2018) used a Mask R-CNN and its applications for detection and segmentation in the domain of oral pathology. However, since oral diseases are numerous (such as thrush, lichen planus, and leukoplakia), their method was limited to the two common conditions of herpes labialis and aphthous ulcer, which are popularly known as cold sores and canker sores, respectively. Their framework is illustrated in Fig. 8.9.

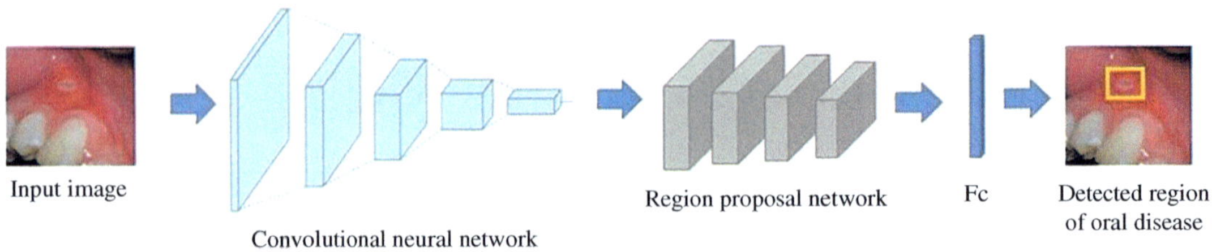

FIG. 8.9 The working flow of an R-CNN for detection of oral disease (Anantharaman et al., 2018).

One of the most important issues in hyperspectral imaging is the classification of cancer images based on spatial information. Several classification techniques have used dimensionality reduction for features, although CNN approaches have remained limited in this regard. To tackle this issue, Jeyaraj et al. (2019) proposed a technique that merged the characteristic spectral and spatial information for the purposes of classification. They developed a deep classification framework based on the ResNet architecture for the detection of cancer regions. In the domain of oral cancer, the detection of esophageal cancer is relatively poor and patients are often diagnosed at a late stage, which usually means that treatment is not effective. However, the recent advances in deep learning have made remarkable strides toward overcoming this problem, and CNNs have yielded promising results with improved accuracy. For example, Horie et al. (2019) developed a CNN-based method to tackle this problem by collecting data from 384 patients. The Internet of Things (IoT) is also an emerging aspect of modern health systems that can provide high-quality treatment and care with lower computational complexity. With this in mind, Rajan et al. (2020) proposed a method that used vesselness measurements with a deep CNN for the identification of oral cancer in an IoT-based care system. The vesselness scheme was applied to handle the issue of noise and to preserve the structure, while the CNN helped in deblurring the region of interest by integrating it with multi-dimensional information obtained from the feature maps. These feature maps were extracted from each component of the region and used as input to train the CNN.

Oral lesions are also considered in oral cancer diagnosis, which has enormous growth in the research community. These oral lesions can be differentiated on the basis of their characteristics, such as their color, location, size, and surface morphology. However, most oral lesions only persist for a short time and can be removed with simple clinical treatment, while some take a long time to heal and may lead to cancer. In these cases, a complete history and a thorough examination by a physician are highly recommended. Recently, Shamim et al. (2019) presented a comparative study of six deep CNN models that used a transfer learning approach for the diagnosis of oral cavity cancers at an early stage. The results of their experiments showed that the VGG19 model achieved the best accuracy in differentiating between benign and pre-cancerous lesions. A similar method proposed in Ming-Jer et al. (2019) used Raman spectroscopy as a non-invasive way of screening for oral cancer using sub-site-wise differentiation. Linear and quadratic discriminant analyses were used with principal component analysis for sample classification. The method presented in Guo et al. (2021) classified images of oral ulcers using a variant of the residual network framework. Several enhancements and preprocessing techniques were applied to enrich the data and overcome the problem of data overfitting. The transfer learning approach was introduced, and residual blocks were used to obtain accurate classification results.

Chan et al. (2019) combined a texture map with a DCNN to detect cancerous regions and to mark these areas automatically. Their network consisted of two main branches: the upper branch was used to detect oral cancers, while the lower branch performed semantic segmentation. The upper branch extracted the cancerous regions that are more precise than the lower branch. Similarly, the method presented in Fu et al. (2020) used a deep learning-based cascaded CNN for OCSCC detection from photographic images. This scheme used pathological reports as a standard for identifying OCSCC. Dental caries in the oral cavity are another important aspect that should be considered, and Lee et al. (2018) employed a DCNN to detect and diagnose these. They used GoogleNet and an Inception V3 CNN as a pre-trained model and applied a transfer learning strategy, classifying the oral images into those with dental caries and those without.

Results and discussion

This section gives a detailed summary of the results obtained for oral cancer diagnosis, including the overall accuracies, the validation methods used, and descriptions of the data. Deep learning models for oral cancer diagnosis have used histological, clinical, genomic, imaging, epidemiological and demographic data, or combinations of these. For instance, the study in Exarchos et al. (2012) addresses the issue of predicting recurrence in OSCC by introducing a multi-parametric decision support system. Heterogeneous source data (imaging, clinical, and genomic) were used to predict relapses. However, the present chapter only discusses deep models that are based on imaging data. The results obtained from the methods surveyed here are summarized in Table 8.2 and due to a lack of a common dataset for evaluation, we also include descriptions of the data in this comparison.

The imaging data used for the diagnosis of oral cancer can be obtained from various sources, such as tissue samples, CT scans, PET scans, laser endomicroscopy, or camera images of samples. The last column of Table 8.2 shows the types of images used to train the deep models. For instance, Ariji et al. (2020) trained an AlexNet model over 300 epochs on 703 CT images of 51 patients, 178 of which contained an extranodal extension and 525 did not. They cropped these CT images to an arbitrary size to show the lymph and surrounding tissues and split the dataset into 80% for training and 20% for validation. The results yielded scores of 84%, 66.9%, and 89.7% for the accuracy, sensitivity, and specificity, respectively. Similarly, the 3D CNN proposed in Xu et al. (2019) used 7000 CT images of the full oral cavity and achieved scores of 75.4%, 81.8%, and 73.9% for the accuracy, sensitivity, and specificity, respectively, for a binary classification task (benign and malignant). In Panigrahi and Swarnkar (2019), tissue slide images were used for the classification of oral squamous cells with a four-layered CNN model; an accuracy of 96.77% was achieved using K-fold validation and 98.05% using the hold-out validation method with data augmentation.

To classify the normal and cancerous regions in hyperspectral images, Jeyaraj et al. (2019) applied a transfer learning strategy using a ResNet model and achieved scores of 94.8%, 97.8%, and 96.5% for accuracy, sensitivity, and specificity. Similarly, in Folmsbee et al. (2018),

TABLE 8.2 Comparative analysis of studies from the literature in terms of their results, descriptions of the data, and the evaluation criteria.

Ref.	Method	Results				Validation method	Data			
		Accuracy (%)	Sensitivity (%)	Specificity (%)	Area under curve (AUC)		No. of samples	Training / testing split (%)	No. of patients	Type
Xu et al. (2019)	3D CNN	75.4	81.8	73.9	0.79	Holdout validation	7000	–	–	CT
Bhandari et al. (2020)	CNN	96.5	–	–	–	K-fold cross-validation	3045	K=7	230	MRI
Panigrahi and Swarnkar (2019)	CNN	98.05	–	–	–	Holdout validation	1000	70/30	750	Slide images
Jeyaraj et al. (2019)	2D CNN	94.8	97.8	96.5	–	Holdout validation	1400	–	–	Hyperspectral images
Rajan et al. (2020)	2D CNN	96.8	92	97.5	–	Holdout validation	1500	–	–	PET scan
Shamim et al. (2019)	2D CNN	97	89	97	0.98	Holdout method	300	80/20	–	Sample images
Guo et al. (2021)	2D CNN	98.7	98.2	99.2	–	Holdout method	360	80/10	78	Sample images
Aubreville et al. (2017)	CNN	87	90.7	83.8	0.96	Leave-one-patient-out cross-validation	7894	–	12	Laser endomicroscopy images
Ariji et al. (2020)	AlexNet	84	66.9	89.7	0.81	Holdout method	703	80/20	51	CT images
Folmsbee et al. (2018)	CNN +active learning	96.44	–	–	–	Holdout method	143	–	–	Tissue slides
Chan et al. (2019)	DCNN		96.8	71.2		Holdout method	80	80/20	–	Cancerous and normal images
Fu et al. (2020)	CNN		94·9	88·7	0·98	Holdout method	44,409	–	–	Clinical oral photographs
Lee et al. (2018)	DCNN	89.0	92.3	94.0	–	Holdout method	3000	80/20	–	Periapical radiographic images

active and random learning strategies for tissue classification were compared based on 143 digitized images of eosin- and hematoxylin-stained oral cavity cancers. Using active learning, an accuracy of 96.44% was achieved for the classification of seven classes: stroma, keratin pearls, mucosa, tumor, blood, background/adipose, and lymphocytes. Likewise, the study in Bhandari et al. (2020) used MRI data for the classification of oral tumors with a modified loss function to minimize the overall error and time complexity. These authors achieved an accuracy of 96.5% and reduced the time complexity by 30–40ms. Confocal laser endomicroscopy (CLE) images have also been used (Aubreville et al., 2017) and yielded scores of 88.3% for accuracy and 0.96 for AUC.

An IoT-assisted framework presented in Rajan et al. (2020) achieved values of 96.8%, 92%, and 97.5% for the accuracy, sensitivity, and specificity, respectively, for three classes (cancerous, non-cancerous, and pre-cancerous regions). Six different deep models (AlexNet, SqueezeNet, VGG19, Inceptionv3, ResNet50, and GoogleNet) were explored, and a transfer learning strategy was used to classify tongue lesions into pre-cancerous and benign types. Images were downloaded from the Internet via an image search engine and manually labeled as different types with the help of a certified physician. For a binary classification problem (benign or pre-cancerous), VGG19 achieved a maximum accuracy of 98%, with scores of 89% for sensitivity and 97% for specificity. Multi-class classification was also performed on five types of tongue lesions: hairy tongue, fissured tongue, geographic tongue, strawberry tongue, and oral hairy leukoplakia, and ResNet50 achieved an accuracy of 97%.

Current challenges and future research direction

In this section, we discuss the challenges that remain in relation to deep learning-based oral cancer diagnosis methods and possible solutions to these challenges. Despite the considerable attention that has been paid to these problems and the great efforts of researchers in this domain, there are certain aspects that have not yet reached maturity, and several hurdles still exist. These challenges are discussed below, and solutions are suggested in the subsequent sections.

Lack of oral cancer staging/classification methods

Staging or classifying cancer involves describing the extent of cancer, i.e., how many cancerous cells there are and where they are in the body. This classification is based on the size of the tumor, the parts of the organ affected, and the spread (metastasis). This classification helps specialists to study and compare similar cases in clinical trials, to determine the chance of recovery (prognosis), and to plan the overall treatment. There are various deep learning models that can perform detection and segmentation tasks for cancerous tissues/cells as part of a classification model (for binary classification tasks, i.e., normal and cancerous cells). There are four basic stages of oral cancers that are defined based on the tumor nodal and distant metastases (TNM) (i.e., stages T1–T4) having their own appearance. For instance, the method presented in Sajjad et al. (2019) classifies brain tumors into four stages. The current literature on oral cancer diagnosis lacks deep models that can both detect and stage cancer, and these would be valuable in terms of better results and understanding.

Lack of common large-scale datasets for oral cancer diagnosis

In most cases of medical imaging analysis, deep learning techniques suffer from the SSL problem for specific diseases. There are two main reasons for this absence of annotated datasets, which are related to data confidentiality and a lack of proper annotation. There is no doubt that recent advanced imaging technologies have produced huge volumes of data, but access to this data requires several precautions even for research purposes, due to the need for patient confidentiality. Similarly, annotation of the samples is a highly specialized task and needs in-depth domain knowledge that puts extra enormous pressure on biomedical experts. Fu et al. (2020) recently published the most highly organized and well-annotated dataset for oral cavity cancer diagnosis, consisting of 44,409 samples from different categories, which can be used for research purposes on request.

Active and incremental learning for newly added samples and classes

The continuous advancements in every field of science have given rise to many other challenges and solutions. In the field of medical sciences, and especially in relation to cancer diagnosis, each patient case differs from the others, and entirely new samples are sometimes encountered. Updating a trained model with new classes and samples is one of the challenges faced in the automation of cancer diagnosis. On the other hand, machine learning techniques also gained overwhelming attractions where the domain knowledge is transferred into machines. Active and incremental learning techniques are promising paradigms that can be used for sample- and class-wise learning, while ensuring that previous knowledge is unaltered. In active learning, the model intelligently queries to label new data samples with the desired outputs. Hence, in current AL terminology, the human-to-artificial-intelligence (AI) interface is minimal and opaque to visually view the entire data processing and change in data points, while in incremental learning, the model is updated by adding or removing classes, without altering existing knowledge. In order to maintain high accuracy and diversity in new samples and classes, deep models should, therefore, apply active and incremental learning strategies.

Use of explainable artificial intelligence (XAI) for understanding data

Recent dramatic developments in machine learning have resulted in the production of thousands of autonomous systems and many applications that can learn, decide, and act on their own. However, the effectiveness of these systems is limited in terms of explaining their decisions to users. For instance, DNNs rely on complex interconnected hierarchical architectures, and interpreting or understanding their predictions becomes very difficult, making this a "black box" problem (Castelvecchi, 2016). In an important domain such as the diagnosis of oral cancer, medical specialists, therefore, face a dilemma in terms of trusting predictions based on features that they do not understand. To address this problem, the AI community has introduced Explainable AI with the aim of improving interpretability (Montavon et al., 2018) and reducing the gaps between computer science and other domains, in order to increase their trustworthiness by finding new ways to measure sensitivity and visualize features.

Conclusion

Recent advancements in deep learning and medical imaging technologies have made it possible to diagnose cancer in its early stages. Great efforts have been made by researchers to transfer domain knowledge to computers to allow for the intelligent diagnosis of cancers from imaging data using identification, classification, and segmentation tasks. In this chapter, we have described the basic concepts of deep learning as applied to medical imaging data and have given an overview of the deep models that have been developed for oral cancer diagnosis along with their achievements and drawbacks. The main aim of this chapter is to present a short tutorial that will allow medical experts to understand the basic working of deep learning models, and which will help new researchers in this domain explore the recent literature on the diagnosis of oral cancer. We have also discussed the promising results that have been achieved thus far and have highlighted some current challenges in oral cancer diagnosis. Finally, we have tried to provide some recommendations for future research directions, such as addressing the lack of properly annotated datasets, the development of staging/classification-based deep models, and the exploration of active and incremental learning for new data. We believe that this chapter can help readers identify the current challenges faced by the oral cancer diagnosis research community.

Acknowledgment

This work is supported by the Institute for Information & communication Technology Planning & evaluation (IITP, Korea) funded by the Ministry of Science and ICT (MSIT, Korea) 1711117081.

References

Anantharaman, R., Velazquez, M., & Lee, Y. (2018). Utilizing mask R-CNN for detection and segmentation of oral diseases. In *2018 IEEE international conference on bioinformatics and biomedicine (BIBM)* (pp. 2197–2204).

Antipov, G., Berrani, S. A., Ruchaud, N., & Dugelay, J. L. (2015). Learned vs hand-crafted features for pedestrian gender recognition. In *MM 2015 - Proceedings of the 2015 ACM multimedia conference* (pp. 1263–1266). Association for Computing Machinery, Inc. https://doi.org/10.1145/2733373.2806332.

Ariji, Y., Sugita, Y., Nagao, T., Nakayama, A., Fukuda, M., Kise, Y., Nozawa, M., Nishiyama, M., Katumata, A., & Ariji, E. (2020). CT evaluation of extranodal extension of cervical lymph node metastases in patients with oral squamous cell carcinoma using deep learning classification. *Oral Radiology, 36*(2), 148–155. https://doi.org/10.1007/s11282-019-00391-4.

Aubreville, M., Knipfer, C., Oetter, N., Jaremenko, C., Rodner, E., Denzler, J., Bohr, C., Neumann, H., Stelzle, F., & Maier, A. (2017). Automatic classification of cancerous tissue in laser endomicroscopy images of the oral cavity using deep learning. *Scientific Reports, 7*(1). https://doi.org/10.1038/s41598-017-12320-8.

Bhandari, B., Alsadoon, A., Prasad, P. W. C., Abdullah, S., & Haddad, S. (2020). Deep learning neural network for texture feature extraction in oral cancer: Enhanced loss function. *Multimedia Tools and Applications*, *79*(37–38), 27867–27890. https://doi.org/10.1007/s11042-020-09384-6.

Castelvecchi, D. (2016). Can we open the black box of AI? *Nature News*, *538*.

Chan, C. H., Huang, T. T., Chen, C. Y., Lee, C. C., Chan, M. Y., & Chung, P. C. (2019). Texture-map-based branch-collaborative network for oral cancer detection. *IEEE Transactions on Biomedical Circuits and Systems*, *13*(4), 766–780. https://doi.org/10.1109/TBCAS.2019.2918244.

Das, N., Hussain, E., & Mahanta, L. B. (2020). Automated classification of cells into multiple classes in epithelial tissue of oral squamous cell carcinoma using transfer learning and convolutional neural network. *Neural Networks*, *128*, 47–60. https://doi.org/10.1016/j.neunet.2020.05.003.

Deng, J., Dong, W., Socher, R., Li, L.-J., Li, K., & Fei-Fei, L. (2009). Imagenet: A large-scale hierarchical image database. In *2009 IEEE conference on computer vision and pattern recognition* (pp. 248–255).

Erickson, B. J., Korfiatis, P., Kline, T. L., Akkus, Z., Philbrick, K., & Weston, A. D. (2018). Deep learning in radiology: Does one size fit all? *Journal of the American College of Radiology*, *15*(3), 521–526. https://doi.org/10.1016/j.jacr.2017.12.027.

Esteva, A., Kuprel, B., Novoa, R. A., Ko, J., Swetter, S. M., Blau, H. M., & Thrun, S. (2017). Dermatologist-level classification of skin cancer with deep neural networks. *Nature*, *542*(7639), 115–118. https://doi.org/10.1038/nature21056.

Exarchos, K. P., Goletsis, Y., & Fotiadis, D. I. (2012). Multiparametric decision support system for the prediction of oral cancer reoccurrence. *IEEE Transactions on Information Technology in Biomedicine*, *16*(6), 1127–1134. https://doi.org/10.1109/TITB.2011.2165076.

Folmsbee, J., Liu, X., Brandwein-Weber, M., & Doyle, S. (2018). Active deep learning: Improved training efficiency of convolutional neural networks for tissue classification in oral cavity cancer. In (Vols. 2018). *Proceedings - International symposium on biomedical imaging* (pp. 770–773). IEEE Computer Society. https://doi.org/10.1109/ISBI.2018.8363686.

Frangioni, J. V. (2008). New technologies for human cancer imaging. *Journal of Clinical Oncology*, *26*(24), 4012–4021. https://doi.org/10.1200/JCO.2007.14.3065.

Fu, Q., Chen, Y., Li, Z., Jing, Q., Hu, C., Liu, H., Bao, J., Hong, Y., Shi, T., Li, K., Zou, H., Song, Y., Wang, H., Wang, X., Wang, Y., Liu, J., Liu, H., Chen, S., Chen, R., … Xiong, X. (2020). A deep learning algorithm for detection of oral cavity squamous cell carcinoma from photographic images: A retrospective study. *EClinicalMedicine*, *27*. https://doi.org/10.1016/j.eclinm.2020.100558.

Genomic Data Commons Portal, n.d.," https://portal.gdc.cancer.gov/repository.

Gulshan, V., Peng, L., Coram, M., Stumpe, M. C., Wu, D., Narayanaswamy, A., Venugopalan, S., Widner, K., Madams, T., Cuadros, J., Kim, R., Raman, R., Nelson, P. C., Mega, J. L., & Webster, D. R. (2016). Development and validation of a deep learning algorithm for detection of diabetic retinopathy in retinal fundus photographs. *JAMA: The Journal of the American Medical Association*, *316*(22), 2402–2410. https://doi.org/10.1001/jama.2016.17216.

Guo, J., Wang, H., Xue, X., Li, M., & Ma, Z. (2021). Real-time classification on oral ulcer images with residual network and image enhancement. *IET Image Processing*. https://doi.org/10.1049/ipr2.12144.

Horie, Y., Yoshio, T., Aoyama, K., Yoshimizu, S., Horiuchi, Y., Ishiyama, A., Hirasawa, T., Tsuchida, T., Ozawa, T., Ishihara, S., Kumagai, Y., Fujishiro, M., Maetani, I., Fujisaki, J., & Tada, T. (2019). Diagnostic outcomes of esophageal cancer by artificial intelligence using convolutional neural networks. *Gastrointestinal Endoscopy*, *89*(1), 25–32. https://doi.org/10.1016/j.gie.2018.07.037.

Hussain, T., Muhammad, K., Ser, J. D., Baik, S. W., & De Albuquerque, V. H. C. (2020). Intelligent embedded vision for summarization of multiview videos in IIoT. *IEEE Transactions on Industrial Informatics*, *16*(4), 2592–2602. https://doi.org/10.1109/TII.2019.2937905.

Jeyaraj, P. R., Samuel Nadar, E. R., & Panigrahi, B. K. (2019). ResNet convolution neural network based hyperspectral imagery classification for accurate cancerous region detection. In *2019 IEEE conference on information and communication technology, CICT 2019* Institute of Electrical and Electronics Engineers Inc. https://doi.org/10.1109/CICT48419.2019.9066215.

Krizhevsky, A., Sutskever, I., & Hinton, G. E. (2017). ImageNet classification with deep convolutional neural networks. *Communications of the ACM*, *60*(6), 84–90. https://doi.org/10.1145/3065386.

Lecun, Y., Bengio, Y., & Hinton, G. (2015). Deep learning. *Nature*, *521*(7553), 436–444. https://doi.org/10.1038/nature14539.

Lee, J. H., Kim, D. H., Jeong, S. N., & Choi, S. H. (2018). Detection and diagnosis of dental caries using a deep learning-based convolutional neural network algorithm. *Journal of Dentistry*, *77*, 106–111. https://doi.org/10.1016/j.jdent.2018.07.015.

Lin, T. Y., Maire, M., Belongie, S., Hays, J., Perona, P., Ramanan, D., Dollár, P., & Zitnick, C. L. (2014). Microsoft COCO: Common objects in context. In *8693. Lecture notes in computer science (including subseries lecture notes in artificial intelligence and lecture notes in bioinformatics)* (pp. 740–755). Springer Verlag. https://doi.org/10.1007/978-3-319-10602-1_48. Issue 5.

"MIDAS Dataset, n.d.," http://insight-journal.org/midas/.

Ming-Jer, J., Mukta, S., Lokesh, S., Ting-Yu, C., Shiang-Fu, H., Liann-Be, C., Shih-Lin, W., & Lee, C. (2019). Raman spectroscopy analysis for optical diagnosis of oral cancer detection. *Journal of Clinical Medicine, 1313*. https://doi.org/10.3390/jcm8091313.

Montavon, G., Samek, W., & Müller, K. R. (2018). Methods for interpreting and understanding deep neural networks. *Digital Signal Processing: A Review Journal, 73*, 1–15. https://doi.org/10.1016/j.dsp.2017.10.011.

Panigrahi, S., & Swarnkar, T. (2019). Automated classification of oral cancer histopathology images using convolutional neural network. In *Proceedings - 2019 IEEE international conference on bioinformatics and biomedicine, BIBM 2019* (pp. 1232–1234). Institute of Electrical and Electronics Engineers Inc. https://doi.org/10.1109/BIBM47256.2019.8982979.

Poplin, R., Varadarajan, A. V., Blumer, K., Liu, Y., McConnell, M. V., Corrado, G. S., Peng, L., & Webster, D. R. (2018). Prediction of cardiovascular risk factors from retinal fundus photographs via deep learning. *Nature Biomedical Engineering, 2*(3), 158–164. https://doi.org/10.1038/s41551-018-0195-0.

Rajan, J. P., Rajan, S. E., Martis, R. J., & Panigrahi, B. K. (2020). Fog computing employed computer aided cancer classification system using deep neural network in internet of things based healthcare system. *Journal of Medical Systems, 44*(2). https://doi.org/10.1007/s10916-019-1500-5.

Rajpurkar, P., Irvin, J., Ball, R. L., Zhu, K., Yang, B., Mehta, H., Duan, T., Ding, D., Bagul, A., Langlotz, C. P., Patel, B. N., Yeom, K. W., Shpanskaya, K., Blankenberg, F. G., Seekins, J., Amrhein, T. J., Mong, D. A., Halabi, S. S., Zucker, E. J., … Lungren, M. P. (2018). Deep learning for chest radiograph diagnosis: A retrospective comparison of the CheXNeXt algorithm to practicing radiologists. *PLoS Medicine, 15*(11). https://doi.org/10.1371/journal.pmed.1002686.

Rossenblatt, F. (1957). The perceptron—A perceiving and recognizing automaton, Cornell Aeronautical Laboratory, Report 85-460-1.

Russakovsky, O., Deng, J., Su, H., Krause, J., Satheesh, S., Ma, S., Huang, Z., Karpathy, A., Khosla, A., Bernstein, M., Berg, A. C., & Fei-Fei, L. (2015). ImageNet large scale visual recognition challenge. *International Journal of Computer Vision, 115*(3), 211–252. https://doi.org/10.1007/s11263-015-0816-y.

Sajjad, M., Khan, S., Muhammad, K., Wu, W., Ullah, A., & Baik, S. W. (2019). Multi-grade brain tumor classification using deep CNN with extensive data augmentation. *Journal of Computational Science, 30*, 174–182. https://doi.org/10.1016/j.jocs.2018.12.003.

Shamim, M. Z. M., Syed, S., Shiblee, M., Usman, M., & Ali, S. (2019). Automated detection of oral pre-cancerous tongue lesions using deep learning for early diagnosis of oral cavity cancer. *arXiv preprint arXiv:1909.08987*.

Siuly, S., & Zhang, Y. (2016). Medical big data: Neurological diseases diagnosis through medical data analysis. *Data Science and Engineering, 1*(2), 54–64. https://doi.org/10.1007/s41019-016-0011-3.

"Tongue Dataset, n.d.," http://biogps.org/dataset/tag/tongue/.

Ullah, A., Muhammad, K., Haq, I. U., & Baik, S. W. (2019). Action recognition using optimized deep autoencoder and CNN for surveillance data streams of non-stationary environments. *Future Generation Computer Systems, 96*, 386–397. https://doi.org/10.1016/j.future.2019.01.029.

Ullah, F. U. M., Ullah, A., Haq, I. U., Rho, S., & Baik, S. W. (2019). *Short-term prediction of residential power energy consumption via CNN and multilayer bi-directional LSTM networks*. IEEE Access.

Welikala, R. A., Remagnino, P., Lim, J. H., Chan, C. S., Rajendran, S., Kallarakkal, T. G., Zain, R. B., Jayasinghe, R. D., Rimal, J., Kerr, A. R., Amtha, R., Patil, K., Tilakaratne, W. M., Gibson, J., Cheong, S. C., & Barman, S. A. (2020). Automated detection and classification of Oral lesions using deep learning for early detection of oral cancer. *IEEE Access, 8*, 132677–132693. https://doi.org/10.1109/ACCESS.2020.3010180.

WHO. (2020). *Human papillomavirus (hpv) and cervical cancer*. http://www.who.int/mediacentre/factsheets/fs380/en/.

Xu, S., Liu, Y., Hu, W., Zhang, C., Liu, C., Zong, Y., Chen, S., Lu, Y., Yang, L., Ng, E. Y. K., Wang, Y., & Wang, Y. (2019). An early diagnosis of oral cancer based on three-dimensional convolutional neural networks. *IEEE Access, 7*, 158603–158611. https://doi.org/10.1109/ACCESS.2019.2950286.

Yan, H., Yu, M., Xia, J., Zhu, L., Zhang, T., Zhu, Z., & Sun, G. (2020). Diverse region-based CNN for tongue squamous cell carcinoma classification with Raman spectroscopy. *IEEE Access, 8*, 127313–127328. https://doi.org/10.1109/ACCESS.2020.3006567.

CHAPTER

9

An intelligent deep learning approach for colon cancer diagnosis

Pemmada Suresh Kumar[a], *K. Anisha Kumari*[b], *and Uttam Ghosh*[c]

[a]Department of Computer Science and Engineering, GITAM School of Technology, GITAM (Deemed to be University), Visakhapatnam, Andhra Pradesh, India [b]Department of Artificial intelligence, Northeastern University, Boston, MA, United States [c]Department of CS & DS, Meharry Medical College, Nashville, TN, United States

Introduction

Cancer is characterized as a genetic mutation of normal cells, damaging the deoxyribonucleic acid (DNA) and affecting the cells' life cycle, which causes uncontrollable growth of abnormal cells. The growth of these abnormal cells in the colon causes colon cancer, also known as colorectal cancer or bowel cancer (Al-Rajab et al., 2017). It has a significant effect on human health with a high mortality rate. According to the World Health Organization (WHO) in 2019, out of 183 countries, in 112 countries, cancer is placed first or second as a cause of death, and in other 23 countries, it is placed at third or fourth as a cause of death for people younger than 70 years. Moreover, in 2020, around 19.3 million new cancer cases and 10.0 million deaths due to cancer cases were estimated worldwide. Also, 10.0% of new cases are colorectal cancer cases and 9.4% deaths due to colorectal cancer (Sung et al., 2021). If detected on time, then about 90% of colorectal cases are cured. Therefore, it is a major challenge for physicians to diagnose and researchers to develop tools that can help physicians in their diagnosis.

Colon cancer is a common type of disease found both in men and women, has a significant effect on human life, and has a high mortality rate. Colon cancer begins as a polyp (non-cancerous growth) developed in the mucosal layer of the colon part of the body. Polyps are commonly detected in average-risk individuals of 50 years of age or older age groups and commonly found in men as than in women. The modifiable risk factors of colon cancer

https://doi.org/10.1016/B978-0-323-85240-1.00014-6

include physically inactivity, high consumption of processed red meat, long-term smoking, heavy alcohol consumption, excessive weight loss, and low intake of calcium, whole-grain fiber, and fruits and vegetables. Medical and genetic/hereditary factors increasing risk includes inherited genetic syndrome, family or personal history of colorectal cancer, personal history of type-2 diabetes, and chronic inflammatory bowel disease. The stages of colon cancer include bleeding from the rectum, dark or black stools, blood in toilet after having a bowel movement, changes in bowel habit, discomfort, pain and cramping in the lower abdomen, urge to have movement in bowel with empty bowel, diarrhea or constipation lasting for few days, decrease in appetite, and unintentional loss in weight. Excessive blood loss due to cancer may also lead to anemia, causing excessive fatigue, weakness and sometimes shortness of breath (American Cancer Society: Cancer Facts and Figures 2021, 2021).

Diagnosis of cancer using gene expression profiles is generally based on clinical stages, irrational characteristics, tumor tissues, and differentiation. Evolution in biotechnology helps molecular biology to acquire data from genes, especially microarray analysis, which provides discoveries in medicine and allows diagnostic tool advancement. The microarray gene expression data are essential to obtain data regarding gene regulation and interaction (Bouazza et al., 2015; Chen et al., 2015). This process considers thousands of genes, monitors each gene under multiple conditions, and records their expression levels (Arockia et al., 2015). Microarray cancer data incorporate high dimensionality and small sample size. Generally, medical datasets have overfitting issues and low classification accuracy due to the existence of typical noise, various feature values, and class imbalance problems. The microarray classification approach allows predicting new observations by identifying genes, that is contributing to specific biological outcomes and developing an accurate model to predict cancer (Basavegowda & Dagnew, 2020).

Models for classification of cancer using machine learning techniques such as k-nearest neighbor (KNN) (Ayyad et al., 2019), support vector machine (SVM) (Jimoh et al., 2018), random forest (RF) (Geetha et al., 2019), decision tree (DT) (Venkatesan & Velmurugan, 2015), and artificial neural networks (ANN) (Agrawal & Agrawal, 2015) are implemented. Among these, artificial neural networks have been used immensely on gene expression data. These machine learning classifiers come up with their limited advantages and disadvantages as follows: (i) KNN: its implementation is simple, easy, and training is done quickly. However, it requires large memory space and is very slow during testing. (ii) Naïve Bayes (NB): compared to other sophisticated methods, it is fast, and training requires huge data. But, it does not integrate feature data, and the performance is sensitive while working in skewed data. (iii) RF: it solves the classification and regression problems, and it automatically handles the missing values. But, it requires more time to build the tree's structure and requires more storage space to train deep roots. Hence, it is expensive. (iv) DT: it does not need scaling or normalization of data, thus requiring less effort during preprocessing of data. But, it takes more time to train the model and is expensive to compute. (v) SVM: it is efficient for high-dimensional spaces. But, it is not efficient for the dataset with a large number of instances and overlapping of target classes. (vi) Multi-layer Perceptron (MLP): It is capable of parallel processing and computation. But, it is hardware-dependent and challenging to handle the number of layers of neurons. (vii) Stochastic Gradient Descent (SGD): it is efficient and easy to implement. However, the selection of input parameters impacts accuracy. It needs more memory space and is thus expensive (Pradhan & Chawla, 2020). In medical practices, deep learning techniques are effectively used for diagnosis with success and found to be superior to classic machine learning methods. Deep learning methods are expected to help in transforming current medical practices to assist patients to treat with better accuracy (Pacal et al., 2020). These techniques use efficient and more abstract representation of input data to perform a specific task, whereas machine learning methods use hand-craft features to represent input data. Deep learning uses the behavior of input data features, and it learns from important features in training phase and predicts the class of unseen data (Skrede et al., 2020) testing phase. Incorporating domain-specific data in standard deep neural networks can be problematic, and training is computationally unintuitive and intensive. Based on the architecture, deep learning techniques can be divided into supervised, unsupervised, and hybrid type. A hybrid deep network has much efficient performance as its structure is designed by combining different deep learning classifiers.

In this work, a hybrid model is formed by combining the GA with deep neural network for the effective classification of colon cancer. Training the deep neural network with a GA finds the global solution to a problem as they evaluate multiple candidate solutions simultaneously in the given search space. They do not use derivative information and only require scalar performance measures by evaluating objective function. It provides a set of optimal solutions from which the most appropriate one could be accepted to avoid local optima (Javadi et al., 2005; Juang, 2004).

The major contribution of the chapter includes

1. A deep neural network (DNN)-based model has been proposed to classify colorectal survival.
2. The DNN is optimized with a GA for effective selection of weights.
3. A microarray cancer dataset is used for identification of color cancer.
4. The proposed approach has been compared with various machine learning algorithms such as KNN, SGD, DT, RF, MLP, and DNN to validate the performance.
5. Performance metrics such as accuracy, true positive rate, false positive rate, true negative rate, precision, and f-1 score are used to test the proposed and contrasted methods' efficiency.

The remaining article is structured as follows: "Literature study" section discusses literature review on colon cancer and machine learning model. "Proposed method" section describes the proposed methodology. "Experimental setup" section explains the experimental setup including dataset description, preprocessing procedure of the data, and simulation environment. "Performance measures" section is an analysis of the result, and "Conclusion" section concludes the article.

Literature study

Microarray data analysis is challenging due to noisy data, variance in feature values, small sample size, high dimensionality, and imbalance class problem. A recent study (Basavegowda & Dagnew, 2020) proposed a 7-layered deep feed-forward neural network to classify microarray data for subsequent diagnosis. In this work, the input featured value is scaled using the min-max approach in the pre-processing phase. Here, the performance metrics such as precision, f1-measure, accuracy, log-loss, confusion matrix, recall, and receiver operating characteristic curve are used to validate the performance of the proposed method. The experiment results proved that this method is significantly better than state-of-art methods and existing studies.

Roadknight et al. (2015) investigated a single classification technique and feature selection on a 5-year survival rate of stage 2 and 3 of TNM (where T is the tumor's size, N is nearby lymph nodes, and M is metastasis) patient. The experiment is conducted using simple CART, support vector machine, logistic regression, and J48 methods. Here, the SVM is trained with the 10 best and worst attributes. This approach shows better results at eight attributes with 64.58 accuracy. This ensemble model outperforms and gives better accuracy among the compared models.

A classification model has been proposed (Jimoh et al., 2018) for colon cancer prediction using Support Vector Machine-Poly kernel (SVM-PK) and Support Vector Machine-Gaussian kernel (SVM-GK). The experimentation is conducted on a dataset obtained from the UCI machine learning repository, and performance is evaluated using accuracy, precision, recall, F1-measure, specificity, and sensitivity. The results showed that SVM-PK has high efficiency in accuracy, precision, F1-measure, and specificity, whereas SVM-GK performs better in recall and sensitivity. Overall, these methods performed better than other classification methods.

(Vuong et al., 2020) implemented a multi-task deep learning neural network for the prediction of colon cancer grading. A DenseNet121 model is employed with four dense blocks repeated 6, 12, 24, and 16 times. The dataset is collected from six tissue microarrays which contains three whole slide images stained with hematoxylin and eosin (H&E). This approach has an accuracy of 85.91% and could be used to analyze other tumors in histopathological images.

In another study (Gedeon et al., 2020), the deep convolutional neural network with a transfer learning method has been modeled and experimented to predict colon cancer. The experiment is conducted on a dataset consisting of 100,000 images from H&E stained, consisting of achieving nine class sizes greater than 97%. Five CNN models, namely, Inception v3, DenseNet, SqueeseNet, Resnet, and VGG, are trained, validated and tested with 70%, 15%, and 15% of the dataset, respectively. It is reported that the VGG outperforms other models with 98.86% accuracy. The model is used to segment whole slide images and investigates cohort TCGA colon cancer patients with TILAB and TSR scores. Some more literature related to colon cancer are presented in Table 9.1.

Materials and methods

This section discusses the implementation of Deep Neural network and Genetic Algorithm and its functionalities. The details of a hybrid model of a deep neural network trained with a genetic algorithm is presented and explained.

Deep neural network

A DNN can be described as an artificial neural network composed of multiple hidden layers with a complex structure to enable high performance and complex computation of

TABLE 9.1 The literature on colon cancer.

Intelligent method	Compared methods	Datasets (source)	Evaluation factors	Limitation	Ref
ResNet50	AlexNet, VGG16, VGG19, DenseNet121, InceptionV3	CRCHistoPhenotypes	Precision, Recall, F1-measure, AUC, Cross-entropy loss	The discriminative power of the model is compromised when used on the complex dataset with high intra-class heterogeneity, procedural artifacts, and low resolution	Tripathi and Singh (2020)
Nested Genetic Algorithm	GA-SVM, GA-NNW, K-Nearest Neighbor, Random Forest	TCGA Gene Expression data, GEO Gene Expression data, DNA Methylation	Accuracy	The experiment is conducted with a reduced number of genes from the dataset	Sayed et al. (2019)
ResNet50	ResNet101, Xception, VGG19, and Inception V3	Urduliz and Biodonostia Hospital	Accuracy, Precision, Recall, F1-score, Specificity	The performance of the model is limited to the quality of the data	Bour et al. (2019)
Convolutional neural network	Inception, Densenet, SE-RX50, VGG-16	University Hospital Schleswig-Holstein in Lübeck	Accuracy, Sensitivity, Specificity, F1-Score, and AUC	The proposed model is limited to the transferability of features	Gessert et al. (2019)
Convolutional neural network	Logistic Regression, K-Nearest Neighbor	Synthetic data	Accuracy, Precision, Recall, and F1 Score	The model proposed is limited to few patient samples and the number of patients	Mobilia et al. (2019)
RCCNet	Softmax CNN_IN27, softmaxCNN, AlexNet, CIFAR-VGG, GoogleNet, and WRN	CRCHistoPhenotypes	Accuracy, weighted average F1 score	The methods are modified to fit low-resolution images	Shabbeer Basha et al. (2018)
Long Short-Term Memory, Gated Recurrent Units	CART, Logistic Regression, Support Vector Machine, Random Forest	Julius Center	AUC	The dataset used is only 6 months before diagnosis	Amirkhan et al. (2018)
T-ladder CNN	Baseline CNN, Fine-tuned CNN	H&*E*-stained prostate biopsies	Precision, recall, F1 score, and AUPR	The experiment is conducted with limited labeled data	Cardoso et al. (2017)

Continued

TABLE 9.1 The literature on colon cancer—cont'd

Intelligent method	Compared methods	Datasets (source)	Evaluation factors	Limitation	Ref
Particle Swarm Optimization and Support Vector Machine	Naïve Bayes, Genetic Algorithm, and Decision Tree	DNA Microarray	Accuracy	The dataset used is reduced to a selected subset of features	Al-Rajab et al. (2017)
Bilinear Convolutional neural network	CNN, CNN-H, CNN-E	H&*E*-stained histology image	Accuracy, Sensitivity, Specificity	The feature used for the proposed method has high dimensionality	Wang et al. (2017)
Ensembled Deep neural network	PangNet, Fully Convolutional Net, DeconvNet	H&E-stained histology image	Accuracy, Precision, Recall, F1 score, IU, Performance	The dataset does not consider morphological variability, which may occur in real cases	Naylor et al. (2017)
SC-SNN with Neighboring Ensemble Predictor	Softmax CNN+SSPP, superpixel descriptor, CRImage	H&*E*-stained histology image	Precision, Recall, F1 Score, and Median Distance	The experiment is limited to a specific size of nuclei in the dataset	Sirinukunwattana et al. (2016)
Convolutional neural network	AlexNet, GoogleNet	H&E-stained tissue	Jaccard Index, Dice Coefficient	The starting learning rate is reduced for ImageNet training	Li et al. (2016)
Convolutional neural network	GoogleNet, CNN-S, AlexNet MCN.	CC-i-Scan	Accuracy	Only one performance measure is used to evaluate the performance	Ribeiro et al. (2017)
Convolutional Neural Network with featured learning	Support vector Machine Kernel	PWH database, Two publicly available non-medical	Accuracy, Precision, Recall, and F1 score	This study is limited to the manual selection of the region of interest for every polyp	Zhang et al. (2017)
SC-CNN	SSAE algorithm, SC-CNN	H&E-stained	Precision, Recall, F1 score	The dataset used for the experiment is small size	Kashif et al. (2016)

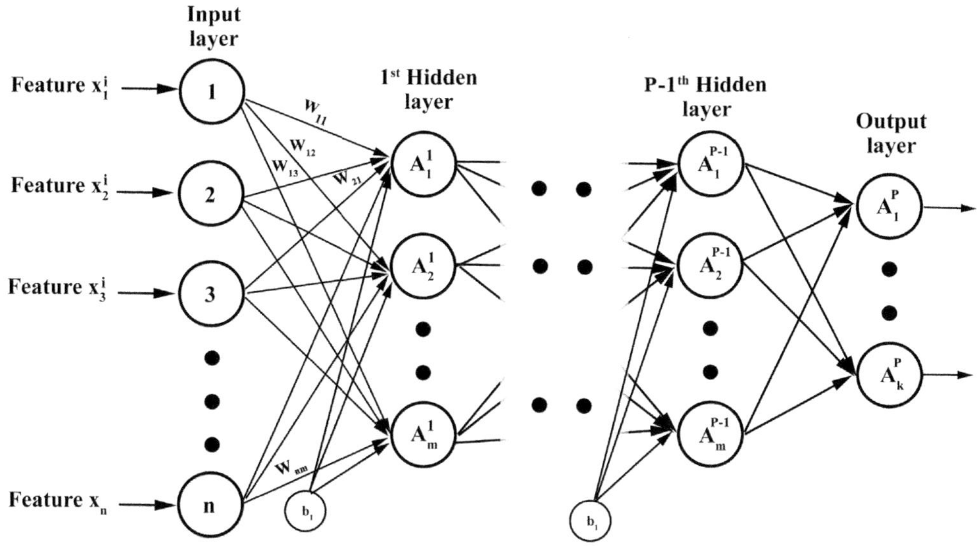

FIG. 9.1 Deep neural network architecture.

linear and non-linear problems. DNN takes a set of inputs, applying mathematical principles to turns the input into the desired output. At each layer, the probability of the output is calculated and passed to the forward layer. Initially, the random weights are initialized and gradually updated according to the output response. The weights are updated using a learning algorithm until the desired output is obtained. The inputs are associated with different weight values, which are summed together on a non-linear function to give the final output (Schmidhuber, 2015).

Fig. 9.1 presents the process of workflow of a deep neural network. Here, each layer may contain different initialization methods, different activation functions, different number of neurons, and different number of outputs. The selection of these parameter affects the performance of the DNN. The number of nodes and hidden layers influences the output, and its value depends on the data fed to the model. Setting the optimal values to the parameter takes multiple attempts and searching in a large search space. In this chapter, the features' optimization uses the GA (Koo et al., 2019; Martin et al., 2017).

Genetic algorithm

GA is an optimization method and stochastic search derived from evolutionary computation. This algorithm adopts the natural selection process where the candidates/individuals of a population are selected to reproduce next-generation offspring with improved and better fitness. For a specific problem, each candidate solution is considered a chromosome or individual representing a point in the search space, and a population is a collection of such individuals. Each individual is assigned to a fitness value through an evaluating mechanism. A new population referred as 'generation' is generated iteratively based on the fitness value and by undergoing genetic operations. The genetic operation mainly consists

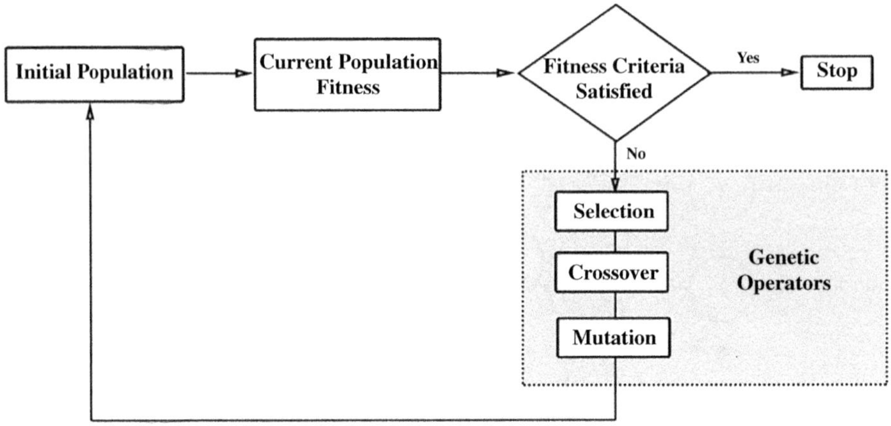

FIG. 9.2 GA workflow.

of reproduction, crossover, and mutation. Reproduction is the process of reproducing highly rated individuals of the current generation to a new generation. Crossover is a process where two new offspring are reproduced by exchanging alternative pairs of selection between the crossings sites selected uniformly. The mutation process is an alternation of some gene values randomly in an individual. This process is iterated as a specified number of generation-producing populations of individuals in each generation to maximize the fitness value. The optimal values of the parameter are identified based on the chromosomes with the best fitness value (Alonso et al., 2021).

Fig. 9.2 presents the process of workflow in the GA. The first initial population is created, and the fitness value of the current population is determined. Then, it is compared with the fitness criteria and checked if the fitness criteria are satisfied or not. The new population is created using genetic operations, including selection, crossover, and mutation. This process is repeated until the stopping criteria are satisfied, and the best fitness value is obtained.

Proposed method

The proposed method implemented on the microarray gene dataset consists of a small number of observation and thousands of genes are there in each observation. Among these thousands of genes, only a few numbers of genes may be strongly correlated to detection of colon cancer. This small number of genes is called as biomarker genes, which can identify the biological relationship between the genes and the target disease. The selection of the genes is processed by converting the input space into high-dimensional space and classifying the input data. Use of biomarker genes in detection of colon cancer increases the classification accuracy and reduce the computational errors (Sayed et al., 2019).

The deep neural network is first fed with the input data with various input features. At first, random weights and bias are allocated to each neuron in every layer of the network.

The output of the network is calculated by applying the activation function on the neuron. GA is used as a learning algorithm to determine the parameters for the DNN. An initial population is created with the predefined number of individuals, where each individual represents a possible solution. The fitness value is calculated for each of the individual solution in the current population. Based on the fitness of the current population, the reproduction, crossover, and mutation operation are performed. The optimal solution is obtained once the stopping criteria are reached. The DNN's final output is calculated based on the optimal value and the updated parameters.

Dataset $X=\{x_i, t_i\}_{i=1}^{n}$ where x_i is microarray-based gene expression data and t_i represents the target class labels tumor and normal. Initially, $P_i=\{w_i, b_i\}$ (i^{th} chromosome) in $P=\{P_i\}_{i=1}^{M}$ are evaluated in the deep neural network model for recognition of colon cancer (Algorithm 9.2). The accuracy obtained from DNN corresponding to P_i has been recorded as fitness f_i (Algorithm 9.2). Similarly, $f=\{f_1,f_2,\ldots,f_n\}$ has been computed for all P_i, $i=1, 2, \ldots, n$. The crossover pool is created using route wheel selection strategy, where the fittest solutions are given higher chance to pass on to the successive generation. Update the population by comparing the fitness. Check the eligibility for mutation operation as per the mutation rate and apply mutation on P if it is found eligible. Finally return the best P_i from P.

The detailed procedure of the proposed DNN with GA is presented in the Algorithms 9.1 and 9.2. The working of the proposed approach has been presented in Fig. 9.3.

ALGORITHM 9.1

Proposed deep neural network with genetic algorithm.

1. Initialize population P with M number of individual solutions $P=\{P_1, P_2, \ldots, P_M\}$, where $P_i=\{w_i, b_i\}$ represents the ith solution
2. Set the number of generations g
3. Repeat for $i=1$ to g
4. Compute the fitness of each $P_i=\{w_i, b_i\}$ in DNN with dataset $X=\{x_i, t_i\}_{i=1}^{n}$
5. For $i=1$ to M repeat $f_i=DNN(P_i, X)$ and get $f=\{f_i\}_{i=1}^{M}$
6. Select K number of P_i using route wheel selection and construct a crossover pool by P'_{Pool}
7. Generate new population P'_{Pool} by applying crossover on P'_{Pool}
8. Update the population by comparing fitness of P and P'_{cross}
9. Check the eligibility for mutation operation as per the mutation rate and apply mutation on P if it is found eligible.
10. if (stopping condition is reached), then go to step 11, Else go to step 3
11. Return best P_i from P

ALGORITHM 9.2

Evaluation of fitness of P_i in DNN with $\mathcal{X}$.

1. Initialize training epoch, number of layers is l
2. For all x_j=1, j=1 to n, compute output Z_j: $Z_j = w_j x_j + b_j$
3. Calculate net output A_i: $A_j = Relu(Z_j)$
4. Propagate the output to layer$l$and calculate the net output: $A_{l-1} = Relu(Z_{l-1})$and $Z_l = w_l A_{l-1} + b_l$
5. Output prediction is calculated as $o_j = \sigma(Z_j)$
6. Compute error: $e_j = \| t_j - o_j \|$
7. Compute total root mean square error: $RMSE = \sqrt{\frac{\sum_{j=1}^{n}(t_j - o_j)^2}{n}}$
8. Calculate the fitness of: $f_i = \frac{1}{RMSE_{P_i}}$

Return f_i

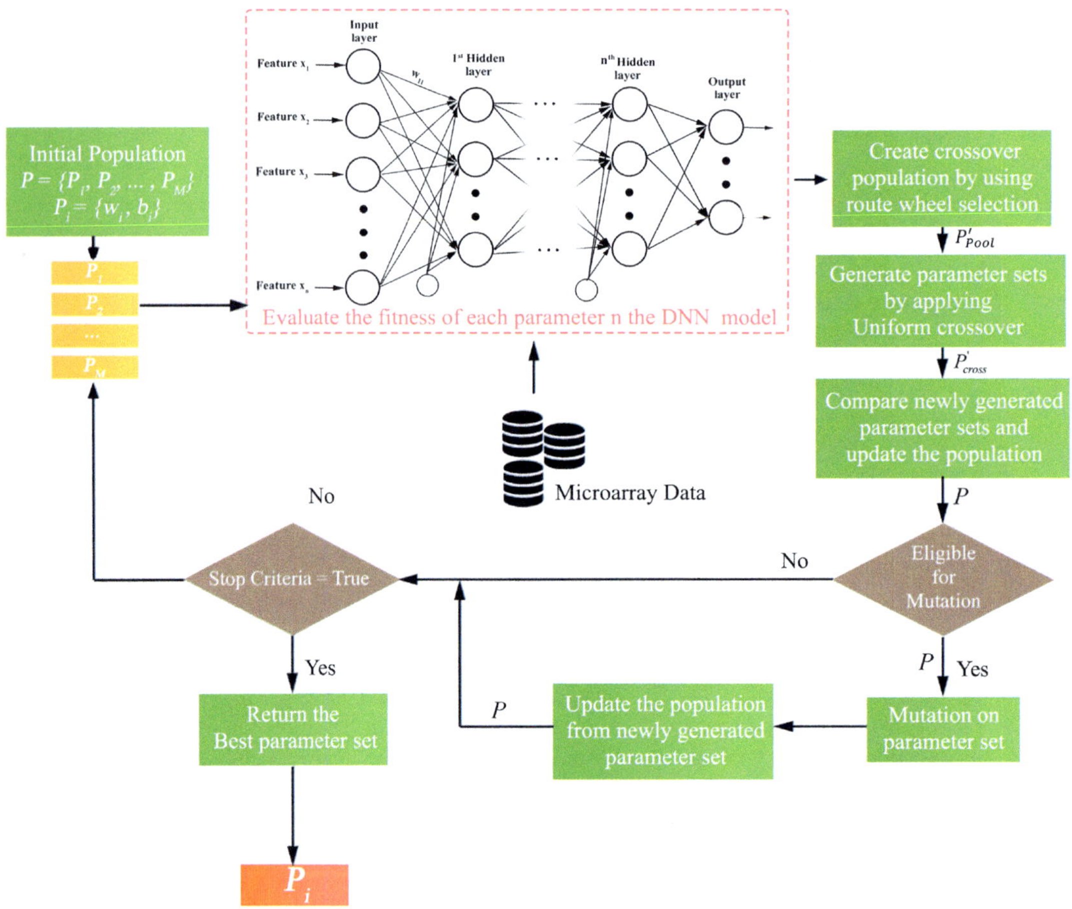

FIG. 9.3 Deep neural network trained with genetic algorithm workflow.

Experimental setup

This section discusses the dataset and its attributes on which the experiment is conducted. Also, this section presents data preprocessing, the performance measures of proposed method, and comparison with other machine learning methods.

Dataset preparation

Analysis and diagnosis of colon cancer use a microarray-based gene expression dataset widely. The DNA microarray dataset consists of thousands of genes and allows studying the expression level of multiple genes simultaneously. It consists of about 65,000 features, each with $\approx 10^7$ Strands of DNA 25-per oligonucleotide. Humann cDNAs of about 3200 full length and 3400 ESTs similar to other eukaryotic genes are presented as a set of four chips with 20 feature pairs representing each EST and each feature containing 25-bp sequence (Åkerrén, 1951). In this work, the data related to colon cancer contain 62 adenocarcinoma specimens of different patients classified into two classes: tumor and normal. Among which, 40 belongs to the tumor (positive patients) and 22 belongs to normal (negative patient). The gene expression dataset is expressed with 2000*62, which is the expression results of 2000 genes obtained across 62 patients with the highest minimal intensities (Jingjing, 2017).

Data preprocessing in the experimentation includes data normalization, thresholding, filtering, and logarithmic transformation, and it is an essential step to be carried out before feeding data to model for classification. Preprocessing of data is carried out using log normalization, quantile normalization, and standardization. Log normalization rescales the data to match its neighboring data based on logical reasoning, and it eliminates systematic errors from log curves. Applying log normalization on the microarray dataset will give normally distributed data. Quantile normalization replaces the dataset's value with the variable of the same rank in the reference distribution. Each array's data is given a rank first, and then data of the same rank are distributed across the array. It removes unwanted technical variation and reduces bias and mean square errors. Standardization is a process of rescaling the values to be centered around a mean with a unit standard deviation, that is, to have a mean value between 0 and a standard derivation of 1 (Wu et al., 2005). After the normalization and standardization of data, the dataset is fed to the model. While preprocessing of data, 89 duplicate genes are removed.

Simulation environment

The experiment has been conducted on Windows 10 Pro, 64-bit operating system, Intel(R) Core (TM) i5- 6300 U CPU @ 2.40GHz, ~2.5GHz processor, and 8 GB RAM. The proposed and comparison models have been simulated using a python programming environment. The python programming setup includes the Numpy and Pandas framework used for data analysis, the sklearn framework used for machine learning classification models and data preprocessing, Matplotlib and seaborn frameworks data visualization, and pyGAD used for building and optimization of proposed DNN trained by genetic algorithms. The

TABLE 9.2 Proposed and comparative models parameter setting.

Method	Parameter
K-Nearest Neighbor (KNN)	n_neighbors=15,algorithm='kd_tree'
Stochastic Gradient Descent (SGD)	loss='modified_huber',random_state=1,max_iter=600
Random Forest (RF)	n_estimators=50,max_depth=5,random_state=1
Decision Tree (DT)	max_depth=4,criterion='entropy',splitter='best',.
Multi-Layer Perceptron (MLP)	activation='identity', batch_size=20, random_state=2
Deep Neural Network (DNN)	num_neurons_hidden_layers=[20, 10],num_neurons_output=1,hidden_activations=[" sigmoid ", "sigmoid"],output_activation=" sigmoid "
Proposed Model (DNN +GA)	num_solutions=15, num_neurons_input=35,num_neurons_hidden_layers=[30, 20, 15, 10],num_neurons_output=2,hidden_activations=["relu", "relu", "relu", "relu"], output_activation="softmax".

parameter setting of the proposed model and comparative ensemble and machine learning methods are presented in Table 9.2.

Performance measures

The experiment is carried out on proposed methods along with several other machine learning classifiers. The proposed method and other machine learning classifier methods are validated using different performance measures such as Accuracy, True Positive, False Positive, True Negative, False Negative, True Positive Rate, False Positive Rate, True Negative rate, precision, and F1 score.

Accuracy

Accuracy can be defined as the measure of closeness between the detected value and the actual true value. It can be represented as Eq. (9.1)

$$\text{Accuracy} = \frac{\text{TP} + \text{TN}}{\text{TP} + \text{FP} + \text{TN} + \text{FN}} \tag{9.1}$$

True positive rate (TPR)

True Positive Rate is the proportion of true positive, which are classified correctly. It is represented in Eq. (9.2).

$$\text{TPR} = \frac{\text{TP}}{\text{TP} + \text{FN}} \tag{9.2}$$

False positive rate (FPR)

False Positive rate is the ratio of the number of negative cases categorized as positive to the total number of negative cases. It is represented in Eq. (9.3).

$$\mathrm{FPR} = \frac{\mathrm{FP}}{\mathrm{FP} + \mathrm{TN}} \tag{9.3}$$

True negative rate (TNR)

True Negative Rate is the proportion of negative cases which are classified correctly. It is represented in Eq. (9.4).

$$\mathrm{TNR} = \frac{\mathrm{TN}}{\mathrm{TN} + \mathrm{FP}} \tag{9.4}$$

Precision

Precision is defined as the fraction of correct results to the total results returned. It is represented in Eq. (9.5).

$$\mathrm{Precision} = \frac{\mathrm{TP}}{\mathrm{TP} + \mathrm{FP}} \tag{9.5}$$

F1-score

F1-Score is defined as the weighted average of precision and recall and can be represented as Eq. (9.6)

$$F1 - \mathrm{Score} = 2 * \frac{\mathrm{Precision} \times \mathrm{Recall}}{\mathrm{Precision} + \mathrm{Recall}} \tag{9.6}$$

AUC

AUC stands for "Area under ROC Curve," a measure of a classifier's ability to distinguish between classes. The higher the AUC, the better the performance of the model at distinguishing between the classes. It can be represented as Eq. (9.7).

$$\mathrm{AUC} = \frac{\mathrm{TPR} + \mathrm{TNR}}{2} \tag{9.7}$$

Result analysis

The method proposed in this chapter to classify the colon cancer patients through a hybrid deep learning-based model. Experimentation has been conducted on a microarray colon cancer dataset. PyGAD library is utilized to build the model deep neural network trained by a genetic algorithm.

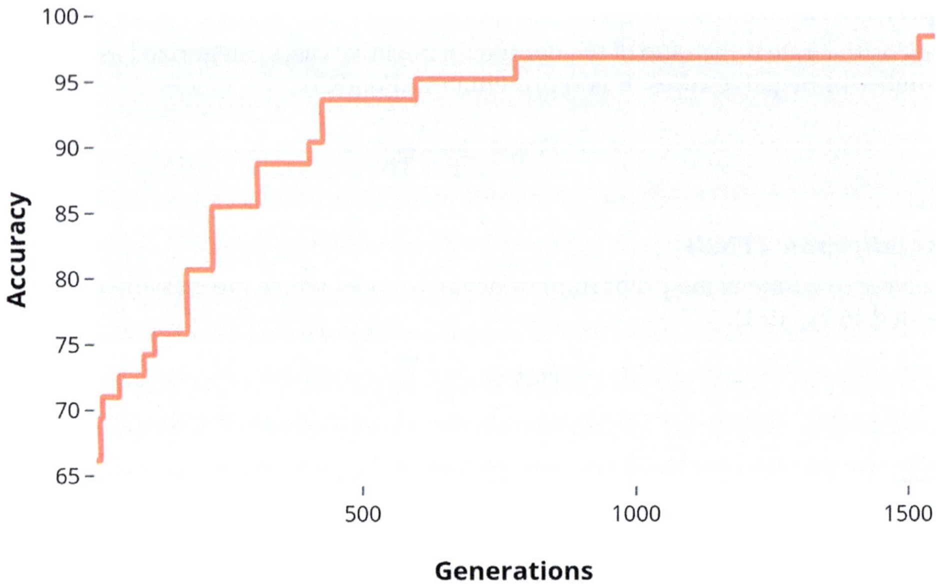

FIG. 9.4 Evolving of fitness value for each generation iteration.

Fig. 9.4 shows the fitness generation in contrast to the number of epochs. The fitness value of the best solution is recorded as 98.39, and it reached its best solution after the 1521st generation.

As the number of generations passes, the accuracy value increases, and the error value decreases. Fig. 9.5 shows the reduction in error value for each generation iteration in the GA. The least error value is recorded as 1.61, and it is reached after the 1521st generation.

The proposed model's performance is compared with that of other machine learning classifiers such as a k-nearest neighbor, stochastic gradient descent, random forest, decision tree, multi-layer perception, and deep neural network and validated using performance measurements accuracy, true positive rate, false positive rate, precision, true negative rate, f1-score, and AUC. The proposed method's performance and that of other comparative methods with each performance measure are shown in Table 9.3.

Based on the above table results, the proposed methods' performance shows an accuracy of 98.39%. The classification of data using the proposed method shows significant results with a true positive rate of 1.00, false positive rate of 0.05, and true negative rate of 0.95. The proposed method's precision value is 0.98, f1-score is 0.99, and the AUC value is 0.98. Accuracy alone cannot be trusted to select a well-performed model as it does not include the incorrect classified cases due to the class imbalance issue in the considered data. The performance measures such as precision, f1-score, TPR, FPR, TNR, and AUC help measure the model's exactness and completeness and keep a balance between them. So, using these performance measurements altogether gives a better understanding of a well-performed model.

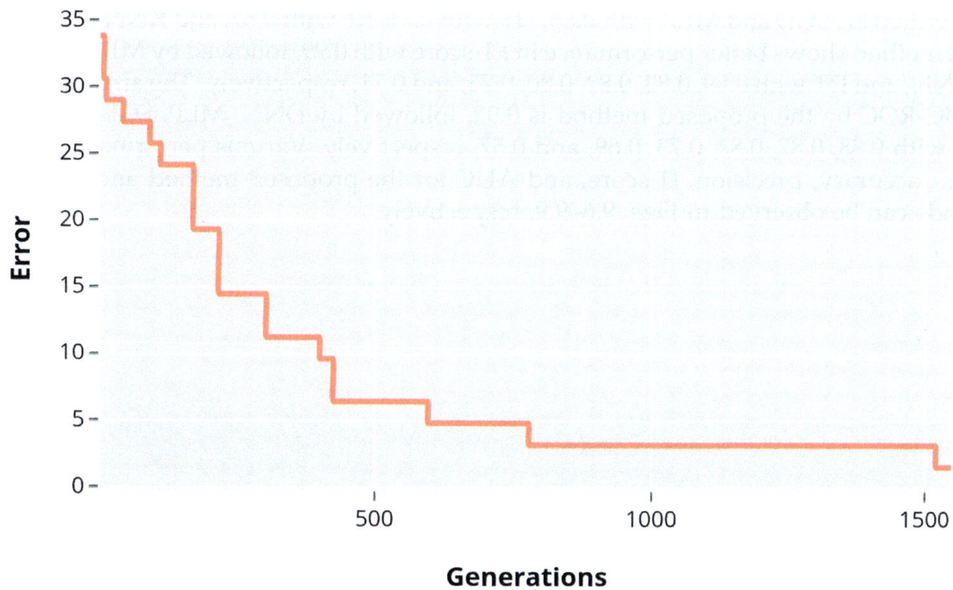

FIG. 9.5 Plot showing error value for each generation iteration.

TABLE 9.3 The performance measure of the proposed method and other comparative methods.

Technique	Accuracy	TPR	FPR	Precision	TNR	*F*1-score	AUC
DNN+GA	98.39	1.00	0.05	0.98	0.95	0.99	0.98
DNN	89.47	0.88	0.12	0.83	0.87	0.83	0.88
KNN	62.50	1.00	0.86	0.60	0.14	0.75	0.57
SGD	81.25	0.78	0.14	0.88	0.86	0.82	0.82
RF	75.00	0.89	0.43	0.73	0.57	0.80	0.73
DT	68.75	0.67	0.29	0.75	0.71	0.71	0.69
MLP	87.50	0.89	0.14	0.89	0.86	0.89	0.87

The proposed method shows better results than the comparative methods in case of all performance measures used in this experiment. The proposed method is with significant accuracy of 98.39, followed by DNN, MLP, SGD, RF, DT, and KNN with an accuracy of 89.47, 87.50, 81.25, 75.00, 68.75, and 62.50, respectively. In the case of the TPR, the proposed method and KNN perform equally with an outcome of 1.00, followed by RF and MLP with 0.89, DNN with 0.88, and SGD with 0.78, and DT 0.67. In terms of FPR (falsely classified positive cases), the proposed method scored minimum error among the comparative methods with the value of 0.05, followed by DNN with 0.12, MLP and SGD scoring equally with 0.14, DT with 0.29, RF with 0.43, and KNN with 0.89. The proposed method performs well in precision with 0.98 followed by MLP, SGD, DNN, DT, RF, and KNN with values 0.89, 0.88, 0.83, 0.75, 0.73, and 0.60, respectively. The proposed method's TNR value is better with 0.95, followed by

DNN with 0.87, SGD and MLP with 0.86, DT with 0.71, RF with 0.57, and KNN 0.14. The proposed method shows better performance in $F1$-score with 0.99, followed by MLP, DNN, SGD, RF, KNN, and DT with 0.89, 0.83, 0.82, 0.80, 0.75, and 0.71, respectively. The area accumulated in AUC-ROC by the proposed method is 0.98, followed by DNN, MLP, SGD, RF, DT, and KNN with 0.88, 0.87, 0.82, 0.73, 0.69, and 0.57, respectively. Various performance measures such as accuracy, precision, f1-score, and AUC for the proposed method and comparative methods can be observed in Figs. 9.6–9.9, respectively.

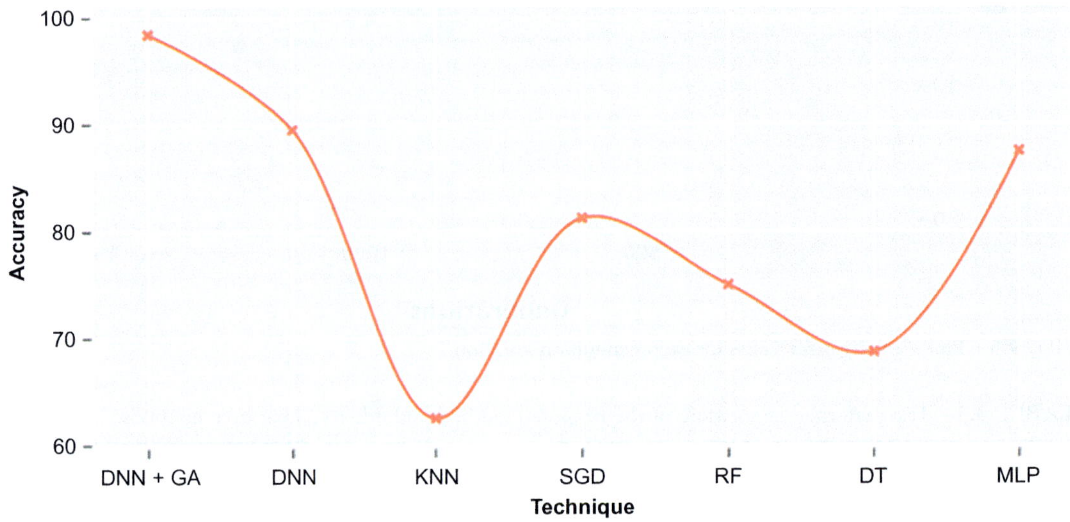

FIG. 9.6 Analysis of accuracy among all methods.

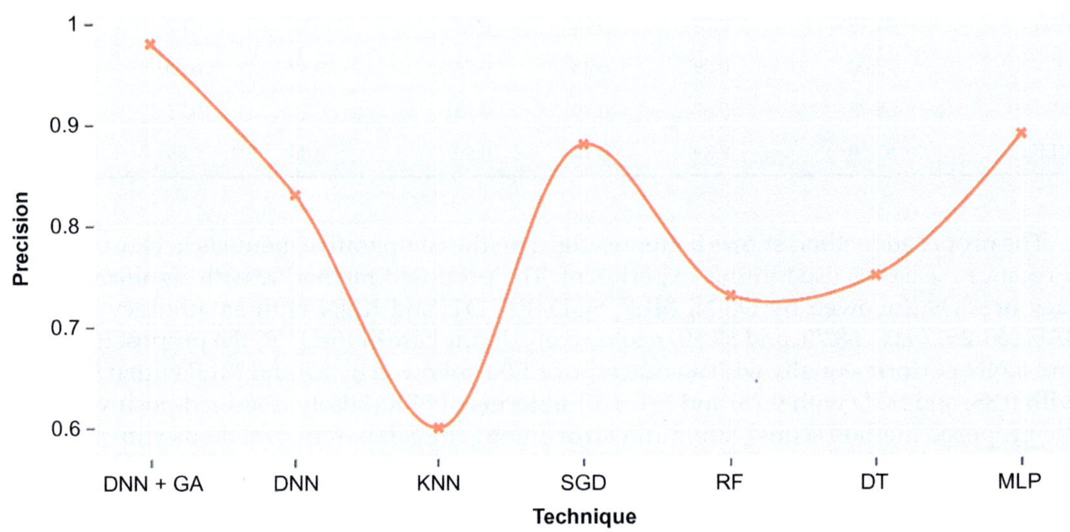

FIG. 9.7 Analysis of precision among all methods.

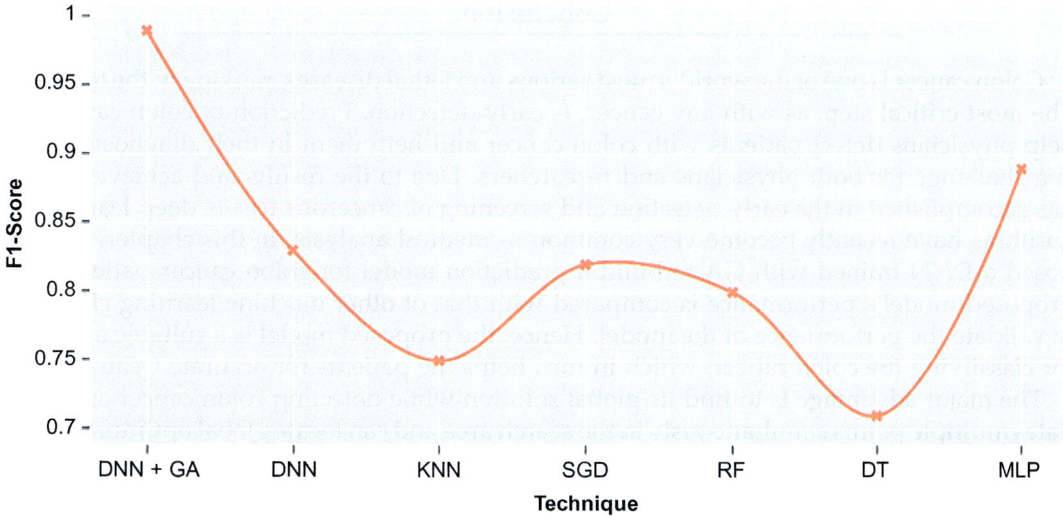

FIG. 9.8 Analysis of *F*1-score among all methods.

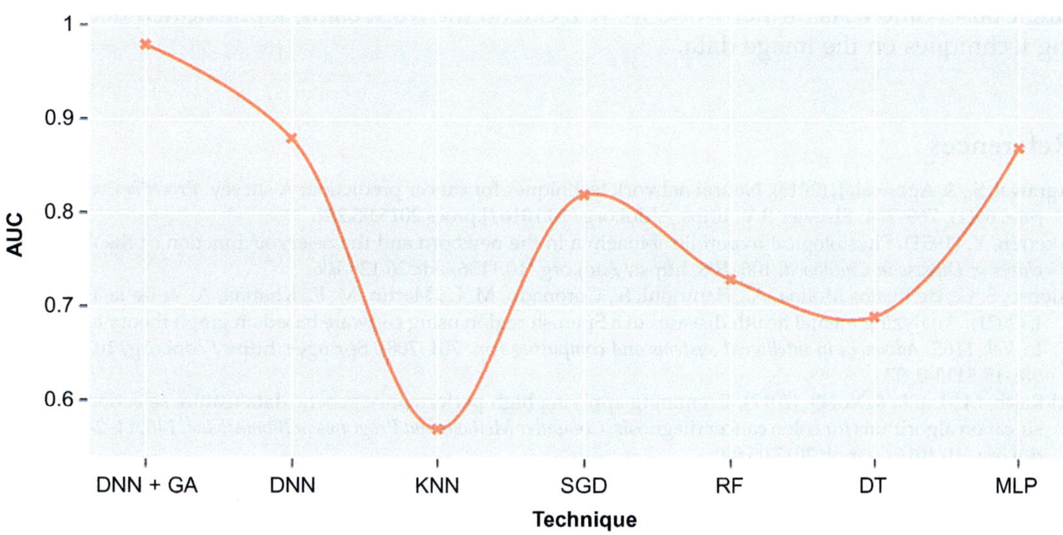

FIG. 9.9 Analysis of AUC among all methods.

From the results of proposed and compared models, it is proven that the proposed model outperforms the other comparative methods as the comparative methods have scored lower performance in various measures such as accuracy, TPR, FPR, TNR, precision, f1-score, and AUC on the microarray dataset for colon cancer classification.

Conclusion

Colon cancer is one of the world's most serious and lethal diseases, ranking in the top three. The most critical step, as with any cancer, is early detection. Prediction of colon cancer can help physicians detect patients with colon cancer and help them in their diagnosis, and it is a challenge for both physicians and researchers. Due to the results and achievements it has accomplished in the early detection and screening of cancerous tissue, deep learning algorithms have recently become very common in medical analysis. In this chapter, we proposed a DNN trained with GA to build a prediction model for colon cancer patients. The proposed model's performance is compared with that of other machine learning classifiers to validate the performance of the model. Hence, the proposed model is a suitable approach for classifying the colon cancer, which in turn helps the patients for accurate treatment.

The major advantage is to find its global solution while detecting colon cancer as it evaluates multiple points simultaneously in the search area and can locate global optimum as they create and search for a population of points instead of single points. This method provides a set of optimal solutions from which the most appropriate one is accepted. This experiment is limited to number of patient cases as the used dataset contains only few sample records of colon cancer. This work does not include colonoscopy images, videos, and medical image analysis. The challenge faced in this work is computation efficiency of the microarray dataset due to its complex and high-dimensional data structure consisting of thousands of genes for a single observation. As further work, we will extend the work using sophisticated deep learning techniques on the image data.

References

Agrawal, S., & Agrawal, J. (2015). Neural network techniques for cancer prediction: A survey. *Procedia Computer Science, 60*(1), 769–774. Elsevier B.V. https://doi.org/10.1016/j.procs.2015.08.234.

Åkerrén, Y. (1951). Physiological hyperbilirubinaemia in the newborn and the reservoir function of the spleen. *Archives of Disease in Childhood*, 106–108. https://doi.org/10.1136/adc.26.126.106.

Alonso, S. G., de Bustos Molina, A., Hamrioui, S., Coronado, M. L., Martín, M. F., Khanna, A., & de la Torre Díez, I. (2021). Analyzing mental health diseases in a Spanish region using software based on graph theory algorithms. In *Vol. 1165. Advances in intelligent systems and computing* (pp. 701–708). Springer. https://doi.org/10.1007/978-981-15-5113-0_57.

Al-Rajab, M., Lu, J., & Xu, Q. (2017). Examining applying high performance genetic data feature selection and classification algorithms for colon cancer diagnosis. *Computer Methods and Programs in Biomedicine, 146*, 11–24. https://doi.org/10.1016/j.cmpb.2017.05.001.

American Cancer Society: Cancer Facts and Figures 2021 (pp. 13–15). (2021).

Amirkhan, R., Hoogendoorn, M., Numans, M. E., & Moons, L. (2018). Using recurrent neural networks to predict colorectal cancer among patients. In (Vols. 2018). *2017 IEEE symposium series on computational intelligence, SSCI 2017 - Proceedings* (pp. 1–8). Institute of Electrical and Electronics Engineers Inc. https://doi.org/10.1109/SSCI.2017.8280826.

Arockia, V. C. D., Devaraj, D., & Venkatesulu, M. (2015). Gene expression data classification using support vector machine and mutual information-based gene selection. *Procedia Computer Science*, 13–21. https://doi.org/10.1016/j.procs.2015.03.178.

Ayyad, S. M., Saleh, A. I., & Labib, L. M. (2019). Gene expression cancer classification using modified K-nearest neighbors technique. *Biosystems, 176*, 41–51. https://doi.org/10.1016/j.biosystems.2018.12.009.

Basavegowda, H. S., & Dagnew, G. (2020). Deep learning approach for microarray cancer data classification. *CAAI Transactions on Intelligence Technology, 5*(1), 22–33. https://doi.org/10.1049/trit.2019.0028.

Bouazza, S. H., Hamdi, N., Zeroual, A., & Auhmani, K. (2015). Gene-expression-based cancer classification through feature selection with KNN and SVM classifiers. In *2015 Intelligent systems and computer vision, ISCV 2015* Institute of Electrical and Electronics Engineers Inc. https://doi.org/10.1109/ISACV.2015.7106168.

Bour, A., Castillo-Olea, C., Garcia-Zapirain, B., & Zahia, S. (2019). Automatic colon polyp classification using convolutional neural network: A case study at Basque Country. In *2019 IEEE 19th international symposium on signal processing and information technology, ISSPIT 2019* Institute of Electrical and Electronics Engineers Inc. https://doi.org/10.1109/ISSPIT47144.2019.9001816.

Cardoso, M. J., Arbel, T., Lee, S. L., Cheplygina, V., Balocco, S., Mateus, D., Zahnd, G., Maier-Hein, L., Demirci, S., Granger, E., et al. (2017). Intravascular imaging and computer assisted stenting, and large-scale annotation of biomedical data and expert label synthesis. In M. J. Cardoso, T. Arbel, S.-L. Lee, V. Cheplygina, S. Balocco, D. Mateus, G. Zahnd, L. Maier-Hein, S. Demirci, E. Granger, L. Duong, M.-A. Carbonneau, S. Albarqouni, & G. Carneiro (Eds.), CVII-STENT and Second International Workshop, LABELS (Vol. 10552). Springer International Publishing. https://doi.org/10.1007/978-3-319-67534-3.

Chen, H., Zhao, H., Shen, J., Zhou, R., & Zhou, Q. (2015). Supervised machine learning model for high dimensional gene data in colon cancer detection. In *Proceedings - 2015 IEEE international congress on big data, BigData congress 2015* (pp. 134–141). Institute of Electrical and Electronics Engineers Inc. https://doi.org/10.1109/BigDataCongress.2015.28.

Gedeon, R., Nagar, A. K., & Naguib, R. (2020). Using convolutional neural networks to predict colon cancer patients survival. In *Vol. 1139. Advances in intelligent systems and computing* (pp. 47–55). Springer. https://doi.org/10.1007/978-981-15-3287-0_4.

Geetha, R., Sivasubramanian, S., Kaliappan, M., Vimal, S., & Annamalai, S. (2019). Cervical cancer identification with synthetic minority oversampling technique and PCA analysis using random forest classifier. *Journal of Medical Systems, 43*(9). https://doi.org/10.1007/s10916-019-1402-6.

Gessert, N., Bengs, M., Wittig, L., Drömann, D., Keck, T., Schlaefer, A., & Ellebrecht, D. B. (2019). Deep transfer learning methods for colon cancer classification in confocal laser microscopy images. *International Journal of Computer Assisted Radiology and Surgery, 14*(11), 1837–1845. https://doi.org/10.1007/s11548-019-02004-1.

Javadi, A. A., Farmani, R., & Tan, T. P. (2005). A hybrid intelligent genetic algorithm. *Advanced Engineering Informatics, 19*(4), 255–262. https://doi.org/10.1016/j.aei.2005.07.003.

Jimoh, K., Yakub, S., & Alimi, M. (2018). *Colon cancer classification and patients' survival detection using support vector machine kernels*. Proceedings of the 14th iSTEAMS International Multidisciplinary Conference, AlHikmah University, Ilorin, Nigeria

Jingjing, W. (2017). Feature selection for cancer classification using microarray gene expression data. *Biostatistics and Biometrics Open Access Journal*. https://doi.org/10.19080/BBOAJ.2017.01.555557.

Juang, C. F. (2004). A hybrid of genetic algorithm and particle swarm optimization for recurrent network design. *IEEE Transactions on Systems, Man, and Cybernetics, Part B: Cybernetics, 34*(2), 997–1006. https://doi.org/10.1109/TSMCB.2003.818557.

Kashif, M. N., Raza, S. E. A., Sirinukunwattana, K., Arif, M., & Rajpoot, N. (2016). Handcrafted features with convolutional neural networks for detection of tumor cells in histology images. In (Vols. 2016). *Proceedings - International symposium on biomedical imaging* (pp. 1029–1032). IEEE Computer Society. https://doi.org/10.1109/ISBI.2016.7493441.

Koo, Y. D., An, Y. J., Kim, C. H., & Na, M. G. (2019). Nuclear reactor vessel water level prediction during severe accidents using deep neural networks. *Nuclear Engineering and Technology, 51*(3), 723–730. https://doi.org/10.1016/j.net.2018.12.019.

Li, W., Manivannan, S., Akbar, S., Zhang, J., Trucco, E., & McKenna, S. J. (2016). Gland segmentation in colon histology images using hand-crafted features and convolutional neural networks. In (Vols. 2016). *Proceedings - International symposium on biomedical imaging* (pp. 1405–1408). IEEE Computer Society. https://doi.org/10.1109/ISBI.2016.7493530.

Martin, A., Fuentes-Hurtado, F., Naranjo, V., & Camacho, D. (2017). Evolving deep neural networks architectures for android malware classification. In *2017 IEEE congress on evolutionary computation, CEC 2017 - Proceedings* (pp. 1659–1666). Institute of Electrical and Electronics Engineers Inc. https://doi.org/10.1109/CEC.2017.7969501.

Mobilia, S., Sirkeci-Mergen, B., Deal, J., Rich, T. C., & Leavesley, S. J. (2019). Classification of hyperspectral colon cancer images using convolutional neural networks. In *2019 IEEE data science workshop, DSW 2019 - Proceedings* (pp. 232–236). Institute of Electrical and Electronics Engineers Inc. https://doi.org/10.1109/DSW.2019.8755582.

Naylor, P., Lae, M., Reyal, F., & Walter, T. (2017). Nuclei segmentation in histopathology images using deep neural networks. In *Proceedings - International symposium on biomedical imaging* (pp. 933–936). IEEE Computer Society. https://doi.org/10.1109/ISBI.2017.7950669.

Pacal, I., Karaboga, D., Basturk, A., Akay, B., & Nalbantoglu, U. (2020). A comprehensive review of deep learning in colon cancer. *Computers in Biology and Medicine, 126*. https://doi.org/10.1016/j.compbiomed.2020.104003.

Pradhan, K., & Chawla, P. (2020). Medical internet of things using machine learning algorithms for lung cancer detection. *Journal of Management Analytics, 7*(4), 591–623. https://doi.org/10.1080/23270012.2020.1811789.

Ribeiro, E., Uhl, A., Wimmer, G., & Häfner, M. (2017). Transfer learning for colonic polyp classification using off-the-shelf CNN features. In *Vol. 10170. Lecture notes in computer science (including subseries lecture notes in artificial intelligence and lecture notes in bioinformatics)* (pp. 1–13). Springer Verlag. https://doi.org/10.1007/978-3-319-54057-3_1.

Roadknight, C., Suryanarayanan, D., Aickelin, U., Scholefield, J., & Durrant, L. (2015). An ensemble of machine learning and anti-learning methods for predicting tumour patient survival rates. In *Proceedings of the 2015 IEEE international conference on data science and advanced analytics, DSAA 2015* Institute of Electrical and Electronics Engineers Inc. https://doi.org/10.1109/DSAA.2015.7344863.

Sayed, S., Nassef, M., Badr, A., & Farag, I. (2019). A nested genetic algorithm for feature selection in high-dimensional cancer microarray datasets. *Expert Systems with Applications, 121*, 233–243. https://doi.org/10.1016/j.eswa.2018.12.022.

Schmidhuber, J. (2015). Deep learning in neural networks: An overview. *Neural Networks, 61*, 85–117. https://doi.org/10.1016/j.neunet.2014.09.003.

Shabbeer Basha, S. H., Ghosh, S., Kishan Babu, K., Ram Dubey, S., Pulabaigari, V., & Mukherjee, S. (2018). RCCNet: An efficient convolutional neural network for histological routine colon cancer nuclei classification. In *2018 15th international conference on control, automation, robotics and vision, ICARCV 2018* (pp. 1222–1227). Institute of Electrical and Electronics Engineers Inc. https://doi.org/10.1109/ICARCV.2018.8581147.

Sirinukunwattana, K., Raza, S. E. A., Tsang, Y. W., Snead, D. R. J., Cree, I. A., & Rajpoot, N. M. (2016). Locality sensitive deep learning for detection and classification of nuclei in routine colon cancer histology images. *IEEE Transactions on Medical Imaging, 35*(5), 1196–1206. https://doi.org/10.1109/TMI.2016.2525803.

Skrede, O. J., De Raedt, S., Kleppe, A., Hveem, T. S., Liestøl, K., Maddison, J., Askautrud, H. A., Pradhan, M., Nesheim, J. A., Albregtsen, F., Farstad, I. N., Domingo, E., Church, D. N., Nesbakken, A., Shepherd, N. A., Tomlinson, I., Kerr, R., Novelli, M., Kerr, D. J., & Danielsen, H. E. (2020). Deep learning for prediction of colorectal cancer outcome: A discovery and validation study. *The Lancet, 395*(10221), 350–360. https://doi.org/10.1016/S0140-6736(19)32998-8.

Sung, H., Ferlay, J., Siegel, R. L., Laversanne, M., Soerjomataram, I., Jemal, A., & Bray, F. (2021). Global cancer statistics 2020: GLOBOCAN estimates of incidence and mortality worldwide for 36 cancers in 185 countries. *CA: a Cancer Journal for Clinicians*. https://doi.org/10.3322/caac.21660.

Tripathi, S., & Singh, S. K. (2020). Ensembling handcrafted features with deep features: An analytical study for classification of routine colon cancer histopathological nuclei images. *Multimedia Tools and Applications, 79*(47–48), 34931–34954. https://doi.org/10.1007/s11042-020-08891-w.

Venkatesan, E., & Velmurugan, T. (2015). Performance analysis of decision tree algorithms for breast cancer classification. *Indian Journal of Science and Technology*. https://doi.org/10.17485/ijst/2015/v8i1/84646.

Vuong, T. L. T., Lee, D., Kwak, J. T., & Kim, K. (2020). Multi-task deep learning for colon cancer grading. In *2020 International conference on electronics, information, and communication, ICEIC 2020* Institute of Electrical and Electronics Engineers Inc. https://doi.org/10.1109/ICEIC49074.2020.9051305.

Wang, C., Shi, J., Zhang, Q., & Ying, S. (2017). Histopathological image classification with bilinear convolutional neural networks. In *Proceedings of the annual international conference of the IEEE engineering in medicine and biology society, EMBS* (pp. 4050–4053). Institute of Electrical and Electronics Engineers Inc. https://doi.org/10.1109/EMBC.2017.8037745.

Wu, W., Dave, N., Tseng, G. C., Richards, T., Xing, E. P., & Kaminski, N. (2005). Comparison of normalization methods for CodeLink bioarray data. *BMC Bioinformatics, 6*. https://doi.org/10.1186/1471-2105-6-309.

Zhang, R., Zheng, Y., Mak, T. W. C., Yu, R., Wong, S. H., Lau, J. Y. W., & Poon, C. C. Y. (2017). Automatic detection and classification of colorectal polyps by transferring low-level CNN features from nonmedical domain. *IEEE Journal of Biomedical and Health Informatics, 21*(1), 41–47. https://doi.org/10.1109/JBHI.2016.2635662.

CHAPTER

10

Effect of COVID-19 on cancer patients: Issues and future challenges

H. Swapnarekha[a] *and Janmenjoy Nayak*[b]

[a]Department of Information Technology, Aditya Institute of Technology and Management (AITAM), Tekkali, Andhra Pradesh, India [b]P.G. Department of Computer Science, Maharaja Sriram Chandra Bhanja Deo University, Baripada, Odisha, India

Introduction

In December 2019, the local hospitals in Wuhan were reported to have unidentified pneumonia cases. Most of these cases have exposure to the Huanan wholesale seafood live market (Swapnarekha, Sekhar, Janmenjoy, & Bighnaraj, 2020). On January 7, 2020, the virus was identified as a novel coronavirus caused by SARS-CoV-2 (severe acute respiratory syndrome coronavirus-2) after performing a swab analysis of the throat (Hongzhou, Stratton, & Yi-Wei, 2020). Later, the disease was renamed COVID-19 by the World Health Organization on February 11, 2020 (WHO, 2020). Since its origin, the number of COVID-19 cases is increasing rapidly and has infected 132,507,403 million people all over the world as of April 6, 2021 (https://www.worldometers.info/coronavirus/). The most common symptoms of COVID-19 include fever, cough, shortness of breath, and muscle pain. It is reported that 33% of patients infected with COVID-19 infection suffer from severe complications such as pneumonia, septic shock, syndrome of severe respiratory distress, and renal failure (Al-Quteimat & Amer, 2020).

As COVID-19 is a new disease, the availability of data is limited. However, from the epidemiological and clinical data, it was observed that patients with comorbidities enhance the risk of infection (CDC, 2020). Hypertension, diabetes, and cardiovascular disorders are the most common comorbidities reported in COVID-19 patients until now (Zhou et al., 2020). From the clinical data, it was also observed that COVID-19 patients suffering from hypertension, lung disease, obesity, cardiovascular disorder, and diabetes experience the worst prognosis and mostly wind up with degenerating outcomes such as acute respiratory distress syndrome (ARDS) and pneumonia. Moreover, elderly people having long-term care facilities

https://doi.org/10.1016/B978-0-323-85240-1.00010-9

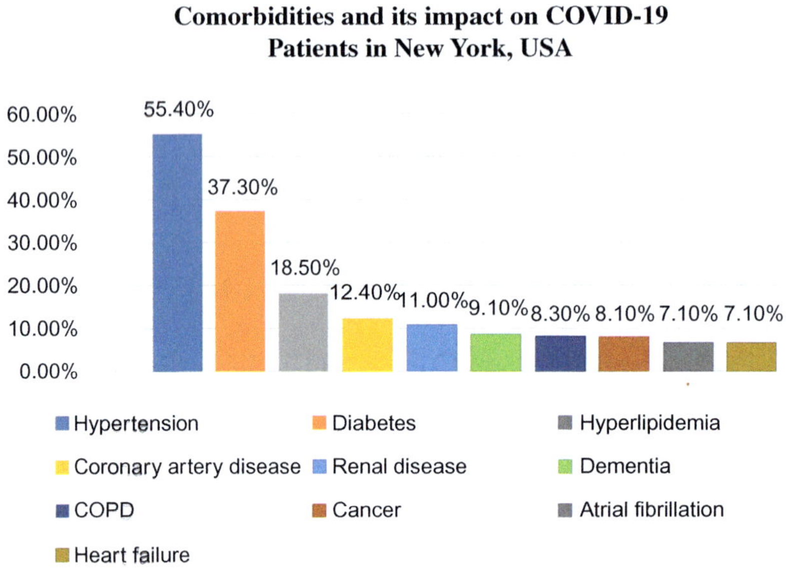

FIG. 10.1 Comorbidities and their impact on COVID-19 inmates in New York, USA (Franki, 2020).

and people of any age with medical conditions are at greater risk of COVID-19 infection when compared with other people (BCCDC, 2020) (Fig. 10.1).

Cancer is a major health problem and is the second leading cause of death all over the world. According to the WHO, cancer is responsible for approximately 10 million deaths all over the world. The present healthcare emergency is having special concern for the oncology community, as COVID-19 disease causes negative consequences in the diagnosis and treatment of cancer patients (Siegel, Miller, Fuchs, & Jemal, 2021). For instance, the suspension of healthcare systems leads to the postponement of diagnosis and treatment of disease which may decline the health condition of advanced-stage patients, resulting in increased mortality. Patients suffering from cancer have a high risk of developing an infection when compared to the general population because of the immunosuppressed status of cancer patients (Mengyuan et al., 2020). To investigate why cancer patients are more vulnerable to a COVID-19 infection, a study on 105 cancer patients and 536 non-cancer patients was performed by the authors (Mengyuan et al., 2020). From the study, it was observed that the admission rate to the ICU, usage of invasive mechanical ventilation, critical symptoms, and the mortality rate are higher in patients suffering from cancer when compared with non-cancer patients. In addition, patients undergoing various treatments for cancer such as immunotherapy or surgery have a high chance of acquiring critical symptoms when compared with patients receiving radiotherapy treatment. A retrospective study performed by the authors (Graziano, Giovanni, & Silvio, 2020) on 355 COVID-19 infected patients reveals that of 355 patients, 25% of patients are suffering from cancer, 36% of patients have diabetes, and 30% of patients are suffering from heart diseases., of the 355 COVID-19 infected patients, only 8% of patients do not suffer from any type of disease. A study on 28 cancer patients having a history of antitumor treatment of the 1276 COVID-19 infected patients admitted to three hospitals in Wuhan city of China was done by Zhang et al. (2020). It is revealed from the study that lung cancer is the most common cancer identified in the COVID-19 infected patients. After lung cancer, esophageal and breast cancer are the most common cancers identified in the COVID-19 infected inmates. Furthermore, to minimize the risk of COVID-19 infection in cancer patients, several measures have been recommended by various studies all over the world (Shankar & Saini, 2021; Zhijie, Jie, & Jie, 2020).

The objective of the paper is to provide a systematic analysis of the impact of COVID-19 infection on different types of cancers along with the case studies representing the scenario of cancer patients in various countries of the world during the COVID-19 pandemic. Further, the paper highlights the major challenges encountered by the oncology community in the management of cancer care during the pandemic period. Finally, it highlights the measures necessary for the efficient management of cancer care during the COVID-19 pandemic.

The rest of the paper is organized as follows. "Effect of COVID-19 on various types of cancer" section outlines the consequences of COVID-19 infection in different types of cancer. Different case studies representing the scenario of cancer patients in various countries of the world during the COVID-19 pandemic have been represented in "Scenario of the impact of COVID-19 on cancer patients in various countries" section. "Major challenges of oncology community and cancer patients during the COVID-19 pandemic" section highlights the major challenges encountered by the oncology community during the COVID-19 pandemic, whereas "Discussion and future directions" section discusses the future directions for the efficient management of cancer care in the present as well as future pandemics. Finally, the paper ends with a conclusion and future scope in "Conclusion" section.

Effect of COVID-19 on various types of cancer

Though medical oncologists decided to provide safe and benevolent care to cancer patients all over the world, the COVID-19 pandemic has forced the current healthcare system to change its approach to managing various types of cancer. This section describes the impact of COVID-19 infection on various types of cancer such as lung cancer, colorectal cancer, breast cancer, head and neck cancer, and thyroid cancer and their perspectives on the efficient management of cancer care during the COVID-19 pandemic. The impact of COVID-19 on different types of cancer has been depicted in Fig. 10.2.

Impact of COVID-19 on lung cancer

Lung cancer is the most common type of cancer in men, while in women, it is the third most commonly occurring cancer (https://www.wcrf.org/dietandcancer/cancer-trends/lung-cancer-statistics). Several studies reveal that patients with lung cancer have a high risk of hospitalization and have a high chance of developing critical symptoms due to the COVID-19 infection, which is basically a respiratory infection (Leora et al., 2020; Wenhua et al., 2020). To estimate the severity of COVID-19 infection in lung cancer inmates, a study on 102 lung cancer patients diagnosed with the COVID-19 infection during the period March 12, 2020, to May 6, 2020, has been performed by Luo et al. (2020). The study reveals that the

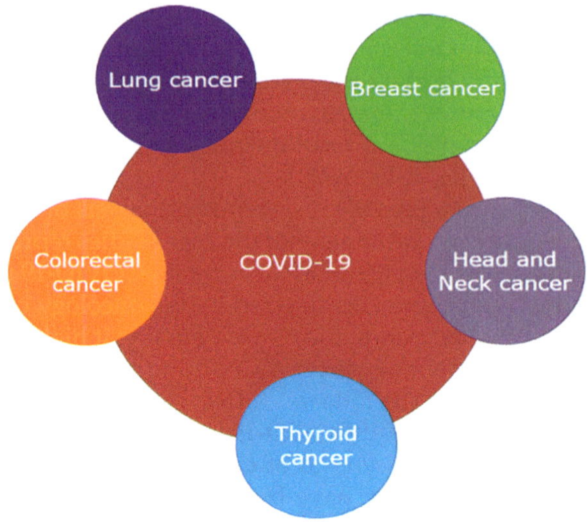

FIG. 10.2 Impact of COVID-19 on various types of cancer.

hospitalization ratio of lung cancer patients infected with the COVID-19 infection was 62%, while the death ratio of lung cancer patients infected with COVID-19 is 25%. Moreover, it reveals that the severity of COVID-19 infection in lung cancer patients is due to patient-specific factors such as smoking and chronic obstructive pulmonary disease rather than cancer-specific factors such as surgery, radiation, and systemic treatments. To investigate the consequences of COVID-19 infection on the screening and diagnosis of pulmonary cancer, a study on 2153 patients was conducted by Van Haren et al. (2021). The study reveals that there was a significant reduction in the total monthly mean $\pm$SD low-dose CTs (146 ± 31 vs 39 ± 40; $p < 0.01$) and monthly low-dose CTs of new patients (56 ± 14 vs 15 ± 17; $p < 0.01$) at the time of COVID-19 pandemic. In addition, a significant increase was observed in patients having lung nodules suspicious of malignancy after resuming the screening (8% vs 29%; $p < 0.01$). To assess the impact of the COVID-19 pandemic on the scheduling of pulmonary cancer therapy, a systematic review of lung cancer patients having anticancer treatment at the National Hospital Organization Kyoto Medical in Kyoto, Japan, between the period March 1, 2020, and May 31, 2020, has been carried out by Fujita et al. (2020). From the study, it was observed that a total of 9.1% of patients faced a delay in the treatment of pulmonary cancer during the COVID-19 infection due to anxiety. In addition, more immune checkpoint inhibitor (ICI) monotherapy was obtained by inmates with treatment delay rather than inmates without treatment delay. For the efficient screening of pulmonary cancer patients at the time of the COVID-19 pandemic at the European Institute of Oncology, Italy, proactive management measures such as social media, telemedicine, and telephone prioritization and containment measures have been suggested by de Marinis et al. (2020). In addition, these measures assist in the identification of COVID-19-infected inmates, which further helps in restricting the admission of suspected COVID-19 patients to the cancer center.

Impact of COVID-19 on colorectal cancer

Colorectal cancer affects the colon and the rectum. It is also known as bowel cancer, colon cancer, or rectal cancer. As per the American Cancer Society, an estimate of 1 in 21 men and 1 in 23 women has the chance of developing colorectal cancer during their lifetime in the United States (https://www.medicalnewstoday.com/articles/155598#, 2021). To analyze the data of the colorectal cancer screening (CRCS) colonoscopies during the lockdown period that was from March 9, 2020, to May 4, 2020, with the data of the same duration as 2019 (control group), a retrospective controlled cohort study was carried out by D'Ovidio et al. (2020) in "COVID-free" hospitals of Italy. In addition, a survey of pre-/post-endoscopic sanitation for COVID-19 disease was also conducted for patients and sanitary staff of the hospitals. From the analysis, it is observed that during lockdown 60 CRCS colonoscopies were performed. While in the control group, 238 CRCS colonoscopies were performed. Moreover, a greater number of colorectal cancer and high-risk adenoma detection rate was observed in the lockdown period when compared with the control group. The survey of pre-/post-endoscopic sanitation for COVID-19 disease reveals that no patients and staff were infected with COVID-19 infection. Therefore, the cohort study concludes that even though the number of CRCS colonoscopies decreased during the lockdown period, the effectiveness and safety of CRCS colonoscopies continued. To assess the consequences of COVID-19 infection on the detection and management of colorectal cancer in England, a study was conducted by Morris et al. (2021) by extracting the data from January 1, 2019, to October 31, 2020, from four population-based datasets related to colorectal cancer in England. From the findings, it was observed that there was a reduction of 63% in the number of two-week referrals for suspected cancer, 22% in the number of cases recommended for treatment, and 31% in the number of surgeries in April 2020 when compared to the monthly average for 2019. Because of the usage of short-course regimens, a relative increase of 44% was observed in the utilization of neoadjuvant radiotherapy for rectal cancer in April 2020. Although a reduction has been seen in April 2020, the target has returned to 2019 levels by October 2020. A survey of a total of 1051 respondents from 84 countries was carried out by Santoro et al. (2021) assess the impact of the COVID-19 pandemic on elective surgery and preoperative and postoperative management of colorectal cancer patients. In this survey, the respondents are basically divided into two groups known as the "delay" group where colorectal cancer is affected by the COVID-19 pandemic and the "no delay" group where colorectal cancer care is unaltered by the COVID-19 pandemic. From the findings, it was observed that factors such as cancellation of meetings of multidisciplinary teams, quarantine or relocation of staff members to COVID-19 units, complete dedication of cancer care units to COVID-19 disease care, and unavailability of a sufficient number of PPE kits are responsible for the delays associated with the endoscopy, surgery, prolonged chemotherapy, and radiology of colorectal cancer care. It is also observed that 48.9% and 26.3% of respondents from the delay groups reported an alteration in the surgical plan and a move from elective to urgent surgery respectively. Finally, it can be concluded that changes in the management of colorectal care occurred because of the discrepancies in the delivery system and preparedness of healthcare units, availability of resources, and predominance of COVID-19 diseases instead of geographical factors. To investigate the impact of

COVID-19 infection on colorectal diagnosis in the state of Spain, a study was conducted by Javier et al. (2021)) by analyzing the data of newly diagnosed colorectal cancer patients for the period March 14, 2020, to June 20, 2020, with the data of patients diagnosed with colorectal cancer during the same period in 2019. From the analysis, it is observed that fewer patients were diagnosed with CRC and referred to the colorectal care multidisciplinary unit because of the temporary closing of colorectal screening programs.

Impact of COVID-19 on breast cancer

Breast cancer is the most common and second leading cancer diagnosed in women. According to the American Cancer Society, an estimate of 281,550 new cases of invasive breast cancer and 49,290 new cases of noninvasive (in situ) breast cancer are predicted to be diagnosed in women in 2021 in the United States (https://www.cancer.net/cancer-types/breast-cancer/statistics, 2021). A systematic analysis of the consequences of the COVID-19 pandemic on the management and emphasizing care of breast cancer in Singapore has been presented by Chan et al. (2020). To emphasize radiotherapy and systemic treatments of cancer during the pandemic of COVID-19, a conceptual framework proposed by Hanna, Evans, and Booth (2020) has been adopted by the authors for the continuity of breast oncology practice in Singapore. Moreover, the authors have recommended transformations in the established clinical practices and processes over various specialties of breast oncology in the public healthcare sector of Singapore. Further, these recommendations provide guidelines for the management of breast cancer in the present pandemic as well as public health emergencies in the future. To overcome the consequences of COVID-19 on breast cancer, Giuseppe et al. (2020) suggested standard procedures such as selection criteria for providing the services and prioritization of surgical, radiation, and medical treatment for the efficient management of breast cancer patients during the pandemic. A systematic study has been performed by the authors (Gianluca et al., 2020) to determine the anxiety caused among breast cancer patients due to the rapid dissemination of COVID-19 infection. To carry out the study, 82 patients with suspected breast tumors and 78 patients with breast cancer characteristics such as age, marital status, the diameter of a breast lesion, clinical stage, personal and family history of breast cancer, and molecular subtype were considered. Further, both the suspected and breast cancer patients were categorized as pre-COVID-19 and post-COVID-19 patients. In addition, procedure refusal and surgical refusal were also considered for the study. From the results, it is observed that both post-COVID-19-suspected and breast cancer patients exhibited a higher rate of procedure and surgical refusal. Therefore, the results indicate that anxiety caused by the spread of COVID-19 infection affects the decision-making process of breast cancer patients. Swainston, Chapman, Grunfeld, and Derakshan (2020) proposed a systematic study on the effect of the COVID-19 pandemic on the intellectual and emotional health of breast cancer patients which may further assist the healthcare sector in the pandemic preparedness plans. From the findings of the study, it is noticed that breast cancer patients are more likely to develop psychological disorders such as anxiety and depression at the time of the COVID-19 pandemic. Moreover, the study also highlights the significance of mental health interferences that are delivered remotely.

Impact of COVID-19 on head and neck cancer

Among various types of cancer, head and neck cancer is the eighth most common cancer in the world in 2018 and is responsible for 1.5% of deaths in the United States (https://www.rgare.com/knowledge-center/media/articles/head-and-neck-cancers-update-2020, 2021). The risk related to the investigation, biopsy, and therapy of head and neck cancer originating in the upper part of the aerodigestive tract is considered one of the significant challenges of head and neck cancer at the time of the COVID-19 pandemic (Shuman & Campbell, 2020). To balance the imperils of COVID-19 vulnerabilities with the risks related to dysphagia of head and neck cancer, clinical practice guidelines have been suggested by Ku et al. (2020). These guidelines include securing instrumental evaluations for emergency cases only, optimization of noninstrumental swallowing assessments, proper usage of personal protective equipment (PPE), and implementation of telemedicine when suitable. Further, alterations in the existing setup and deployment of newer technologies can provide safe and proper dysphagia care regardless of the limitations in clinical services during the COVID-19 pandemic. During the COVID-19 pandemic, the changes in the surgical patterns, alterations in oncology treatment, and issues associated with the multidisciplinary care of head and neck cancer have been reported by Kiong et al. (2020) by comparing the data of the same interval in 2019. From the findings, it was observed that a decrease of 46.7% occurred in the outpatient visits and a decrease of 46.8% has been found in the surgical practice patterns. Further, the factors associated with the decision-making process have been discussed to stabilize the oncologic and safety requirements of head and neck cancer patients during the COVID-19 pandemic. Thomson et al. (2020) conducted a survey that included the panel of the American Society of Radiation Oncology (ASTRO) and the European Society for Radiotherapy and Oncology (ESTRO) to provide guidelines for radiation oncologists to treat head and neck cancer patients in the face of COVID-19 with limited resources and intense risk for cancer inmates and staff. Using Rand methodology, the cutoffs for strong agreement (≥80%) and agreement (≥66%) were chosen. In addition, two scenarios of the pandemic such as risk mitigation known as the early scenario and severely reduced radiation therapy resources known as the late scenario were assessed. The objective of this study was to provide high-quality head and neck cancer therapies in the face of the COVID-19 pandemic. To predict the risk of head and neck squamous cell carcinoma (HNSCC) patients dying because of the COVID-19 infection and also to suggest some mitigation schemes, an online tool based on a mathematical model known as COVID-19 cancer impact calculator (COCIC) has been developed by Matos et al. (2021). The results of the model show that the more cancer care is provided during the pandemic and mitigation period, the less will be the risk of HNSC patients dying because of the time-to-time therapy.

Impact of COVID-19 on thyroid cancer

According to global cancer statistics, thyroid cancer occupies 9th rank among cancers for incidence in the year 2020 and is responsible for a total of 586,000 cases all over the world. The global occurrence rate in women is 10.1 per 100,000 which means 3-fold higher than that of men (Freddie et al., 2018). To assess the impact of COVID-19 infection on surgery for thyroid cancer in Italy, a retrospective study on a total of 1570 patients with thyroid cancer from March 9, 2020, to August 31, 2020, was carried out by Medas et al. (2021). In addition, the study was compared with the data of the same period in 2019 to find the differences. From the findings, it was observed that there was a reduction of 27.1% in the number of operations when compared with the number of operations in 2019. Further, they have also analyzed the size of the tumor, extent of lymph node metastases, and the extension of extrathyroidal and multicentricity to identify the differences between the two groups and results indicating that no differences were found between the two groups. To describe the predominance of COVID-19 infection in thyroid cancer patients, a study on 1464 patients affected by constant thyroid cancer was conducted by Alessandro et al. (2021) from March to September 2020. From the study, it was identified that COVID-19 infection was observed in 0.3% of cases of patients having constant thyroid cancer. Among the active surveillance and multikinase inhibitors (MKIs), a higher predominance of COVID-19 infection was found in the group of thyroid cancer inmates treated with MKI. During MKI treatment, patients suffering from mild COVID-19 do not require discontinuing treatment. While patients with severe COVID-19, the decision of continuing or stopping the treatment will be done by the multidisciplinary team after assessing the benefits and risks. A systematic analysis of the consequences of risk mitigation approaches of COVID-19 on thyroid cancer diagnosis and treatment and the effect of thyroid cancer on COVID-19 has been performed by Tsang, Matti, Anthony, Roderick, and Robinson (2020) to provide guidelines for thyroid cancer patients and multidisciplinary teams of oncology community in the face of COVID-19 pandemic.

Scenario of the impact of COVID-19 on cancer patients in various countries

This section describes the case studies of the consequences of COVID-19 on cancer patients in various regions of the world.

Case study 1: Assessment of mortality rate of COVID-19 in cancer patients of the United Kingdom undergoing chemotherapy or other anticancer treatments

On March 18, 2020, Lee, Cazier, Angelis, et al. (2020) have started the United Kingdom Coronavirus Cancer Monitoring Project (UKCCMP) to assess the mortality rate of COVID-19 in cancer patients undergoing chemotherapy or other anticancer treatments. This study considers active cancer patients who were tested positive for COVID-19 infection. Analysis of throat and nose swabs using RT-PCR is the approach used for the determination of COVID-19 infection. Moreover, patients with a clinical diagnosis of COVID-19 with negative RT-PCR tests were not considered in the study. Therefore, to assess the mortality rate of COVID-19 in cancer patients undergoing chemotherapy or other anticancer treatments, the study has analyzed a total of 800 patients having cancer and symptomatic COVID-19 infection from March 18, 2020, to April 26, 2020. To carry out the assessment, two-sided Welch's t and Fisher's exact tests were used. The following is Table 10.1.

The analysis of the usage of anticancer treatments within 4 weeks of COVID-19 diagnosis was depicted using the forest plot as shown in Fig. 10.3.

TABLE 10.1 Multivariate regression analysis and odds of deaths depending on the anticancer treatment and cytotoxic chemotherapy of patients (Lee, Cazier, Angelis, et al., 2020).

	Odds ratio (95% CI)	*p* value
Anticancer treatment		
Chemotherapy vs no chemotherapy	1.18 (0.81–1.72)	0.380
Hormone therapy vs no hormone therapy	0.90 (0.49–1.68)	0.744
Immunotherapy vs no immunotherapy	0.59 (0.27–1.27)	0.177
Radiotherapy vs no radiotherapy	0.65 (0.36–1.18)	0.159
Targeted treatment vs no targeted treatment	0.83 (0.45–1.54)	0.559
Cytotoxic chemotherapy		
Non-palliative chemo vs palliative chemo	0.40 (0.17–0.96)	0.040
Palliative first-line chemotherapy vs other line	0.84 (0.36–1.98)	0.690
Palliative chemotherapy vs no chemotherapy	1.48 (0.93–2.36)	0.102
Palliative chemotherapy vs no treatment	1.05 (0.63–1.76)	0.854

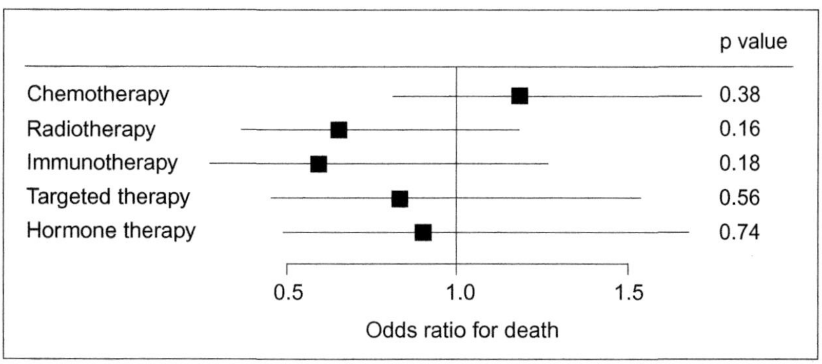

FIG. 10.3 Forest plot displaying the impact of anticancer treatments and mortality from COVID-19 (Lee, Cazier, Angelis, et al., 2020).

From the multivariate analysis of COVID-19-diagnosed inmates who had taken chemotherapy in recent times after modifying age, gender, and comorbidities, it was noticed from Table 10.1 that there was a reduced odds of deaths in cancer inmates taking non-palliative chemotherapy in comparison with those patients taking palliative chemotherapy, i.e., 16% versus 35%. It was also noticed that there was no significant difference in the odds of deaths in inmates getting palliative chemotherapy compared with inmates not receiving anticancer treatment or chemotherapy in recent times. Moreover, no significant difference in the mortality of inmates was found between patients taking first-line palliative chemotherapy and patients taking later lines of palliative treatment. From Fig. 10.3 also, it was noticed that patients who were on immunotherapy, radiotherapy, hormonal, and targeted therapy for cancer were not found at any additional risk of death when compared with patients who were not on cancer therapies.

Case study 2: Assessment of the clinical impact of COVID-19 infection on cancer patients

To determine the prognostic factors of fatality and acute illness of cancer patients having COVID-19 infection, a study on 1035 records of patients having acute or previous malignancy and COVID-19 infection collected from the United States, Spain, and Canada COVID-19 and Cancer Consortium (CCC19) database has been done by Kuderer et al. (2020) during the period March 17, 2020, to April 16, 2020. Patients having noninvasive cancers such as in situ carcinoma, hematological lumps, and non-melanomatous skin carcinoma are not considered for assessment in this study. Further, logistic regression analysis was used to determine the relationship between the outcome of the study and prognostic factors. The factors related to the 30-day all-cause fatality have been depicted using forest plot as shown in Figs. 10.4 and 10.5 represents the prevalence of primary and secondary outcomes by type of cancer, status of cancer and subgroups defined by anticancer treatment types.

Case study 3: Investigating the risk of COVID-19 in cancer patients in the United Kingdom based on the tumor subtype and patient demographics

To investigate the risk of COVID-19 in cancer patients in the United Kingdom based on the tumor subtype and patient demographics, a study was carried out by Lee, Cazier, Starkey, et al. (2020) on adult patients having cancer by collecting the data from the UK Coronavirus Cancer Monitoring Project (UKCCMP) cohort from March 18, 2020, to May 8, 2020. Simultaneously, the non-COVID-19 United Kingdom cancer control population data from the UK Office for National statistics were also collected. The study aimed to determine the impact of tumor subtype, age, and sex on the occurrence of COVID-19 infection and mortality rate at the time of hospital admission. The risk of COVID-19 in cancer patients in the United Kingdom based on the tumor subtype and patient demographics has been analyzed using univariable and multivariable models (Table 10.2). Fig. 10.6 depicts the age and all-cause case-mortality rate of patients with cancer after having COVID-19 infection in the UKCCMP.

Major challenges of oncology community and cancer patients during the COVID-19 pandemic

Since December 2019, the outbreak of the COVID-19 pandemic has caused a serious threat to the healthcare system all over the world due to the shortage of medical resources. As cancer patients are considered, COVID-19 is having a profound impact on the entire oncology community because of the immunosuppressed status of cancer patients. This section presents the major challenges encountered by the healthcare practitioners and health system of the oncology community and cancer inmates during the COVID-19 pandemic.

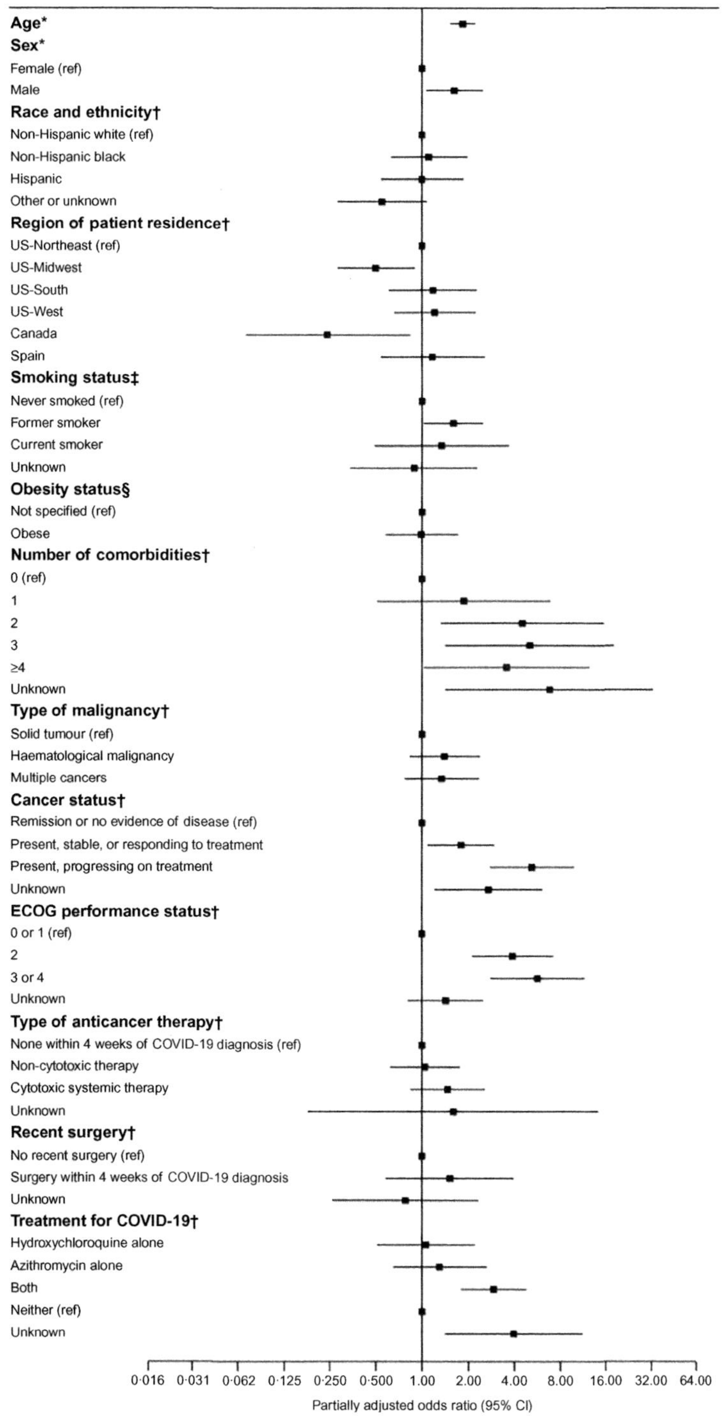

FIG. 10.4 Forest plot depicting the factors related to the 30-day all-cause fatality (Kuderer et al., 2020).

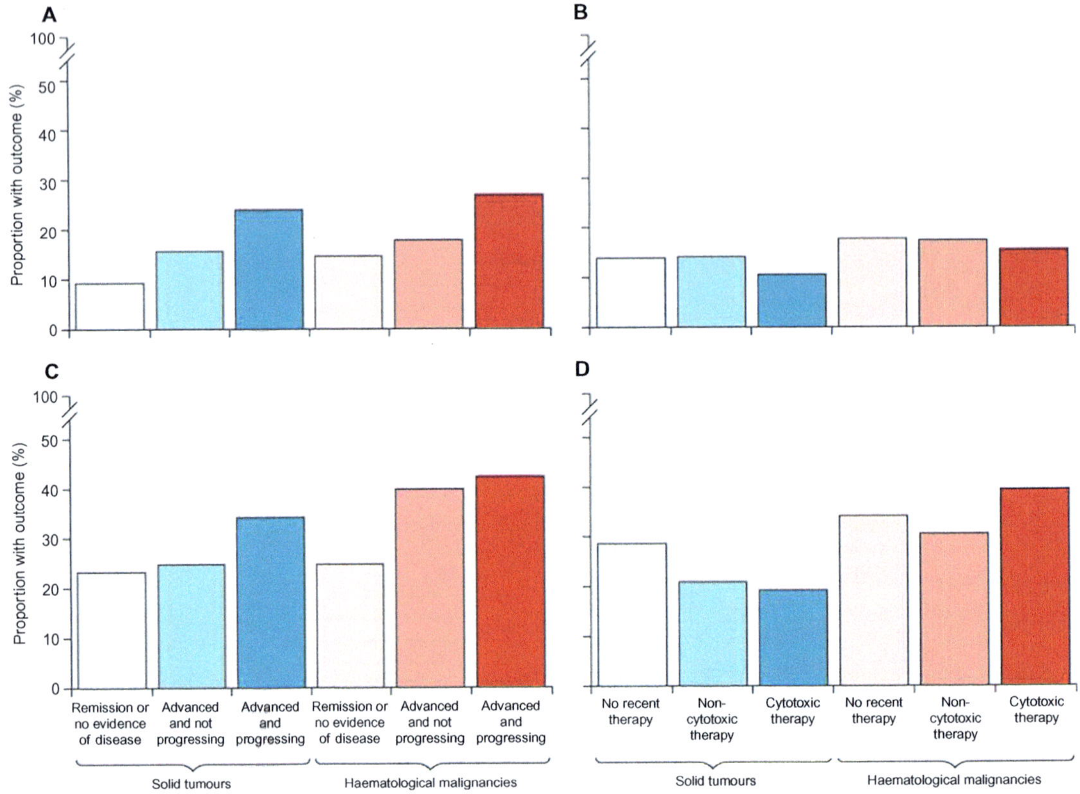

FIG. 10.5 Prevalence of primary and secondary outcomes by type of cancer, status of cancer, and anticancer treatment (Kuderer et al., 2020).

Healthcare practitioners

In the present pandemic, healthcare practitioners are one of the most affected communities by COVID-19 infection (Desideri et al., 2020; Tait, Jonathan, & Mickey, 2020). Healthcare practitioners have encountered the stress of losing jobs due to declining patient visits to hospitals because of the fear caused by the COVID-19 infection. The COVID-19 pandemic also caused a considerable disturbance to the infrastructure of cancer healthcare, such as disruption in the visits of cancer patients to the hospital, postponement or cancellation of medical surgery, and restrained therapeutic procedures (Desideri et al., 2020; Segelov et al., 2020). These unexpected disruptions in the cancer healthcare sector force healthcare practitioners to adapt to new technologies such as telemedicine tools. However, a lack of adequate training on new technologies causes additional stress among healthcare practitioners. In addition, the fear of being vulnerable to

TABLE 10.2 Mortality rate after presentation with COVID-19 by tumor subtype (Lee, Cazier, Starkey, et al. (2020)).

	Deaths	Mortality rate	Univariable OR (95% CI)	*p*-value	Multivariable adjusted OR (95% CI)	*p*-value
Prostate (C61)	49	0.43	2.14 (1.17–3.96)	0.014	1.09 (0.51–2.33)	0.82
Lung (C34)	43	0.387	1.62 (0.89–3.00)	0.118	1.41 (0.75–2.67)	0.29
Mesothelial and soft tissue (C45–C49)	6	0.375	1.18 (0.37–3.51)	0.772	1.52 (0.43–5.30)	0.51
Urinary tract (C64–C68)	23	0.299	1.08 (0.54–2.13)	0.834	0.87 (0.41–1.81)	0.72
Colorectal (C18–C21)	35	0.282	1.03 (0.56–1.90)	0.934	0.85 (0.44–1.64)	0.63
CNS (C69–C72)	7	0.28	1.15 (0.39–3.18)	0.797	1.87 (0.57–6.05)	0.29
Respiratory organs and intrathoracic organs (not lungs; C30–C39)	3	0.273	0.84 (0.17–3.29)	0.813	0.96 (0.18–4.10)	0.95
Lips, oral cavity, and pharynx (C00–C14)	8	0.242	0.75 (0.28–1.85)	0.542	0.77 (0.25–2.27)	0.64
Breasts (C50)	26	0.182	0.53 (0.28–1.00)	0·049	0.97 (0.40–2.52)	0.94
Female genital organs (C51–C58)	7	0.125	0.36 (0.13–0.87)	0.031	0.79 (0.24–2.63)	0.70
Myeloma (C90)	16	0.432	1.85 (0.81–4.22)	0.142	1.65 (0.71–3.85)	0.24
Leukemia (C91–C95)	33	0.418	2.03 (1.04–3.97)	0.038	2.25 (1.13–4.57)	0.023
Lymphoma (C81–C85)	25	0.316	1.60 (0.80–3.19)	0.184	1.72 (0.81–3.68)	0.16
Other hematological (C86, C88, C96)	7	0.241	0.81 (0.28–2.12)	0.675	0.81 (0.26–2.33)	0.70
Digestive organs, non-colorectal (C15–C17, C22–C26)	28	0.295	1 (ref)		1 (ref)	

Where OR is the odds ratio.

COVID-19 infection among healthcare practitioners also causes additional stress, which in turn causes adverse effects on the performance of healthcare practitioners in the management and treatment of cancer patients.

Patients

In general, cancer patients are susceptible to psychological distress related to their health condition. As 10%–20% of cancer patients feel chronic psychological distress, it is essential to detect and prevent psychological distress of cancer patients (Fielding, 2019). Since cancer

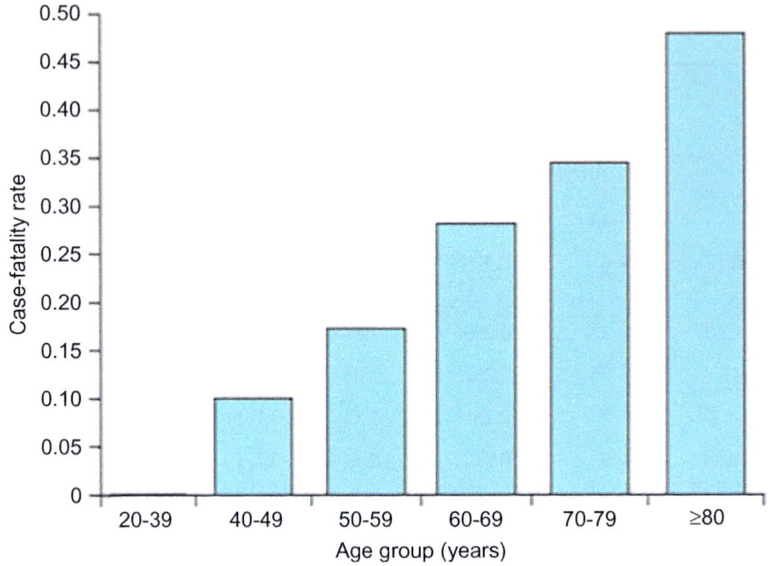

FIG. 10.6 Age and all-cause case-mortality rate of patients with cancer after having COVID-19 infection in the UKCCMP (Lee, Cazier, Starkey, et al., 2020).

patients share 20% of the mortality rate of the COVID-19 pandemic (Graziano et al., 2020), the following are some of the challenges for cancer inmates that add further psychological burden to their pre-existing distress.

- Cancer patients have specific guidelines during the COVID-19 pandemic as they are considered a high-risk group. For instance, to avoid respiratory droplets from other people, social distancing is implemented in many countries, particularly cancer patients. Though social distancing reduces pollution, it may negatively impact the emotions of cancer patients and may add psychological distress.
- Urgent requirement for therapy at hospitals adds psychological distress to cancer patients as it contravenes the social distancing measure (Gregucci et al., 2020).
- Lockdown restrictions on traveling during pandemics increased the psychological distress of cancer inmates who are seeking their treatment in foreign countries with advanced technologies because of the postponement of the treatment plans.
- In general, most cancer patients are admitted to ICU for their respiratory difficulties (Taccone et al., 2009). During the COVID-19 pandemic, the ICUs were overwhelmed by the COVID-19-infected patients (Xiya & Dominique, 2020). The lack of beds in ICUs forced medical experts to allocate beds to COVID-19 and cancer patients. Among cancer patients, the news of less priority for older patients caused psychological distress among cancer inmates.

The concerns expressed by the cancer patients due to the fear of COVID-19 infection have been depicted in Fig. 10.7.

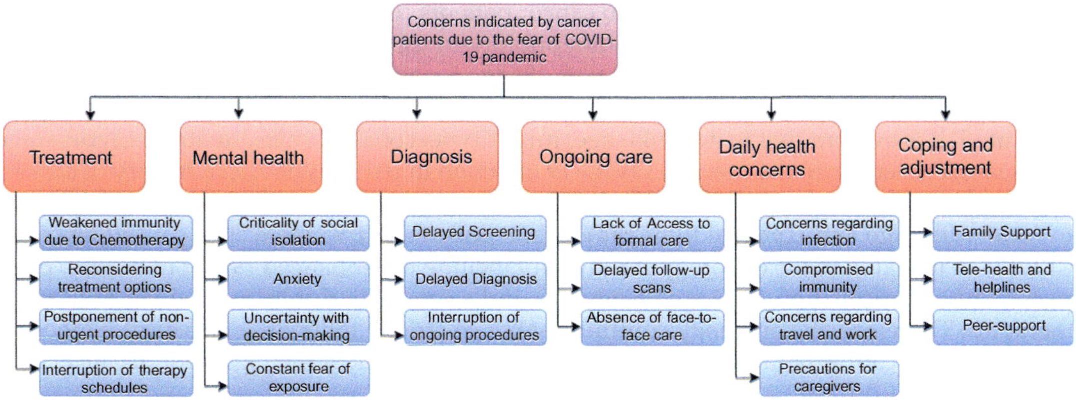

FIG. 10.7 Concerns expressed by the patients due to the fear of COVID-19 infection.

Therefore, the COVID-19 infection has created a feeling of uncertainty and psychological distress among cancer patients about the management of cancer during the pandemic.

Health systems

The outbreak of COVID-19 has affected the health system in terms of loss of patients, healthcare practitioners, revenue, and operational activity. It is approximated that a total of 2,324,069 elective cancer surgeries were delayed or canceled because of the interruptions caused by the COVID-19 pandemic during the 12-week peak period (Nepogodiev et al., 2020). In the United Kingdom, it is predicted that COVID-19 will lead to excess deaths of 10,000 from breast cancer and colorectal cancer patients (Sharpless, 2020). Due to the declining demand for healthcare services caused by the fear of the COVID-19 pandemic, healthcare practitioners are losing their jobs. It is approximated that 1.4 million healthcare practitioners have lost their jobs since January 2020 due to the outbreak of COVID-19 (US Bureau of Labor Statistics, 2021). Though it is difficult to estimate the impact of COVID-19 on the health system in terms of loss of patients and healthcare practitioners, it is easy to estimate the loss in terms of revenue and operational activity because of disrupted healthcare services. As stated by the American Hospital Association, it is estimated that US hospitals and health systems encountered a loss of 202.6 billion dollars between February 2020 and June 2020 due to the outbreak of COVID-19 infection (American Hospital Association, 2021).

Discussion and future directions

The rapid spread of the novel coronavirus has drastically impacted the healthcare system all over the world. The present healthcare emergency is having special concern for the oncology community, as COVID-19 disease causes negative consequences in the diagnosis and treatment of cancer patients. From the studies, it is evident that cancer patients and survivors are at greater risk of having COVID-19 infection when compared to the general population.

The prevalence of COVID-19 infection in cancer patients could be higher than is reported. As cancer patients are at a higher rate of susceptibility to COVID-19 infection, early diagnosis, screening, therapy, isolation, and education of cancer patients play a crucial role in the efficient management of cancer care during the COVID-19 pandemic. From the studies presented, it is evident that factors like age, tumor type, comorbidities, treatment type, stage of disease, etc., enhance the risk of COVID-19 infection in patients having a history of cancer. However, it is also evident from studies that the mortality rate in cancer patients is mainly due to factors like age and comorbidities rather than anticancer therapies. The consequences of COVID-19 infection in the oncology community include the enhanced rate of mortality, postponement or cancellation of cancer diagnosis and surgery, and interruptions in anticancer therapies. Therefore, the oncology community has to strictly follow the measures suggested by the relevant cancer organizations to provide safe and efficient management of cancer care in the face of COVID-19.

Telemedicine

Due to the rapid advancement of science and technology, the deployment of telemedicine in cancer care plays a crucial role in the effective management of cancer care in the face of the COVID-19 pandemic (Fertleman et al., 2018; Joseph & Maria, 2018). According to (Strehle & Shabde, 2006), telemedicine aims at enhancing population health by providing timely and cost-effective medical services to patients (Telemedicine: Opportunities and developments in member states, 2010). Telemedicine is considered the best solution to provide efficient healthcare services at the time of a pandemic as it provides social distancing and self-isolation measures (Lewis, Hatch, Wiederhold, & Swanson, 2020; Lopez, 2019; Ochs et al., 2019). Telemedicine facilitates training for healthcare practitioners from remote locations because of the presence of virtual reality devices (Mendu et al., 2018). It is possible to facilitate cancer screening remotely because of emerging technologies like artificial intelligence (AI) (Lindsey et al., 2016). For the efficient implementation of telemedicine in cancer care and primary care, it is necessary to provide education and training to both patients and healthcare practitioners. A study on the current situation and obstructions to international telemedicine conferencing in Japan has been presented by Kudo et al. (2021). The study was conducted by sending a questionnaire on characteristics, activities, technical resources, and obstruction to international telemedicine conferencing at the hospitals to the Internationalization Project Team (I-PT), medical staff in-charge of internationalization (MI), and technical staff in telemedicine (TT). The study reveals that attaching MI and TT to an I-PT is not enough for efficient telemedicine implementation. In addition, enhanced cooperation between MI and TT at each university hospital is required. Therefore, the organization of a structured international telemedicine center at each university hospital is recommended to enhance telemedicine activities in Japan (Kudo et al., 2021). Therefore, telemedicine tools and training programs need to be updated frequently to assure that healthcare practitioners can handle the present and future pandemic challenges efficiently and ensure safe treatment for patients (Bidgoli, 2018).

Enhancement of cancer screening through social media campaigns

To enhance screening awareness among patients, the healthcare systems need to choose integrated marketing services such as social media campaigns. Social media campaigns use social media platforms to communicate health strategies to the targeted audience to change their behavior toward health. As social media campaigns are evidence-based, they often generate desirable outcomes (Key & Czaplewski, 2017; Noar, Harrington, & Aldrich, 2009). Because of the advantages of social media campaigns such as cost-behavior effectiveness and scalability, social media are useful for promoting the services of cancer screening to the risk population (Wakefield, Loken, & Hornik, 2010). Evidence shows that social media campaigns are done through Google and Facebook for promoting lung cancer screening awareness to the risk population producing higher rates than the industry standards. Therefore, to reduce the cancer disparities produced by the COVID-19 pandemic, healthcare practitioners can make use of social media campaigns to promote screening to the risky population.

Conclusion

The novel COVID-19 pandemic has drastically affected the healthcare system throughout the world. Due to the rapid spread of the COVID-19 infection, the structure of the healthcare system including the oncology community has been altered to address the infectious disease. Though the oncology community is committed to providing safe care and treatment to cancer inmates all over the world, the present pandemic has resulted in a major shift in the approaches to cancer management due to the overwhelming healthcare system with COVID-19 patients and the unavailability of medical resources. Due to the COVID-19 pandemic, the oncology community has encountered challenges like postponement or cancellation of surgeries, diagnosis and screening, and interruption in anticancer treatment, which results in enhanced mortality rates and severe illness among cancer patients. This study provides a systematic analysis of the impact of COVID-19 infection on various types of cancers along with patient case studies representing the scenario of COVID-19 infection on cancer patients in various regions of the world. The study also highlights the challenges and opportunities for the oncology community, which provides guidelines for the efficient management of cancer care during the COVID-19 pandemic.

References

Alessandro, P., Marco, F., Valeria, B., Carlotta, G., Giusy, T., Laura, A., Antonio, M., Virginia, C., Laura, V., Loredana, L., Elisa, M., Eleonora, M., & Rossella, E. (2021). Thyroid cancer and COVID-19: Experience at one single thyroid disease referral center. *Endocrine*. https://doi.org/10.1007/s12020-021-02650-z.

Al-Quteimat, O. M., & Amer, A. M. (2020). The impact of the COVID-19 pandemic on cancer patients. *American Journal of Clinical Oncology*, *43*(6), 452–455. https://doi.org/10.1097/coc.0000000000000712.

American Hospital Association. (2021). *Hospitals and health systems face unprecedented financial pressures due to COVID-19*. https://www.aha.org/guidesreports/.

BCCDC. (2020). *COVID-19 vulnerable populations*. British Columbia Centre for Disease Control. http://www.bccdc.ca/health-info/diseases-conditions/covid-19/vulnerable-populations.

Bidgoli, H. (2018). Successful integration of information technology in healthcare: Guides for managers. *Journal of Strategic Innovation and Sustainability, 13*.

CDC. (2020). *Coronavirus (COVID-19): Symptoms of coronavirus*. Centers for Disease Control and Prevention. https://www.cdc.gov/coronavirus/2019-ncov/symptoms-testing/symptoms.html.

Chan, J. J., Sim, Y., Ow, S. G. W., Lim, J. S. J., Kusumawidjaja, G., Zhuang, Q., Wong, R. X., Wong, F. Y., Tan, V. K. M., & Tan, T. J. Y. (2020). The impact of COVID-19 on and recommendations for breast cancer care: The Singapore experience. *Endocrine-Related Cancer, 27*(9), R307–R327. https://doi.org/10.1530/ERC-20-0157.

de Marinis, F., Attili, I., Morganti, S., Stati, V., Spitaleri, G., Gianoncelli, L., Signore, E. D., Catania, C., Rampinelli, C., Salè, E. O., Spaggiari, L., Mastrilli, F., & Passaro, A. (2020). Results of multilevel containment measures to better protect lung cancer patients from COVID-19: The IEO model. *Frontiers in Oncology, 10*.

Desideri, I., Pilleron, S., Battisti, N. M. L., Gomes, F., de Glas, N., Neuendorff, N. R., Liposits, G., Paredero-Pérez, I., Lok, W. C. W., Loh, K. P., DuMontier, C., Mian, H., & Soto-Perez-de-Celis, E. (2020). Caring for older patients with cancer during the COVID-19 pandemic: A young International Society of Geriatric Oncology (SIOG) global perspective. *Journal of Geriatric Oncology, 11*(7), 1175–1181. https://doi.org/10.1016/j.jgo.2020.05.001.

D'Ovidio, V., Lucidi, C., Bruno, G., Lisi, D., Miglioresi, L., & Bazuro, M. E. (2020). Impact of COVID-19 pandemic on colorectal cancer screening program. *Clinical Colorectal Cancer, 20*, e5–e11.

Fertleman, C., Aubugeau-Williams, P., Sher, C., Lim, A. N., Lumley, S., Delacroix, S., & Pan, X. (2018). A discussion of virtual reality as a new tool for training healthcare professionals. *Frontiers in Public Health, 6*. https://doi.org/10.3389/fpubh.2018.00044.

Fielding, R. (2019). Developing a preventive psycho-oncology for a global context. The International Psycho-Oncology Society 2018 Sutherland Award Lecture. *Psycho-Oncology, 28*(8), 1595–1600. https://doi.org/10.1002/pon.5139.

Franki, R. (2020). Comorbidities the rule in New York's COVID-19 deaths. In *Hospitalist*. https://www.the-hospitalist.org/hospitalist/article/220457/coronavirus-updates/comorbidities-rule-new-yorks-covid-19-deaths.

Freddie, B., Jacques, F., Isabelle, S., Siegel, R. L., Torre, L. A., & Ahmedin, J. (2018). Global cancer statistics 2018: GLOBOCAN estimates of incidence and mortality worldwide for 36 cancers in 185 countries. *CA: A Cancer Journal for Clinicians*, 394–424. https://doi.org/10.3322/caac.21492.

Fujita, K., Ito, T., Saito, Z., Kanai, O., Nakatani, K., & Mio, T. (2020). Impact of COVID-19 pandemic on lung cancer treatment scheduling. *Thoracic Cancer, 11*(10), 2983–2986. https://doi.org/10.1111/1759-7714.13615.

Gianluca, V., Marco, M., Marco, P., Sara, I., Maurizio, R., Francesca, S., Maria, C., Jonathan, C., Anna, M., Adriana, P. C., & Claudio, B. O. (2020). Breast cancer and COVID-19: The effect of fear on patients' decision-making process. *In Vivo*, 1651–1659. https://doi.org/10.21873/invivo.11957.

Giuseppe, C., Joao, C. M., Philip, P., Oreste, G., Gabriella, P., Ketti, M., Nehmat, H., Olivia, P., Elzbieta, S., & Fatima, C. (2020). Recommendations for triage, prioritization and treatment of breast cancer patients during the COVID-19 pandemic. *The Breast*, 8–16. https://doi.org/10.1016/j.breast.2020.04.006.

Graziano, O., Giovanni, R., & Silvio, B. (2020). Case-fatality rate and characteristics of patients dying in relation to COVID-19 in Italy. *Journal of the American Medical Association*. https://doi.org/10.1001/jama.2020.4683.

Gregucci, F., Caliandro, M., Surgo, A., Carbonara, R., Bonaparte, I., & Fiorentino, A. (2020). Cancer patients in Covid-19 era: Swimming against the tide. *Radiotherapy and Oncology, 149*, 109–110. https://doi.org/10.1016/j.radonc.2020.04.002.

Hanna, T. P., Evans, G. A., & Booth, C. M. (2020). Cancer, COVID-19 and the precautionary principle: Prioritizing treatment during a global pandemic. *Nature Reviews. Clinical Oncology, 17*(5), 268–270. https://doi.org/10.1038/s41571-020-0362-6.

Hongzhou, L., Stratton, C. W., & Yi-Wei, T. (2020). Outbreak of pneumonia of unknown etiology in Wuhan, China: The mystery and the miracle. *Journal of Medical Virology*, 401–402. https://doi.org/10.1002/jmv.25678.

(2021). https://www.cancer.net/cancer-types/breast-cancer/statistics.

(2021). https://www.medicalnewstoday.com/articles/155598#.

(2021). https://www.rgare.com/knowledge-center/media/articles/head-and-neck-cancers-update-2020.

(2021). https://www.worldometers.info/coronavirus/.

Javier, S., Elena, M., Ana, G., Gloria, J., Marta, M., Fernando, A., Ciga, M. A., Eugenia, U., María, E., Beñat, A., Matilde, L., Asunción, S., & Ruth, V. (2021). Impact of the COVID-19 pandemic during Spain's state of emergency on the diagnosis of colorectal cancer. *Journal of Surgical Oncology, 123*(1), 32–36. https://doi.org/10.1002/jso.26263.

Joseph, S. S., & Maria, L. A. (2018). Telemedicine in cancer care. *American Society of Clinical Oncology Educational Book*, 540–545. https://doi.org/10.1200/EDBK_200141.

Key, T. M., & Czaplewski, A. J. (2017). Upstream social marketing strategy: An integrated marketing communications approach. *Business Horizons, 60*(3), 325–333. https://doi.org/10.1016/j.bushor.2017.01.006.

Kiong, K. L., Guo, T., Yao, C. M. K. L., Gross, N. D., Hanasono, M. M., Ferrarotto, R., Rosenthal, D. I., Myers, J. N., Hanna, E. Y., & Lai, S. Y. (2020). Changing practice patterns in head and neck oncologic surgery in the early COVID-19 era. *Head & Neck*, 1179–1186. https://doi.org/10.1002/hed.26202.

Ku, P. K. M., Holsinger, F. C., Chan, J. Y. K., Yeung, Z. W. C., Chan, B. Y. T., Tong, M. C. F., & Starmer, H. M. (2020). Management of dysphagia in the patient with head and neck cancer during COVID-19 pandemic: Practical strategy. *Head & Neck*, 1491–1496. https://doi.org/10.1002/hed.26224.

Kuderer, N. M., Choueiri, T. K., Shah, D. P., Shyr, Y., Rubinstein, S. M., Rivera, D. R., ... & Loaiza-Bonilla, A. (2020). Clinical impact of COVID-19 on patients with cancer (CCC19): A cohort study. The Lancet, 395(10241), 1907–1918. https://doi.org/10.1016/S0140-6736(20)31187-9.

Kudo, K., Isobe, N., Ueda, S., Tomimatsu, S., Moriyama, T., & Shimizu, S. (2021). Barriers to international telemedicine conferencing: A survey of the National University Hospital Council of Japan. *Telemedicine and E-Health*. https://doi.org/10.1089/tmj.2021.0046.

Lee, L. Y. W., Cazier, J. B., Angelis, V., Arnold, R., Bisht, V., Campton, N.A., … Middleton, G. (2020). COVID-19 mortality in patients with cancer on chemotherapy or other anticancer treatments: A prospective cohort study. The Lancet, 395(10241), 1919–1926. https://doi.org/10.1016/S0140-6736(20)31173-9.

Lee, L. Y. W., Cazier, J. -B., Starkey, T., Briggs, S. E. W., Arnold, R., Bisht, V., Booth, S., Campton, N. A., Cheng, V. W. T., Collins, G., Curley, H. M., Earwaker, P., Fittall, M. W., Gennatas, S., Goel, A., Hartley, S., Hughes, D. J., Kerr, D., Lee, A. J. X.,... Kerr, R. (2020). COVID-19 prevalence and mortality in patients with cancer and the effect of primary tumour subtype and patient demographics: A prospective cohort study. The Lancet Oncology, 21(10), 1309–1316. https://doi.org/10.1016/s1470-2045(20)30442-3.

Leora, H., Whisenant, J. G., Torri, V., Huang, L. C., Trama, A., Paz-Ares, L. G., Felip, E., Pancaldi, V., De Toma, A., Tiseo, M., Garrido, P., Genova, C., Cadranel, J., Michielin, O., Dingemans, A. M. C., Van Meerbeeck, J. P., Barlesi, F., Wakelee, H. A., Peters, S., & Garassino, M. C. (2020). Thoracic cancers international COVID-19 collaboration (TERAVOLT): Impact of type of cancer therapy and COVID therapy on survival. *Journal of Clinical Oncology*, LBA111. https://doi.org/10.1200/jco.2020.38.18_suppl.lba111.

Lewis, G. D., Hatch, S. S., Wiederhold, L. R., & Swanson, T. A. (2020). Long-term institutional experience with telemedicine services for radiation oncology: A potential model for long-term utilization. *Advances in Radiation Oncology*, 780–782. https://doi.org/10.1016/j.adro.2020.04.018.

Lindsey, M., Anna, S., Natalie, P., Ivette, T., Sarah, S., Brad, P., & Gail, T. (2016). Reaching high-risk underserved individuals for cancer genetic counseling by video-teleconferencing. *The Journal of Community and Supportive Oncology, 14*(4), 162–168. https://doi.org/10.12788/jcso.0247.

Lopez, A. M. (2019). Telemedicine, telehealth, and e-health technologies in cancer prevention. In *Fundamentals of cancer prevention* (4th ed., pp. 333–352). Springer International Publishing. https://doi.org/10.1007/978-3-030-15935-1_10.

Luo, J., Rizvi, H., Preeshagul, I. R., Egger, J. V., Hoyos, D., Bandlamudi, C., McCarthy, C. G., Falcon, C. J., Schoenfeld, A. J., Arbour, K. C., Chaft, J. E., Daly, R. M., Drilon, A., Eng, J., Iqbal, A., Lai, W. V., Li, B. T., Lito, P., Namakydoust, A., … Hellmann, M. D. (2020). COVID-19 in patients with lung cancer. *Annals of Oncology, 31*(10), 1386–1396. https://doi.org/10.1016/j.annonc.2020.06.007.

Matos, L. L., Forster, C. H. Q., Marta, G. N., Junior, G. C., Ridge, J. A., Hirata, D., Miranda-Filho, A., Hosny, A., Sanabria, A., Gregoire, V., Patel, S. G., Fagan, J. J., D'Cruz, A. K., Licitra, L., Mehanna, H., Hao, S.-P., Psyrri, A., Porceddu, S., Galloway, T. J., … Kowalski, L. P. (2021). The hidden curve behind COVID-19 outbreak: The impact of delay in treatment initiation in cancer patients and how to mitigate the additional risk of dying—The head and neck cancer model. *Cancer Causes & Control*. https://doi.org/10.1007/s10552-021-01411-7.

Medas, F., Ansaldo, G. L., Avenia, N., Basili, G., Bononi, M., Bove, A., Carcoforo, P., Casaril, A., Cavallaro, G., Conzo, G., De Pasquale, L., Del Rio, P., Dionigi, G., Dobrinja, C., Docimo, G., Graceffa, G., Iacobone, M., Innaro, N.,

Lombardi, C. P., … Pezzolla, A. (2021). Impact of the COVID-19 pandemic on surgery for thyroid cancer in Italy: Nationwide retrospective study. *British Journal of Surgery*. https://doi.org/10.1093/bjs/znab012.

Mendu, S., Boukhechba, M., Gordon, J. R., Datta, D., Molina, E., Arroyo, G., Proctor, S. K., Wells, K. J., & Barnes, L. E. (2018). Design of a culturally-informed virtual human for educating hispanicwomen about cervical cancer. In *ACM international conference proceeding series* (pp. 360–366). Association for Computing Machinery. https://doi.org/10.1145/3240925.3240968.

Mengyuan, D., Dianbo, L., Miao, L., Fuxiang, Z., Guiling, L., Zhen, C., Zhian, Z., Hua, Y., Meng, W., Qichao, Z., Yong, X., Huihua, X., Chun, W., Changchun, C., Fei, X., Yan, Z., Yaqin, P., Siping, G., Bo, Z., … Hongbing, C. (2020). Patients with cancer appear more vulnerable to SARS-COV-2: A multi-center study during the COVID-19 outbreak. *Cancer Discovery*, CD-20-0422. https://doi.org/10.1158/2159-8290.cd-20-0422.

Morris, E. J. A., Raphael, G., Enti, S., Marion, M., Finan, P. J., Jon, S., Mike, R., Katie, S., Jonathan, E., Sam, H., Paula, C., Dominic, G., David, S.-M., Chris, C., Rutter, M. D., Nicholson, B. D., Jem, R., Martin, L., Rory, C., … Colin, B. (2021). Impact of the COVID-19 pandemic on the detection and management of colorectal cancer in England: A population-based study. *The Lancet Gastroenterology & Hepatology*, *6*(3), 199–208. https://doi.org/10.1016/s2468-1253(21)00005-4.

Nepogodiev, D., Omar, O. M., Glasbey, J. C., Li, E., Simoes, J. F. F., Abbott, T. E. F., Ademuyiwa, A. O., Biccard, B. M., Chaudhry, D., Davidson, G. H., Di Saverio, S., Gallo, G., Ghosh, D., Harrison, E. M., Hutchinson, P. J., Kamarajah, S. K., Keller, D. S., Lawani, I., … Minaya-Bravo, A. (2020). Elective surgery cancellations due to the COVID-19 pandemic: Global predictive modelling to inform surgical recovery plans. *British Journal of Surgery*, *107*(11), 1440–1449. https://doi.org/10.1002/bjs.11746.

Noar, S. M., Harrington, N. G., & Aldrich, R. S. (2009). The role of message tailoring in the development of persuasive health communication messages. *Annals of the International Communication Association*, *33*, 73–133.

Ochs, M., Mestre, D., de Montcheuil, G., Pergandi, J. M., Saubesty, J., Lombardo, E., Francon, D., & Blache, P. (2019). Training doctors' social skills to break bad news: Evaluation of the impact of virtual environment displays on the sense of presence. *Journal on Multimodal User Interfaces*, *13*(1), 41–51. https://doi.org/10.1007/s12193-018-0289-8.

Santoro, G. A., Ugo, G., Sthela, M.-R., Nunoo-Mensah, J. W., Anders, M., Luca, D. T. G., Gaetano, G., Charles, T., & Wexner, S. D. (2021). DElayed COloRectal cancer care during COVID-19 pandemic (DECOR-19): Global perspective from an international survey. *Surgery*, *169*(4), 796–807. https://doi.org/10.1016/j.surg.2020.11.008.

Segelov, E., Underhill, C., Prenen, H., Karapetis, C., Jackson, C., Nott, L., Clay, T., Pavlakis, N., Sabesan, S., Heywood, E., Steer, C., Lethborg, C., Gan, H. K., Yip, D., Karanth, N., Karikios, D., & Raina MacIntyre, C. (2020). Practical considerations for treating patients with cancer in the COVID-19 pandemic. *JCO Oncology Practice*, *16*(8), 467–482. https://doi.org/10.1200/OP.20.00229.

Shankar, A., & Saini, D. (2021). Cancer care delivery amidst the COVID-19 pandemic: Common challenges and counter measures. *Cancer Research, Statistics, and Treatment*, *4*.

Sharpless, N. E. (2020). COVID-19 and cancer. *Science*, 1290. https://doi.org/10.1126/science.abd3377.

Shuman, A. G., & Campbell, B. H. (2020). Ethical framework for head and neck cancer care impacted by COVID-19. *Head and Neck*, *42*(6), 1214–1217. https://doi.org/10.1002/hed.26193.

Siegel, R. L., Miller, K. D., Fuchs, H. E., & Jemal, A. (2021). Cancer statistics, 2021. *CA: A Cancer Journal for Clinicians*, *71* (1), 7–33. https://doi.org/10.3322/caac.21654.

Swainston, J., Chapman, B., Grunfeld, E. A., & Derakshan, N. (2020). COVID-19 lockdown and its adverse impact on psychological health in breast cancer. *Frontiers in Psychology*, *11*. https://doi.org/10.3389/fpsyg.2020.02033.

Swapnarekha, H., Sekhar, B. H., Janmenjoy, N., & Bighnaraj, N. (2020). Role of intelligent computing in COVID-19 prognosis: A state-of-the-art review. *Chaos, Solitons & Fractals*, *138*, 109947. https://doi.org/10.1016/j.chaos.2020.109947.

Taccone, F. S., Artigas, A. A., Sprung, C. L., Moreno, R., Sakr, Y., & Vincent, J.-L. (2009). Characteristics and outcomes of cancer patients in European ICUs. *Critical Care*, *R15*. https://doi.org/10.1186/cc7713.

Tait, S., Jonathan, R., & Mickey, T. (2020). Understanding and addressing sources of anxiety among health care professionals during the COVID-19 pandemic. *JAMA*, *2133*. https://doi.org/10.1001/jama.2020.5893.

Telemedicine: Opportunities and developments in member states. (2010). http://apps.who.int/ iris/bitstream/handle/10665/44497/9789241564144_eng.pdf.

Thomson, D. J., Palma, D., Guckenberger, M., Balermpas, P., Beitler, J. J., Blanchard, P., Brizel, D., Budach, W., Caudell, J., Corry, J., Corvo, R., Evans, M., Garden, A. S., Giralt, J., Gregoire, V., Harari, P. M., Harrington, K., Hitchcock, Y. J., Johansen, J., … Yom, S. S. (2020). Practice recommendations for risk-adapted head and neck cancer radiation therapy during the COVID-19 pandemic: An ASTRO-ESTRO consensus statement. *International Journal of Radiation Oncology Biology Physics*, *107*(4), 618–627. https://doi.org/10.1016/j.ijrobp.2020.04.016.

Tsang, V. H. M., Matti, G., Anthony, G., Roderick, C.-B., & Robinson, B. G. (2020). Thyroid cancer in the age of COVID-19. *Endocrine-Related Cancer*, R407–R416. https://doi.org/10.1530/erc-20-0279.

US Bureau of Labor Statistics. (2021). *The employment situation—September 2020*. https://www.bls.gov/news.release/pdf/empsit.pdf.

Van Haren, R. M., Delman, A. M., Turner, K. M., Waits, B., Hemingway, M., Shah, S. A., & Starnes, S. L. (2021). Impact of the COVID-19 pandemic on lung cancer screening program and subsequent lung cancer. *Journal of the American College of Surgeons*, *232*(4), 600–605. https://doi.org/10.1016/j.jamcollsurg.2020.12.002.

Wakefield, M. A., Loken, B., & Hornik, R. C. (2010). Use of mass media campaigns to change health behaviour. *The Lancet*, 1261–1271. https://doi.org/10.1016/s0140-6736(10)60809-4.

Wenhua, L., Weijie, G., Ruchong, C., Wei, W., Jianfu, L., Ke, X., Caichen, L., Qing, A., Weixiang, L., Hengrui, L., Shiyue, L., & Jianxing, H. (2020). Cancer patients in SARS-CoV-2 infection: A nationwide analysis in China. *The Lancet Oncology*, *21*(3), 335–337. https://doi.org/10.1016/s1470-2045(20)30096-6.

WHO. (2020). *Laboratory testing for coronavirus disease (COVID-19) in suspected human cases: Interim guidance*.

Xiya, M., & Dominique, V. (2020). Critical care capacity during the COVID-19 pandemic: Global availability of intensive care beds. *Journal of Critical Care*, 96–97. https://doi.org/10.1016/j.jcrc.2020.04.012.

Zhang, L., Zhu, F., Xie, L., Wang, C., Wang, J., Chen, R., Jia, P., Guan, H. Q., Peng, L., Chen, Y., Peng, P., Zhang, P., Chu, Q., Shen, Q., Wang, Y., Xu, S. Y., Zhao, J. P., & Zhou, M. (2020). Clinical characteristics of COVID-19-infected cancer patients: A retrospective case study in three hospitals within Wuhan, China. *Annals of Oncology*, *31*(7), 894–901. https://doi.org/10.1016/j.annonc.2020.03.296.

Zhijie, W., Jie, W., & Jie, H. (2020). Active and effective measures for the care of patients with cancer during the COVID-19 spread in China. *JAMA Oncology*, *631*. https://doi.org/10.1001/jamaoncol.2020.1198.

Zhou, F., Yu, T., Du, R., Fan, G., Liu, Y., Liu, Z., Xiang, J., Wang, Y., Song, B., Gu, X., Guan, L., Wei, Y., Li, H., Wu, X., Xu, J., Tu, S., Zhang, Y., Chen, H., & Cao, B. (2020). Clinical course and risk factors for mortality of adult inpatients with COVID-19 in Wuhan, China: A retrospective cohort study. *The Lancet*, *395*(10229), 1054–1062. https://doi.org/10.1016/S0140-6736(20)30566-3.

CHAPTER

11

Empirical wavelet transform-based fast deep convolutional neural network for detection and classification of melanoma

Bhanja Kishor Swain, Mrutyunjaya Sahani, and Renu Sharma

Shiksa 'O' Anusandhan deemed to be University, Bhubaneswar, Odisha, India

Introduction

Skin cancer is a physical condition defined by an abnormal growth of skin cells, and the most common reason is the exposure to sun light, but it can also develop in parts of skin that generally do not get exposed to sun light. Skin cancer can be broadly categorized as keratinocyte carcinoma and melanoma. Squamous cell skin cancers and basal skin cancers come under the category of keratinocyte carcinoma, and they usually develop on the skin areas of the head and neck because of the increased exposure to sunlight. Squamous cell and basal skin cancers usually do not spread, but if they are left untreated, they may spread and grow larger to become life-threatening. Melanoma comes under another category of skin cancer which develops from melanocytes (Satheesha et al., 2017). The cells of melanocytes are responsible to provide skin color. It may get developed at any skin area, and more observably it starts developing on the skin areas of chest and back in case of men and on the legs in case of women. If identified at early stage and treated properly, then the melanoma can be cured, but if left untreated, they may spread and grow larger to become even more life-threatening than squamous cell and basal skin cancers. The severity of melanoma cancer is defined on the basis of different stages which depends upon the depth of penetration (Siegel et al., 2017). The early detection of melanoma at its prodromal stage is very crucial to prevent its advancement to different parts of the body. Melanoma can be very dangerous at its later stage as it has a high mortality rate; however, it accounts for 1% of all skin cancers. At the advanced stage, the

Computational Intelligence in Cancer Diagnosis
https://doi.org/10.1016/B978-0-323-85240-1.00006-7

survival rate is 13%, whereas when it is detected at its early stage, the survival rate is 95% (Siegel et al., 2017). Hence, the early detection of melanoma at its prodromal stage is very important.

Many researchers have proposed many methods such as the ABCDE method (Stolz et al., 1994), 7-point checklist (Argenziano et al., 1998), texture analysis (Maglogiannis & Doukas, 2009), and pattern analysis (Sáez et al., 2014) by comparing the texture and symmetry of the normal skin with the skin of the subject. The shape and color of the skin under study are also compared with normal skin. But in epiluminescence microscopy, otherwise called as dermoscopy, the dermal features are captured from the skin surface by applying efficient digital imaging techniques called as dermatoscopy. Dermatoscopy, a non-invasive method, captures the distinguished dermal features from the skin surface which cannot be seen with naked eyes and thus contributes significantly in the process of identifying melanoma.

In computer-aided diagnosis (CAD) systems, image acquisition is the first step and then the acquired images are pre-processed to eliminate unnecessary information. Then, the distinguished features are extracted from the segmented skin region to feed the classifier. Fig. 11.1 represents the complete process of a CAD, which helps in increasing the efficiency of clinicians and dermatologists by reducing unnecessary biopsies. The pre-processing techniques such as illumination enhancement (Barata et al., 2012), median filtering (Celebi et al., 2007), and DullRazor hair removal algorithm (Lee et al., 1997) are applied to remove the spurious noise. In segmentation the surrounding skin is separated from the lesion to analyze the lesion and to extract the features. For manual feature extraction, generally the ABCD rule is followed which examines the asymmetry border irregularities, color variation, and diameter of the lesion. The statistical features usually extracted in CAD are mean, standard deviation, color variance, color histogram, color asymmetry, gradient histogram, gray level co-occurrence etc.

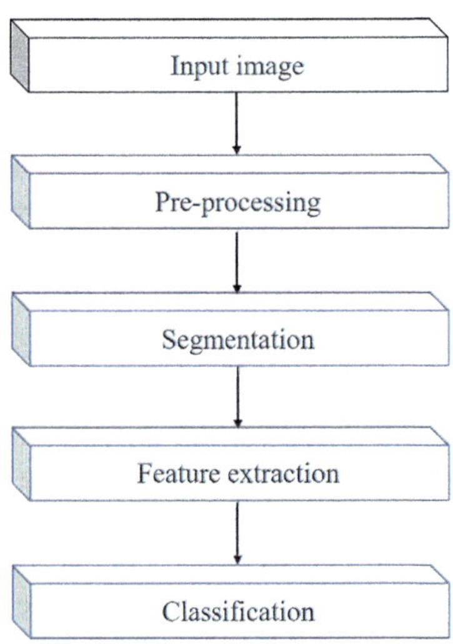

FIG. 11.1 Block diagram of computer-aided detection system for melanoma.

In literature, various conventional classifiers are proposed for detection of melanoma. For instance, the SVM is used to create a hyperplane between the positive and negative classes (Mocellin et al., 2006). KNN is used for the skin lesion classification based on the color feature and texture of the dermoscopy images (Ballerini et al., 2013). Although KNN alone is not able to produce good classification accuracy, a combination of KNN with SVM and AdaBoost produces better accuracy by concentrating on the extraction of local as well as global features from the skin lesion digital images (Barata et al., 2014). Similarly, the logistic regression (Blum et al., 2004) and Naïve Bayes classifier (Maglogiannis & Doukas, 2009) along with some variants of SVM (Korotkov & Garcia, 2012; Mishra & Celebi, 2016) are used to classify the malignant melanoma. Yu et al. (2017) proposed a two-stage frame, one for segmentation and other for classification. A fully convolutional residual network (FCRN) is used for image segmentation, and a very deep residual network is used for image classification. The process of segmentation allows the feature extraction of only the lesion areas, and such features carry the maximum information. For classification purpose, the softmax and SVM are used to obtain an average precision of 0.637. A multi-tract CNN is proposed by Kawahara et al. to classify the skin lesion (Kawahara & Hamarneh, 2016). This method provides an accuracy of 77.3% with auxiliary loss and 75.5% without auxiliary loss. Then, for further improvement, the data augmentation method is used to achieve an accuracy of 79.5%. Codella et al. (2015) proposed a method consisting of parallel paths, one path uses CNN with transfer learning and the other one uses sparse coding. The maximum accuracy achieved is 93.1% by using non-linear SVM with histogram intersection kernel. The accuracy for classification of melanoma and non-melanoma is 93.1% and that of melanoma and atypical nevus is 73.9%. Codella et al. (2017) proposed a segmentation method using a fully convolutional neural network, and the non-linear SVM is applied for classification purpose. This method produced a precision of 0.649 by combining some specific features like sparse coding, edge histogram, and color histogram.

Dataset

In this paper, the publicly available PH^2 database is used to detect the early stage of melanoma by classifying the atypical nevi from the dermoscopic images of melanomas and common nevi. The dermoscopic images present in the PH^2 database are acquired at the Dermatology Service of Hospital Pedro Hispano, Matosinhos, Portugal. The dermoscopic images are acquired under the same conditions through the Tuebinger Mole Analyzer system using a magnification of 20×. Each of the dermoscopic images are 8-bit RGB color images with a resolution of 768 × 560 pixels. A total of 200 8-bit RGB color images are there in the database, which includes 80 common nevi, 80 atypical nevi, and 40 melanomas. Fig. 11.2 shows some selected samples of dermoscopy images collected from the PH2 dataset.

Continuous wavelet transform (CWT)

When wavelet transform is applied to a signal $x(t)$, the transformed signal is represented as $X_{wt}(\tau,s)$ where τ is the translation parameter and s is the scale parameter and $X_{wt}(\tau,s)$ is given as

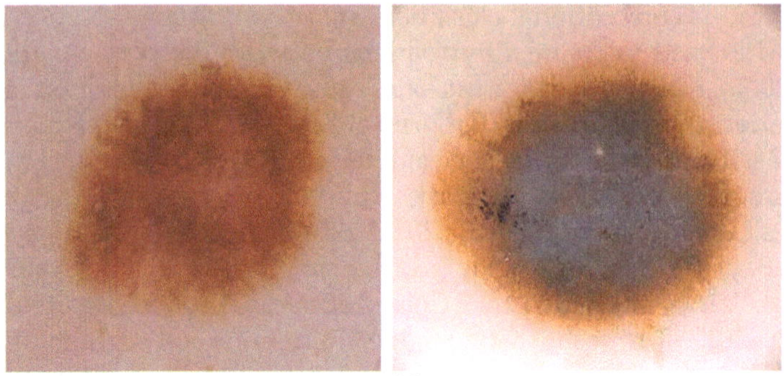

FIG. 11.2 Selected samples from PH2 database. (A) Common nevus. (B) Melanoma.

$$X_{wt}(\tau, s) = \frac{1}{\sqrt{|s|}} \int_{-\infty}^{\infty} x(t)\varphi^{*}\left(\frac{t-\tau}{s}\right) dt \tag{11.1}$$

If a complex wavelet function is used, $\varphi^{*}$is the complex conjugate of φ, and for the normalization of the signal energy at every scale, the term $\frac{1}{\sqrt{|s|}}$ is multiplied. The window length and the central frequency are controlled by the variation of the scale s, which results in the contraction and the dilation of the mother wavelet, and the variation of the translation parameter τ provides the position of the wavelet in time when shifted over the signal. A varying translation τ with constant scale s fills the rows, and a varying scale s with constant translation τ fills the columns of the time scale plane respectively.

Each element obtained from Eq. (11.1) is called wavelet coefficient, and they belong to a scale frequency and to a point in the time domain. Generally, the scaling parameter s and translation parameter τ represented by discrete values to calculate the performance of CWT and the discrete values of the translation parameters provide the wavelet coefficients known as wavelet series.

The quality factor of the filter is denoted as Q and defined as the constant Q property, which is otherwise called as the constant relative frequency resolution of the wavelet analysis. It is defined by the ratio of center-frequency f_c to that of the bandwidth f_b. To analyze the constant Q property of the filter, a dyadic sample-grid is applied for scaling, and the scale parameters are discretized by a dyadic grid on a logarithmic scale, and according to the scale parameter, the time parameter is discretized. The application of the dyadic grid results in the construction of an orthogonal wavelet basis. In terms of the dyadic grid, the wavelet series can be represented as,

$$X_{WT_{m,n}} = \int_{-\infty}^{\infty} x(t)\varphi_{m,n}(t) dt, \tag{11.2}$$

$$\text{With } \varphi_{m,n} = s_0^{-m/2} \varphi\left(s_0^{-m} t - n\tau_0\right) \tag{11.3}$$

In Eqs. (11.2) and (11.3), the wavelet dilation and translation are controlled by the integers m and n, and s_0 is chosen as zero and τ_0 as 1 for a dyadic grid. The chosen discrete dyadic grid wavelets have the property of normalized unit energy, and they are mutually orthogonal. The choice of selecting a discrete dyadic grid wavelet with normalized unit energy and which are mutually orthogonal facilitates the restoration of the original signal as,

$$x(t) = \sum_{m=-\infty}^{\infty} \sum_{n=-\infty}^{\infty} X_{WT_{m,n}} \varphi_{m,n}(t) \tag{11.4}$$

Discrete wavelet transform (DWT)

In CWT, the dilation and contraction of the wavelet function helps in the multiresolution analysis, but in DWT multiresolution, filter banks are used for the same purpose (Swain et al., 2020). A filter bank with various types of filters decomposes the original signal into different frequency bands. Fig. 11.3 shows a two-channel filter bank which is divided into two parts, the analysis bank and the synthesis bank. The analysis bank has one LPF and one HPF, and these filters allow the input signal to pass through it in such a way that the output of the filter will have the same number of samples of the input signal, but the frequency content will be halved. The outputs of the filters present in analysis bank have the same frequency content of the signal $x(k)$ with a doubled amount of data, and therefore downsampling is required to reduce the data redundancy. Synthesis filters are used to reconstruct the original signal by a factor of 2 and allowed to pass through the filters $L_s(z)$ and $H_s(z)$. The synthesis bank's filters are correlated to the filters of analysis bank, and the outputs of $L_s(z)$ and $H_s(z)$ are added to reconstruct the original signal as $y(k)$. In the analysis filter bank, the outputs of $L_a(z)$and $H_a(z)$ are known as sub-bands. But, it is worth mentioning that the downsampling and consequent upsampling of the same signal will not produce the original signal.

As the transpose of downsampling is upsampling, again a set of operations of upsampling and downsampling by first inserting zeros and then removing them, it is possible to obtain the original signal. Mathematically, the operation of DWT is given as,

$$DWT_g^{\phi}(m,n,k) = (P_0{}^m)^{\frac{-1}{2}} \sum\nolimits_n g(n).\phi^* \left[\frac{k - nq_0 P_0{}^m}{P_0{}^m} \right] \tag{11.5}$$

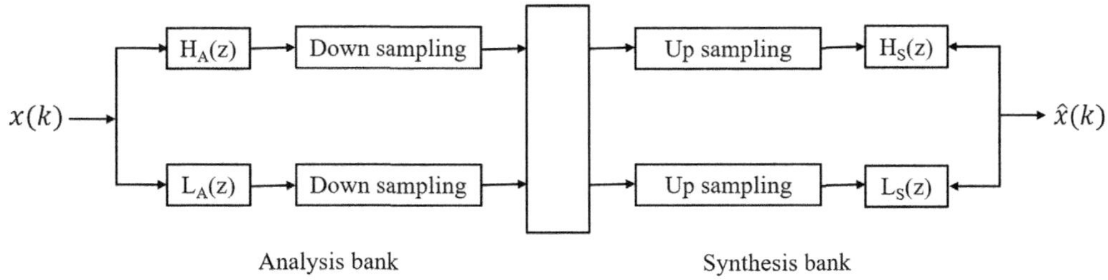

FIG. 11.3 A two-channel filter bank.

where $g(n)$ is the discrete version of $g(t)$which is discretely dilated by scaling parameter $P_o{}^m$ and translated by translational parameter $nq_0P_0{}^m nq_0P_0{}^m$ where q_0 is a constant with m and n are integers. Generally, $P_0 = 2$ and $q_0 = 1$ were chosen for dyadic transform.

Empirical wavelet transform (EWT)

The two-dimensional discrete wavelet transform has produced promising results in the decomposition of non-stationary and complicated signals (Mishra et al., 2015; Sahani, 2019; Swain, Rout, Sahani, & Sharma, 2021), but with a strong mathematical background and the advantages of the combination of wavelet theory and Fourier analysis, the two-dimensional empirical wavelet transform decomposes non-stationary and complicated signals consistently.

The input dermoscopic images $I(t)$ of the selected dataset are decomposed by employing the adaptive wavelet filter bank to obtain their sub-band images. The Fourier spectra of each row and column are determined in the range of $[0,\pi]$ to obtain the mean Fourier spectrum $I(\omega)$. Then, the center frequency is obtained using the mean Fourier support of $I(\omega)$ within a width of $2\tau_n$ of compact frequency support. Then, the boundary detection techniques are performed to obtain the corresponding filter banks with a frequency boundary $[0,\Omega_1]$, $[\Omega_1,\Omega_2]$, $\ldots[\Omega_{N-1,}\pi]$. Eqs. (11.6) and (11.7) represent the scaling function and wavelet function, respectively, and they are defined as follows,

$$S_n(\omega) = \begin{cases} 1, if\, |\omega| \leq \omega_n - \tau_n \\ \cos\left[\frac{\pi}{2}\alpha\left(\frac{1}{2\tau_n}(|\omega| - \omega_n + \tau_n)\right)\right], if\, \omega_n - \tau_n \leq |\omega| \leq \omega_n + \tau_n \\ 0, \text{otherwise} \end{cases} \tag{11.6}$$

$$W_n(\omega) = \begin{cases} 1, if\, \omega_n + \tau_n \leq |\omega \leq \omega_{n+1} - \tau_{n+1}| \\ \cos\left[\frac{\pi}{2}\alpha(\frac{1}{2\tau_{n+1}}(|\omega| - \omega_{n+1} + \tau_{n+1}\right], if\, \omega_{n+1} - \tau_{n+1} \leq |\omega| \leq \omega_{n+1} + \tau_{n+1} \\ \sin\left[\frac{\pi}{2}\alpha\left(\frac{1}{2\tau_n}(|\omega| - \omega_n + \tau_n)\right)\right], if\, \omega_n - \tau_n \leq |\omega| \leq \omega_n + \tau_n \\ 0, \text{otherwise} \end{cases} \tag{11.7}$$

The arbitrary function $S_n(\omega)$ in the tight form is defined as

$$\alpha(x) = \begin{cases} 0, \text{if}\, I \leq 0\, \text{and}\, \alpha(I) + \alpha(1 - I) = 1 \forall I \in [0, 1] \\ 1, \text{if}\, I > 1 \end{cases} \tag{11.8}$$

The approximate coefficients $A(n,t)$ are computed based on the inner product of scaling function $S_n{}^{coloum}(\omega)$ and $S_n{}^{row}(\omega)$ as follows

$$A(n,t) = \langle I, S_n \rangle = \int I(\tau).\overline{S_n(\tau - t)}d\tau \tag{11.9}$$

Similarly, the detail coefficients $D(0,t)$ are computed by the inner product of wavelet function $W_n{}^{coloum}(\omega)$ and $W_n{}^{row}(\omega)$ as follows

$$D(0,t) = \langle I, \omega_n \rangle = \int I(\tau).\overline{\omega_n(\tau - t)}d\tau \tag{11.10}$$

So, total (N_c+1) with (N_R+1) dimensions of images as sub-band images are generated.

Fast deep convolutional neural network (fast-DCNN)

Being receptive to shift deviation and translation has been a major disadvantage of the artificial neural network (ANN) especially in the field of pattern recognition for digital images as input. In this regard, the deep convolutional neural network (DCNN) has been a better alternative for achieving competitive classification accuracy (Sahani et al., 2021; Swain, Rout, Sahani, Kumari, et al., 2021). In this paper, the authors have proposed a variation in the existing DCNN called as fast deep convolutional neural network (fast-DCNN). The fast-DCNN structure has the following layers.

Convolutional layer

In all deep learning techniques, the prime objective is to extract the distinguished features from the input signal. In fast-DCNN, the convolutional layer yields the feature map contained with the distinguished features of the input signal. In this work, the resized dermoscopic images are used as the input signal, and the size and number of filters are predefined. Mathematically, the feature map is represented as the following equation.

$$C_m = \sum\nolimits_{n=0}^{N-1} f_n k_{m-n} \tag{11.11}$$

where k and f are the input signal and the filter selected for convolution operation, respectively. C_m is the mth output element, and the total number of data points is represented by N.

Rectified liner unit (ReLU) layer

The ReLU layer performs thresholding operation to reduce non-linearity by assigning 0 to the negative values and constant value to the positive values and is given as follows

$$f(m) = \begin{cases} M; \; m \geq 0 \\ 0; \; \text{Otherwise} \end{cases} \tag{11.12}$$

As a compensation of over-fitting problem, the output of ReLU undergoes the normalization operation called as batch normalization (BN).

Pooling layer

The smooth processing of distinguished features without losing its vitality is imperative, and this is performed by a pooling layer. The most popular maximum pooling (MP) operation is performed in this research work, and the stride step is selected as 2.

Fully connected layer

The feature map is flattened after the dimensionality reduction to construct the fully connected (FC) layer. Subsequently, rest of the FC layers are constructed with the last FC having two outputs. The output of the fully connected layer is the sum of the product of input with the weights w and bias b and is represented in Eq. (11.11).

$$x_i = \sum_j w_{ji} y_j + b_i \tag{11.13}$$

where x represents the output of the current layer and$y$ represents the output of the previous layer.

Softmax layer

Finally, two outputs of the last FC are fed as input to the softmax layer to predict the class of the input signal. The softmax activation function is given as

$$P_i = \frac{e^{x_i}}{\sum_m^k e^{x_i}} \tag{11.14}$$

where $P(x_i)$ belongs to the class of $x_i$and represents its probability within a range of 0–1.

Result and discussion

In this work, the dermoscopic images of the PH^2 dataset is used. The fast deep convolutional neural network (fast-DCNN), a modified version of the DCNN, is proposed by deciding the number of filters with a predefined size. Fig. 11.4 depicts the complete architecture of the proposed fast-DCNN. After the convolution layer, the Rectified Linear Unit (ReLU) layer performs thresholding operation to reduce non-linearity by assigning 0 to the negative values and a constant value to the positive values. For smooth processing of the distinguished features, the max pooling operation is performed with a selected stride step of 2. After three sets of convolutional layers, the feature map is flattened to construct the fully connected (FC) layer. Finally, two outputs of the last FC layer are given to the softmax layer to categorize the normal nevus and melanoma.

Max pooling operation is performed after every set of convolution operations to minimize the problem of overfitting. Table 11.1 provides the detail information about the number of filters and the size of filters used in the architecture of fast-DCNN. The dataset contains the images with variable size, and to avoid the ambiguity, the images are resized to $600 \times 800 \times 3$ before applying them as input to the proposed classifier. In the fast-DCNN method, the resized images of the PH^2 dataset are used to achieve the maximum classification accuracy with minimum training loss as shown in Fig. 11.5. The MATLAB/Simulink software is used to carry out this research work, and the system used has a processor of Intel core(R) of 2.6GHz clock speed along with a RAM size of 16GB.

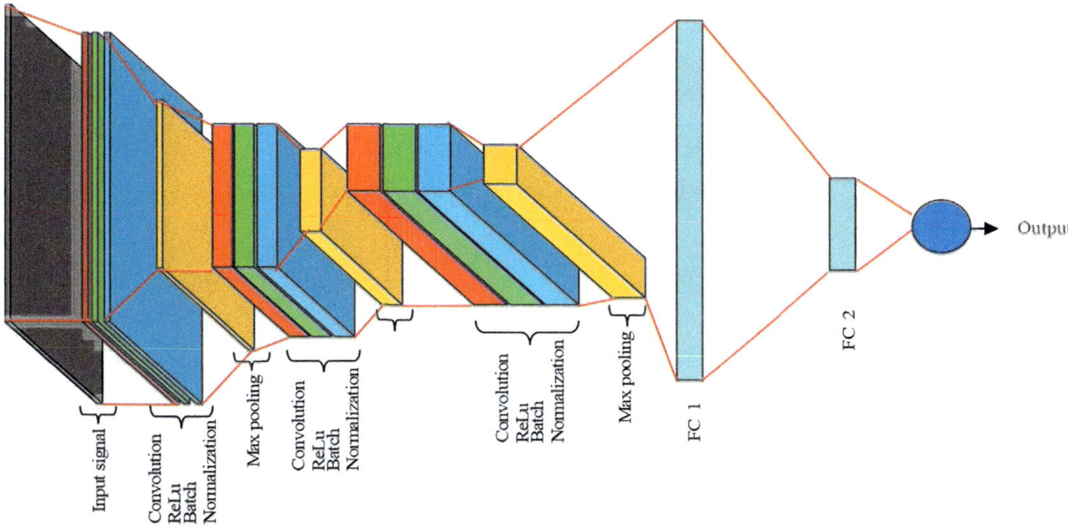

FIG. 11.4 The architecture of fast deep convolutional neural network (fast-DCNN).

TABLE 11.1 The architecture of the fast deep convolutional neural network (fast-DCNN).

Layer	Name	Output nodes	Filter size	Stride
0–1	Convolution	[598 × 798 × 3] × 8	3	1
1–2	ReLU	[598 × 798 × 3] × 8	1	1
2–3	BN	[598 × 798 × 3] × 8	10	10
3–4	MP	[299 × 399 × 3] × 8	2	2
4–5	Convolution	[295 × 395 × 3] × 5	5	1
5–6	ReLU	[295 × 395 × 3] × 5	1	1
6–7	BN	[295 × 395 × 3] × 5	10	10
7–8	Max-pooling	[147 × 197 × 3] × 5	2	2
8–9	Convolution	[138 × 188 × 3] × 5	10	1
9–10	ReLU	[138 × 188 × 3] × 5	1	1
10–11	BN	[138 × 188 × 3] × 5	10	10
11–12	MP	[69 × 94 × 3] × 5	2	2
12–13	FC	100 × 1	0	0
13–14	FC	2 × 1	0	0

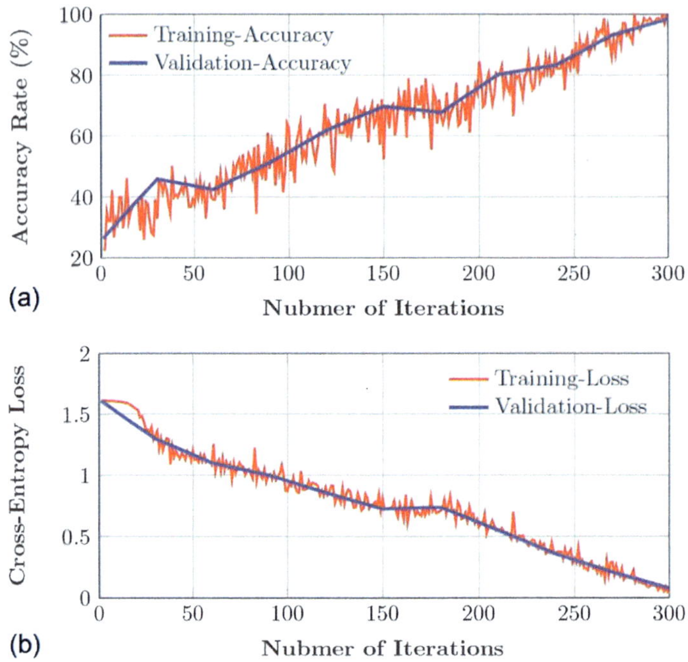

FIG. 11.5 Performance of the proposed fast DCNN classifier, (A) accuracy rate, (B) cross-entropy loss.

For the operation of DWT, the Daubechies wavelet is used as the mother wavelet. The application of such mother wavelets provides the translated version of the signal in terms of the shifted and scaled version of the mother wavelet. As the input digital image is a two-dimensional (2-D) signal, the low-pass filter (LPF) and high-pass filter (HPF) are applied on the rows and the columns of the original input image. The LPF preserves the low-frequency components, and the HPF preserves the high-frequency components. The HPF provides the detailed, and the LPF provides the approximation value. In this way, by applying the LPF, we get the horizontal approximation. Again, we apply the LPF and HPF to the columns of horizontal approximation. The application of two successive LPFs, one to the rows of original image and next to the columns of the horizontal approximation provide an approximate image. Vertical details are obtained by applying a HPF to the horizontal approximation. The application of two successive filters, one LPF to the rows of original input image and one HPF to the horizontal approximation, provides vertical details of the original image. Similarly, the horizontal details and diagonal details are obtained from the input image as shown in Fig. 11.6.

Furthermore, the empirical wavelet transform (EWT) is applied on the gray images of the dermoscopic images after confirming that the images are resized again and they are squared. Multiple sub-images are obtained as the output images of EWT with respect to various

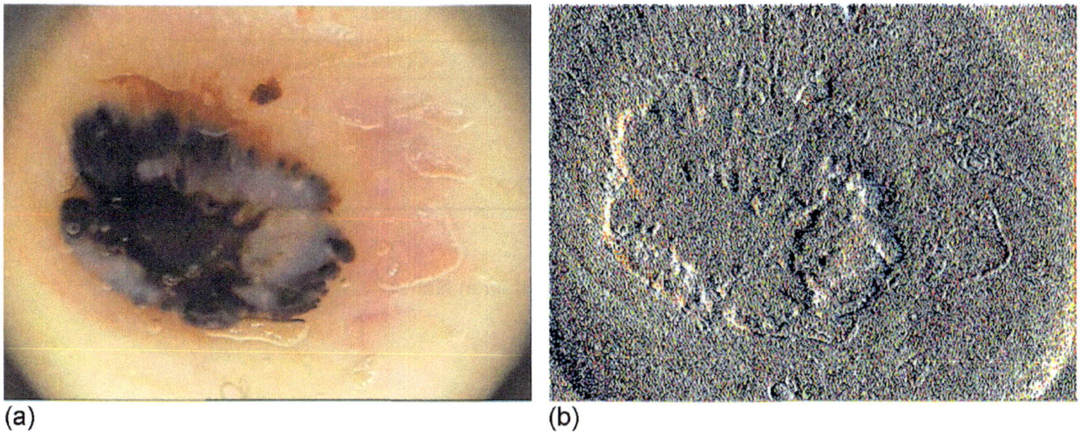

FIG. 11.6 The output of DWT of a selected sample of melanoma and its horizontal details. (A) Melanoma sample. (B) Horizontal details.

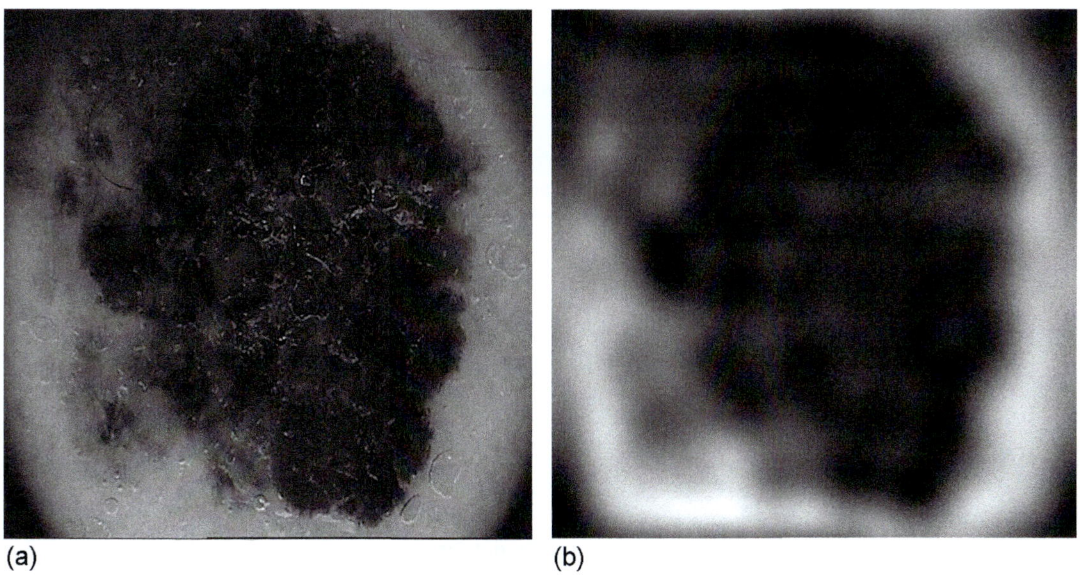

FIG. 11.7 The first sub-band image of a selected sample of melanoma. (A) Melanoma sample. (B) First sub-band image.

frequency bands. In this work, the first sub-band images are selected as the input to the proposed classifier. Fig. 11.7 shows a selected sample of melanoma and its first sub-band image.

Table 11.2 represents the obtained classification accuracy of all the three methods, i.e., the fast-DCNN, DWT based fast-DCNN, and the EWT based fast-DCNN methods. Fig. 11.8 represents the confusion matrix for the binary classification of normal nevus and melanoma.

TABLE 11.2 The performance of proposed classifiers.

Serial number	Proposed methods	Classification accuracy (%)
1	Fast-DCNN	92.76
2	DWT based fast-DCNN	93.88
3	EWT based fast-DCNN	95.00

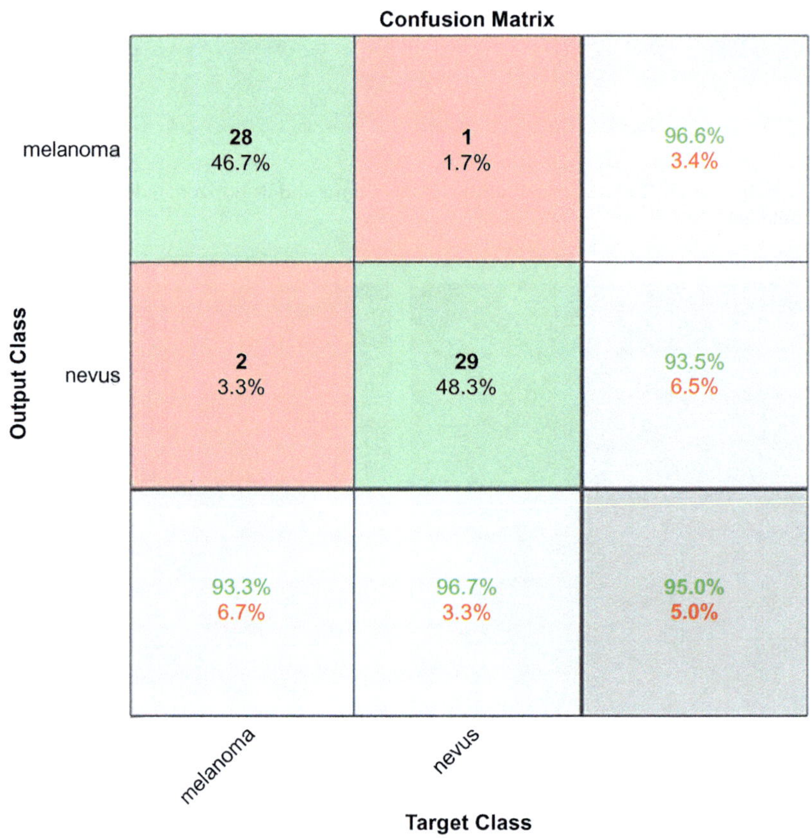

FIG. 11.8 The confusion matrix of the classification performance of EWT based fast-DCNN.

Conclusion

The binary classification of nevus and melanoma is performed by using the resized dermoscopic images of the PH^2 dataset. A novel fast deep convolutional neural network (fast-DCNN), consisting of three convolutional layers, two fully connected layers, and the probability density function based softmax layer is proposed. The resized dermoscopic images are decomposed to extract the horizontal details and the first sub-band images by applying DWT and EWT respectively and used as input to the fast-DCNN. The performance of the binary classification of common nevus and melanoma is compared. Finally, in terms of the cross-entropy loss and accuracy rate along with the processing time, the EWT-based fast-DCNN method outperforms the fast-DCNN and the DWT based fast-DCNN.

References

Argenziano, G., Fabbrocini, G., Carli, P., De Giorgi, V., Sammarco, E., & Delfino, M. (1998). Epiluminescence microscopy for the diagnosis of doubtful melanocytic skin lesions: Comparison of the abcd rule of dermatoscopy and a new 7-point checklist based on pattern analysis. *Archives of Dermatology, 134*(12), 1563–1570.

Ballerini, L., Fisher, R. B., Aldridge, B., et al. (2013). A color and texture based hierarchical K-NN approach to the classification of non-melanoma skin lesions. *Color Medical Image Analysis, 6*, 63–86.

Barata, C., Marques, J. S., & Rozeira, J. (2012). A system for the detection of pigment network in dermoscopy images using directional filters. *IEEE Transactions on Biomedical Engineering, 59*(10), 2744–2754.

Barata, C., Ruela, M., Francisco, M., et al. (2014). Two systems for the detection of melanomas in dermoscopy images using texture and color features. *IEEE Systems Journal, 8*(3), 965–979.

Blum, A., Luedtke, H., Ellwanger, U., Schwabe, R., Rassner, G., & Garbe, C. (2004). Digital image analysis for diagnosis of cutaneous melanoma. Development of a highly effective computer algorithm based on analysis of 837 melanocytic lesions. *The British Journal of Dermatology, 151*(5), 1029–1038.

Celebi, M. E., Kingravi, H. A., Uddin, B., Iyatomi, H., Aslandogan, Y. A., Stoecker, W. V., & Moss, R. H. (2007). A methodological approach to the classification of dermoscopy images. *Computerized Medical Imaging and Graphics, 31*(6), 362–373.

Codella, N., Cai, J., Abedini, M., Garnavi, R., Halpern, A., & Smith, J. R. (2015). Deep learning, sparse coding, and SVM for melanoma recognition in dermoscopy images. In L. Zhou, L. Wang, Q. Wang, & Y. Shi (Eds.), *Vol. 9352. MLMI 2015. LNCS* (pp. 118–126). Cham: Springer. https://doi.org/10.1007/978-3-319-24888-2 15.

Codella, N. C., Nguyen, Q. B., Pankanti, S., Gutman, D., Helba, B., Halpern, A., & Smith, J. R. (2017). Deep learning ensembles for melanoma recognition in dermoscopy images. *IBM Journal of Research and Development, 61*(4), 1–5.

Kawahara, J., & Hamarneh, G. (2016). Multi-resolution-tract CNN with hybrid pretrained and skin-lesion trained layers. In L. Wang, E. Adeli, Q. Wang, Y. Shi, & H.-I. Suk (Eds.), *Vol. 10019. MLMI 2016. LNCS* (pp. 164–171). Cham: Springer. https://doi.org/10.1007/978-3-319-47157-0 20.

Korotkov, K., & Garcia, R. (2012). Computerized analysis of pigmented skin lesions: A review. *Artificial Intelligence in Medicine, 56*(2), 69–90.

Lee, T., Ng, V., Gallagher, R., Coldman, A., & McLean, D. (1997). DullrazorR: A software approach to hair removal from images. *Computers in Biology and Medicine, 27*(6), 533–543.

Maglogiannis, I., & Doukas, C. N. (2009). Overview of advanced computer vision systems for skin lesions characterization. *IEEE Transactions on Information Technology in Biomedicine, 13*(5), 721–733.

Mishra, N. K., & Celebi, M. E. (2016). *An overview of melanoma detection in dermoscopy images using image processing and machine learning*. arXiv preprint arXiv:1601.07843.

Mishra, M., Rout, P. K., & Routray, P. (2015). High impedance fault detection in radial distribution system using wavelet transform. In *Annual IEEE India conference (INDICON)* (pp. 1–6). IEEE.

Mocellin, S., Ambrosi, A., Montesco, M. C., et al. (2006). Support vector machine learning model for the prediction of sentinel node status in patients with cutaneous melanoma. *Annals of Surgical Oncology, 13*(8), 1113–1122.

Sáez, A., Acha, B., & Serrano, C. (2014). Pattern analysis in dermoscopic images. In J. Scharcanski, & M. Celebi (Eds.), *Computer vision techniques for the diagnosis of skin cancer* (pp. 23–48). Heidelberg: Springer. https://doi.org/10.1007/978-3-642-39608-3 2.

Sahani, M. (2019). Detection and classification of power quality events using empirical wavelet transform and error minimised extreme learning machine. *International Journal of Power and Energy Conversion, 10*(4).

Sahani, M., Swain, B. K., & Dash, P. K. (2021). FPGA-based favourite skin colour restoration using improved histogram equalization with variable enhancement degree and ensemble extreme learning machine. *IET Image Processing, 15*(6), 1247–1259.

Satheesha, T., Satyanarayana, D., Prasad, M. G., & Dhruve, K. D. (2017). Melanoma is skin deep: A 3D reconstruction technique for computerized dermoscopic skin lesion classification. *IEEE Journal of Translational Engineering in Health and Medicine*, *5*, 1–17.

Siegel, R. L., Miller, K. D., & Jemal, A. (2017). Cancer statistics. *CA: A Cancer Journal for Clinicians*, *67*(1), 7–30.

Stolz, W., Riemann, A., Cognetta, A., Pillet, L., Abmayr, W., Holzel, D., Bilek, P., Nachbar, F., & Landthaler, M. (1994). ABCD rule of dermatoscopy: A new practical method for early recognition of malignant-melanoma. *European Journal of Dermatology*, *4*(7), 521–527.

Swain, B. K., Rout, S. K., Sahani, M., Kumari, P., & Sharma, R. (2021). Effectiveness of data augmentation for classification of melanoma using deep convolutional neural network. In *2021 International conference in advances in power, signal, and information technology (APSIT)* (pp. 1–5). IEEE.

Swain, B. K., Rout, S. K., Sahani, M., & Sharma, R. (2021). Unsupervised reduced deep convolutional neural network of process empirical wavelet transform data for recognition of the early stage of Alzheimer's disease. In *Green technology for smart city and society* (pp. 55–67). Singapore: Springer.

Swain, B. K., Sahani, M., & Sharma, R. (2020). Automatic recognition of the early stage of Alzheimer's disease based on discrete wavelet transform and reduced deep convolutional neural network. In *Innovation in electrical power engineering, communication, and computing technology* (pp. 531–542). Singapore: Springer.

Yu, L., Chen, H., Dou, Q., Qin, J., & Heng, P. A. (2017). Automated melanoma recognition in dermoscopy images via very deep residual networks. *IEEE Transactions on Medical Imaging*, *36*(4), 994–1004.

PART 3

Advance computational intelligence paradigms

CHAPTER

12

Convolutional neural networks and stacked generalization ensemble method in breast cancer prognosis

Tahmina Akter Tisha[a], *Mir Moynuddin Ahmed Shibly*[b], *Kowshik Ahmed*[a], *and Shamim H. Ripon*[a]

[a]Department of Computer Science and Engineering, East West University, Dhaka, Bangladesh
[b]Department of Computer Science and Engineering, United International University, Dhaka, Bangladesh

Introduction

Breast cancer is one of the main reasons for cancer deaths for women in the USA (DeSantis et al., 2019). From breast tissue, this cancer develops gradually. Although it is possible that a woman can be diagnosed with breast cancer without having any lump or tumor, the majority of women have breast cancer from some sort of tumor. A breast tumor can be either benign or malignant. Benign tumors usually do not spread to other tissues. If this type of tumor is found, there is nothing to worry about. But if the tumors are malignant or cancerous, they will attack other tissues, and eventually, they will spread. These types of tumors are responsible for breast cancer. Breast cancer is treatable if it is found in an earlier stage. Therefore, early detection of malignant tumors can be very helpful to treat a breast cancer patient. The earlier the malignancy of a tumor is detected, the more the chance of reducing breast-cancer-related death.

A reliable way to decide whether a patient has breast cancer is to perform a biopsy (Gupta & Bhavsar, 2017). Through histopathological biopsy, even the most infallible cancer can be diagnosed. Thin sections of breast tissues are extracted in slides using a fine needle during the biopsy, and after that, those slides are analyzed under the microscope with different levels of magnifications to highlight the regions of interest (ROIs). The pathologists examine the textures, pattern, and morphology of the extracted tissue to do clinical examination finding clues

https://doi.org/10.1016/B978-0-323-85240-1.00008-0

for better prognosis (Bayramoglu, Kannala, & Heikkila, 2016). This needs the process of zooming, panning, focusing, and scanning through the entire image which is time-consuming and expensive, and to perform these experiments on the images, a pathologist must have certain expertise (Benhammou, Achchab, Herrera, & Tabik, 2020). Another drawback is this costly process in many cases can be hampered through the different interpretations of different pathologists. A more improved data-driven and trained deep-learning model has been introduced to overcome all these stumbling blocks in detecting breast cancer and save many precious lives. Proper training of such a model can lead to obtaining a decision support system that can detect breast cancer from related biomedical images with very high accuracy. So, this study aims to develop appropriate deep-learning models that can detect breast cancer from biomedical breast images efficiently.

There are many examples of automated machine learning predictors for breast cancer detection. In this era of technological advancements, a machine-made prediction can be more accurate than a doctor-made prediction. Researchers have shown that an artificially intelligent model can outperform expert doctors in breast cancer screening (McKinney et al., 2020). There is a great chance that a doctor can miss cancer for various reasons in a real-world environment while screening the images. This inability to detect a potential breast cancer patient in the early stage can be life-threatening. To avoid such cases, a trained model can assist a doctor in a meaningful way (Houssami, Lee, Buist, & Tao, 2017). However, developing such machine learning and artificially intelligent models that can screen cancer patients from various imaging can be challenging.

The computer-aided prognosis of breast cancer is an area of research interest among many researchers (Chan, Samala, & Hadjiiski, 2020). Different researchers have applied different methods to create efficient and accurate models for breast cancer detection based on different types of images like histopathology, mammography, thermography, or based on data other than images. For example, the support vector machine has been used to a tabular dataset for breast cancer prognosis (Ferroni et al., 2019). However, detecting breast cancer from images is a more widely used prognosis mechanism. Neural network methods can come to aid in breast cancer prognosis from images (Alan, 2017). The majority of researchers have used various neural networks for breast cancer detection like convolutional neural networks (Han et al., 2017), artificial neural networks (Kaymak, Helwan, & Uzun, 2017), and extreme learning (Zhiqiong et al., 2019). This scientific base has guided this study to use the convolutional neural network (CNN)-based methods for breast cancer prognosis. Moreover, the ensemble of convolutional neural networks has not been used much for the detection of cancer from breast images. The ensemble of multiple models can enhance the performance of accurate prognosis rates drastically. Using an ensemble of typical shallow machine learning models has been proved to be an effective method for classification as opposed to a single classifier. But, the deep learning-based ensemble methods are rarely seen in the field of image classification. This study also aims to introduce an ensemble method for the detection task to develop more enhanced and robust predictors. Additionally, this work has explored the portability of deep learning models for the detection task.

The major contribution of this chapter includes the following:

- Detecting breast cancer from histopathological images using advanced convolutional neural networks.
- Improving the performance of detection by introducing an ensemble of convolutional neural networks.

In the following sections of this chapter, an extensive literature review for breast cancer prognosis from images is presented followed by an elaborate description of the proposed methodology of this work. After that, the experimental setup of this study is described. Besides, the outcomes of the conducted experiments are presented with appropriate analysis and discussion. Finally, the key research findings are summarized with some concluding remarks.

Literature study

Nahid, Mehrabi, and Kong (2018) have classified biomedical images into malignant and benign classes. To extract the statistical and structural information from images, they have proposed three deep neural network models—a CNN, a neural network with long short-term memory (LSTM), and another model that has combined both CNN and LSTM. They have used both the support vector machine (SVM) and the SoftMax activation for the classification. The experiments have shown that the CNN model has worked better than the other two proposed methods. And for the classification layer, the SoftMax activation has performed better than SVM. A survey has been conducted by Mehdy, Ng, Shair, Saleh, and Gomes (2017) to find out the gap of knowledge in early breast cancer detection by evaluating recent literature with special methods. They have identified ultrasound, thermal, mammogram, and MRI as four major sources of breast imaging. Different types of neural networks such as feed-forward backpropagation, which is the simplest neural network, and cascade forward, the CNN, have been applied for classification. The majority of the learning methods for breast cancer detection have been hybrid models like a self-organizing map (SOM), bootstrap with a neural network, SVM with a neural network, and so on.

Considering the difficulty of tuning a large number of parameters for CNNs, Erkan et al. (2018) have introduced transfer learning and the process of deep feature extraction to detect breast cancer from histopathological images. The experiment has been conducted by arranging the BreakHis (Spanhol, Oliveira, Petitjean, & Heutte, 2016) dataset into fivefolds. CNN models such as AlexNet and VGG16 have been adapted for extracting the feature, and SVM has been used to classify the benign and malignant class from each fold. At the same time, introducing transfer learning, pretrained AlexNet has been implemented for further fine-tuning which eventually has produced the better result having an accuracy of more than 90% in every fold.

Jiang, Chen, Zhang, and Xiao (2019) have proposed a SE-ResNet module that has been built upon the convolution operation to reduce the training parameter so that the risk of overfitting could be mitigated. The core module has coalesced with the residual block and the squeeze-and-excitation block of the ResNet. Using this method, they have been able to do similar performances with fewer parameters. Additionally, a new scheduler for the learning rate has been built. In another study Celik, Talo, Yildirim, Karabatak, and Acharya (2020), a deep transfer learning-based model has been used to detect invasive ductal carcinoma (IDC) from histopathological images. Pretrained ResNet50 and DenseNet161 have been used for the detection task. Only the last few layers of the models have been trained for IDC detection.

The best accuracy of the study has been 91.57% using the DenseNet-161 model, and the best F1-score has been 94.11% using the ResNet-50 model.

Chan et al. (2020) have improved the evaluation of histological breast images by developing a quantitative measure that is well-linked to the severity of cancer and by using an automated model to make objective calculations of such measures. The authors of this study (Wei, Han, He, & Yin, 2017) have proposed a biCNN model for classifying breast histopathological images. The biCNN model has adopted transfer learning and data augmentation to overcome the overfitting problem for the lack of instances in the dataset. Their experimental setup has been able to achieve a maximum of 97.50% test accuracy. Song, Zou, Chang, and Cai (2017) have used CNN to extract features from pathological images. Then, the features have been represented using the Fisher vector (FV) encoding. An adaptation layer has also been introduced that has transformed the FV descriptor. With the descriptor from the adaptation layer, an SVM classifier has been built. Another study on the histopathological breast image classification has been conducted by Bhuiyan, Shamsujjoha, Ripon, Proma, and Khan (2019). They have used four pretrained convolutional neural networks—ResNet50, InceptionV3, InceptionResNetV2, and Xception as feature extractors. On the extracted features, principal component analysis has been applied for dimensionality reduction. On the reduced features, three typical classifiers have been trained. The combination of ResNet50 as a feature extractor and logistic regression as a classifier has yielded a maximum accuracy of 96.24%. Class structure-based deep convolutional neural network (CSDCNN), a comprehensive multiclassification approach based on deep learning method, has newly been proposed in another study (Han et al., 2017) to focus on more quantitative analysis of breast cancer research, rather than just binary classification.

Imbalance in the dataset degrades the performance of classifier models as the classification becomes biased to a class with more instances. It has been a key concern for the experimenters to ensure the balance of the class distribution. Thus, researchers have come up with various data augmentation techniques, e.g., generative adversarial networks (GAN), deep convolution generative adversarial network (DCGAN), etc. These models can be applied to the minority class from time to time to tackle the problem. Saini and Susan (2020) have introduced a novel architecture based on deep transfer learning using DCGAN as a data augmentation technique to keep the balance of the minor benign class of the BreakHis dataset.

Breast cancer has also been detected based on thermography. Acharya, Ng, Tan, and Sree (2012) have used SVM and the texture features extracted from images for the classification. As the cancerous cells have a higher temperature than normal cells, they can be identified easily on infrared images. As data preprocessing, they have used four texture features for the thermogram representation. These features have reflected the variation of pixel and image contours. After feature extraction, the SVM classifier has been fitted, and the model has obtained 88.10% test accuracy. Breast cancer can also be detected from breast mammography images. The authors in Becker et al. (2017) have used a deep artificial neural network for cancer detection from mammogram images. They have made the developed classifier to compete against three expert radiologists. The results have shown that the deep neural network has been able to perform almost as well as human experts. However, in terms of sensitivity, artificial decision-makers have been better, and human experts have been better in terms of specificity. Another study of breast cancer detection using mammography has been conducted by Tan, Sim, and Ting (2018). They have used a CNN-based classifier and have achieved 82.71% accuracy. The summary of a few state-of-the-art works on breast cancer detection is shown in Table 12.1.

TABLE 12.1 Research map of the state-of-the-art in breast cancer detection.

Work	Method	Dataset	Finding	Limitation/gap
Nahid et al. (2018)	Deep neural network based on CNN and LSTM	BreakHis (Spanhol et al., 2016)	CNN works better than LSTM	Lack of exploring other CNNs, lack of guiding the model by cell nuclei direction and position
Jiang et al. (2019)	SE-ResNet	BreakHis (Spanhol et al., 2016)	The combination of a newly proposed scheduler and SE-ResNet has shown excellent performance	Work with uneven color distribution and cell overlap in the pathological images
Celik et al. (2020)	DenseNet161 and ResNet50	A dataset from Cancer Institute of New Jersey and BreakHis (Spanhol et al., 2016)	ResNet50 model has a better performance than DenseNet161	Use of only two CNNs and lack of using ensemble methods
Chan and Tuszynski (2016)	SVM with fractal dimension	BreakHis (Spanhol et al., 2016)	Maximum F1-score of 96.4%	Not having benign breast tumor images as a control group
Wei et al. (2017)	BiCNN	BreakHis (Spanhol et al., 2016)	Accuracy of 97.5%	Work with uneven color distribution and cell overlapping phenomena in the pathological images causes difficulty in the classification task
Song et al. (2017)	CNN, Fisher vector encoding	BreakHis (Spanhol et al., 2016)	Maximum accuracy of 93.2%	Use of only one CNN and potential of improving performance
Saini and Susan (2020)	DCGAN, CNN	BreakHis (Spanhol et al., 2016)	Balancing dataset and maximum accuracy of 96.6%	DCGAN failed to generate better-quality images
Acharya et al. (2012)	SVM with texture feature extraction using CNN	Thermography breast image dataset	A cell to be cancerous depends on the temperature of the image	The performance was not up to the mark
Becker et al. (2017)	Deep artificial neural network	Breast mammography dataset	The developed model performed almost as well as expert radiologists	The model was not good enough in terms of specificity
Tan et al. (2018)	CNN	Breast mammography dataset	The model has achieved 82.71% accuracy	The parameters tuning have not been appropriate
Bhuiyan et al. (2019)	Four transfer learning models as feature extractors	BreakHis (Spanhol et al., 2016)	ResNet50 has extracted better-quality features	Reducing dimensionality using PCA harmed classifier performance

Methodology

A convolutional neural network (CNN) is an artificial neural network that contains at least one convolutional layer. It is a type of deep neural network that is widely used in image classification. Theoretically, a CNN can have as many convolutional layers as possible. Convolutional layer-based neural networks are often called convnets. In addition to the convolutional layer, some other layers are found in a convnet like the pooling layer, dropout layer, batch normalization layer, fully connected dense layer, etc. To construct a convnet, after the input layer, different combinations of those layers are added repeatedly. And, in the end, an output layer that gives the predicted class is added. In this study, seven convnets have been used for the detection of breast cancer from histopathological images. Those convnets are VGG16, VGG19, ResNetV2, Inception, InceptionResNetV2, Xception, and DenseNet. In the subsequent sections, these architectures are described.

VGG16 and VGG19

VGG16 and VGG19 are two convolutional neural network models that have similar sequential structures. Before these two, AlexNet was one of the very first convnet architectures that utilize a graphical performance unit (GPU) for processing images (Krizhevsky, Sutskever, & Hinton, 2017). The VGG16 and VGG19 architectures have been developed by Simonyan and Zisserman (2015). They have an AlexNet-like structure. However, they are deeper than AlexNet. VGG16 and VGG19 are 16- and 19-layer architectures, respectively, as opposed to eight-layer AlexNet. In the convolutional layers of these two architectures, filters of size 3×3 with strides 1×1 have been used. And, in the pooling layers, a pool size of 2×2 with strides 2×2 has been used. In Figs. 12.1 and 12.2, block diagrams of two architectures have been shown. As depicted in the figures, both models start with two convolutional layers followed by a max-pooling layer. These two convolutional layers followed by the max-pooling layer repeat two times. After that, VGG16 has three convolutional layers with a max-pooling layer while VGG19 has four convolutional layers with a max-pooling layer. For both architectures, their last corresponding combination of convolutional layers and the pooling layer repeats three times. Moreover, both models have three dense layers on top of their last respective max-pooling layer. There is an optional dropout in between two dense layers. Lastly, the number of neurons in the last layer of networks of this study is one. The neuron fires if an input image is malignant and does not fire for a benign image. If the number of convolutional layers and dense layers in VGG16 is counted, it would be 16 (13 convolutional

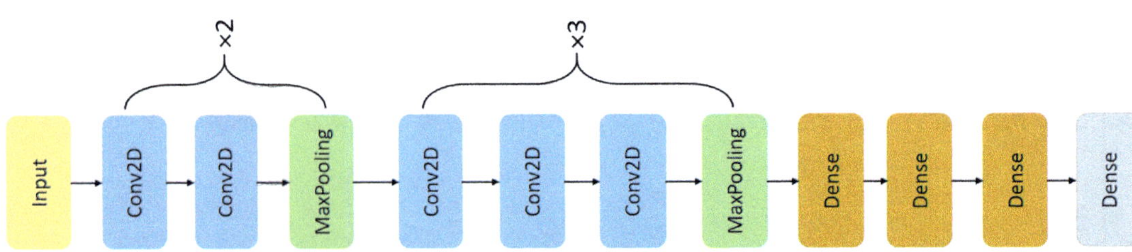

FIG. 12.1 VGG16 architecture. This figure illustrates the block diagram of VGG16 architecture.

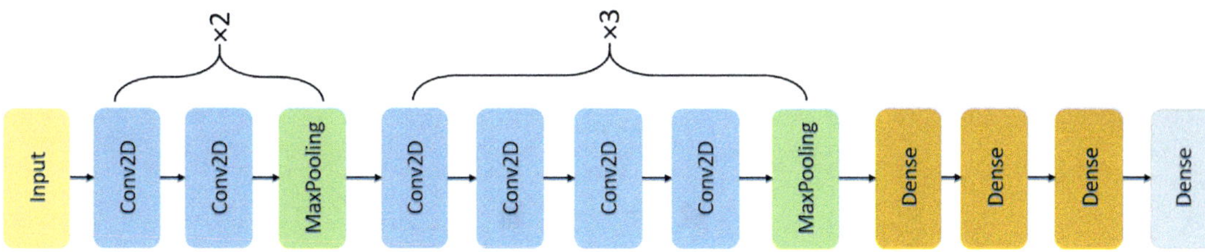

FIG. 12.2 VGG19 architecture. This figure illustrates the block diagram of VGG19 architecture.

layers and three dense layers). Based on the number of layers, these two architectures are known as VGG16 and VGG19, respectively. In this study, both architectures have been utilized to accomplish the classification task of breast cancer detection.

ResNetV2

The third convnet that has been used in this study is ResNetV2. It has been developed by He, Zhang, Ren, and Sun (2016a, 2016b) with slight modification on the original ResNet (He et al., 2016a). The two major building blocks of the original ResNet are the convolutional block and the identity block. Another important feature of ResNet is using residual connections. ResNets can have 18, 34, 50, 101, and 152 layers. All the ResNets start with a 7×7 convolutional layer with 64 filters and strides of 2 followed by a 3×3 max-pooling layer having strides of 2. ResNet also uses batch normalization after every few layers to prevent the model from gradient explosion. The modification of ResNetV2 over the original ResNet lies in the identity block. In the identity block of the original ResNet, there are three convolutional layers—each of which is followed by a batch normalization layer. But in the identity block of ResNetV2, the batch normalization layer comes before the convolution layer.

In Fig. 12.3, ResNet50V2 architecture has been illustrated. From the figure, it can be seen that the architecture starts with a convolutional layer, followed by batch normalization and a max-pooling layer. After that, there is a convolutional block followed by two identity blocks, then another convolutional block followed by three identity blocks. This pattern continues two more times except with five identity blocks and two identity blocks, respectively. One of the building blocks of ResNetV2 is the convolutional block, and its structure is given in Fig. 12.4. The input of this block goes through a 1×1 convolutional layer followed by batch normalization, and the input also passes through three convolutional layers. These three convolutional layers have filter sizes of 1×1 , 3×3, and 1×1,respectively. Each of the convolutional layers is followed by a batch normalization layer. Two branches are added

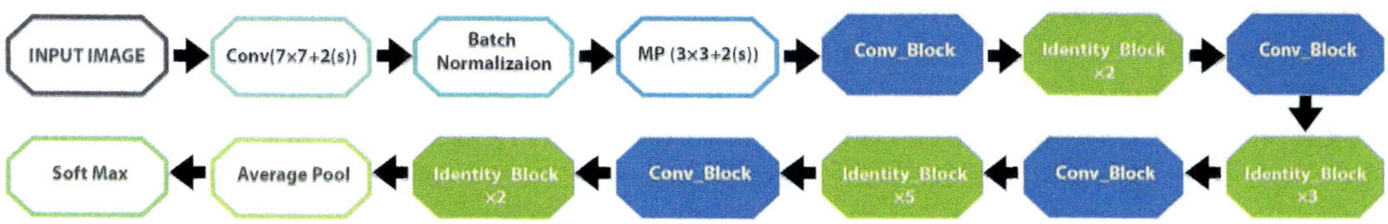

FIG. 12.3 ResNet50V2 architecture. This is a block diagram of ResNet50V2 architecture.

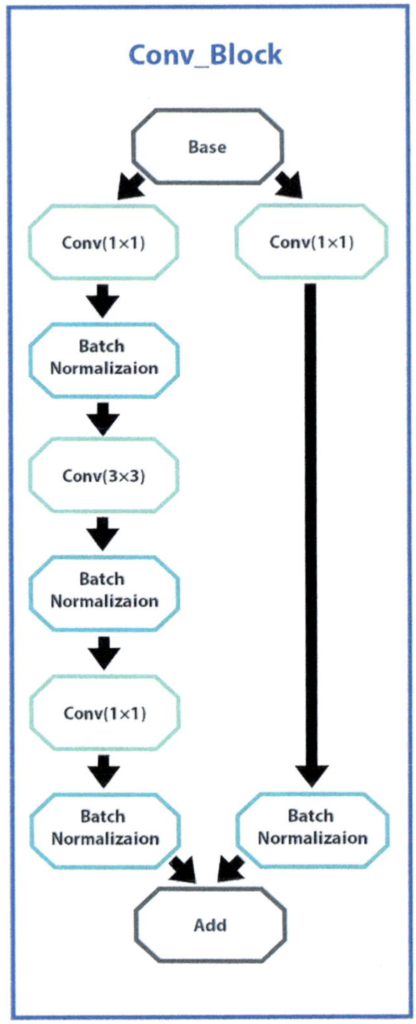

FIG. 12.4 Convolutional block. This is a block diagram of convolutional block—one of the building blocks of ResNetV2.

at the end of a convolutional block to get the final output. The identity block is the other building block of a ResNetV2 whose structure is shown in Fig. 12.5. To get the final output of an identity block, the identity of the input is combined with the output that has been obtained by passing the input through three convolutional layers. In ResNet50V2, after the last identity block, the outputs are reduced with average pooling, and finally, a SoftMax classification layer with one neuron has been added. An input to the ResNet50V2 network can pass through at most 50 convolutional layers. That is why this network is named ResNet50V2.

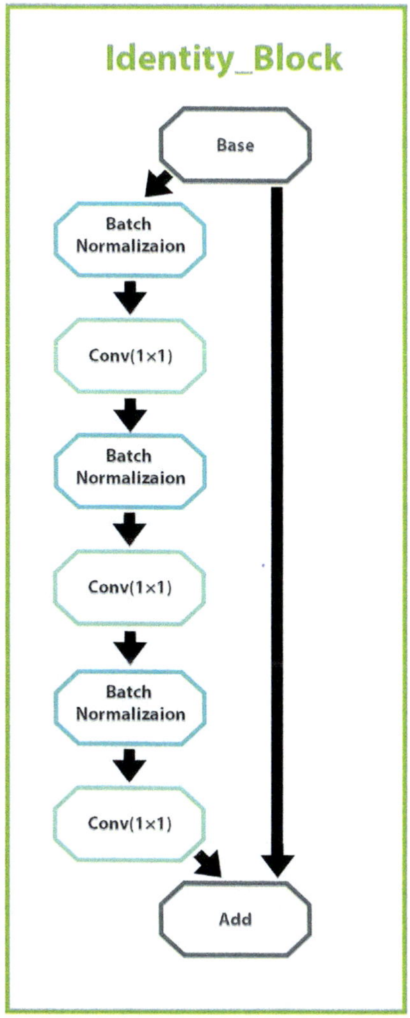

FIG. 12.5 Identity block. This is a block diagram of identity block—one of the building blocks of ResNetV2.

Inception

The fourth convnet that has been used for breast histopathological image classification is Inception. This architecture has been developed by Szegedy et al. (2015). The major building block of this network is the Inception block. The Inception block has been shown in Fig. 12.6. Utilizing this building block, the Inception network has been developed. The Inception network has been presented in Fig. 12.7. After the input layer, the network starts with a 7×7 convolutional layer followed by a max-pooling layer. After that, the network has two back-to-back convolutional layers followed by a max-pooling layer, then come two Inception blocks followed by a max-pooling layer. After that, there are five consecutive Inception blocks

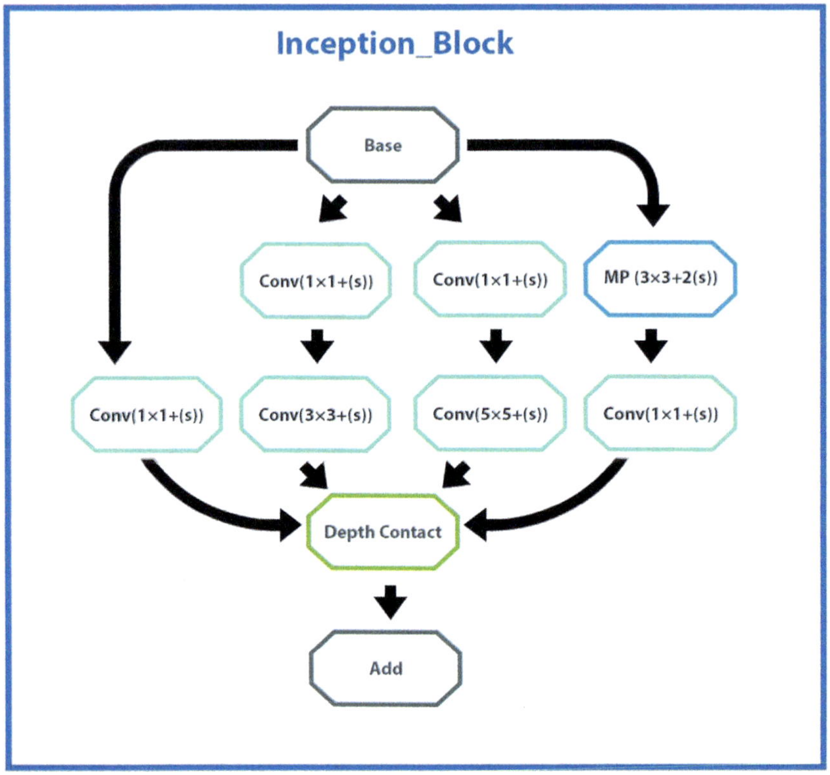

FIG. 12.6 Inception block. This is a block diagram of Inception block—the building block of Inception architecture.

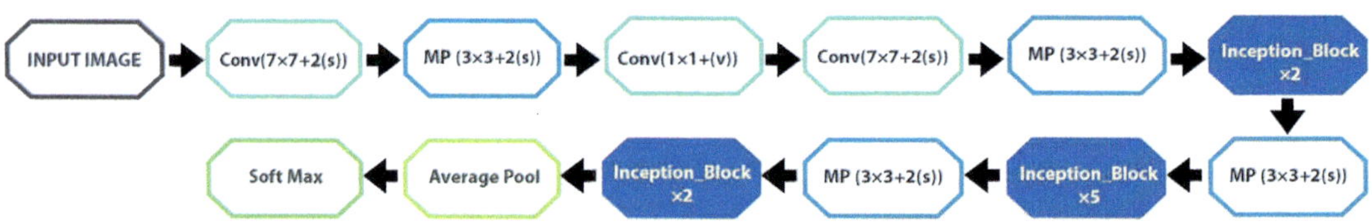

FIG. 12.7 Inception architecture. This is a block diagram of Inception architecture.

followed by another max-pooling layer. The convolutional operation of the architecture finishes with two Inception blocks followed by an average pooling layer. The network ends with a classification layer with SoftMax activation.

InceptionResNetV2

InceptionResNetV2 has been introduced by Szegedy, Ioffe, Vanhoucke, and Alemi (2017). It had a state-of-the-art performance on the 2015 ILSVRC challenge. The fundamental idea of this network is to combine the Inception network with the residual connection. In Figs. 12.8

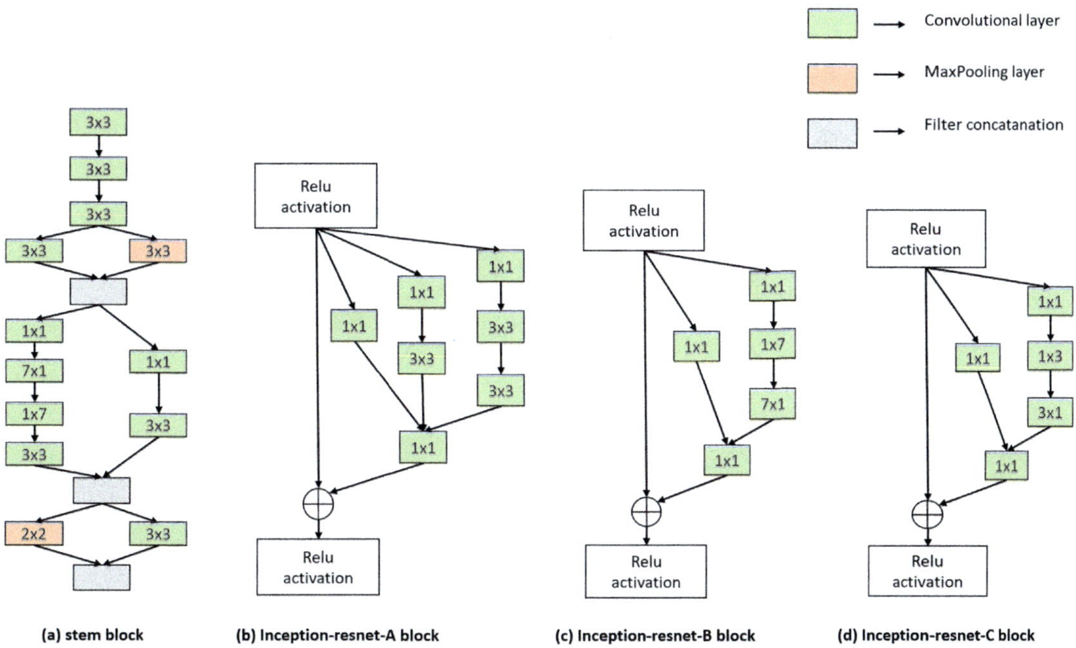

FIG. 12.8 Building blocks of InceptionResNetV2. This figure illustrates the building blocks of InceptionResNetV2 architecture: (A) stem block, (B) InceptionResNet-A block, (C) InceptionResNet-B block, and (D) InceptionResNet-C block.

and 12.9, the building blocks for InceptionResNetV2 have been presented. After the input layer, a stem block, which is described in Fig. 12.8A, has to be added to create this network. After the stem block, five consecutive Inception-ResNet-A blocks are added. The structure of an Inception-ResNet-A block is given in Fig. 12.8B. After the last Inception-ResNet-A block, a Reduction-A block from Fig. 12.9A is stacked. After that, 10 consecutive Inception-ResNet-B blocks have to be added followed by a Reduction-B block. Inception-ResNet-B and Reduction-B blocks are explained in Figs. 12.8C and 12.9B. The feature extraction process of InceptionResNetV2 architecture ends with five consecutive Inception-ResNet-C blocks followed by an average pooling layer. The architecture ends with a 40% dropout layer and a classification layer having SoftMax activation. In this chapter, this architecture has been used for breast cancer detection.

Xception

Like ResNets, Xception architecture has residual connections too. It has been developed by Chollet (2017). This convnet is inspired by the Inception network. Inception modules of the Inception network have been replaced with a depthwise separable convolutional layer. The depthwise convolutional layer has the same output as a typical convolutional layer, but it achieves the same result with fewer computations. In a typical convolutional layer, the kernels have the same number of channels that the input image has. If an input matrix has

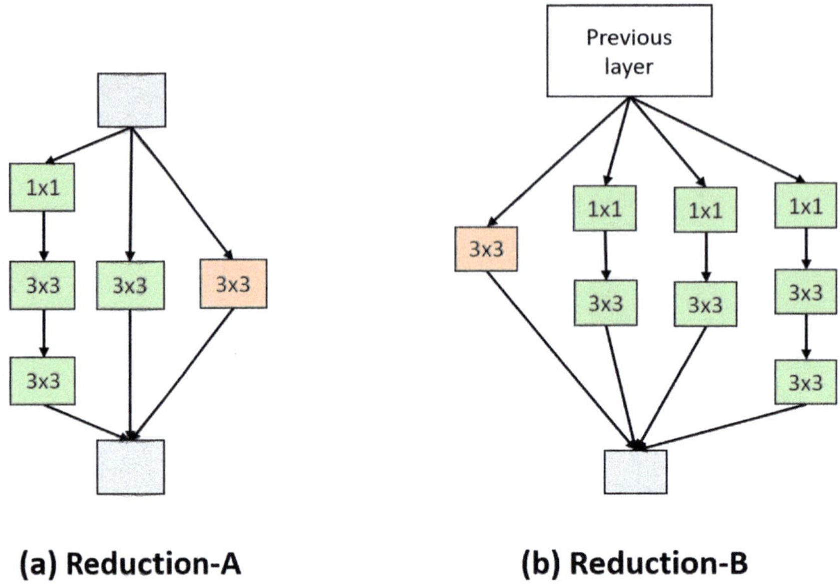

FIG. 12.9 Reduction blocks. This figure illustrates two reduction blocks of InceptionResNetV2 architecture: (A) Reduction-A and (B) Reduction-B.

$n \times n \times d$ dimensions (where n is height and width of the input and d is depth), and the kernels should be of $f \times f \times d$ size where f is the kernel height and width. If we want to have an output of $m \times m \times t$ dimensions (where $m = n - f + 1$), we need t kernels with $f \times f \times d$ dimension.

Here, the total number of multiplications needed to get the output is $f \times f \times d \times m \times m \times t$. The same output can be obtained by applying a depthwise separable convolution followed by a pointwise separable convolution. On the input image, a kernel of size $f \times f$ is applied to individual depths of the inputs separately. If the input image has $n \times n \times d$ dimension just like before, after applying depthwise convolution, an output of $m \times m \times d$ will be found. After that, pointwise convolution with kernel size $1 \times 1 \times t$ will be applied to the output of depthwise convolution, and the final output with $m \times m \times t$ dimension will be obtained. If we perform convolution separately in this way, we need $d \times f \times f \times m \times m$ multiplications for the depthwise convolution and $t \times 1 \times 1 \times d \times m \times m$ multiplications for pointwise convolution. Therefore, the total number of multiplications is significantly less than the multiplications needed in regular convolution. Xception utilizes this method and improves training time.

In Fig. 12.10, Xception architecture has been shown. Xception starts with two regular convolutional layers with 32 and 64 filters, respectively, with a filter size of 3×3. After that, there are five Xception blocks. Each input of an Xception block passes through two separable convolutional layers followed by a max-pooling layer and passes through a pointwise convolution via residual connection parallelly except the fourth Xception block. The fourth Xception block consists of only one separable convolutional layer, and this block repeats eight times. After the last Xception block, there are two separable convolutional layers followed by a global average pooling layer. After that, the fully connected layers are added with the last

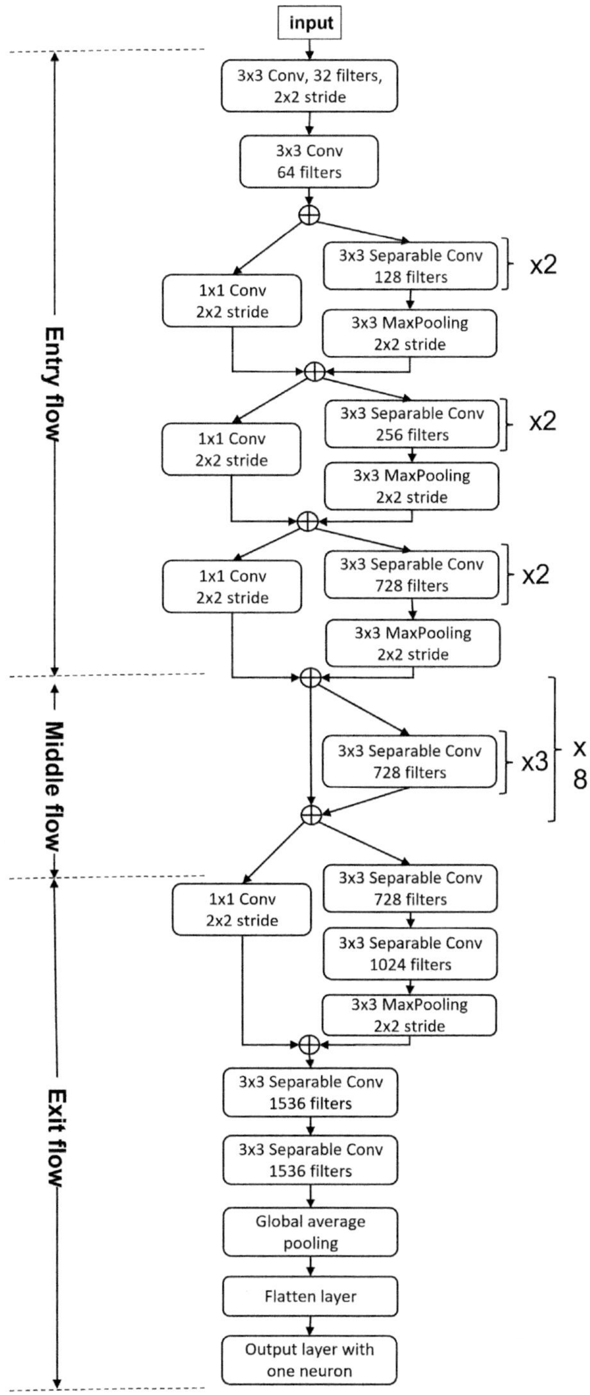

FIG. 12.10 Xception architecture. This is a block diagram of Xception architecture.

one having a single neuron with SoftMax activation. The Xception network has been adopted in this study to detect malignant breast tumors from histopathological images.

DenseNet

The last convnet that has been used is DenseNet (Huang, Liu, Van Der Maaten, & Weinberger, 2017). The intuition behind this model is that every convolutional layer in a certain block is connected to every other layer of that block. Every layer not only processes the input forwarded by the immediately preceding layer but also processes the input combination forwarded by all the previous layers. In Fig. 12.11, a structure of a DenseNet is demonstrated. A DenseNet model consists of a few dense blocks. Before every dense block, there is a convolutional layer followed by an average pooling layer except the first dense block. There is a convolutional layer with a filter size of 7×7 followed by a max-pooling layer with a pool size of 3×3 before the first dense block. The other transitional convolutional layers have a filter size of 1×1. After the last dense block, there is a global average pooling layer and few customizable classification layers. The network ends with a dense layer having neurons equal to the number of classes in a classification task.

Stacked generalization ensemble method

Apart from using convnets as classifiers and feature extractors, the classification task has been carried out using an ensemble technique as well. In ensemble learning, a prediction has been made based on multiple learning algorithms rather than single learning algorithms (Opitz & Maclin, 1999). It is a technique used in machine learning fields where more than one model is trained for the same task as opposed to a typical machine learning technique where a single model is developed for solving a particular task. In ensemble learning, hypotheses from different models are combined to create a more generalized, accurate, and robust model to solve a specific problem. Few ensemble techniques can develop a generalized model from multiple models. Regarding this, the most commonly used ensemble techniques are majority voting, weighted majority voting, Borda count, bagging, boosting, stacked generalization, and so on (Cha & Yunqian, 2012). In this study, the stacked generalization ensemble technique has been used to detect breast cancer.

Stacked generalization follows a very general process. First of all, the training dataset is divided into two sets naming A and B. Throughout the experiments of this work, the dataset has been split into a training set, validation set, and test set. Training, validation, and test sets have 75%, 10%, and 15% images, respectively. For the stacked generalization method, the training set and validation set have acted as set-A and set-B. On the first set, k different first-level classifiers are fitted. There are many methods to obtain these first-level classifiers (Aggarwal, 2015).

In this study, a fivefold cross-validation method has been employed on set A, and five first-level classifiers have been developed. For each first-level classifier, fourfolds have acted as a training set while the remaining fold has acted as a validation set chronologically. If the whole training set is used to build for each first-level classifier, there is a possibility that the final classifier might be overfitted. To avoid this, five different sets of training instances are created

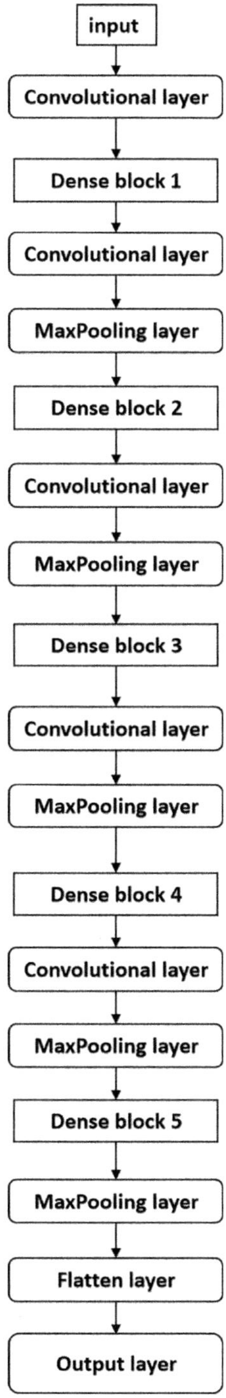

FIG. 12.11 DenseNet architecture. This is a block diagram of DenseNet architecture.

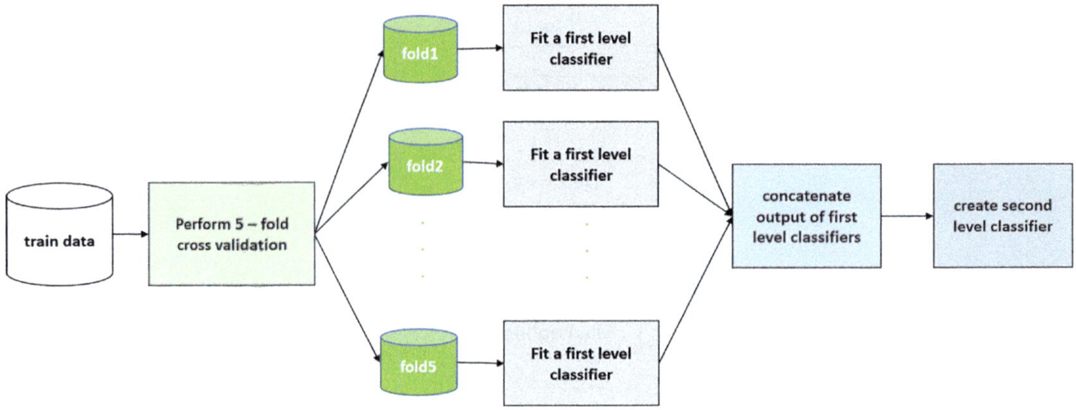

FIG. 12.12 Stacked generalization ensemble method overview. This figure illustrates the staked generalization ensemble method.

with a fivefold cross-validation technique. Then, five convolutional neural network models are trained on each set with different image augmentation techniques. This allows the first-level classifiers to recognize various patterns and shapes of the training images appropriately.

After obtaining first-level classifiers, they are stacked, and their outputs are concatenated. After the concatenation, three dense layers have been added. The last layer of the second-level stacked model is the output layer. After creating the stacked model, it has been trained with set B, where the output of each first-level classifier works as the new dataset for training the second-level classifier. This synthesized dataset has five features as 5 first-level classifiers have been used. As set B is completely unknown to the first-level classifiers, using it to train the second-level classifier adds more robustness to the final model. Using set A for the training of second-level classifier does not add significant information to the final model because the first-level classifiers are already familiar with the instances of set A. That is why the training and validation sets are not combined to train the first-level classifiers. A block diagram of the stacked generalization method is presented in Fig. 12.12.

The description of how the first-level classifiers are built is given as follows:

- First-level classifier on fold1—DenseNet model
- First-level classifier on fold2—InceptionResNetV2 model
- First-level classifier on fold3—ResNetV2 model
- First-level classifier on fold4—VGG16 model
- First-level classifier on fold5—Xception model

Experimental setup

This study aims at detecting breast cancer from histopathological breast images. The working procedure of this study is largely divided into three stages. In the first stage, seven classifiers have been developed using seven popular convolutional neural networks. In the second stage, the stacked generalization ensemble method has been used to improve the performance of the breast cancer detector. In the final stage of the study, to check the portability of the classifiers, the developed models have been tested on the other dataset having different magnification factors. For creating the convnets, Keras on top of TensorFlow, a python library, has been used. For a few of the experimentations, the sci-kit-learn library of python has also been utilized. The models have been trained on a computer having Ryzen 5, 1600 CPU, 8 GB ram, and Nvidia GeForce 1050 TI GPU with Linux Mint 19.3 operating system.

Dataset

Different types of datasets have been used for breast cancer detection like thermography datasets (Arora et al., 2008; Wishart et al., 2010), MRI dataset (Meyer et al., 2015), an ultrasound image dataset (Al-Dhabyani, Gomaa, Khaled, & Fahmy, 2020), and mammogram dataset (Huang et al., 2017). But in this study, a widely used breast histopathology dataset—BreakHis (Spanhol et al., 2016)—has been used. This dataset consists of pathological images with four magnification factors—40 ×, 100 ×, 200 ×, and 400 ×. The dataset contains images labeled with two classes—benign and malignant. This dataset is imbalanced, i.e., the malignant class has more images than the benign class. Images were collected through a clinical study from January 2014 to December 2014 from P&D Laboratory, Brazil. This dataset has a total of 7909 images. Among them, 2480 images are labeled as benign and 5429 images are labeled as malignant. These images were collected from 82 individuals among which 58 were diagnosed with breast cancer and 24 were not diagnosed with breast cancer. In Table 12.2, the detailed description of the BreakHis dataset has been shown. In Figs. 12.13 and 12.14, an image from each class of each magnification factor-based dataset is demonstrated.

Training, testing, and validating

The whole dataset has been divided into three sets—train, validation, and test. The train set has 75%, the validation set has 10%, and the test set has 15% of the data. 1496 of 1995 images from the 40 × dataset are used for training. And for validation and test, the number of images is 191 and 308, respectively, for the 40 × dataset. For the 100 ×, 200 ×, and 400 × datasets, the

TABLE 12.2 BreakHis dataset details.

Magnification factor	Number of malignant images	Number of benign images	Total images
40 ×	1370	625	1995
100 ×	1437	644	2081
200 ×	1390	623	2013
400 ×	1232	588	1820
Total	5429	2480	7909

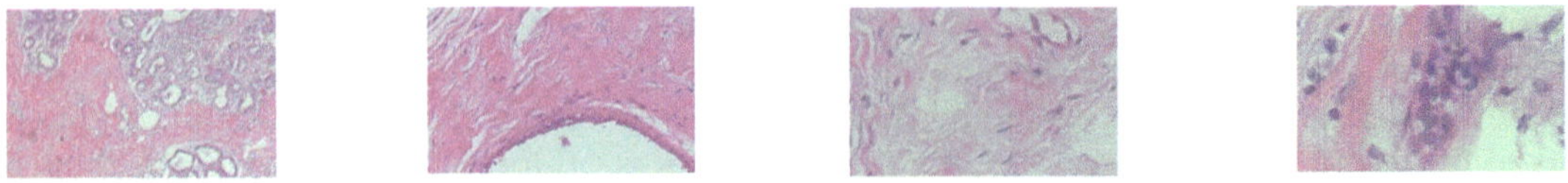

(a) 40X benign image **(b) 100X benign image** **(c) 200X benign image** **(d) 400X benign image**

FIG. 12.13 Few benign images from BreakHis dataset. This figure presents four images from the benign class from four magnification factor-based datasets—(A) a benign image from the 40 × dataset, (B) a benign image from the 100 × dataset, (C) a benign image from the 200 × dataset, and (D) a benign image from 400 × dataset.

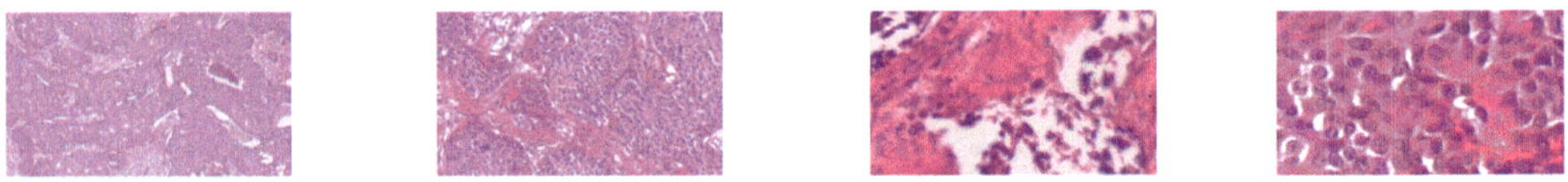

(a) 40X malignant image **(b) 100X malignant image** **(c) 200X malignant image** **(d) 400X malignant image**

FIG. 12.14 Few malignant images from BreakHis dataset. This figure presents four images from the malignant class from four magnification factor-based datasets—(A) a malignant image from the 40 × dataset, (B) a malignant image from the 100 × dataset, (C) a malignant image from the 200 × dataset, and (D) a malignant image from the 400 × dataset.

images have been distributed in three sets in a similar fashion. For training the first-level classifiers of the stacked generalization model, the training set has been divided into fivefolds using fivefold cross-validation and a classifier on each fold has been fitted. After combining the first-level classifiers, the second-level classifier has been trained with the validation set of the original dataset. The reason behind not using the training set again is that the first-level classifiers have already known the training set. That is why the original validation set which is completely unknown for the first-level classifiers has been used. After training the second-level classifier, it has been tested against the original test images—the same test images that have been used for testing throughout the study.

Hyperparameters settings

The classification problem of this study is binary—either malignant or benign. The single neuron at the output layer of the neural networks activates if the input is a positive class instance, i.e., malignant image, or deactivates if the input instance is negative class, i.e., benign image. In the output layers of the neural networks, the SoftMax activation with a single neuron has been used which is technically a sigmoid activation. Throughout all the experiments, a stochastic gradient descent optimizer with a learning rate of 0.001 and a momentum of 0.9 has been used. As the loss function, binary cross-entropy has been used for the binary nature of the classification task. The input size of the images has been kept fixed at 224 × 224 for all the experiments. And due to the limitation of computational resources, a minimum batch size of 8 and the maximum batch size of 16 have been used. For the deeper models like DenseNet and InceptionResNetV2, the experiments have been carried out with a batch size of 8, and for the less deep networks like VGG16 and VGG19, a batch size of 16 has been utilized.

The datasets that have been used contain three-channeled RGB images. To keep the RGB characteristics of the images intact, they have been loaded with the hyperparameter of the color mode set to "RGB." Additionally, few image augmentation techniques have been used for the better generalizability of the deep learning models. 10% height and width shift range have been used. The width shift range of 0.1 adds extra training images having 10% left or right shift than the original images. Similarly, the height shift range of 0.1 adds a few more training images that are shifted about 10% of the original images. The other augmentation technique, i.e., horizontal flip, allows the training images to be augmented upside down.

Evaluation metrics

Six evaluation metrics have been used throughout the whole working process of this study: precision, recall, F1-score, accuracy, sensitivity, and specificity. If TP , TN , FP , FN , P , and N are true positive, true negative, false positive, false negative, the total number of positive instances, and the total number of negative instances, respectively, then, precision is the portion of the correct predictions made by the classifier with respect to total positively predicted classes.

$$Precision = \frac{TP}{TP + FP} \tag{12.1}$$

The recall is the portion of the correct predictions made by the classifier regarding the total existing accurate classes.

$$Re\,call = \frac{TP}{TP + FN} \tag{12.2}$$

F1-score is the harmonic mean of precision and recall, i.e., both precision and recall are given equal importance while evaluating the performance of a classifier.

$$F1\_score = \frac{2 \times precision \times recall}{precision + recall} \tag{12.3}$$

Accuracy is known as the overall recognition rate. And accuracy is defined as follows:

$$Accuracy = \frac{TP + TN}{TP + FP + TN + FN} \tag{12.4}$$

Sensitivity is the true positive rate. It is defined as follows:

$$Sensitivity = \frac{TP}{P} \tag{12.5}$$

Specificity is the true negative rate. It is defined as follows:

$$Specificity = \frac{TN}{N} \tag{12.6}$$

Results

The experimental results of this study have proved the acceptability of various convolutional neural network-based architectures in breast cancer prognosis. The results have also proved that the ensemble methods combining the convnets can produce more robust and accurate classifiers in the field of cancer diagnosis and prognosis. In this section of the chapter, the results of this study from various points of view have been described and explained.

Validation accuracy and loss

The validation accuracy after completion of each epoch has been measured against the validation set for the developed models. Doing this has helped to observe whether a model is being trained properly or not. If performance loss is observed in the validation accuracy of a model in consecutive 10 epochs, further training may become useless. Although this study has not faced such constant performance loss among the models, all the convnet models have been trained for 100 epochs.

In Fig. 12.15, the validation accuracy of the models on each epoch on the dataset having a 40× magnification factor has been shown. DenseNet, Inception, InceptionResNetV2, and Xception models have gained validation accuracy as high as 1.0. On the other hand, the other three convnets—VGG16, VGG19, and ResNetV2—have not been able to reach the highest possible validation accuracy. Among the four highest validation accuracy-achieving models on the 40× dataset, the Inception model has reached the peak faster than the other models on the 35th epoch while the Xception model has come close by reaching the peak validation accuracy on the 46th epoch. The InceptionResNetV2 and DenseNet have gained their first perfect validation accuracy on 53rd and 54th epochs, respectively. The ResNetV2 model has been

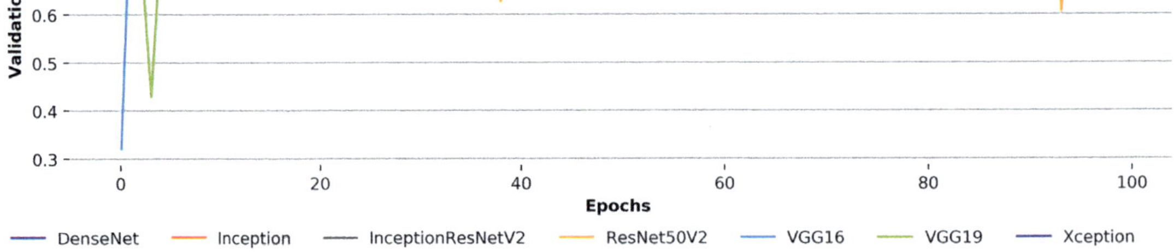

FIG. 12.15 Validation accuracy vs epochs comparison for the 40× dataset. This figure describes the validation accuracy of the seven convnets over the number of epochs for 40× dataset.

more unstable than the others during training in terms of validation accuracy. The yellow line in Fig. 12.15 depicts this convnet. It has reached the highest validation accuracy of 98.33% at the 49th epoch. The other two lower-performing convnets—VGG16 and VGG19—have gained their highest validation accuracies of 98.74% and 97.90% at 74th and 57th epoch, respectively. From Fig. 12.15, it can also be predicted that the models with higher validation performance would perform better than the models with lower performance in the testing phase. Moreover, for 40 × dataset, the comparison of validation losses for the developed convolutional neural network models is shown in Fig. 12.16.

For the 100 × magnification breast histopathological images, all seven models' epochwise performance comparison is given in Fig. 12.17. From the figure, it can be seen that only DenseNet and Xception models have demonstrated perfect validation accuracy during training. The first one has achieved 1.00 accuracy on the validation set on the 42nd epoch while the latter achieved it on the 95th epoch. Among the other five models, the InceptionResNetV2 model has performed better with a validation accuracy of 99.54%, and the VGG16 model has performed worse with 96.79% accuracy on the validation set. The other three models have validation performances in between 99.54% and 96.79%. The highest epochwise performances are 98.63%, 98.17%, and 98.17%, respectively, for Inception, VGG19, and ResNetV2. Another noticeable aspect from Fig. 12.17 is that the ResNetV2 model for the 100 × dataset is more unstable in training just like its instability for the 40 × dataset. Moreover, for the 100 × dataset, the comparison of validation losses for the developed convolutional neural network models is shown in Fig. 12.18.

The validation accuracy comparison of the seven models for the 200 × dataset has been shown in Fig. 12.19. Only the DenseNet model has the perfect validation accuracy for this dataset on the last epoch. Despite having only one model with a validation accuracy of 1.0, the average highest accuracies of the models are higher than the previous two datasets. Inception, ResNetV2, and InceptionResNetV2 all three models have the highest validation accuracy of 99.53% while the Xception model has 99.54% validation accuracy. The validation

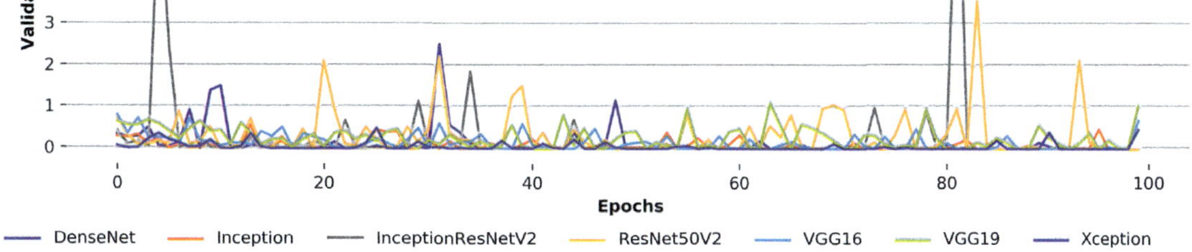

FIG. 12.16 Validation loss vs epochs comparison for the 40 × dataset. This figure describes the validation loss of the seven convnets over the number of epochs for 40 × dataset.

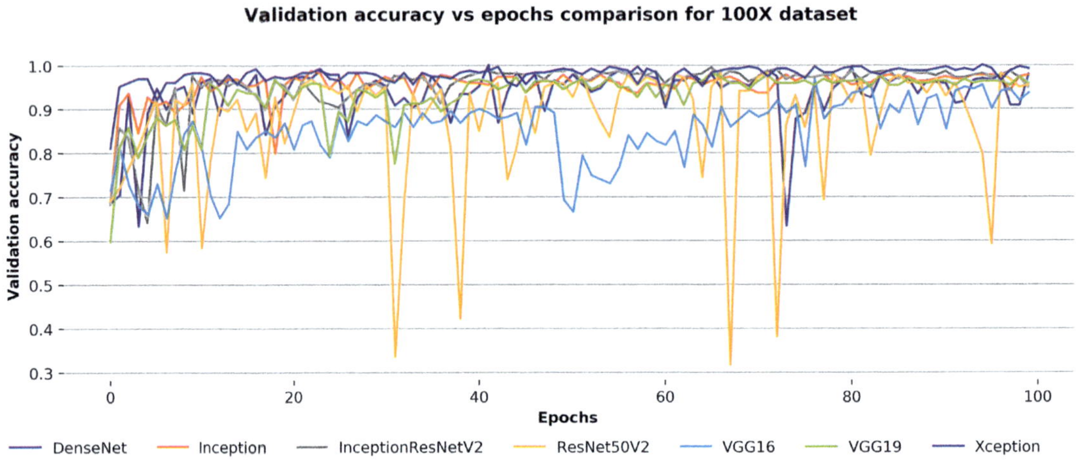

FIG. 12.17 Validation accuracy vs epochs for the 100 × dataset. This figure describes the validation accuracy of the seven convnets over the number of epochs for 100 × dataset.

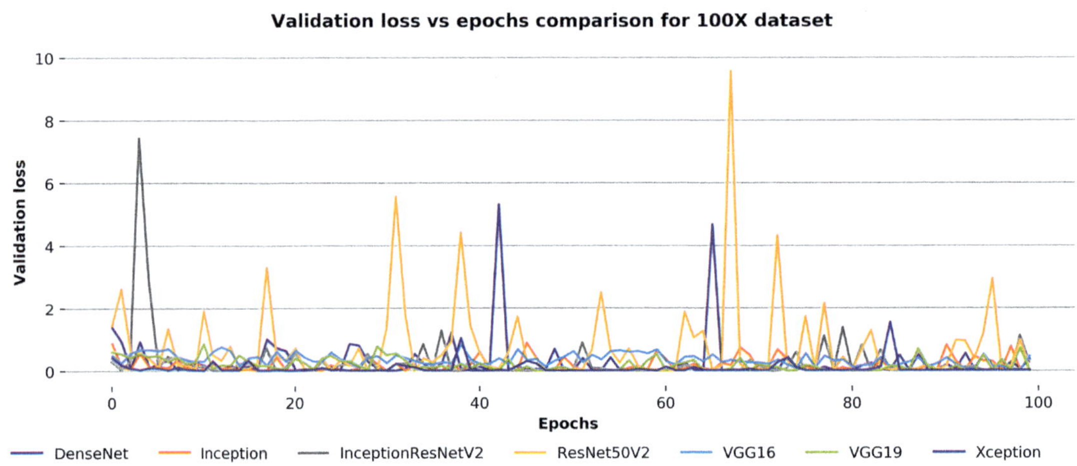

FIG. 12.18 Validation loss vs epochs comparison for the 100 × dataset. This figure describes the validation loss of the seven convnets over the number of epochs for 100 × dataset.

performance of the ResNetV2 model has increased most significantly from the previous two datasets. Two VGG models have demonstrated similar lower validation performance in comparison with the other models' performances. The instability pattern of the models' performance is different for the 200 × dataset than the 40 × and 100 × datasets. The VGG16 model has been most unstable in this case as opposed to ResNetV2. Moreover, for the 200 × dataset, the comparison of validation losses for the developed convolutional neural network models is shown in Fig. 12.20.

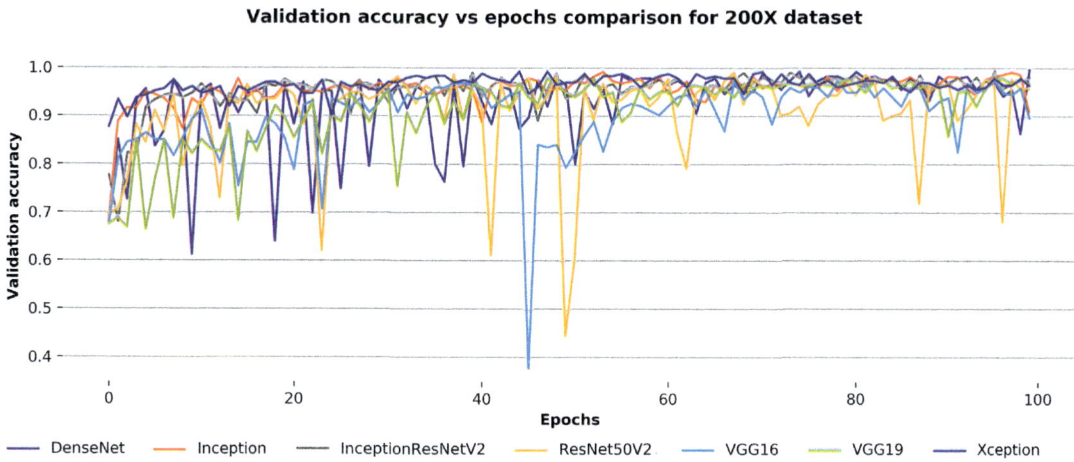

FIG. 12.19 Validation accuracy vs epochs comparison for the 200 × dataset. This figure describes the validation accuracy of the seven convnets over the number of epochs for 200 × dataset.

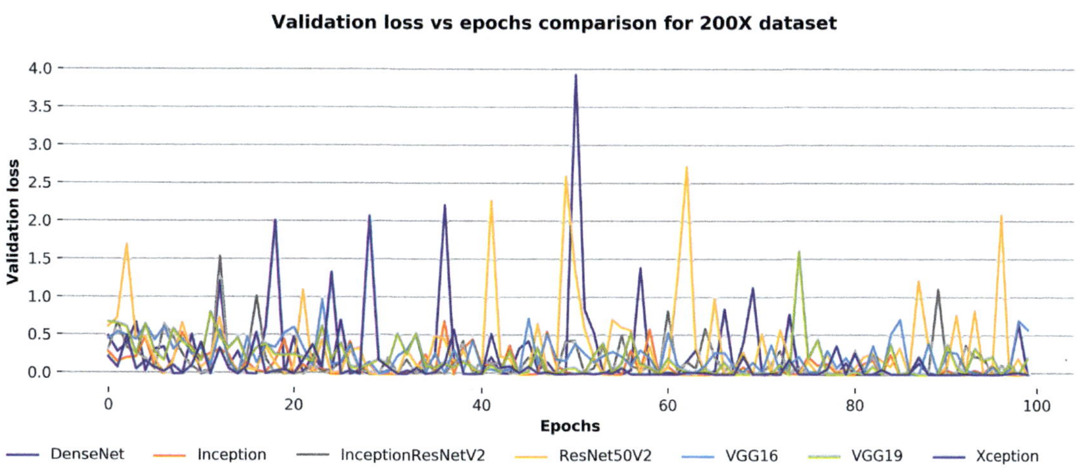

FIG. 12.20 Validation loss vs epochs comparison for the 200 × dataset. This figure describes the validation loss of the seven convnets over the number of epochs for 200 × dataset.

The last image dataset for breast histopathology has 400 × magnification factors. For the previous three magnification factors datasets, the DenseNet model has been able to show 1.0 validation accuracy while training. But for this magnification factor, it has failed to show a similar result. In Fig. 12.21, the validation accuracies of seven convnets against the number of epochs are presented. The figure depicts that the DenseNet and the Xception have better results than other convnets with 99.51% validation accuracy while ResNetV2 has worse performance than others with 96.91% accuracy on the validation set. The poor-performing VGG19 model so far has surprisingly demonstrated a comparatively better result of

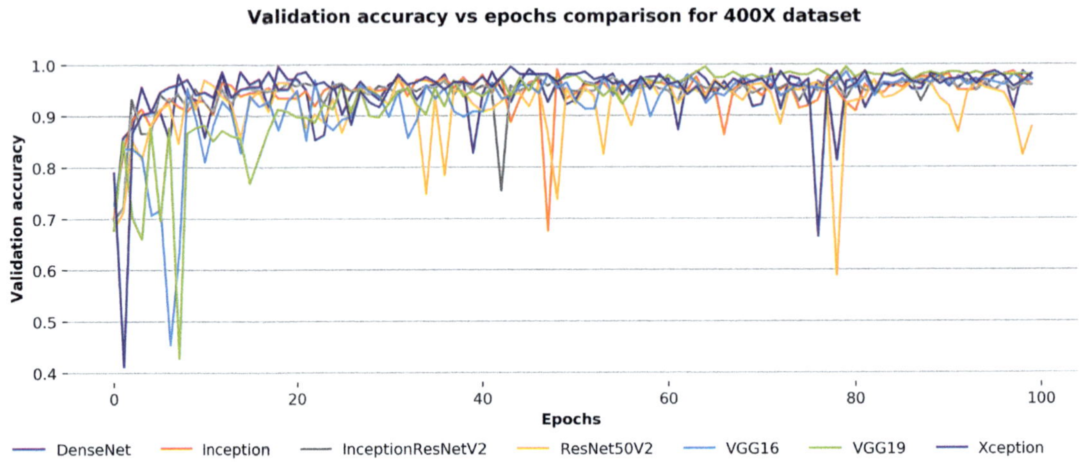

FIG. 12.21 Validation accuracy vs epochs for the 400 × dataset. This figure describes the validation accuracy of the seven convnets over the number of epochs for 400 × dataset.

99.49% validation accuracy. VGG19 and Inception models have similar validation accuracy for this dataset. Lastly, the InceptionResNetV2 model has generally been performing efficiently for the previous three magnification factors, but in this case, it has shown the worst performance among the seven convnets. Moreover, for the 400 × dataset, the comparison of validation losses for the developed convolutional neural network models is shown in Fig. 12.22.

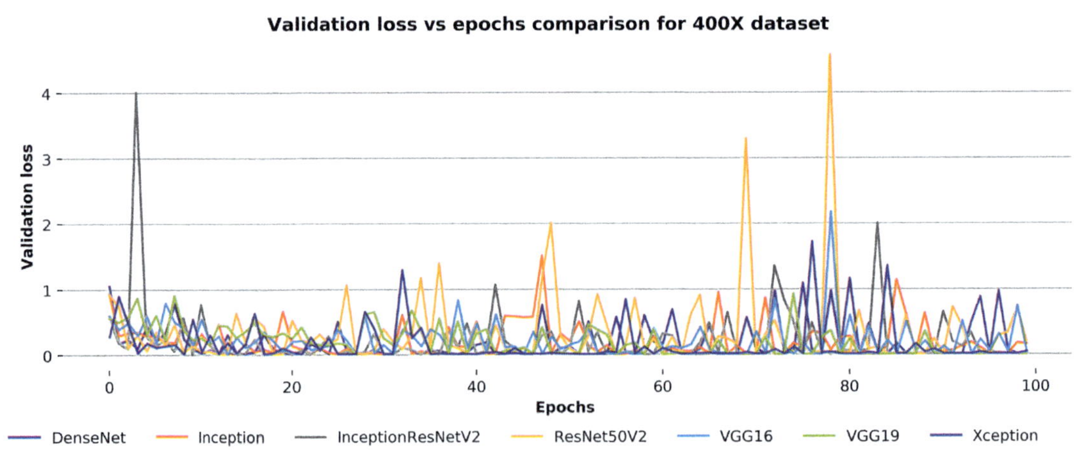

FIG. 12.22 Validation loss vs epochs comparison for the 400 × dataset. This figure describes the validation loss of the seven convnets over the number of epochs for 400 × dataset.

Test performance

The overall test performances of the convnets have been good for the four datasets with different magnification factors. After training seven convnets for each magnification factor-based dataset, they have been tested with the holdout test images. Then, the predictions of each classifier have been evaluated with respect to some evaluation metrics after constructing a confusion matrix considering "Malignant" as a positive class and "Benign" as a negative class. In Table 12.3, the performances of convnets have been presented. From the table, it is apparent that different convnets have not performed exactly in the same way in terms of different evaluation metrics. But the better-performing convnets have been consistent throughout the four datasets. For example, the InceptionResNetV2 model has been the best performer in each dataset in terms of test accuracy. On the other hand, there has not been a single model that has performed worst in each dataset. Instead, four different convnets have shown worse performances in terms of the evaluation metrics.

For the 40 × magnification factor dataset, Inception and VGG19 have jointly been worse-performing, for 100 × and 200 ×, it has been VGG16, and for the 400 × dataset, ResNetV2 has turned out to be the worse-performing convnet. Table 12.3 also shows datasetwise results of the convnets. For the 40 × dataset, apart from the highest- and lowest-performing models, DenseNet, ResNetV2, VGG16, and Xception models have gained test accuracy better than 98%. For the 100 × dataset, the DenseNet model has been able to perform as well as the InceptionResNetV2 model. Inception and Xception models have respected performances with 98.32% and 98.04% test accuracy, respectively. The remaining models have not been able to perform exceptionally. They even could not reach 97% test accuracy. For the 200 × dataset, the test accuracy pattern of the models is different from the previous one. Apart from VGG16, the models have demonstrated similar test accuracies ranging from 97.40% to 98.27%. The models' performances for the 400 × dataset have been poorer than those for the datasets with other magnification factors. Without the better- and worse-performing models for this dataset, the test accuracies of the average models have varied from 95% to 96.88%.

Along with the overall test accuracies, the other evaluation metrics are also important since the classification task of this study is related to cancer prognosis. Recognizing malignant breast tumors as malignant is more important than recognizing benign breast tumors as benign. Because, if malignant is recognized as benign, the patient may not go through again with the cancer test until it is too late. On the other hand, if a tumor is falsely identified as malignant, she may go through again with the tests for a second opinion. Therefore, the models need to be as accurate as possible to classify true positive instances. Precision, recall, sensitivity, and specificity determine whether the performance of a model toward a specific class (in this case, malignant and benign) is good enough or not. From Table 12.3, it can be seen that these evaluation metrics have generally been good for the 40 × dataset for all the convnets. For the 100 × dataset, the models have been better in terms of these metrics except VGG16. It has a good enough precision of 97.87% but a poor recall of 92.74%. A similar trend has been repeated for the 200 × dataset. VGG16 model has achieved 93.80% recall while the other models have achieved at least 97.52% recall. Lastly, for the 400 × dataset, the ResNetV2 model has been comparatively poorer than the others. It has 93.64% recall while others have a minimum of 95.91% recall. All the models for all the datasets have demonstrated decent enough performance in terms of precision. Recognizing benign breast tumors as benign

TABLE 12.3 All convnets' performances for four magnification factor-based datasets.

Magnification factor	Convnets	Precision (%)	Recall (%)	F1 score (%)	Sensitivity (%)	Specificity (%)	Accuracy (%)
40 ×	DenseNet	98.68	98.68	98.68	98.68	97.17	98.20
	Inception	98.24	97.81	98.02	97.81	96.23	97.31
	InceptionResNetV2	100.00	99.56	99.78	99.56	100.00	99.70
	ResNetV2	98.28	100.00	99.13	100.00	96.23	98.80
	VGG16	98.26	99.12	98.69	99.12	96.22	98.20
	VGG19	97.82	98.25	98.03	98.25	95.28	97.31
	Xception	98.69	99.12	98.91	99.12	97.17	98.50
100 ×	DenseNet	99.19	98.79	98.99	98.79	98.18	98.60
	Inception	98.79	98.79	98.79	98.79	97.27	98.32
	InceptionResNetV2	98.41	99.60	99.00	99.60	96.36	98.60
	ResNetV2	96.79	97.18	96.98	97.17	92.72	95.81
	VGG16	97.87	92.74	95.24	92.74	95.45	93.58
	VGG19	99.17	96.37	97.75	96.37	98.18	96.93
	Xception	98.01	99.19	98.60	99.19	95.45	98.04
200 ×	DenseNet	98.75	97.93	98.34	97.93	97.11	97.69
	Inception	99.16	97.52	98.33	97.52	98.08	97.69
	InceptionResNetV2	98.76	98.76	98.76	98.76	97.12	98.27
	ResNetV2	99.16	97.93	98.54	97.93	98.08	97.98
	VGG16	97.01	93.80	95.38	93.80	93.27	93.64
	VGG19	96.41	100.00	98.17	100.00	91.35	97.40
	Xception	98.35	98.35	98.35	98.35	96.15	97.69
400 ×	DenseNet	96.83	97.72	97.27	97.72	93.00	96.24
	Inception	96.82	96.82	96.82	96.82	93.00	95.63
	InceptionResNetV2	99.07	97.27	98.17	97.27	98.00	97.50
	ResNetV2	95.81	93.64	94.71	93.64	91.00	92.81
	VGG16	97.24	95.91	96.57	95.91	94.00	95.31
	VGG19	96.79	95.91	96.35	95.91	93.00	95.00
	Xception	99.07	96.36	97.70	96.36	98.00	96.88

may not be as important as recognizing malignant breast tumors as malignant but accurate classification of benign tumors helps to provide better and hassle-free healthcare for the individuals. This can be ensured by achieving better specificity of the machine learning models. In this study, the specificities of the models may have not been as good as precision and recall (for reference, see Table 12.3), but they have not been too bad either. For the 40 × dataset, they have been decent. Then, for the 100 × dataset, specificities have been good enough except for the ResNetV2 model. For 200 × and 400 × datasets, the VGG19 and ResNetV2 models have shown poor specificities. The rest of the models have performed decently in terms of specificity.

Stacked generalization result

The stacked generalization ensemble method has been employed to obtain better performances than the individual convnets. Fivefold cross-validation has been applied to the training set. To train models on these fivefolds of the four datasets based on magnification factor, top-performing five individual convnets for the 40 × dataset have been used. These five convnets are DenseNet, InceptionResNetV2, ResNetV2, VGG16, and Xception. These individual models are known as first-level classifiers. After developing the classifier on each of the fold, they are combined to develop the second-level classifier. The second-level classifier has been trained with the validation set of the original dataset. Both the first-level and second-level classifiers are tested against the test images. Their datasetwise performances have been presented in Tables 12.4–12.7.

From Table 12.4, it is clear that the stacked model has performed better than the individual convnets by achieving higher test accuracy for the 40 × dataset. The second-level classifier has obtained 99.41% accuracy and F1-score. Though improvement in performance is not that significant for the second level for the 40 × dataset, the second-level classifiers for the other three datasets have performed comparatively better than first-level classifiers. For the 100 × dataset, the stacked model has achieved 98.88% test accuracy. And for 200 × and 400 × datasets, the test accuracies of the second-level classifiers have been 99.42% and

TABLE 12.4 First-level and second-level classifiers performance for the 40 × dataset.

Fold	Model	Precision (%)	Recall (%)	F1-score (%)	Accuracy (%)
1	DenseNet	98.82	98.80	98.80	98.80
2	InceptionResNetV2	98.80	98.80	98.80	98.80
3	ResNetV2	96.24	96.11	96.17	96.11
4	VGG16	97.90	97.90	97.90	97.90
5	Xception	99.40	99.40	99.40	99.40
Stacked	Combination of 5	99.41	99.41	99.41	99.41

TABLE 12.5 First-level and second-level classifiers performance for the 100 × dataset.

Fold	Model	Precision (%)	Recall (%)	F1-score (%)	Accuracy (%)
1	DenseNet	98.34	98.32	98.33	98.32
2	InceptionResNetV2	98.05	98.04	98.04	98.04
3	ResNetV2	94.67	94.69	94.67	94.69
4	VGG16	96.12	96.09	96.09	96.09
5	Xception	98.32	98.32	98.32	98.32
Stacked	Combination of 5	98.90	98.88	98.89	98.88

TABLE 12.6 First-level and second-level classifiers performance for the 200 × dataset.

Fold	Model	Precision (%)	Recall (%)	F1-score (%)	Accuracy (%)
1	DenseNet	98.31	98.27	98.25	98.27
2	InceptionResNetV2	99.13	99.13	99.13	99.13
3	ResNetV2	97.44	97.40	97.40	97.40
4	VGG16	99.14	99.13	99.13	99.13
5	Xception	98.01	97.98	97.98	97.98
Stacked	Combination of 5	99.43	99.42	99.42	99.42

TABLE 12.7 First-level and second-level classifiers performance for the 400 × dataset.

Fold	Model	Precision (%)	Recall (%)	F1-score (%)	Accuracy (%)
1	DenseNet	95.43	95.33	95.38	95.33
2	InceptionResNetV2	95.96	95.94	95.95	95.94
3	ResNetV2	91.94	91.8	97.91	91.88
4	VGG16	95.61	95.63	95.62	95.63
5	Xception	97.18	97.19	97.18	97.19
Stacked	Combination of 5	97.49	97.50	97.49	97.50

97.5%, respectively. The performance of the stacked model for the 40 × dataset has not beaten the best-performing individual convnet. From Table 12.3, the best performer convnet has been InceptionResNetV2 with 99.70% as opposed to the accuracy of 99.41% for the second-level classifier. The performance of the best-performing convnet for the 400 × dataset and second-level classifier has been the same with 97.5% accuracy. On the other hand, the performances of the second-level classifiers have improved from the best-performing convnets for 100× and 200× datasets. For the 100× dataset, the test accuracy has improved from 98.60% to 98.88%. The accuracy improvement has been most significant for the 200× dataset. The stacked model has shown 99.42% accuracy as opposed to 98.27% achieved by InceptionResNetV2. Tables 12.4–12.7 also show the first- and second-level classifiers' precision, recall, and F1-score. The pattern of test accuracy performance improvement has been similar for these evaluation metrics too.

Cross-dataset performance

Another objective of this study is to analyze how does a model trained on a dataset with a specific magnification factor perform on the datasets with different magnification factor. To do that, the best-performing convnet for each of the datasets has been tested with the test set from the remaining datasets. InceptionResNetV2 has been the top performer across all four datasets. Its cross-dataset accuracies have been presented in Fig. 12.23. A model trained on a specific magnification factor-based dataset has not performed with the same accuracy on the different magnification factor-based datasets. Test accuracy has been better for its own test set than the test sets from other datasets. For example, InceptionResNetV2 trained on the 40× dataset has a test accuracy of 99.70% while it has only 92.46% test accuracy for the 100× dataset. This models' test accuracies for the rest of the two datasets are worse than 92.46%. InceptionResNetV2 models trained on the other three datasets have similar cross-dataset results.

A similar pattern in the cross-dataset accuracy has also been observed for the stacked generalization models. In Fig. 12.24, the cross-dataset accuracies of the stacked generalization models have been displayed. Just like individual InceptionResNetV2, the stacked model trained on the 40× dataset has achieved the highest test accuracy of 99.41% on the test set from the 40× dataset while it has only been able to achieve a maximum of 93.02% accuracy among the test sets from other three datasets. Similar downward test performance on the other datasets has been observed for the stacked model trained on the other three datasets too.

Tested with

Trained with	Dataset	40X	100X	200X	400X
	40X	**99.70%**	92.46%	84.97%	75.94%
	100X	88.62%	**98.60%**	94.22%	86.56%
	200X	80.54%	88.55%	**98.27%**	93.12%
	400X	84.73%	86.87%	91.62%	**97.5%**

FIG. 12.23 Cross-dataset accuracy for the best-performing convnet (InceptionResNetV2). This figure presents the cross-dataset accuracy for the best-performing convnet (InceptionResNetV2).

Tested with

Trained with: Dataset	40X	100X	200X	400X
40X	**99.41%**	93.02%	82.95%	75.63%
100X	90.12%	**98.88%**	95.07%	89.38%
200X	85.03%	91.06%	**99.42%**	95.63%
400X	77.25%	82.40%	94.22%	**97.5%**

FIG. 12.24 Cross-dataset accuracy for the stacked generalization model. This figure presents the cross-dataset accuracy for the stacked generalization model.

Training and testing time

In this study, the training and testing time of the deep learning models have also been measured. Time is an important metric to measure the performance of a deep learning model. Due to the lack of a graphical processing unit, training a deep learning model can be a tedious job. Thus, it is necessary to know about the training time. Prediction time is also important when the deep learning models are deployed in a web platform. Table 12.8 outlines the average time to process an image during training by the convnets across the four datasets. The average processing time of an image during testing is also presented in the table for the individual convnets. These values are measured based on our experimental setup and are subject to be a change in a different environment.

DenseNet, InceptionResNetV2, and Xception models have been on the slower side in terms of training time while the Inception model has been trained faster than other models. The other models have an average training time per image ranging from 25.78 to 37.50ms. The models have consistent training time across the four datasets except for VGG19 for the 200 × dataset. During training on this dataset, VGG19 has required 97.31ms per image which can be considered as an anomaly with respect to other measured values. From Table 12.8, it is also observed that the models that have been trained faster have also faster average testing time. However, it is an intriguing matter that all the models have taken less amount of time to process an image in the testing phase than in the training phase.

Discussion

In this study, the histopathological images of breasts have been classified into benign or malignant using various methods. First, seven convnets have been developed for classification challenges. After that, the stacked generalization ensemble method has been used to increase the classifiers' performances. The BreakHis (Spanhol et al., 2016) dataset is divided into four subsets based on magnification factors. The experiments have been conducted for every subset. The cross-dataset performance of the developed models has also been measured to determine the portability of the models. The performances of the developed classifiers have been generally good. For the 40 × dataset, the best-performing model has been the individual

TABLE 12.8 Training and testing time of the convnets.

Dataset	Convnets	Average training time per image (ms)	Average testing time per image (ms)
40 ×	DenseNet	55.96	50.9
	Inception	21.72	13.84
	InceptionResNetV2	60.17	26.95
	ResNetV2	26.24	13.03
	VGG16	32.26	14.88
	VGG19	37.5	12.56
	Xception	49.73	17.13
100 ×	DenseNet	54.08	15.59
	Inception	21.49	12.44
	InceptionResNetV2	47.66	28.03
	ResNetV2	25.98	12.34
	VGG16	31.24	11.66
	VGG19	37.02	13.84
	Xception	49.55	15.94
200 ×	DenseNet	54.43	31.25
	Inception	21.55	16.25
	InceptionResNetV2	47.74	28.16
	ResNetV2	25.78	13.6
	VGG16	31.19	17.56
	VGG19	97.31	25.53
	Xception	49.72	18.13
400 ×	DenseNet	55.08	29.94
	Inception	22.88	11.53
	InceptionResNetV2	48.55	14.38
	ResNetV2	25.84	13.69
	VGG16	31.13	18.56
	VGG19	37.48	19.81
	Xception	50.02	12.22

InceptionResNetV2 model. For 100× and 200× datasets, a stacked model combining five first-level classifiers has outperformed all the other classifiers. And, for the 400× dataset, both InceptionResNetV2 and the stacked model have the same performance. Since for the majority of cases, the ensemble classifiers have outperformed the individual convnets; therefore, the use of the described ensemble method has been justified. The main reason behind the better performance by the stacked model is that each first-level classifier has been built by following different convolutional neural networks. Five first-level classifiers have been trained using five different convnets, and those convnets are selected from a pool of seven convnets. Five of the better-performing convnets on the BreakHis datasets have been utilized to develop the first-level classifiers. After obtaining these first-level classifiers, a new combined model has been developed which has reduced bias from the predictions. The validation set which is completely unknown to the first-level classifiers has been used for training the second-level classifier. By following these, the second-level stacked models have been able to demonstrate better performances than individual convnets.

Among the individual classifiers, InceptionResNetV2 has outperformed the other six convnets. It has been consistent in giving better performances throughout the four datasets. The reason behind such good performance is that it blends the residual connections and Inception structures. While the individual ResNetV2 and InceptionV3 could not be the top performer at any of the datasets, the combination of residual and Inception has been able to be the better convnet in terms of various evaluation metrics. DenseNet has demonstrated performances close to InceptionResNetV2 for having the dense blocks in its architecture. Being deeper than other convnets has also helped DenseNet to be one of the good performers. Another consistent performing convnet has been Xception. It has been able to show better accuracies throughout the study than its counterpart—InceptionV3. Having those depthwise separable convolutional layers instead of the typical convolutional layer has helped Xception to improve its performances and to reduce training times. Although they have not shown top-of-the-line performances in classifying histopathological breast images, the two VGG models have shown surprisingly good performances despite having fewer deeper architectures. With 16 and 19 layers, they have been able to demonstrate performances compatible with other convnets having larger and deeper architectures.

Another aspect to discuss is the length of the training phase of this study. All the individual convnets of this study have been trained for 100 epochs. Validation accuracies during training have provided important insights. Different models have achieved their best validation accuracies at different epoch. For example, the DenseNet model for the 200× dataset has gained the best validation accuracy at the last epoch. On the other hand, the DenseNet model for the 400× dataset has gained the best validation performance at the 19th epoch, i.e., after the 19th epoch, no improvement in validation performance whatsoever. This is to explain the uncertain behavior of the models during training. Determining when to stop the training and the weights of the model on which epoch would give the best prediction on the test set is a very difficult task. To tackle this issue, after each epoch, a model's validation accuracy has been observed. If it is better than the previous epoch, the model has been saved. This process has been repeated for all the 100 epochs. And, after the completion of training, the model has also been saved. This is the model containing all the updated weights after the last epoch. Then, both the saved model during training and the model after the last epoch are tested with the test images. The higher-performing model has been kept, and the other one has been discarded. In this way, the best possible model has been preserved.

In this study, the cross-dataset performance of the models has also been observed. A model does not tend to perform well in a dataset other than the own dataset on which the model has been trained. This poor performance indicates that the developed classifiers lack portability. To predict breast histopathology images with a specific magnification factor, the models trained on the dataset with the same magnification factor should be used. For example, if an image with a 100× magnification factor is to be classified, a model that has been trained on 100× images has to be used. A new insight has also been found with these cross-dataset experiments. A model trained on images having a specific magnification factor performs relatively better for the images that have a closer magnification factor than that for the images with a distant magnification factor. For example, if a model has been trained on a 400× dataset, its performance for the 200× dataset will be better than its performance for the 100× and 40× datasets, and its performance for the 40× dataset will be worse than its performance for the 100× dataset (see Figs. 12.23 and 12.24).

Performance comparison

The performances of this study have been outstanding and have beaten many existing recent works. In Table 12.9, a performance comparison of this study with a few top-performing works on the BreakHis dataset has been presented. From the table, it is clear that this study has superior performances. The study by Saini and Susan (2020) has been able to perform well after balancing the dataset using DCGAN. But, the methods used in this study have been able to perform better without balancing. In this study, state-of-the-art convolutional networks have been trained from the scratch for image classification tasks in the biomedical domain. Training convnets from scratch has been proven to be better than transfer learning as this study has demonstrated better performances than the studies that use pretrained convnets

TABLE 12.9 Test accuracy comparison of this study with other existing works on BreakHis dataset.

Works	Method	40× (%)	100× (%)	200× (%)	400× (%)
Nahid et al. (2018)	CNN and LSTM	90.00	90.00	91.00	90.00
Erkan et al. (2018)	Transfer learning with AlexNet, VGG16	93.57	92.50	93.24	91.87
Song et al. (2017)	CNN, Fisher vector encoding	87.00	86.20	85.20	82.90
Saini and Susan (2020)	DCGAN, CNN	96.50	94.00	95.50	93
Bhuiyan et al. (2019)	Transfer learning	96.24	92.81	94.90	92.86
Proposed work	InceptionResNetV2	99.70	98.60	98.27	97.50
Proposed work	DenseNet	98.20	98.60	97.69	96.24
Proposed work	Stacked generalization ensemble	99.41	98.88	99.42	97.50

on the ImageNet dataset, e.g., Erkan et al. (2018) and Bhuiyan et al. (2019). Transfer learning could introduce a significant improvement in the model performances for breast cancer detection if the models are pretrained on a large dataset from the medical domain rather than pretrained on a general dataset.

Conclusion

A set of deep learning models have been developed in this study for breast cancer prognosis. The models have been able to perform with high efficiency, and they have achieved outstanding results. Among various experiments, InceptionResNetV2 has been one of the top performers for the detection task. However, the stacked generalization ensemble method has achieved better or equal performances than InceptionResNetV2 in three out of four datasets that have been used in this study. Using this ensemble method has also been one of the first of its kind that has been used for breast cancer prognosis. The performance of this study has also been better than most of the state-of-the-art works in this field of breast cancer image analysis. Another aspect of this study has been to explore the portability of the machine learning models. The experimental results indicate that for better performance on specific biomedical images, a specialized model on those images is needed to be trained. If training data come from a different dataset, the models' performance decreases.

Despite getting good performances in the experimental results, few issues have been observed in this study. For example, the BreakHis dataset has fewer benign images than malignant. The ratio of benign to malignant images is about 1:3. If this imbalance can be resolved, even better performances can be obtained. Besides, histopathology images have been used for the experiments, and these images are subject to very complex structures. Although convnets have been proven to be very good at extracting complex features from images, even better convnets and modified convnets can increase models' performances. Another room for improvement is to use other ensemble methods. Here in this study, only one ensemble method has been used so far. Finally, works can be done to increase the explainability and interpretability of machine learning models in the biomedical domains.

References

Acharya, U. R., Ng, E. Y. K., Tan, J. H., & Sree, S. V. (2012). Thermography based breast cancer detection using texture features and support vector machine. *Journal of Medical Systems, 36*(3), 1503–1510. https://doi.org/10.1007/s10916-010-9611-z.

Aggarwal, C. C. (2015). *Data mining: The textbook* (1st ed.). Springer International Publishing. https://doi.org/10.1007/978-3-319-14142-8.

Alan, G. J. (2017). Deep learning algorithms for detection of lymph node metastases from breast cancer. *JAMA, 2184*. https://doi.org/10.1001/jama.2017.14580.

Al-Dhabyani, W., Gomaa, M., Khaled, H., & Fahmy, A. (2020). Dataset of breast ultrasound images. *Data in Brief, 28*. https://doi.org/10.1016/j.dib.2019.104863.

Arora, N., Martins, D., Ruggerio, D., Tousimis, E., Swistel, A. J., Osborne, M. P., & Simmons, R. M. (2008). Effectiveness of a noninvasive digital infrared thermal imaging system in the detection of breast cancer. *American Journal of Surgery, 196*(4), 523–526. https://doi.org/10.1016/j.amjsurg.2008.06.015.

Bayramoglu, N., Kannala, J., & Heikkila, J. (2016). Deep learning for magnification independent breast cancer histopathology image classification. In *Vol. 0. Proceedings—International conference on pattern recognition* (pp. 2440–2445). Institute of Electrical and Electronics Engineers Inc. https://doi.org/10.1109/ICPR.2016.7900002.

Becker, A. S., Marcon, M., Ghafoor, S., Wurnig, M. C., Frauenfelder, T., & Boss, A. (2017). Deep learning in mammography diagnostic accuracy of a multipurpose image analysis software in the detection of breast cancer. *Investigative Radiology, 52*(7), 434–440. https://doi.org/10.1097/RLI.0000000000000358.

Benhammou, Y., Achchab, B., Herrera, F., & Tabik, S. (2020). BreakHis based breast cancer automatic diagnosis using deep learning: Taxonomy, survey and insights. *Neurocomputing, 375*, 9–24. https://doi.org/10.1016/j.neucom.2019.09.044.

Bhuiyan, M. N. Q., Shamsujjoha, M., Ripon, S. H., Proma, F. H., & Khan, F. (2019). Transfer learning and supervised classifier based prediction model for breast cancer. In N. Dey, H. Das, B. Naik, & H. S. Behera (Eds.), *Advances in ubiquitous sensing applications for healthcare* Big Data Analytics for Intelligent Healthcare Management. https://doi.org/10.1016/B978-0-12-818146-1.00004-0.

Celik, Y., Talo, M., Yildirim, O., Karabatak, M., & Acharya, U. R. (2020). Automated invasive ductal carcinoma detection based using deep transfer learning with whole-slide images. *Pattern Recognition Letters, 133*, 232–239. https://doi.org/10.1016/j.patrec.2020.03.011.

Cha, Z., & Yunqian, M. (2012). *Ensemble machine learning* (1st ed.). Springer Nature. https://doi.org/10.1007/978-1-4419-9326-7.

Chan, A., & Tuszynski, J. A. (2016). Automatic prediction of tumour malignancy in breast cancer with fractal dimension. *Royal Society Open Science, 3*(12). https://doi.org/10.1098/rsos.160558.

Chan, H. P., Samala, R. K., & Hadjiiski, L. M. (2020). CAD and AI for breast cancer-recent development and challenges. *The British Journal of Radiology, 93*(1108), 20190580. https://doi.org/10.1259/bjr.20190580.

Chollet, F. (2017). Xception: Deep learning with depthwise separable convolutions. In (Vols. 2017). *Proceedings—30th IEEE conference on computer vision and pattern recognition, CVPR 2017* (pp. 1800–1807). Institute of Electrical and Electronics Engineers Inc. https://doi.org/10.1109/CVPR.2017.195.

DeSantis, C. E., Ma, J., Gaudet, M. M., Newman, L. A., Miller, K. D., Goding Sauer, A., Jemal, A., & Siegel, R. L. (2019). Breast cancer statistics, 2019. *CA: a Cancer Journal for Clinicians, 69*(6), 438–451. https://doi.org/10.3322/caac.21583.

Erkan, D., Abdulkadir, Ş., Zehra, K., Yanhui, G., Varun, B., & Ümit, B. (2018). Transfer learning based histopathologic image classification for breast cancer detection. *Health Information Science and Systems*. https://doi.org/10.1007/s13755-018-0057-x.

Ferroni, P., Zanzotto, F. M., Riondino, S., Scarpato, N., Guadagni, F., & Roselli, M. (2019). Breast cancer prognosis using a machine learning approach. *Cancers, 11*(3). https://doi.org/10.3390/cancers11030328.

Gupta, V., & Bhavsar, A. (2017). Breast cancer histopathological image classification: Is magnification important? In (Vols. 2017). *IEEE computer society conference on computer vision and pattern recognition workshops* (pp. 769–776). IEEE Computer Society. https://doi.org/10.1109/CVPRW.2017.107.

Han, Z., Wei, B., Zheng, Y., Yin, Y., Li, K., & Li, S. (2017). Breast cancer multi-classification from histopathological images with structured deep learning model. *Scientific Reports, 7*(1). https://doi.org/10.1038/s41598-017-04075-z.

He, K., Zhang, X., Ren, S., & Sun, J. (2016a). Deep residual learning for image recognition. In (Vols. 2016). *Proceedings of the IEEE computer society conference on computer vision and pattern recognition* (pp. 770–778). IEEE Computer Society. https://doi.org/10.1109/CVPR.2016.90.

He, K., Zhang, X., Ren, S., & Sun, J. (2016b). Identity mappings in deep residual networks. In *Vol. 9908. Lecture notes in computer science (including subseries lecture notes in artificial intelligence and lecture notes in bioinformatics)* (pp. 630–645). Springer Verlag. https://doi.org/10.1007/978-3-319-46493-0_38.

Houssami, N., Lee, C. I., Buist, D. S. M., & Tao, D. (2017). Artificial intelligence for breast cancer screening: Opportunity or hype? *Breast, 36*, 31–33. https://doi.org/10.1016/j.breast.2017.09.003.

Huang, G., Liu, Z., Van Der Maaten, L., & Weinberger, K. Q. (2017). Densely connected convolutional networks. In (Vols. 2017). *Proceedings—30th IEEE conference on computer vision and pattern recognition, CVPR 2017* (pp. 2261–2269). Institute of Electrical and Electronics Engineers Inc. https://doi.org/10.1109/CVPR.2017.243.

Jiang, Y., Chen, L., Zhang, H., & Xiao, X. (2019). Breast cancer histopathological image classification using convolutional neural networks with small SE-ResNet module. *PLoS One, 14*(3). https://doi.org/10.1371/journal.pone.0214587.

Kaymak, S., Helwan, A., & Uzun, D. (2017). Breast cancer image classification using artificial neural networks. In *Vol. 120. Procedia computer science* (pp. 126–131). Elsevier B.V. https://doi.org/10.1016/j.procs.2017.11.219.

Krizhevsky, A., Sutskever, I., & Hinton, G. E. (2017). ImageNet classification with deep convolutional neural networks. *Communications of the ACM, 60*(6), 84–90. https://doi.org/10.1145/3065386.

McKinney, S. M., Sieniek, M., Godbole, V., Godwin, J., Antropova, N., Ashrafian, H., Back, T., Chesus, M., Corrado, G. C., Darzi, A., Etemadi, M., Garcia-Vicente, F., Gilbert, F. J., Halling-Brown, M., Hassabis, D., Jansen, S., Karthikesalingam, A., Kelly, C. J., King, D., … Shetty, S. (2020). International evaluation of an AI system for breast cancer screening. *Nature, 577*(7788), 89–94. https://doi.org/10.1038/s41586-019-1799-6.

Mehdy, M. M., Ng, P. Y., Shair, E. F., Saleh, N. I. M., & Gomes, C. (2017). Artificial neural networks in image processing for early detection of breast cancer. *Computational and Mathematical Methods in Medicine, 2017*. https://doi.org/10.1155/2017/2610628.

Meyer, C. R., Chenevert, T. L., Galbán, C. J., Johnson, T. D., Hamstra, D. A., Rehemtulla, A., & Ross, B. D. (2015). Data from RIDER_Breast_MRI. *The Cancer Imaging Archive*. https://doi.org/10.7937/K9/TCIA.2015.H1SXNUXL.

Nahid, A. A., Mehrabi, M. A., & Kong, Y. (2018). Histopathological breast cancer image classification by deep neural network techniques guided by local clustering. *BioMed Research International, 2018*. https://doi.org/10.1155/2018/2362108.

Opitz, D., & Maclin, R. (1999). Popular ensemble methods: An empirical study. *Journal of Artificial Intelligence Research, 11*, 169–198. https://doi.org/10.1613/jair.614.

Saini, M., & Susan, S. (2020). Deep transfer with minority data augmentation for imbalanced breast cancer dataset. *Applied Soft Computing Journal, 97*. https://doi.org/10.1016/j.asoc.2020.106759.

Simonyan, K., & Zisserman, A. (2015). Very deep convolutional networks for large-scale image recognition. In *Int. Conf. Learn. Represent. ICLR 2015 - Conf. Track Proc.*.

Song, Y., Zou, J. J., Chang, H., & Cai, W. (2017). Adapting fisher vectors for histopathology image classification. In *Proceedings—International symposium on biomedical imaging* (pp. 600–603). IEEE Computer Society. https://doi.org/10.1109/ISBI.2017.7950592.

Spanhol, F. A., Oliveira, L. S., Petitjean, C., & Heutte, L. (2016). Breast cancer histopathological image classification using convolutional neural networks. In (Vols. 2016). *Proceedings of the international joint conference on neural networks* (pp. 2560–2567). Institute of Electrical and Electronics Engineers Inc. https://doi.org/10.1109/IJCNN.2016.7727519.

Szegedy, C., Ioffe, S., Vanhoucke, V., & Alemi, A. A. (2017). Inception-v4, inception-ResNet and the impact of residual connections on learning. In *31st AAAI conference on artificial intelligence, AAAI 2017* (pp. 4278–4284). AAAI Press. https://aaai.org/ocs/index.php/AAAI/AAAI17/index.

Szegedy, C., Liu, W., Jia, Y., Sermanet, P., Reed, S., Anguelov, D., Erhan, D., Vanhoucke, V., & Rabinovich, A. (2015). Going deeper with convolutions. In (Vols. 07–12). *Proceedings of the IEEE computer society conference on computer vision and pattern recognition* (pp. 1–9). IEEE Computer Society. https://doi.org/10.1109/CVPR.2015.7298594.

Tan, Y. J., Sim, K. S., & Ting, F. F. (2018). Breast cancer detection using convolutional neural networks for mammogram imaging system. In (Vols. 2018). *Proceeding of 2017 international conference on robotics, automation and sciences, ICORAS 2017* (pp. 1–5). Institute of Electrical and Electronics Engineers Inc. https://doi.org/10.1109/ICORAS.2017.8308076.

Wei, B., Han, Z., He, X., & Yin, Y. (2017). Deep learning model based breast cancer histopathological image classification. In *2017 2nd IEEE international conference on cloud computing and big data analysis, ICCCBDA 2017* (pp. 348–353). Institute of Electrical and Electronics Engineers Inc. https://doi.org/10.1109/ICCCBDA.2017.7951937.

Wishart, G. C., Campisi, M., Boswell, M., Chapman, D., Shackleton, V., Iddles, S., Hallett, A., & Britton, P. D. (2010). The accuracy of digital infrared imaging for breast cancer detection in women undergoing breast biopsy. *European Journal of Surgical Oncology, 36*(6), 535–540. https://doi.org/10.1016/j.ejso.2010.04.003.

Zhiqiong, W., Mo, L., Huaxia, W., Hanyu, J., Yudong, Y., Hao, Z., & Junchang, X. (2019). Breast cancer detection using extreme learning machine based on feature fusion with CNN deep features. *IEEE Access*, 105146–105158. https://doi.org/10.1109/access.2019.2892795.

CHAPTER

13

Light-gradient boosting machine for identification of osteosarcoma cell type from histological features

Etuari Oram[a], *Pandit Byomakesha Dash*[b], *and Bighnaraj Naik*[a]

[a]Department of Computer Application, Veer Surendra Sai University of Technology, Burla, Odisha, India [b]Department of Information Technology, Aditya Institute of Technology and Management (AITAM), Tekkali, Andhra Pradesh, India

Introduction

Though osteosarcomas (OS) is a less-known type of cancer cell disease in India according to the American Cancer Society and American Institute for Cancer (n.d.), every 1 in 10 osteosarcomas are seen in people aged 60 and above. It is also common in pediatrics, as near about 2% of paediatrics cases is common type of osteosarcomas, 5%–10% of pediatric cancer cases are only found in the US, and every year nearly 1000 new cases are diagnosed. Out of the nearly 100 kinds of cancer types, osteogenic sarcoma is a common bone cancer type, and it rarely occurs in India still it is supposed to be found nearly a lakh in a year. Trillions of cells are generated, mutated, and healed naturally in our everyday life. But, very often extra abnormal cells get uncontrolled, and then subsequently cell formation becomes malignancy. This is what called sarcoma. It originates from primitive transformed cells of mesenchymal origin. Later on, it is exposed to osteoblastic differentiation and turns into malignant osteoid. The cancerous tumor usually grows in the knee area, either the lower part, the upper part of the thigh bone, and shin bone, or the metaphysic area, and is rarely seen at the end of a long bone where there is a maximum chance of getting growth faster. Osteosarcoma originates in the bone cell, disrupts cell growth, and hence bone becomes washy and feeble. Also, it may occur when DNA development gets modified and is not controlled over cell instruction. These cases mostly occur in children adolescents or teenagers of the age group of 10–30. Moreover, it is more common in males. In fact, it starts with the bone tumor, which is either benign

https://doi.org/10.1016/B978-0-323-85240-1.00016-X

(non-cancerous) or malignant (cancerous). These cases mostly occur in children adolescents or teenagers of the age group of 10–30. Moreover, it is more common in males. The main cause of this malignant is not known exactly but often it is expected to be inherited from parents to children. The other common causes are environmental diversity, industrialization, diet habits, and hereditary disorders like Werner, Bloom syndrome, and retinoblastoma Li-Fraumeni syndrome. Other risk factors are Paget's, radiation therapy, and fibrous dysplasia. Various forms of metastases of cancerous cells are generally found as multiple myeloma (MM), osteosarcoma (OS), chondrosarcoma, and Ewing's sarcoma, respectively. Osteosarcoma is an immensely malignant tumor originating from the primitive mesenchymal bone-forming cell. The most common MM tumor often called marrow cell is a general type of tumor that grows in the bone marrow area and gradually metastasizes over other bones it usually affects older adults. The second most common primary type is "OS (also called osteogenic sarcoma), is a highly malignant tumor developing around metaphyseal-diaphyseal. This type basically affects the age group 10–25, though it could be found in any age group. The OS cells secrete malignant osteoid from the main tumor bone cells, and it can be detected through biopsy to specify the cancerous or non-cancerous cell. This type of cancer basically affects the age group of 10–25, in fact, it could be found in any age group.

A malignancy experiences symptoms like redness of near localized area, joint pain, swelling to near bone and tissues, feeling fatigued, and accidental bone injury without proper reason followed by certain symptoms of weight loss (Gupte & Mart, 2013). Later on, the resultant poor mass tumor or immense malignancy tends to metastasize and could invade, mortify, and destroy the other healthy soft bone tissues as well as other body parts. Based on the cancerous symptoms and invasiveness, the (low, i.e., stage I, high, i.e., stage II, regardless of grade, i.e., stage III) and extents of cancer cells are determined (from lower to higher grade) through the diagnosis. Pathological experiments are conducted such as biopsy, blood sample test, or image testing including X-rays, bone scanning, and MRI (Magnetic Resonance Imaging). The process of predicting benign tumors or cancerous osteosarcoma is very crucial for medicos. It is a difficult task to predict as it depends on the proliferation of cells, other health issues, the position of metastases, affected size, age of the patient, surgical strategy, histologic response to chemotherapy, and surgical margin. However, surgical therapy is affected by the heterogeneous extracted features, surgeon grasping experience, and sample size. Preventing common types of osteosarcomas is a very significant and challenging factor for health issues in the 21st century. The basic treatment opted for the concerned malignancy is chemotherapy, surgery, or radiotherapy. But, none of the medications is good for health issues because chemotherapy and radiotherapy cause side effects on the body. Sometimes, it becomes necessary to dispose of the cancerous tissue by excising the affected body part. So, prior to diagnosis, it could be suggested to go through the proper identification and prediction of cancerous or non-cancerous type cells by applying a suitable machine learning model in the related research area.

Some literature studies have explored much important research on the concerned domain. A multi-objective algorithm (Lau, 2009) is applied to detect the antibody content of drugs by collecting their inherent molecular cells. In a study (Zhi Li et al., 2017), SVM, RF, and Logistic Regression are used for the early detection restraint to fatal metastases. The proposed machine learning approaches are found to be efficient in early detection, which is helpful to many patients with a greater perspective. In a study (Arunachalam et al., 2017), tumor cells exhibiting viable and non-viable forms (intends merely harmful) are classified with the high-resolution digital image data set. They applied unsupervised k-means clustering for image isolation, multi-threshold Otsu segmentation for classification, and a flood-fill object-based method for further analysis tasks. The results of the proposed method are highly dominant over the other discussed approaches. In another study (Chan et al., 2010), it is suggested that the patients need to be kept in continuous observation and outlook for assessment to diagnose chemoreception over a long duration. Here, the evolutionary variable neighborhood search (EVNS) and variable neighborhood search (VNS) are used for subset selection and feature extraction of the useful information of combinational interacted gene. These genes are then used to predict the histologic response of chemotherapy in bone cancerous patients. The cross-validation outcome demonstrated that EVNS is performing better than other approaches in the initial classification of biopsy data. A neoadjuvant response (chemo response before surgery) of a patent is a mandatory factor for a diagnostic decision process. In this line, a study has been made on a collection of textural signatures of histological data and PET textural features, and these features are used to build machine learning-based models such as Conditional Random Forest (CRF), KNN, NB (Naïve Bayes), and FDG-PET scan tool (Im et al., 2017) for prediction of perception of neoadjuvant chemotherapy of a pediatric patient. Among these machine learning approaches, the RF method was found to be better in the prediction performance. For the 18F-FDG (18F-uorodeoxyglucose), a scanning tool (Jeong et al., 2019) called positron-emitted tomography (PET) has been used for predicting the survival and chemotherapy progression of patients. Several features of 18F-FDG were selected, and the treatment response is predicted using different machine learning-based models such as SVM, RF, and gradient boosting methods. The authors concluded that the intended machine learning methods are suitable for good prediction of chemotherapy progression and survival. OS inmate's transcriptome data and DNA methylation have been analyzed by Deng et al. (2021) for finding the prognosis of OS (osteosarcoma) patients. Here, the dataset TARGET (Therapeutically Applicable Research to Generate Effective Treatments) database has experimented with machine learning algorithm-iterative Lasso Cox regression, Lasso to correlate the survival rate of patients. The models are then validated with Receiver Operating Characteristic (ROC) curves and Kaplan-Meier curves to reveal the risk level of patients as well as the prognostic capability of signature independently with high accuracy. An extreme gradient boosting (XG-Boost), an artificial intelligence model, has been recommended (Jiang et al., 2019) to predict the survival of 5-year osteosarcoma patients with a high accuracy.

To predict the survival of 5-year osteosarcoma patients with high accuracy, out of the 835 features of (Surveillance, Epidemiology, and End Result Research) SEER, 668 cases were trained and 167 cases were validated. Here main features focuses are local lymphatic metastasis, histological grade, age, extension stage, primary tumor site, tumor size, distant metastasis, surgery, radiation, and chemotherapy. The constructed XG-boost model compared Bayesian network models and support vector machine to reveal better performance in both the training and validation cases of AUC 0.977 and 0.911. In order to determine the metastasis level of cartilaginous bone tumors, a machine learning-based approach has been proposed (Gitto et al., 2020). Here, RF model was applied to extract the features from unenhanced MRI, and the resultant dataset was trained on the AdaboostM1 classifier using 10-fold cross-validation. The experimental results reveal the training and test groups were correctly classified using the AdaboostM1 classifier as 75% and 85.7%, respectively, of the lesions. An approach based on the value of multi-parametric MRI associated with machine learning (Huang et al., 2020) has been used to evaluate necrosis tumors after neoadjuvant chemotherapy (NACT). The proposed approach uses MRI parameter signal intensity values of

T2-weighted imaging (T2WI), apparent diffusion coefficient (ADC) values, and subtract-enhanced T1-weighted imaging (ST1WI) to train the models.

From this experimental results, it is observed that the classifier constructed with ADC obtained an AUC of 0.93, whereas the classifier with multi-parametric MRI performed 0.97 to distinguish between the tumor nonviable and non-cartilaginous tumor. An AUC of 0.83 improved to 0.90 to discriminate tumor survival from tumor nonviability. Similarly, to differentiate nonviability from cartilaginous tumors ADC obtained an AUC of 0.61, while the multi-parametric MRI parameter-based classifier was enhanced to 0.81 and concluded that together both machine learning and multi-parametric MRI can improve the distinguishing ability of tumor components.

To predict OS prognosis, an eight-gene predictive model and nomogram have been proposed (Wu & Zhang, 2020) to segregate samples into low- and high-risk groups. The dataset GEO (Gene Expression Omnibus) databases and TCGA (The Cancer Genome Atlas) are collected to classify into non-metastatic and metastatic samples, and the proposed approach makes use of the SVM (support vector machine) classifier. Additionally, it uses univariate and multivariate analysis to obtain independent OS prognostic genes. The performance of the model has been evaluated using nomogram analysis and Kaplan-Meier (KM). To explore the most considerable prognostic factors in patients with non-metastatic osteosarcoma after having primary site surgery using big data, more precise prognostic models have been suggested (Huang et al., 2019). This study considered 1000 patients histologically diagnosed with osteosarcoma from 2005 to 2014 from the SEER dataset. Survival analysis and prognostic factors are identified by Lasso regression and the machine learning model. It is observed from the experimental results that multivariable nomogram-based models could outperform its prediction probability of death of OS and CSS patients than its performance in the sixth and seventh editions of the AJCC (American Joint Committee on Cancer) staging system.

From the above literature study, it is confirmed that most of the authors have emphasized the detection and prevention of malignant tissues on a broader aspect. However, being such an important medical problem, there should be a continuous improvement of the model's performance as well as learning novel intelligent features from past experience and observations. Moving in this direction, this chapter proposed a light-based gradient-boosting machine model for the prediction of osteosarcoma cell type from histological features.

The major objective of this contribution is as follows: (a) detail study of osteosarcoma cells, signs, symptoms and histological features, and (b) design a light gradient boosting ensemble learning model for the prediction of osteosarcoma cell type from histological features. The other sections are organized as follows: "Background information" section explains the osteosarcoma, osteosarcoma cell types, and histological features; "Proposed model" section represents the proposed model; simulation results are presented in "Result and analysis" section, and then the conclusion is presented in "Conclusion" section with future scope.

Background information

Osteosarcoma (OS) is one of the trivial malignant bone cancers that mostly occurs in adolescents around the whole world with an incidence of 0.2%–0.5% (Noone et al., 2018;

Zambo & Veselý, 2014). Generally, it arises from bone tissue with the most common point being the metaphysis of long bones. Over the past few decades, the considerable development in therapeutic methods such as multi-agent systemic and postoperative neoadjuvant chemotherapy has provided promising results in the prevention and treatment of OS (Bielack et al., 2009; Wakamatsu et al., 2019). However, the statistical analysis shows that each year, the occurrence and mortality rate of OS have been increasing approximately by 1.4% (Pei et al., 2019). The most general histological type is the widespread OS which accounts for 75% of all the cases approximately (Spina et al., 1998). These conventional OS are divided into chondroblastic, osteoblastic, and fibroblastic types (Fox & Trotta, 2013). The high-grade central OSs contain small-cell, giant-cell-rich, telangiectatic, and epithelioid variants. Low-grade central osteosarcoma (LCOS) which has a lower rate of metastasis and greater survival is a type of low-grade intramedullary OS (Klein & Siegal, 2006). Sometimes, OS also arises from the cortex of the periosteum. Periosteal sarcoma is the most general type of juxtacortical lesion. It constitutes approximately 1%–6% of the total OS cases (Fox & Trotta, 2013). These periosteal lesions have a "stuck-on" appearance on the metaphyseal part of long bones. These are slow-growing and low-graded lesions when compared with the conventional OS. Due to the presence of cartilage caps, they are often confused with osteochondromas (Clayer, 2015). Another type of juxtacortical lesion is the periosteal sarcoma which has a "stuck-on" appearance. These lesions are mid-grade lesions that exhibit aggressive characteristics on radiographs (Grimer et al., 2005). The most aggressive type of tumor is the high-grade surface OS which has an appearance similar to classical high-grade OS (Temple et al., 2000). The histologic examination of conventional OS has been depicted in Fig. 13.1. Conventional osteosarcomas have a matrix that is prevailed by cartilaginous, osseous, or fibrous elements. The conventional OS is further subdivided based on which matrix

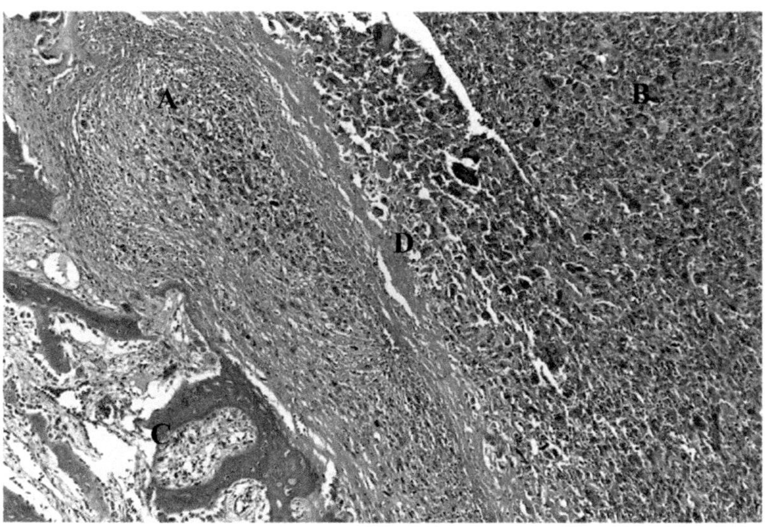

FIG. 13.1 Osteosarcoma histology. *Source: Nephron (https://commons.wikimedia.org/wiki/File:Osteosarcoma_-_intermed_mag.jpg), "Osteosarcoma - intermed mag," Adding marker on the image by adding extra info, https://creativecommons.org/licenses/by-sa/3.0/legalcode.*

cells prevail. The woven micro trabeculae of bone within a bland to moderately cellular fibrous stroma have been demonstrated by the surface OS and lower-grade central. The high-grade morphology has been exhibited by the other types of OS along with areas of abundant giant small cells or epithelioid morphology.

Proposed model

Let $HF = (hf_i, ot_i)_{i=1}^{n}$ be the training data with n no. of samples. hf_i is the ith histological feature profile and ot_i is the osteosarcoma cell type of ith histological feature profile. The construction of the Light-GBM model has five major steps as follows: (i) combining mutually exclusive features of each hf_i in HF using Exclusive feature bundling (Ke et al., 2017), (ii) setting up the initial model and $M_0(HF)$ selecting a base tree model with a minimum loss, (iii) iteratively calculating the gradient g_j of each model M_j in $\{M_i\}_{j=1}^{k}$, (iv) generation of the resampled dataset $SS1$ and $SS2$ from the original data HF by using Gradient-based One-Side Sampling (Ke et al., 2017), (v) calculation of estimated variance information gain of the split $SS1$ and $SS2$ and building a new tree M_k^*, and (vi) updating the model $M_k = M_{k-1} + M_k^*$. The complete working steps have been presented in Algorithm 13.1.

The proposed model makes use of the histogram-based algorithm and the resultant discrete bins (Fig. 13.2). The EFB keeps the exclusive features' value in a bundle and maintains it in different bins. The use of GOSS eliminates the scanning process of the entire dataset to find the best split by selecting subsamples of the dataset based on small and large gradient. Both EFB and GOSS not only make Light-GBM faster but also maintain good accuracy.

Result and analysis

Dataset information

Our study considers the osteosarcoma histological dataset for analysis, which is collected from the Cancer Archive (The Cancer Imaging Archive, n.d.) repository. The collected dataset is uploaded in Dallas by a group of medical scientists at the University of Texas Southwestern Medical Center. It is the collection of diagnosis data of 50 number of child patients diagnosed from the year 1995 to 2015 at Children's Medical Hub, Dallas. After conducting surgical resection of various specimens, four out of 50 treated patients were selected. The generated distributed samples are then recorded as the histological dataset. This dataset consists of 58 features excluding three class labels. The recorded data are distributed as 47% (536) of non-tumor images, 23% (263) of necrotic tumor, and 30% (345) of viable tumor.

Simulation result analysis

The proposed model makes use of histological features of osteosarcoma to predict various cancerous cell metastases. This section represents the simulation results obtained through the experiment by implementing the proposed model (Light GBM) and other studied models such as Decision Tree (DT), Naive Bayes (NB), Random Forest (RF), MLP (Multi-Layer Perceptron), Linear Regression (LR), Bagging, AdaBoost, GBoost, and XGBoost. The

Algorithm 13.1 Light-GBM for identification of osteosarcoma cell type from histological features.

Input: $HF=(hf_i,ot_i)_{i=1}^{n}$ be the training data with n no. of the sample. hf_i is the ith histological feature profile and ot_i is the Osteosarcoma cell type of ith histological feature profile.

Output: Model M for identification of Osteosarcoma cell type

1. Perform exclusive feature bundling for each hf_i in HF, and combine the mutually exclusive.

2. Set $M_0(HF)=\text{argmin}_{BM}(\sum_{i=1}^{n}\ell(ot_i,BM(hf_i)))$, is a base model (usually a decision tree)

3. for $j=1$ to k

3.1. Find the absolute values of gradients $g_j:g_j=\left[\left|\frac{\partial\ell(ot_i,M(hf_i))}{\partial M(hf_i)}\right|_{M_j(HF)=M_{j-1}(HF)}\right]_{i=1}^{n}$

3.2. Perform Light Gradient-based One-Side Sampling for generation of re-sampled dataset from the original data HF:

3.2.1 Rank the training instances according to their absolute values of their gradients $S\leftarrow SortingIndices(abs(g))$ and select top lG % of instances with large gradients

$tN=lG\times len(HF)$

3.2.2 Select top sG% of instances with small gradients

$rN=sG\times len(HF)$

3.2.3 Sample the subset $SS1=S[1:tN]$ and $SS2=PickRandom(S[tN:len(D)],rN)$

3.2.4 $HF^{*}\leftarrow Merge(S1,S2)$

3.3. Calculate estimated variance information gain:

$$V_j(SS1,SS2)=1/N\left(\left(\sum_{hf_i\in SS1_l}g_i+\frac{1-lG}{sG}\sum_{hf_i\in SS2_l}g_i\right)^2\Big/N_l^j+\left(\sum_{hf_i\in SS1_r}g_i+\frac{1-lG}{sG}\sum_{hf_i\in SS2_r}g_i\right)^2\Big/N_r^j\right)$$

3.4. Find the best split and use HF^{*} to build a new decision tree M_j^{*}

3.5. $M_j=M_{j-1}+M_j^{*}$

4. Return M_k

performance metrics in terms of F1-Score, F2-Score, Fbeta-score, Recall, Precision, and AUC-ROC curves are summarized in Table 13.1. The performance of all the corresponding models is evaluated and presented in the consolidated tabular and graphical form. Subsequently, these are analyzed and discussed.

Table 13.1 presents performance measure with various performance metrics such as Precision, Recall, F1 Score, F2 Score, Fbeta Score, and ROC-AUC. It is noticeable that the ensemble learning approaches, particularly boosting techniques, have dominated over other ML models. Although all boosting approaches performed well, the proposed model based on Light-GBM has a comparatively better execution over other models. The Light-GBM based model is not only light weight but also better in prediction accuracy.

The resultant analysis shows that model's classification or prediction performance not only depends upon the dataset but also the model types and approaches. In this study, ensemble

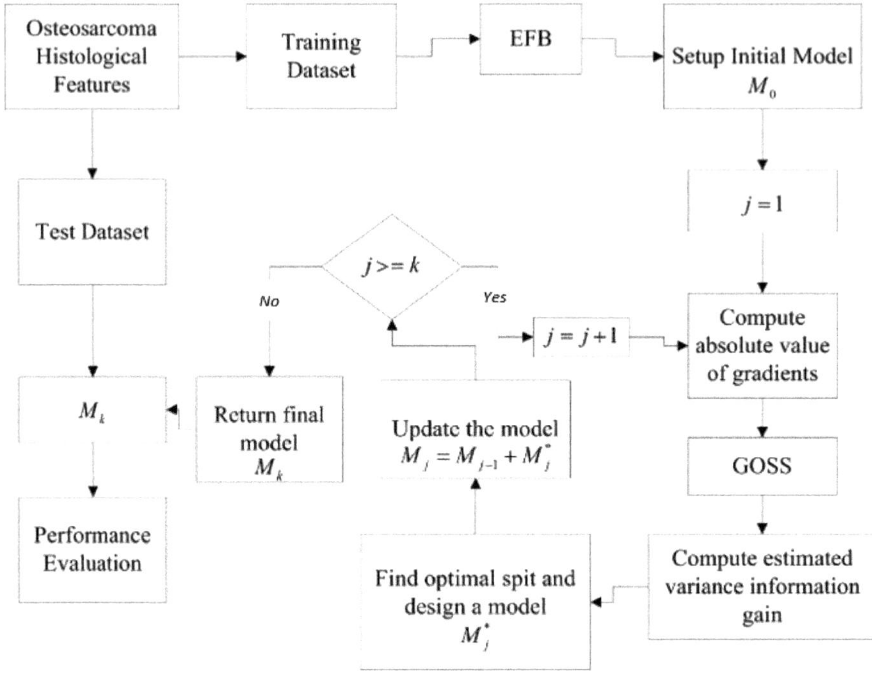

FIG. 13.2 Working sketch of the proposed model.

TABLE 13.1 Results on performance metrics of prediction models.

	Performance metrics					
Prediction models	**Precision**	**Recall**	**F1 score**	**F2 score**	**Fbeta score**	**ROC-AUC**
DT	0.851528	0.851528	0.851528	0.851725	0.853394	0.873722
LDA	0.864628	0.864628	0.864628	0.865024	0.868388	0.900210
MLP	0.877729	0.877729	0.877729	0.877871	0.884803	0.913358
NB	0.746724	0.746724	0.746724	0.748513	0.784775	0.822944
LR	0.746724	0.746724	0.746724	0.748513	0.784775	0.822944
RF	0.976398	0.976398	0.976398	0.976389	0.976387	0.981512
Bagging	0.966783	0.966783	0.966783	0.966755	0.967011	0.973744
AdaBoost	0.969405	0.969405	0.969405	0.969383	0.969382	0.975590
GBoost	0.975524	0.975524	0.975524	0.975518	0.975583	0.980946
XGBoost	0.977272	0.977272	0.977272	0.977262	0.977309	0.981516
Proposed model	0.979020	0.979020	0.979020	0.979022	0.979069	0.984154

learning models and boosting approach are found to be significantly better over the studied machine learning models. This is due to the fact that this approach uses the prediction of multiple additive models which are learned from past prediction error through boosting.

In the Fig. 13.3A–K demonstrating the osteosarcoma cell type prediction performance of all machine learning and ensemble learning techniques. Graph representing the prediction level of each model with histological features versus osteosarcoma cell types. The collected osteosarcoma dataset poses the imbalanced dataset with multiple sub-classes type

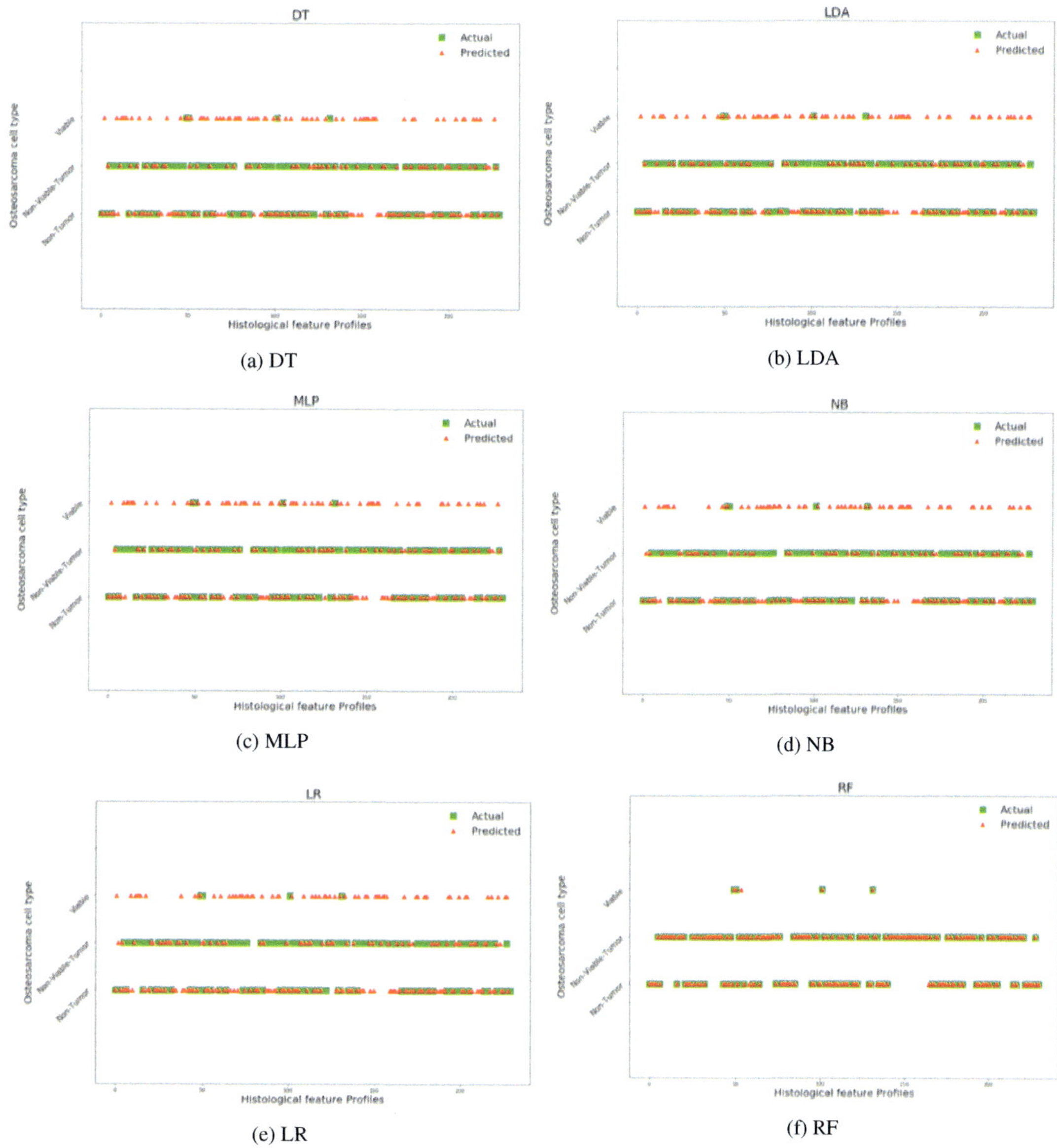

(a) DT (b) LDA (c) MLP (d) NB (e) LR (f) RF

FIG. 13.3 (A–K) Osteosarcoma cell type prediction performance of all the methods. *(Continued)*

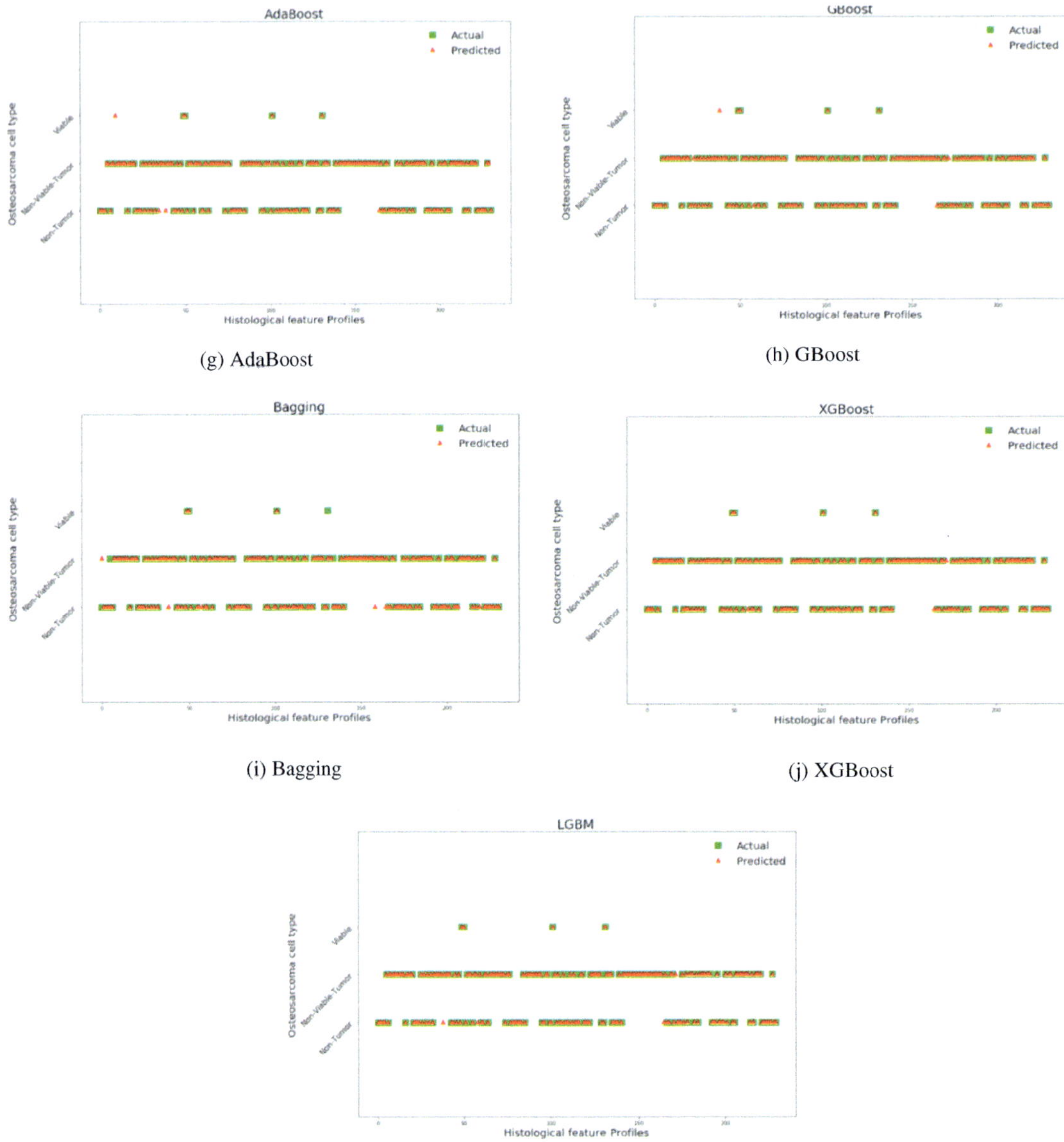

(g) AdaBoost

(h) GBoost

(i) Bagging

(j) XGBoost

(k) Proposed model

FIG. 13.3, Cont'd

such as non-tumor, viable tumor, and non-viable tumor. In Fig. 13.3A, it is clearly visible that actual features and predicted values differ a lot. The performance of the proposed model (Fig. 13.3K) is found to be better than that of the other machine learning (Fig. 13.3A–E) and ensemble learning models (Fig. 13.3F–J). Further, the ROC-AUC analysis on the prediction performances is conducted and demonstrated in Fig. 13.4A–K. The

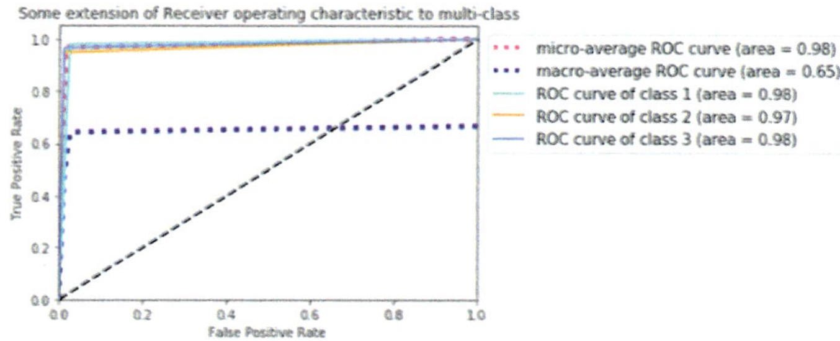

(a) DT

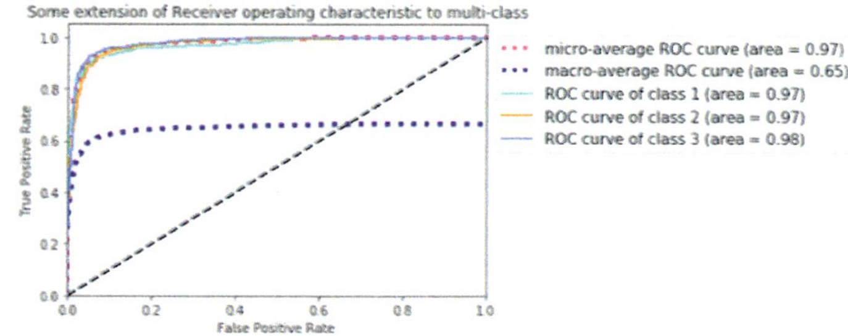

(b) LDA

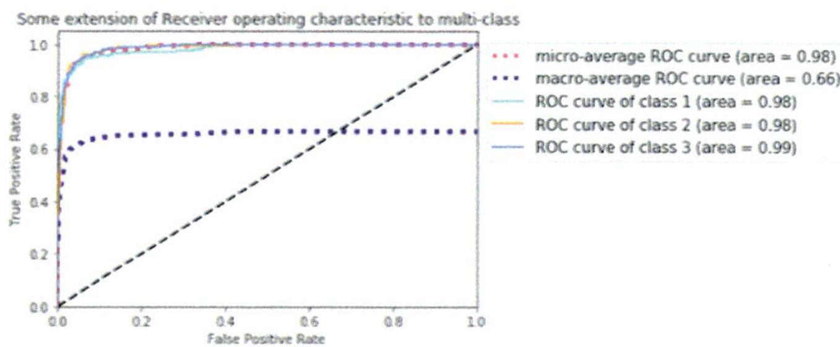

(c) MLP

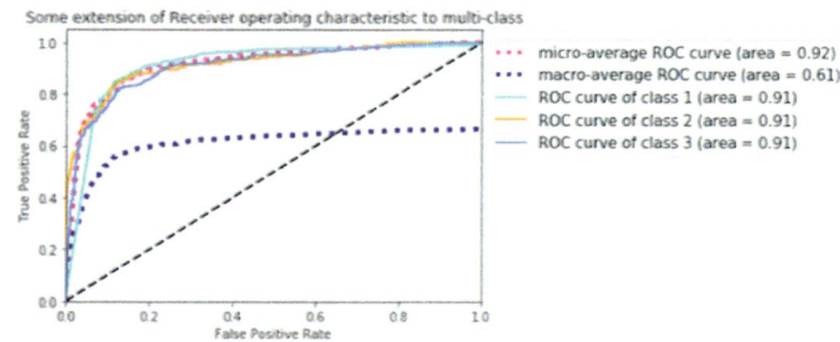

(d) NB

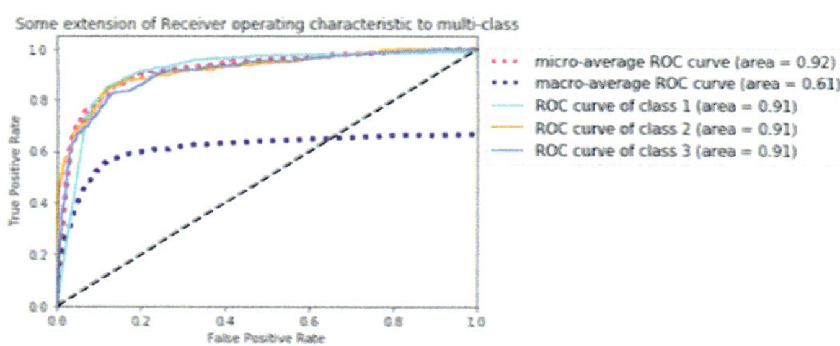

(e) LR

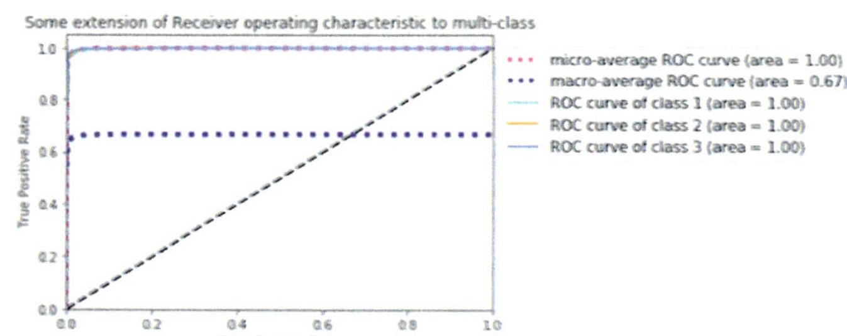

(f) RF

FIG. 13.4 (A–K) AUC-ROC analysis on all the methods. *(Continued)*

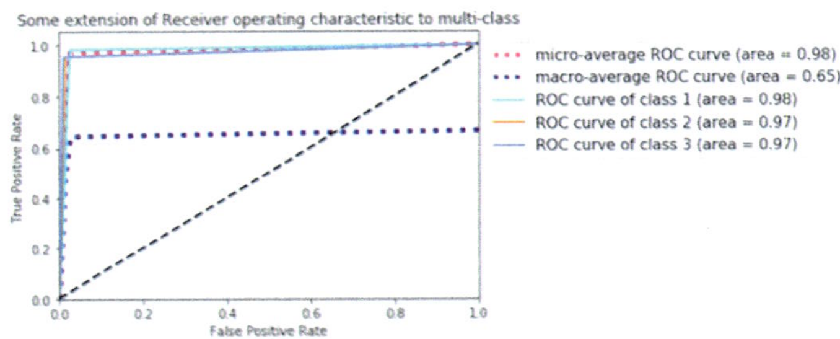

(g) AdaBoost

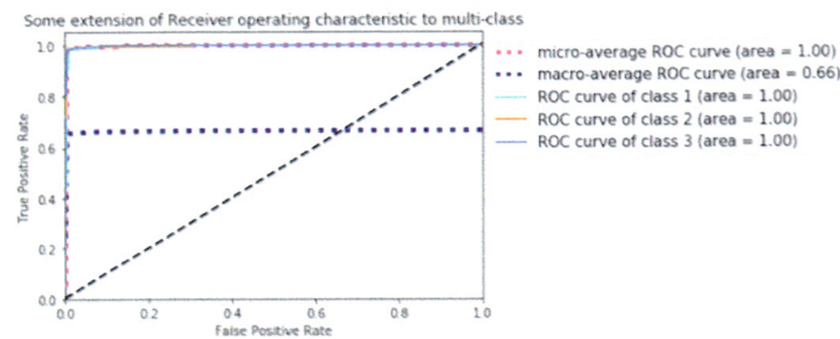

(h) GBT

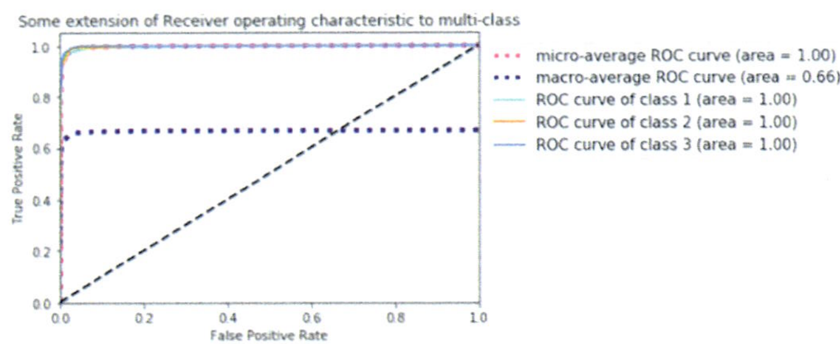

(i) Bagging

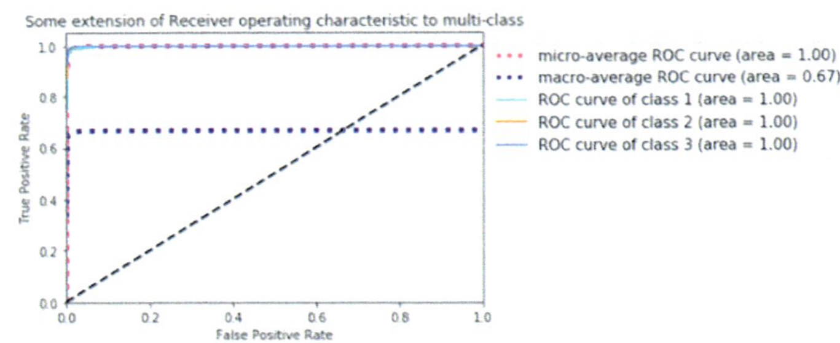

(j) XGBoost

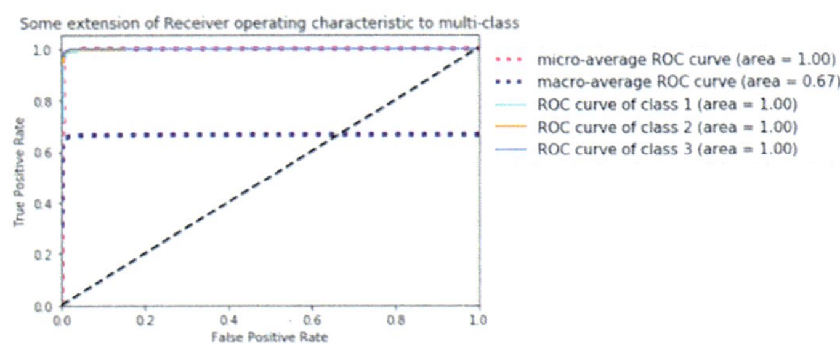

(k) Proposed model

FIG. 13.4, Cont'd

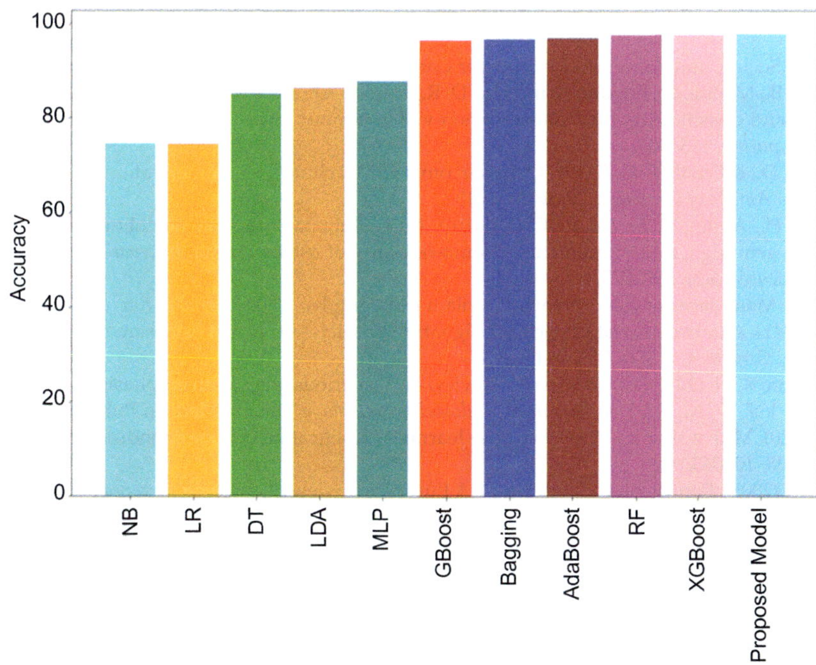

FIG. 13.5 Overall comparison of studied models.

micro-average ROC and macro-average ROC area of the proposed model (Fig. 13.4K) are obtained as 1.00 and 0.67, respectively, which is better than that of other studied models. The overall comparison on prediction accuracies of all the models has been demonstrated in Fig. 13.5.

Conclusion

Early detection of osteosarcoma cells is not only essential for patent health but also helpful for medical practitioners to take decisions. In this line, the proposed model may act like a decision support system for critical decision making related to osteosarcoma. In the close observation, it has been observed that the ensemble learning models have dominated performance over studied machine learning models. However, the AdaBoost, GBoost, XGBoost, and the proposed Light-GBM boosting approach have better performance over other studied ensemble learning approaches such as Bagging and RF. On an overall comparison, the proposed model is found to be better over the other studied models. This work may be used as a decision support system and may provide a framework for study of other types of cancer disease in future. Further, this model can be extended for other types of diseases, which will be certainly helpful for medical practitioners for decision making.

References

American Cancer Society and American Institute for Cancer, n.d. https://www.cancer.org/cancer/osteosarcoma/.

Arunachalam, H. B., Mishra, R., Bogdan Armaselu, D. R., Daescu, O., & Martinez, M. (2017). Computer aided image segmentation and classification for viable and non-viable tumour identification in osteosarcoma. *Pacific Symposium on Biocomputing*, 195–206.

Bielack, S., Carrle, D., & Casali, P. G. (2009). Osteosarcoma: ESMO clinical recommendations for diagnosis, treatment and follow-up. *Annals of Oncology, 20*, iv137–iv139.

Chan, K. Y., Zhu, H., Aydin, M. E., & Lau, C. C. (2010). An evolutionary variable neighborhood search for selecting combinational gene signature in predicting chemo-response of osteosarcoma. *International Journal of Information and Systems Sciences*, *6*(3), 259–275.

Clayer, M. (2015). Many faces of osteosarcoma on plain radiographs. *ANZ Journal of Surgery*, *85*(1–2), 22–26.

Deng, Y., et al. (2021). A four-methylated LncRNA signature predicts survival of osteosarcoma patients based on machine learning. *Genomics*, *113*(1), 785–794.

Fox, M. G., & Trotta, B. M. (2013). Osteosarcoma: Review of the various types with emphasis on recent advancements in imaging. In *Vol. 17. No. 02. Seminars in musculoskeletal radiology* Thieme Medical Publishers.

Gitto, S., et al. (2020). MRI radiomics-based machine-learning classification of bone chondrosarcoma. *European Journal of Radiology*, *128*, 109043.

Grimer, R. J., et al. (2005). Periosteal osteosarcoma—A European review of outcome. *European Journal of Cancer*, *41*(18), 2806–2811.

Gupte, C., & Mart, J.-P. S. (2013). The acute swollen knee: Diagnosis and management. *Journal of the Royal Society of Medicine*, *106*(7), 259–268.

Huang, R., et al. (2019). Evaluating and predicting the probability of death in patients with non-metastatic osteosarcoma: A population-based study. *Medical Science Monitor: International Medical Journal of Experimental and Clinical Research*, *25*, 4675.

Huang, B., et al. (2020). Feasibility of multi-parametric magnetic resonance imaging combined with machine learning in the assessment of necrosis of osteosarcoma after neoadjuvant chemotherapy: A preliminary study. *BMC Cancer*, *20*, 1–9.

Im, H.-J., McIlwain, S., Ong, I., Lee, I., et al. (2017). Prediction of response to neoadjuvant chemotherapy using machine learning algorithm trained by baseline FDG-PET textural parameters in osteosarcoma. *Journal of Nuclear Medicine*, *58*(supplement 1), 44.

Jeong, S. Y., Kim, W., Byun, B. H., Kong, C.-B., Song, W. S., Lim, I., Lim, S. M., & Woo, S.-K. (2019). Prediction of chemotherapy response of osteosarcoma using baseline 18F-FDG textural features machine learning approaches with PC. *Contrast Media & Molecular Imaging*, 3515080. 7 pages https://doi.org/10.1155/2019/3515080.

Jiang, J., et al. (2019). *An artificial intelligence model for predicting the 5-year survival status of osteosarcoma patients based on the SEER database and XGBoost algorithm*. Available at SSRN 3420374.

Ke, G., et al. (2017). Lightgbm: A highly efficient gradient boosting decision tree. *Advances in Neural Information Processing Systems*, *30*, 3146–3154.

Klein, M. J., & Siegal, G. P. (2006). Osteosarcoma: Anatomic and histologic variants. *American Journal of Clinical Pathology*, *125*(4), 555–581.

Lau, C. C. (2009). Molecular classification of osteosarcoma. *Cancer Treatment and Research*, *152*, 459–465.

Noone, A. M., et al. (2018). *SEER cancer statistics review, 1975–2015. 4*. Bethesda, MD: National Cancer Institute.

Pei, Y., et al. (2019). microRNA-211 regulates cell proliferation, apoptosis and migration/invasion in human osteosarcoma via targeting EZRIN. *Cellular & Molecular Biology Letters*, *24*(1), 1–14.

Spina, V., Montanari, N., & Romagnoli, R. (1998). Malignant tumors of the osteogenic matrix. *European Journal of Radiology*, *27*, S98–S109.

Temple, H. T., et al. (2000). Clinical outcome of 38 patients with juxtacortical osteosarcoma. *Clinical Orthopaedics and Related Research (1976–2007)*, *373*, 208–217.

The Cancer Imaging Archive, n.d. https://wiki.cancerimagingarchive.net/pages/viewpage.action?pageId=52756935 [Accessed 5 May 2021].

Wakamatsu, T., et al. (2019). Prognostic implication of adjuvant/neoadjuvant chemotherapy consisting of doxorubicin and ifosfamide in patients with extraskeletal osteosarcoma. *International Journal of Clinical Oncology*, *24*(10), 1311–1319.

Wu, G., & Zhang, M. (2020). A novel risk score model based on eight genes and a nomogram for predicting overall survival of patients with osteosarcoma. *BMC Cancer*, *20*, 1–12.

Zambo, I., & Veselý, K. (2014). WHO classification of tumours of soft tissue and bone 2013: The main changes compared to the 3rd edition. *Ceskoslovenskapatologie*, *50*(2), 64–70.

Zhi Li, S. M., Soroushmehr, R., Hua, Y., Mao, M., Qiu, Y., & Najarian, K. (2017). *Classifying osteosarcoma patients using machine learning approaches*. 978-1-5090-2809-2/17/$31.00 IEEE.

CHAPTER

14

Deep learning-based computer-aided cervical cancer diagnosis in digital histopathology images

Pandia Rajan Jeyaraj[a], *Edward Rajan Samuel Nadar*[a], *and Bijaya Ketan Panigrahi*[b]

[a]Department of Electrical and Electronics Engineering, Mepco Schlenk Engineering College, Sivakasi, Tamilnadu, India [b]Department of Electrical Engineering, Indian Institute of Technology Delhi, New Delhi, Delhi, India

Introduction

Cervical cancer is slow-growing high risk disease with the third highest number of deaths in women. Cervical cancer shows reduced progression since it slow-growing (Li et al., 2020). Hence, early detection is made possible by advanced computer-assisted systems (Li et al., 2019). Early treatment increases the survival rate and reduces the death rate (Kudva & Prasad, 2018). Compared to various imaging techniques like Magnetic Resonance Imaging (MRI), Hyperspectral, Computed Tomography (CT), and ultrasound, histopathological imaging is appropriate gold standard for early cancer detection. Several manual tests like cervical cytology which termed as C-pap test, Cervical Intraepithelial Neoplasia called CIN test and human papillomavirus termed as HPV test are used to detect squamous cells in post-cancerous state. By using histopathological imaging techniques, pre-cancerous cells can be clearly identified (De Fauw et al., 2018; Lu & Mandal, 2015; Siriapisith et al., 2019; Sun et al., 2020; Xiong et al., 2019; Zhang et al., 2020). But, histopathological imaging gives multidimensional information with both spatial and temporal information. It requires computer-aided visual analysis system. Compared to the manual testing method, the computer-assisted detection method is highly accurate, simpler, and time-saving process (Goceri et al., 2016; Jia et al., 2017; Samala et al., 2017, 2019).

The advent of deep learning computation by a convolution neural network makes a feature descriptor. This method does not require annotation, and it can share learned features by

https://doi.org/10.1016/B978-0-323-85240-1.00003-1

patch transfer learning. The generalization of this method for histopathological imaging analysis can be carried out by three simple processes. The first process is domain adaptation for knowing target detection and segmentation. Without having annotation, each image patches obtain domain adaptation. The second process is transfer learning. Transfer learning makes the pre-trained network. In this process, the fine-tuned and pre-trained network predict global features on a large number of image data. In the final process, the histopathological image is multi-view time-lapsed information identified by knowledge transferred pre-trained deep convolutional neural network. This feature extraction and fusion method increases the accuracy of final prediction.

In this chapter, we explain a deep learning framework to detect global feature by transfer learning. Multiple instance learning from a large number of images are fused and transferred to the testing image for image classification. The main contribution of this chapter is as follows;

- To classify histopathological images for early detection of cervical cancer by using global features.
- Transfer learning is used to pre-train and to extract high-level features by ResNet discriminative patch instances extractor.
- Image level fully trained feature descriptors are formed for histopathological image analysis.

Fig. 14.1 shows the overview of the proposed learning and classification scheme. The remaining section of this chapter is structured as follows: "Related work in cervical cancer detection" section summarizes the review of existing cervical cancer detecting techniques. "Proposed methodology" section explains the proposed methodology for classification.

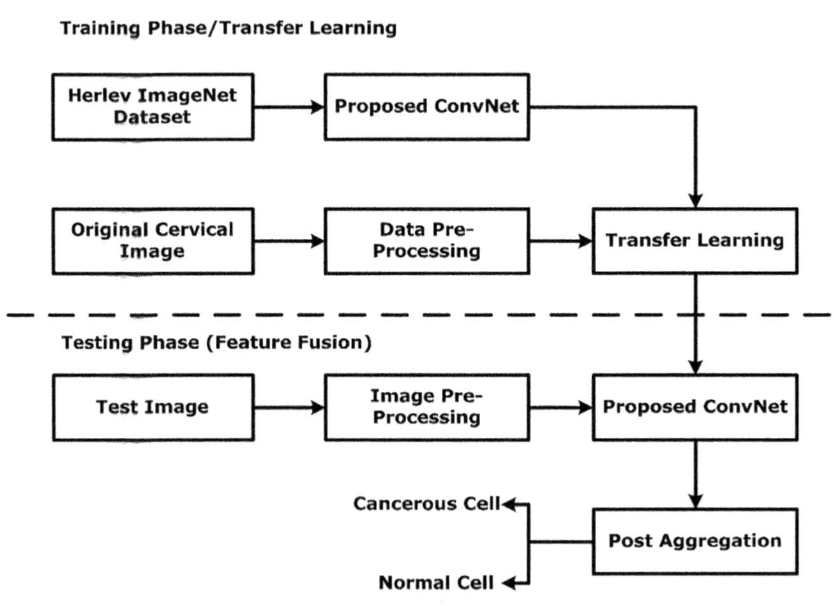

FIG. 14.1 Block diagram of proposed learning and classification scheme.

"Description of dataset for experimental setup" section elaborates on the testing dataset and experimental setting. "Evaluation method and result obtained" section describes the result obtained and evaluation of the proposed framework. Finally, "Conclusion" section draws conclusion of the proposed deep learning framework.

Related work in cervical cancer detection

Many research studies have witnessed the computer-assisted diagnosis system by the deep learning algorithm.

Consider the work; Bernard et al. (2018) designed a deep learning network for liver cancer detection. This method is based on region-based scheme by General Adversial network. Azizi et al. (2017, 2018), Bizopoulos and Koutsouris (2019), Jeyaraj and Nadar (2019), and Shao et al. (2020) used domain generalization with the fully trained network. The unseen image patches have been identified by the feature descriptor. The probabilistic feature descriptor is used in MRI segmentation and CT image annotation removal. Altaf et al. (2019), Hussein et al. (2019), Kallenberg et al. (2016), Koohzadi and Charkari (2017), Saha and Chakraborty (2018), Song et al. (2020), and Zhang et al. (2019) used deep learning network with AlexNet architecture for data augmentation in the CT image to reduce data distribution. Their deep learning model is applied in 2D brain MRI augmentation.

Thakur and Jindal (2020) proposed a CNN to classify oral cancer by a region-based patch identifier. The drawbacks in Jeyaraj and Edward Rajan (2019) and Trivizakis et al. (2019) methods are higher number of training image for paucity of labeled image. Soleimani et al. (2021) used transfer learning for surface crack detection by visual inspection. Similar texture is predicted by the Multiple Instance Learning (MIL) technique. They proposed an autoencoder for combining texture labels and spatial information.

Compared to the work of Balagourouchetty et al. (2020) with Shen et al. (2020) research work, semi-supervised models with high-quality data annotation have been proposed. Feature fusion like the voting method is used for image information extraction and accuracy improvement. Based on overlapping cervical dataset, Hagerty et al. (2019) proposed a deep learning method for cytoplasm segmentation. By optimization of the extracted feature with threefold cross validation, high selectivity of 91.5% and high accuracy of 94.2% was obtained.

Sari and Gunduz-Demir (2019) used a non-linear dimension reduction technique for cervical cell classification. They computed pre-processing for image pixel-level classification. But, this method is suitable for manual screening and not suitable for computer-assisted cancer screening methods. Javed et al. (2020) used redundant information on training and testing image in convolution operation. This method does not represent complex discriminative information.

Recently, Sevakula et al. (2019) developed representation learning for cervical cancer detection and classification. This intermitted patch information is used to train the proposed deep learning method. By representation learning, they achieved high accuracy of 92.1% compared to the SVM model. Wang et al. (2020) used a deep learning model for classification of lung disease with different categories of infection. They used CT imaging and Wang, Tang, et al. (2020) used MRI imaging for brain vessel segmentation. They obtained low accuracy on applying to extract complex feature by transfer learning. Bamunu Mudiyanselage et al. (2020) used high speed system with multi-core process for developing ConvNet for classification of oral cancer. High-speed computation with annotation of image to overcome the knowledge-level learning. Since the data obtained in the infected region are more generic to problem under consideration. To reduce overfitting problem, Jeyaraj and Edward Rajan (2020) used Principal Component Analysis for data dimension reduction. Fine-tuning of the proposed deep neural network allows boosting the performance of large-scale image data.

Zhang, Le, et al. (2017) proposed an automated cell detection system for detecting abnormal nuclei. The cropped block image is used to detect the presence of cytoplasm. The Herlev dataset with pap smear images is used for validation of the proposed computer assisted deep learning model. Bengtsson and Malm (2014) and Zhang, Kong, et al. (2017) reports most of death in low- to medium-income countries are due to lack of early screening. Moreover, manual screening by the acetic acid test reduces the screening accuracy in early stage.

Proposed methodology

The proposed method consists of two-phase processing. Fig. 14.2 shows the block diagram of the proposed learning method. By using global labels, the proposed deep convolution neural network is trained. Each multiple instance learning is obtained by the transfer learning strategy. Multiple instance learning is used to identify discriminative feature by utilizing both negative and positive instances. The proposed method can be explained in the upcoming three sub-sections.

Data processing

In this method, each image is pre-segmented by at least cytoplasm segmentation as indicated by the TBS rule. Each nucleus feature contains cytology abnormalities. Hence, we extract the image patch of size $m \times m$ with nucleus centroid. For extracting nucleus image patches, Laplacian-of-Gaussian used. For increasing accuracy of classification, data augmentation is used for rotationally invariant cervical cell. By using the Herlev dataset, the pre-training of the proposed ConvNet is initiated. The other image patches are avoided by zero padding. In the Herlev dataset, approximately four-time abnormal cells are present compared to normal cells. For getting higher accuracy and increased sensitivity, correct classification of abnormal cells is the practical point of view. Hence, for training, the proposed ConvNet uses the Herlev dataset with a higher number of abnormal patches than normal patches. This Herlev dataset is a whole side image that contains 10000×10000 pixels for direct

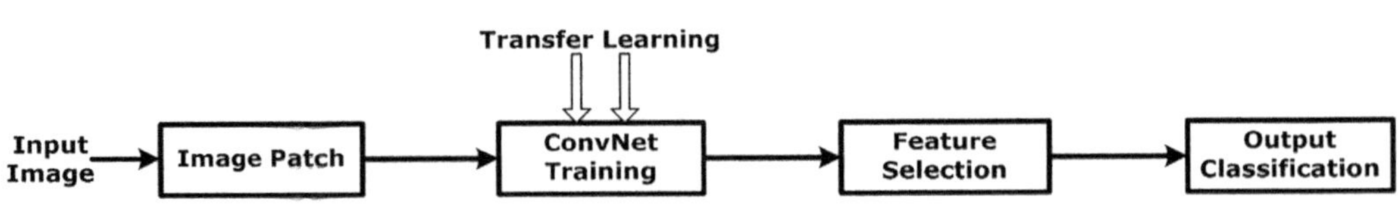

FIG. 14.2 Block diagram of the proposed learning method.

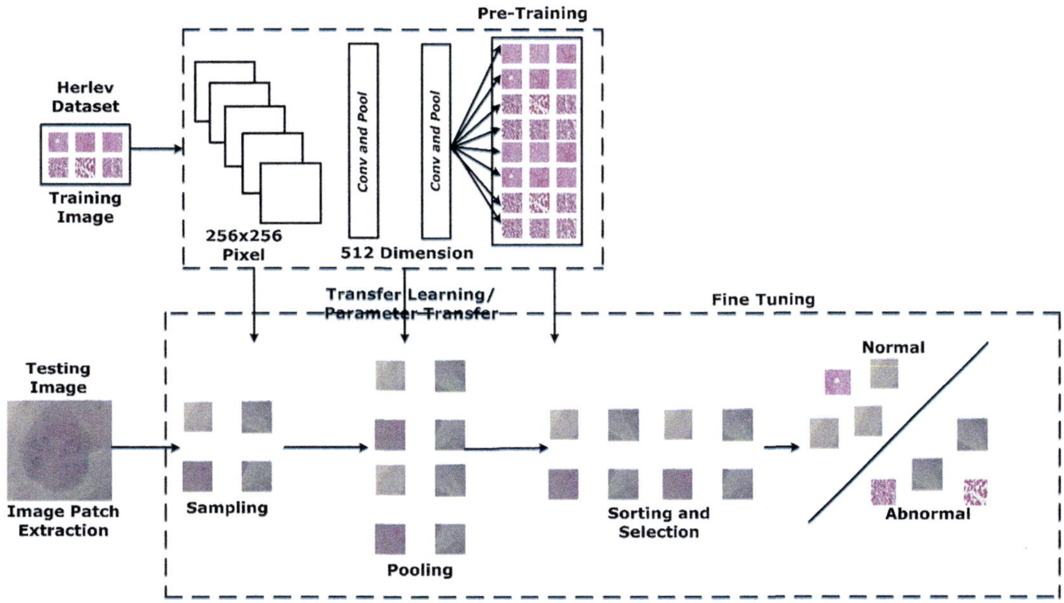

FIG. 14.3 Architecture of proposed deep learning algorithm with pre-training and transfer learning.

analysis. Fig. 14.3 shows the architecture of the proposed deep convolution neural network with transfer learning for processing histopathological images.

Feature extraction

The RGB color images are converted into HSV color space. For each channel of HSV color, Otsu's method is performed for clustering. Higher computation is obtained by color normalization. Each H and E strain has differences by cutting, embedding, and straining process. To extract higher-level features in complex histopathological images, we use ResNet-50 with multilayer perceptron network.

The pre-processed while slide image patch histopathological image patch are presented as input to ResNet-50. The ResNet-50 has already transferred knowledge from the Herlev dataset. From the transfer learning, each aggregated feature is obtained by extracting individual patch-level features. This feature aggregating scheme increases the convolution process dimension, which results in global labels being extracted. Max pooling operation is performed on the entire aggregated feature. This max pooling operation is given by Eq. (14.1);

$$P(t) = \left(\frac{1}{K} \sum_{i=1}^{K} M_i \right)^{\frac{1}{p}} \tag{14.1}$$

where K is the number of aggregation and M_i is input feature vector.

Network training

Gaussian distribution is applied to initialize the weight W. The weight W is updated for every instance by the stochastic gradient descent approach. After the third epoch, the initial learning phase is completed. The overfitting of ConvNet is avoided by fixing the momentum value to 0.6. Hence, the weight values are fine-tuned by transfer learning from the presented Herlev dataset. The convolution and max pooling operation of the ConvNet network are considered as the base network. By this process, more discriminative features are obtained. Multiple instances of the learning strategy eliminate the unrelated features. By selecting and sorting of patch-level features, each aggregated feature separates the normal and abnormal histopathological images.

The sorted vector M_{sort} is obtained by pixelwise separation of entire feature vector M_i.

$$M_{sort} = (M_1, M_2 \ldots M_p) \tag{14.2}$$

where i is the dimension of the feature vector.

Network classification

By the sorting process, each image-level feature is arranged and sorted and selected by Multilayer Perceptron (MP) architecture. This multilayer perceptron architecture for output classification is shown in Fig. 14.4. The MP act as the final classifier. This MP contains 300 layers with 150 neurons and fully connected architecture. The ReLu activation function is used to classify histopathological images. These two layers predict the abnormal cell and normal cell in histopathological images.

Description of dataset for experimental setup

For training, the proposed ConvNet with whole slide image was used. The training standard has abnormal image patches and normal cervical image patch. The Herlev dataset was developed for publicity by different straining methods and different image conditions with

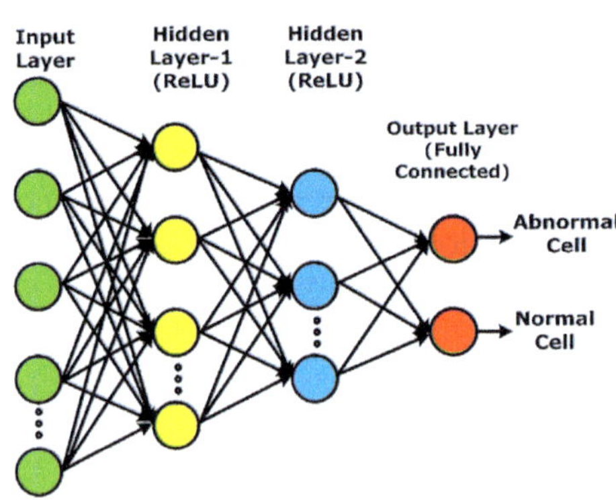

FIG. 14.4 Proposed output classification by perceptron network.

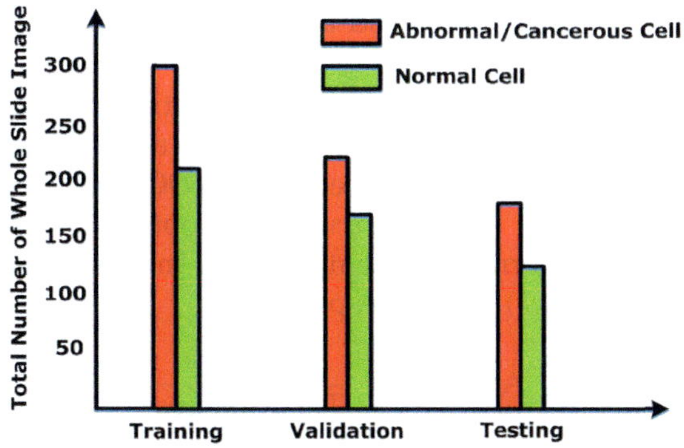

FIG. 14.5 Graphical description of images used for training, validation, and testing.

various slides. In this work, we used the Herlev image available in WWW.Mde-lab.argeam.gr using high resolution of 0.5 μm per pixel digital camera and microscope pap straining with acetic acid are used for strain preparation. A total of 565 cervical cancer patients were found. The scanning level was 0.8 μm/pixel and 48× magnification was carried out. Each image was divided into 0.65 training 0.1 validation and 0.25 testing dataset. Fig. 14.5 shows the graphical description of the dataset used for training proposed deep ConvNet.

For training, the images are regularized to avoid insufficient information. Insufficient images are discarded by L2-regularization. Adam optimizer is adopted to increase the efficiency of the proposed Deep ConvNet. For transfer learning, we kept the learning rate to 0.004, with a batch size of 32. TensorFlow was used to implement the proposed ConvNet. Table 14.1 show the details of training conditions adopted for proposed transfer learning.

Experimental setup

The proposed network was run on Nvidia GFX GPU X-series with 8 GB memory, 1 TB hard disk. For each 512 × 512, training images are cropped to 256 × 256 with the stochastic gradient method for a minimum value of batch size. The proposed deep ConvNet consists of one input layer, five convolution and three max pooling operation, and fully connected output layer.

TABLE 14.1 Herlev training dataset condition.

Category	Image patch condition	Image count
Normal	Squamous carcinoma	45
Normal	Superficial epithelial	65
Abnormal	Severe squamous	115
Abnormal	Dysplasia	125

Evaluation method and result obtained

Three-fold cross-validation is used on training and testing. For each three iterations, two are used for convolution in training and one for validation and remaining one for testing. Performance indexes like accuracy, sensitivity, specificity, ROC, and F1-score were obtained. Receiver Operating Characteristic (ROC) curves are obtained for predicting abnormal histopathological images from normal histopathological images. Accuracy, sensitivity, specificity, and F1 Score are Eqs. (14.3)–(14.6), respectively.

$$\text{Accuracy} = \frac{\text{TP} + \text{TN}}{\text{Total images}} \tag{14.3}$$

$$\text{Sensitivity} = \frac{\text{True positive}}{\text{True positive} + \text{false positive}} \tag{14.4}$$

$$\text{Specificity} = \frac{\text{True positive}}{\text{True positive} + \text{false negative}} \tag{14.5}$$

$$F1 - \text{Score} = \frac{2 \times \text{Sensitivity} \times \text{Specificity}}{\text{Sensitivity} + \text{specificity}} \tag{14.6}$$

From the work of Song et al. (2020) and Shen et al. (2020), the texture classification of the same Herlev image dataset is used for validation. The testing dataset contains 115 images. By transfer learning of ImageNet, the multiple instances learning defect abnormal cervical cell using global labels. The classification score of 0.974 accuracy and 0.962 sensitivity and 0.983 specificity scores are obtained. Fig. 14.6 shows the confusion matrix of high true positive

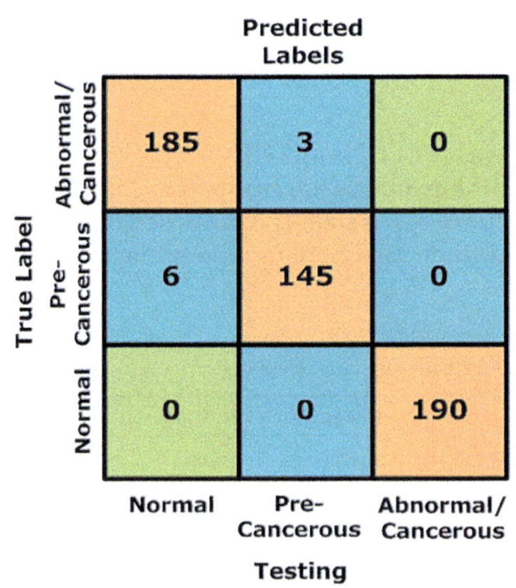

FIG. 14.6 Confusion matrix for testing dataset.

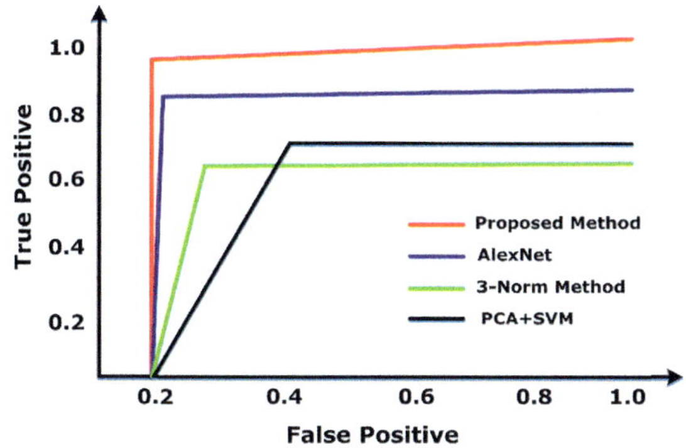

FIG. 14.7 Classification performance comparison by ROC curves under different methods.

TABLE 14.2 Classification of the Herlev dataset by the proposed method with existing methods.

Algorithm used	Reference	Accuracy	Sensitivity	Specificity	*F*1 score	ROC
ANN	De Fauw et al. (2018)	0.852	0.794	0.816	–	0.823
PCA+SVM	Bernard et al. (2018)	00925	0.932	0.923	–	0.901
ConvNet	Sari and Gunduz-Demir (2019)	0.946	0.915	0.947	0.92	0.947
Proposed ConvNet	–	0.974	0.962	0.983	0.97	0.992

for cervical cancer cells. This validates the extraction of higher-level features. The proposed method reduces the false negative rate. The ROC for finding discriminative feature of the proposed method and AlexNet (Shao et al., 2020) is shown in Fig. 14.7.

Table 14.2 gives the summary of the score obtained on the proposed method and other existing methods. The Max pooling method improves the efficiency of detection. In terms of accuracy of abnormal cell classification, our method outperforms other methods, and from ROC viewpoint, high image labels are obtained. This validates the efficiency in detection of abnormal regions. High sensitivity confirms the non-existence of false negative in both normal and abnormal cervical cell classification.

In future, the robustness of our method is increased by including overlapping of nucleus and cytoplasm. This kind of artifact is minor in image samples. Hence, training can be carried out by these images to increase robustness.

Conclusion

In this chapter, cancerous cervical cell classification in histopathological images by transfer learning base convolution neural network is proposed. Using global label and transfer learning approach, robust performance was obtained. The proposed method has reduced annotation and online classification by real-time imaging systems. The transfer learning (pretrained) operation obtained a high accuracy of 0.974 and 0.983 specificity in abnormal cervical cell detection. The fusion of feature produces high sensitivity of 0.962 to detect abnormal cells. Hence, the proposed system provides an automated early screening of cervical cancer by assisting healthcare service providers in real time.

Acknowledgments

The authors would like to thank the Indian academy of science, New Delhi, for providing research fellowship in the Department of Electrical Engineering, Indian Institute of Technology, New Delhi and Department of Electrical and Electronics Engineering, Mepco Schlenk Engineering College, Sivakasi, India, for providing us the necessary research facilities to carry out this work.

References

Altaf, F., Islam, S. M. S., Akhtar, N., & Janjua, N. K. (2019). Going deep in medical image analysis: Concepts methods challenges and future directions. *IEEE Access*, *7*, 99540–99572.

Azizi, S., et al. (2017). Detection and grading of prostate cancer using temporal enhanced ultrasound: Combining deep neural networks and tissue mimicking simulations. *International Journal of Computer Assisted Radiology and Surgery*, *12*(8), 1293–1305.

Azizi, S., et al. (2018). Deep recurrent neural networks for prostate cancer detection: Analysis of temporal enhanced ultrasound. *IEEE Transactions on Medical Imaging*, *37*(12), 2695–2703. https://doi.org/10.1109/TMI.2018.2849959.

Balagourouchetty, L., Pragatheeswaran, J. K., Pottakkat, B., & Ramkumar, G. (2020). GoogLeNet-based ensemble FCNet classifier for focal liver lesion diagnosis. *IEEE Journal of Biomedical and Health Informatics*, *24*(6), 1686–1694.

Bamunu Mudiyanselage, T. K., Xiao, X., Zhang, Y., & Pan, Y. (2020). Deep fuzzy neural networks for biomarker selection for accurate cancer detection. *IEEE Transactions on Fuzzy Systems*, *28*(12), 3219–3228. https://doi.org/10.1109/TFUZZ.2019.2958295.

Bengtsson, E., & Malm, P. (2014). Screening for cervical cancer using automated analysis of pap-smears. *Computational and Mathematical Methods in Medicine*, *2014*, 1–12.

Bernard, O., et al. (2018). Deep learning techniques for automatic MRI cardiac multi-structures segmentation and diagnosis: Is the problem solved? *IEEE Transactions on Medical Imaging*, *37*(11), 2514–2525. https://doi.org/10.1109/TMI.2018.2837502.

Bizopoulos, P., & Koutsouris, D. (2019). Deep learning in cardiology. *IEEE Reviews in Biomedical Engineering*, *12*, 168–193.

De Fauw, J., et al. (2018). Clinically applicable deep learning for diagnosis and referral in retinal disease. *Nature Medicine*, *24*(9), 1342–1350.

Goceri, E., Shah, Z. K., Layman, R., Jiang, X., & Gurcan, M. N. (2016). Quantification of liver fat: A comprehensive review. *Computers in Biology and Medicine*, *71*, 174–189.

Hagerty, J. R., et al. (2019). Deep learning and handcrafted method fusion: Higher diagnostic accuracy for melanoma dermoscopy images. *IEEE Journal of Biomedical and Health Informatics*, *23*(4), 1385–1391. https://doi.org/10.1109/JBHI.2019.2891049.

Hussein, S., Kandel, P., Bolan, C. W., Wallace, M. B., & Bagci, U. (2019). Lung and pancreatic tumor characterization in the deep learning era: Novel supervised and unsupervised learning approaches. *IEEE Transactions on Medical Imaging*, *38*(8), 1777–1787.

Javed, S., Mahmood, A., Werghi, N., Benes, K., & Rajpoot, N. (2020). Multiplex cellular communities in multi-gigapixel colorectal Cancer histology images for tissue phenotyping. *IEEE Transactions on Image Processing*, *29*, 9204–9219.

Jeyaraj, P. R., & Edward Rajan, S. (2019). Deep Boltzmann machine algorithm for accurate medical image analysis for classification of cancerous region. *IET Cognitive Computation and Systems*, *1*(3), 85–90. https://doi.org/10.1049/ccs.2019.0004.

Jeyaraj, P. R., & Edward Rajan, S. (2020). Dynamic image reconstruction and synthesis framework using deep learning algorithm. *IET Image Processing*, *14*(7), 1219–1226. https://doi.org/10.1049/iet-ipr.2019.0900.

Jeyaraj, P. R., & Nadar, E. R. S. (2019). Computer-assisted medical image classification for early diagnosis of oral cancer employing deep learning algorithm. *Journal of Cancer Research and Clinical Oncology*, *145*(4), 829–837. https://doi.org/10.1007/s00432-018-02834-7.

Jia, Z. P., Huang, X. Y., Chang, E. I. C., & Xu, Y. (2017). Constrained deep weak supervision for histopathology image segmentation. *IEEE Transactions on Medical Imaging*, *36*(11), 2376–2388.

Kallenberg, M., et al. (2016). Unsupervised deep learning applied to breast density segmentation and mammographic risk scoring. *IEEE Transactions on Medical Imaging*, *35*(5), 1322–1331. https://doi.org/10.1109/TMI.2016.2532122.

Koohzadi, M., & Charkari, N. M. (2017). Survey on deep learning methods in human action recognition. *IET Computer Vision*, *11*(8), 623–632.

Kudva, V., & Prasad, K. (2018). Pattern classification of images from acetic acid–based cervical cancer screening: A review. *Critical Reviews in Biomedical Engineering*, *46*(2), 117–133.

Li, W., et al. (2019). Path R-CNN for prostate cancer diagnosis and Gleason grading of histological images. *IEEE Transactions on Medical Imaging*, *38*(4), 945–954.

Li, Y., et al. (2020). Computer-aided cervical cancer diagnosis using time-lapsed colposcopic images. *IEEE Transactions on Medical Imaging*, *39*(11), 3403–3415. https://doi.org/10.1109/TMI.2020.2994778.

Lu, C., & Mandal, M. (2015). Automated analysis and diagnosis of skin melanoma on whole slide histopathological images. *Pattern Recognition*, *48*(8), 2738–2750.

Saha, M., & Chakraborty, C. (2018). Her2Net: A deep framework for semantic segmentation and classification of cell membranes and nuclei in breast cancer evaluation. *IEEE Transactions on Image Processing*, *27*(5), 2189–2200. https://doi.org/10.1109/TIP.2018.2795742.

Samala, R. K., Chan, H.-P., Hadjiiski, L. M., Helvie, M. A., Cha, K. H., & Richter, C. D. (2017). Multi-task transfer learning deep convolutional neural network: Application to computer-aided diagnosis of breast cancer on mammograms. *Physics in Medicine and Biology*, *62*(23), 8894–8908.

Samala, R. K., Chan, H.-P., Hadjiiski, L., Helvie, M. A., Richter, C. D., & Cha, K. H. (2019). Breast cancer diagnosis in digital breast tomosynthesis: Effects of training sample size on multi-stage transfer learning using deep neural nets. *IEEE Transactions on Medical Imaging*, *38*(3), 686–696. https://doi.org/10.1109/TMI.2018.2870343.

Sari, C. T., & Gunduz-Demir, C. (2019). Unsupervised feature extraction via deep learning for histopathological classification of colon tissue images. *IEEE Transactions on Medical Imaging*, *38*(5), 1139–1149. https://doi.org/10.1109/TMI.2018.2879369.

Sevakula, R. K., Singh, V., Verma, N. K., Kumar, C., & Cui, Y. (2019). Transfer learning for molecular cancer classification using deep neural networks. *IEEE Transactions on Computational Biology and Bioinformatics*, *16*(6), 2089–2100. https://doi.org/10.1109/TCBB.2018.2822803.

Shao, Y., Wang, J., Wodlinger, B., & Salcudean, S. E. (2020). Improving prostate cancer (PCa) classification performance by using three-player minimax game to reduce data source heterogeneity. *IEEE Transactions on Medical Imaging*, *39*(10), 3148–3158.

Shen, T., Wang, J., Gou, C., & Wang, F.-Y. (2020). Hierarchical fused model with deep learning and type-2 fuzzy learning for breast cancer diagnosis. *IEEE Transactions on Fuzzy Systems*, *28*(12), 3204–3218. https://doi.org/10.1109/TFUZZ.2020.3013681.

Siriapisith, T., Kusakunniran, W., & Haddawy, P. (2019). 3D segmentation of exterior wall surface of abdominal aortic aneurysm from CT images using variable neighborhood search. *Computers in Biology and Medicine*, *107*, 73–85.

Soleimani, M., Vahidi, A., & Vaseghi, B. (2021). Two-dimensional Stockwell transform and deep convolutional neural network for multi-class diagnosis of pathological brain. *IEEE Transactions on Neural Systems and Rehabilitation Engineering*, *29*, 163–172.

Song, R., Li, T., & Wang, Y. (2020). Mammographic classification based on XGBoost and DCNN with multi features. *IEEE Access*, *8*, 75011–75021.

Sun, C., Xu, A., Liu, D., Xiong, Z., Zhao, F., & Ding, W. (2020). Deep learning-based classification of liver cancer histopathology images using only global labels. *IEEE Journal of Biomedical and Health Informatics*, *24*(6), 1643–1651. https://doi.org/10.1109/JBHI.2019.2949837.

Thakur, A., & Jindal, N. (2020). Hybrid deep learning and machine learning approach for passive image forensic. *IET Image Processing*, *14*(10), 1952–1959.

Trivizakis, E., et al. (2019). Extending 2-D convolutional neural networks to 3-D for advancing deep learning cancer classification with application to MRI liver tumor differentiation. *IEEE Journal of Biomedical and Health Informatics*, *23*(3), 923–930. https://doi.org/10.1109/JBHI.2018.2886276.

Wang, X., Tang, F., Chen, H., Luo, L., Tang, Z., Ran, A.-R., Cheung, C. Y., & Heng, P.-A. (2020). UD-MIL: Uncertainty-driven deep multiple instance learning for OCT image classification. *IEEE Journal of Biomedical and Health Informatics*, *24*(12), 3431–3442.

Wang, X., et al. (2020). Weakly supervised deep learning for whole slide lung cancer image analysis. *IEEE Transactions on Cybernetics*, *50*(9), 3950–3962. https://doi.org/10.1109/TCYB.2019.2935141.

Xiong, Z., Fedorov, V. V., Fu, X., Cheng, E., Macleod, R., & Zhao, J. (2019). Fully automatic left atrium segmentation from late gadolinium enhanced magnetic resonance imaging using a dual fully convolutional neural network. *IEEE Transactions on Medical Imaging*, *38*(2), 515–524.

Zhang, L., Kong, H., Liu, S., Wang, T., Chen, S., & Sonka, M. (2017). Graph-based segmentation of abnormal nuclei in cervical cytology. *Computerized Medical Imaging and Graphics*, *56*, 38–48.

Zhang, L., Le, L., Nogues, I., Summers, R. M., Liu, S., & Yao, J. (2017). DeepPap: Deep convolutional networks for cervical cell classification. *IEEE Journal of Biomedical and Health Informatics*, *21*(6), 1633–1643. https://doi.org/10.1109/JBHI.2017.2705583.

Zhang, J., Saha, A., Zhu, Z., & Mazurowski, M. A. (2019). Hierarchical convolutional neural networks for segmentation of breast tumors in MRI with application to radiogenomics. *IEEE Transactions on Medical Imaging*, *38*(2), 435–447.

Zhang, L., et al. (2020). Generalizing deep learning for medical image segmentation to unseen domains via deep stacked transformation. *IEEE Transactions on Medical Imaging*, *39*(7), 2531–2540. https://doi.org/10.1109/TMI.2020.2973595.

CHAPTER

15

Deep learning techniques for hepatocellular carcinoma diagnosis

Dillip Kumar Bishi[a], *Priyadarshini Padhi*[a], *Chhabi Rani Panigrahi*[b], *Bibudhendu Pati*[b], *and Chandi Charan Rath*[c]

[a]Department of Biotechnology, Rama Devi Women's University, Bhubaneswar, Odisha, India
[b]Department of Computer Science, Rama Devi Women's University, Bhubaneswar, Odisha, India
[c]Department of Life Science, Rama Devi Women's University, Bhubaneswar, Odisha, India

Introduction

Hepatocellular carcinoma (HCC) is liver cancer, which predominates in the majority of all primary liver cancers (El-Serag & Rudolph, 2007) and is the fourth key cause of cancer-associated mortality globally (Yang et al., 2019). HCC is caused mostly due to excessive alcohol consumption, chronic viral infections leading to cirrhosis, and ingestion of aflatoxins. Due to the high level of heterogeneity and complex etiological factors, precise interpretation and histopathological grading of HCC and tumor staging require keen visual inspection by experienced pathologists. After decades of research, there is still uncertainty in the clinical diagnosis of cancer including HCC and precise identification of tumor-specific signature markers. Conventional diagnostic methods such as ultrasound, X-ray, computed tomography (CT), positron emission tomography (PET), magnetic resonance imaging (MRI), biopsy, and some serological parameter assessments are widely followed by physicians, though the accuracy in the detection of tumor stage is still not practical in most of the tests. HCC is prevalent, particularly in Eastern Asia and sub-Saharan Africa with an estimate of 85% of all HCC cases worldwide (Tang et al., 2018). The etiologic factors of HCC are mainly hepatic cirrhosis and fibrosis, caused by infection from hepatitis B and C virus and several other nonviral inflammatory factors like steroid drugs, aflatoxins, ochratoxin A, and alcohol (Zimmermann, 2016). Unlike the gradual decrease in all other cancer mortality rates observed worldwide,

https://doi.org/10.1016/B978-0-323-85240-1.00013-4

the mortality for HCC is observed to be increasing each year due to the fact that HCC is detected at a later stage of progression as well as lack of therapeutic advances. In this scenario, for improving long-term survival, precise detection of HCC is required. Much of the focus has been given to diagnostic strategies for the identification of early HCC, based on the size and number of hepatic lesions.

In the generation of big data, there is an upsurge in the new approach involving AI to understand the world and making a decision in various fields, including medicine. The concept of AI arose in the 1950s, but it only a few years ago became a trendsetter in the field of science and technology. AI has a subfield called machine learning (ML), which is further upgraded to a higher version of the technology called deep learning (DL). So, basically, DL is known to be a subset of ML and ML as a subset of AI. In Fig. 15.1, a schematic outline of the relationship between AI, ML, and DL has been shown (Huang et al., 2020). Deep learning models use huge datasets, run on specialized computing hardware, and have predominated many classical ML strategies. Several models based on deep learning principles have accomplished a remarkable accuracy at the clinician level with a myriad of cancer diagnostic tasks. As of now, DL algorithms have succeeded in acquiring high fidelity in classifying some of the tumors, including HCC. Nevertheless, much of the research in DL has emphasized the validation of model performance, while few studies have taken the stride of investigating the implications of DL assistance on the diagnostic performance of physicians.

In the proceeding sections of this chapter, the reader will get basic information about the conventional diagnostic methods to evaluate HCC, the fundamentals of DL, common models of DL in healthcare sectors, and how DL can be used as a tool to support clinicians giving a precise diagnostic evaluation of a range of liver lesions and HCC and about challenges and future perspectives of DL in HCC diagnosis. Specific focus has been made on significant developments of CNN-based deep learning system (DLS) in various proof-of-concept style information that classify and differentiate common and malignant hepatic lesions on a multiphasic MRI. Moreover, the robustness of DL-based models on radiomics and multiomics data of HCC in differentiating survival subpopulations of patients into different cohorts is critically analyzed, which might ultimately help categorize various stages of HCC. Further, CNN-based DLS of histopathological test data has been highlighted in differentiating various genetic mutations to categorize and detect HCC. Finally, some ANN validation models have also been discussed for differentiating various stages of HCC such as early-stage HCC from liver cirrhosis.

Contributions

The main contributions of this chapter are highlighted as follows:

- Conventional methods of diagnostic evaluation of HCC are identified.
- Various DL approaches for the diagnostic evaluation of HCC are presented.
- Challenges involved in the diagnostic evaluation of HCC as well as future prospective research in this direction are highlighted.

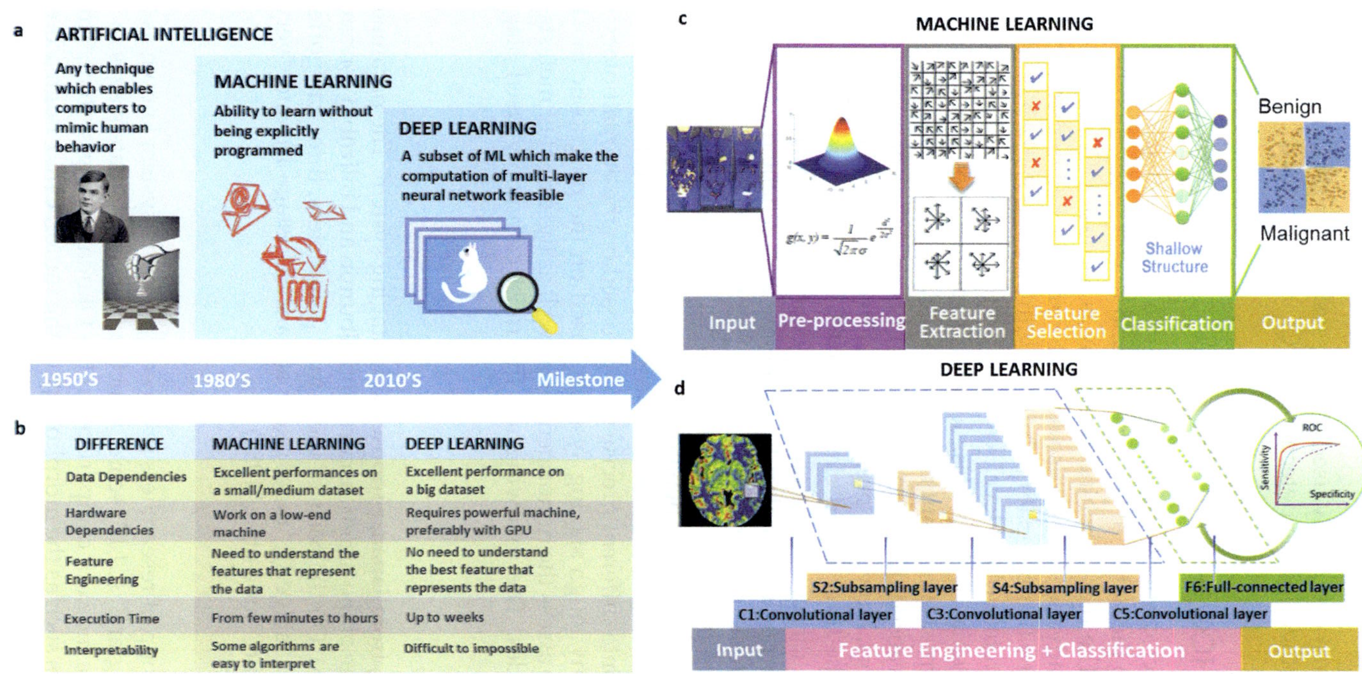

FIG. 15.1 Schematic representation delineating fundamental theories behind artificial intelligence (AI), machine learning (ML), and deep learning (DL) techniques and interrelatedness between them. (A) Definitions of AI, ML, and DL. (B) Major highlights and significant contrasts among ML and DL. (C) Basic sequential algorithms followed for execution of ML innovation in tumor diagnosis. (D) Basic sequential algorithms used by DL models for cancer research and tumor diagnosis. *Adapted with permission from Huang, S., Yang, J., Fong, S., & Zhao, Q. (2020). Artificial intelligence in cancer diagnosis and prognosis: Opportunities and challenges.* Cancer Letters, *471, 61–71. https://doi.org/10.1016/j.canlet.2019.12.007.*

Diagnostic evaluation of HCC

In this section, the most commonly used staging systems for HCC, as well as the conventional screening test method for the detection of HCC are presented.

Clinical features of HCC

Usually, HCC is detected after the manifestation of a tumor with some identifiable clinical features, which come to the notice of both patient and doctor many months after the development of carcinoma. The reason behind this delay is because HCC shows similar clinical features to some other advanced liver diseases like cirrhosis, hepatomegaly, jaundice, viral infections due to hepatitis B and hepatitis C virus, with symptoms like pain in the abdomen, high serum alpha-fetoprotein (AFP) level, vascular bruit in the liver, bleeding inside the gastrointestinal tract, asthenia, weight loss, loss of appetite, anorexia, nausea, and fever (Villanueva, 2019). The most significant risk factors for HCC progression are chronic liver disease and liver cirrhosis caused due to HVB, HVC, and excessive alcohol consumption as well as inherent genetic factors. However, presently, HCC from nonalcoholic fatty liver disease is also not uncommon. The three broad classes of HCC—benign, dysplastic, and malignant lesions—are sometimes difficult to differentiate because of the newly identified subgroups based on the tumor morphology, molecular diversity, and genetic diversity. The most frequent gene mutations in the non proliferative and proliferative class of HCC are attributed to CTNNB1 and TP53 respectively (Villanueva, 2019). Irrespective of the cause and symptoms, diagnosis of different stages of hepatocellular neoplasms is challenging due to the high heterogeneity of the lesions. For accurate diagnosis, prognosis, and treatment, HCC patients need to be categorized into different precise groups based on the etiopathology of the tumor or lesion. Till now, many tough staging systems based on the prognostic markers of HCC have been developed to assess the tumorigenesis as well as the extent of metastasis to distant organs and lymph nodes. The following are the most commonly used staging systems of HCC.

Okuda staging system

In 1984, Okuda et al., from Japan, proposed this staging system. The parameters included in this system are the size of the tumor, ascites, and albumin and bilirubin concentration ($\leq$ or >3 mg/dL in the serum). On the basis of these four parameters, three categories of patients such as "not advanced" (stage I), "moderately advanced" (stage II), and "very advanced" (stage III) are evaluated (Okuda et al., 1984).

CLIP

In 1998, an Italian group of researchers proposed a staging system known as "The Cancer of the Liver Italian Program" (CLIP) score, which aimed to evaluate stages of HCC with a more precise prognostic index in comparison to the Okuda staging system. This system integrates four parameters such as the morphology of the tumor, Child-Pugh score, portal vein thrombosis, and serum level of AFP to categorize patients into seven groups (Manghisi et al., 1998).

BCLC classification

Barcelona clinic liver cancer (BCLC) classification (BCLC) was developed in 1999 (Llovet et al., 1999), categorizing patients into five groups (0, A, B, C, D) on the basis of extension of tumor, physical and functional status of the liver, and symptoms associated with cancer. Stages 0, A, B, C, and D denote very early, early, intermediate, advanced, and terminal stage of HCC respectively. BCLC staging system is regarded as the conventional system for both the American and European Association for the Study of Liver and the European Organization for the Research and Treatment of Cancer and the American Gastroenterology Association (Kinoshita et al., 2015).

GRETCH system

Another staging system of HCC called the "GRETCH system" was proposed by a French group in 1999, which combines five parameters based on a Cox multivariate analytical model to categorize three classes of HCC progression stages in patients. The parameters are bilirubin, AFP, and alkaline phosphatase levels in the serum, Karnofsky index, and ultrasonographic portal hypertension. The three categories of stages of patients are stage A low death risk, stage B moderate death risk, and stage C high death risk (Chevret, 1999).

TNM classification

The TNM system of tumor classification is described as "Tumor, Node, Metastasis," proposed by the American Joint Committee on Cancer and International Union for Cancer Control. As per this system, patients are categorized into four groups based on tumor growth, lymph node involvement, and distant metastasis (Vauthey et al., 2002).

Chines university prognostic index (CUPI)

A Hong Kong group of investigators proposed the staging system CUPI in 2002 (Leung et al., 2002), which combines five features such as ALP, bilirubin, AFP, ascites, and asymptomatic disease on presentation to classify the patients basically into low-, intermediate-, and high-risk categories.

Conventional screening tests for HCC detection

The most common diagnostic protocols for HCC detection include the screening of liver biopsy, serum α-fetoprotein (AFP) as a tumor marker, as well as radiological imaging. Though cytology and histopathology are considered the gold standard for HCC diagnosis, no such universal guidelines for diagnosis exist, somewhat due to the difference in following diagnostic protocols between Western and Eastern institutions (Torzilli et al., 2004). Nonetheless, a few common clinical practices are also followed for prominent distinctions and unvarying conclusions during HCC diagnosis (Bialecki & Di Bisceglie, 2005; Ryder, 2003).

AFP in the diagnosis of HCC

AFP is a glycoprotein, which can represent the size and volume of a tumor or lesion in the liver correlated to its level in the serum. Long ago, AFP was first recognized as a reliable serum biomarker for the preclinical diagnosis of HCC. The yolk sac and liver in the fetal stage generate high levels of AFP, but within 300days after birth, AFP serum level declines to <10ng/dl (Kashyap et al., 2001). Therefore, elevation in the AFP level in the serum indicates an underlying pathology for liver malignancy. However, the principle of AFP elevation in HCC is not universal in all HCC cases (Farinati et al., 2006).

Angiography

Due to the high vascularization of tumors in HCC, angiography has been used commonly for the diagnosis of HCC, but the detection sensitivity of angiography has been questioned, particularly when the tumor size is <2cm in diameter (Bialecki & Di Bisceglie, 2005; Krinsky et al., 2000). Now, the use of angiography is limited to the administration of chemoembolization therapies (Befeler & Di Bisceglie, 2002).

Ultrasound

Ultrasound has been widely used as a screening tool for HCC. But for showing low sensitivity and error-prone prediction value while cirrhosis coexists with HCC, it is now seen to be replaced by two other radiologic imaging methods, CT scan and MRI, for detecting tumors. The use of sonographic contrast agents as well as the application of duplex and color Doppler sonography show promising results in the accuracy of ultrasound imaging for HCC evaluation (Befeler & Di Bisceglie, 2002).

Computerized tomography (CT) scan

Examination of HCC-suspected patients by CT imaging is recommended to be performed using dynamic, multiphasic contrast imaging of the liver. In Far East countries, vascular invasion of the hepatic lesion is evaluated by CT arteriography and portography before surgical intercession (Brancatelli et al., 2003). For improvement of imaging, a spiral scanner is used with an infusion of intravenous contrasting agents. Multidetector CT (MDCT) demonstrates a better speed, thin-section, and high standard imaging with better flexibility and three-dimensional competencies (Szklaruk et al., 2003). However, the not-so-cost-effective nature of this approach illustrates its major limitations.

Magnetic resonance imaging (MRI)

MRI is considered the most favorable and noninvasive choice for HCC imaging and diagnosis at many institutions worldwide, due to its ease of obtaining images in a few seconds (Bialecki & Di Bisceglie, 2005). In comparison to CT, patients are less exposed to radiation and its damaging effects. MRI is also more accurate with respect to more sensitivity and less-negative likelihood than CT (Roberts et al., 2018). The sensitivity of MR imaging is the lowest when tumor sizes <2cm in diameter are evaluated (Krinsky et al., 2001).

Liver biopsy

For a more reliable diagnosis, cytological samples are collected from suspected HCC patients by percutaneous fine needle aspiration (FNA), whereas histologic samples are collected by needle core biopsy (NCB) (VanderLaan, 2016). Both NCB and FNA biopsy methods are generally used to determine preoperative tissue diagnosis which leads to subsequent therapies for several diseases. But cases have been found where these invasive procedures have posed risks of tumor cell implantation. After the diagnosis of HCC, a close examination based on TNM classification is followed for tumor grading to get the exact status regarding the stage and grade of the tumor. The diagnosis methods of HCC state many challenges, which differ with respect to different regions and centers. Further research advancements are needed for exploring novel diagnostic tools and innovative strategies for the early detection of HCC.

Limitations of conventional HCC diagnostic tools

Though there are many regular and modified conventional methods followed by clinicians for diagnosis, HCC is having a high mortality rate due to late detection. The important reasons for the delay in diagnosis are lack of understanding of the symptoms of the disease and the use of old diagnostic procedures as well as an inappropriate decision taken from imaging techniques, sole dependence on serum AFP level, absence of disease confirmation from the histopathological investigation of the lesion, and unpracticed diagnosticians. Specificity of diagnosis from the AFP level is very low when the size of the lesion is below 2cm. The AFP level stays normal during early HCC (Pazgan-Simon et al., 2015). Similarly, angiography also shows less sensitivity when the lesion size is <2cm (Krinsky et al., 2000). Another two diagnostic imaging tools like CT and MRI also have limitations such as the requirement of good contrasting agents for contrast enhancement (e.g., manganese- or gadolinium-based enhancement in MRI), scanning needed at each of the four phases (precontrast, arterial, portal venous, and delayed) in CT, and requirement of specialized expertise (Pazgan-Simon et al., 2015). FNB and NCB possess limitations of very low sensitivity and risks of tumor seeding respectively (VanderLaan, 2016).

Deep learning models

A substantial advancement in statistics and computer science has empowered computer engineers and medical science researchers to work together intently toward diagnosis as well as prognostic betterments using Cox analysis, multifactor analysis, and conventional logistic regression (Huang et al., 2020). These computer-assisted predictions are proven to show significantly higher accuracy than human-run experimental predictions. Human experts may have deep knowledge about a subject in a specific domain, but limitations can be seen when it comes to interpreting data with large variations which are out of a specific subject, and this leads to fatigue while working. To assist and accomplish the task performed by humans to overcome these limitations, the concept of DL is massively being explored (Tobore et al., 2019).

DL can be defined as the learning process executed in deep neural network, which enables the identification of an ideal function for the interpretation of data containing input, hidden, and output layer. Such networks can estimate complex functions by nonlinear transformation in the hidden layers. The final output layer performs as a classifier by exhibiting the possibility of primary input data into either one of the classes (Esteva et al., 2019). A large number of experiments have been performed and validated using various models of DL in healthcare and medicine research.

Many DL models, such as deep neural network (DNN), convolutional neural network (CNN), recurrent neural network (RNN), Boltzmann machine (BM), autoencoder (AE), etc., are developed and used to solve complex problems in various areas. Among them, CNN is reported to be the most convenient tool for HCC diagnosis. Autoencoder (AE) is a simple feed-forward DNN architecture comprising of encoder for input layer functions and decoder for output layer functions. CNNs usually comprise a nested set of feed-forward layers implementing convolutional filters followed by reduction, rectification, or pooling layers. CNN is known to generate high-level abstract features for each layer. Deep Boltzmann machine (DBM) is another interesting DL model, in which connections between neurons are completely undirected in all layers. DL is profound in computational formulations. In this chapter, we have discussed only those models which are used for the diagnostic evaluation of HCC.

Deep neural network

Artificial neural network (ANN) is a computing system developed based on the concept of biological neural networks. ANN is designed as a collection of units or layers and is interconnected with each other using some pattern for allowing communication between the units. Each unit is termed as a neuron or node of the network and is treated as a processor and the nodes are executed in parallel. Each node in the network is connected with other neurons via a connection link. Each connection link is associated with a weight which gives information about the input. The weight of a link is treated as the most useful information for a neuron to solve a problem as the weight helps in simulating the signal that is being communicated through the network. Each neuron has an internal state and is known as an activation signal. Then after combining the input signals and activation rule, the output signals are produced. The multilayer ANN is known as deep neural network. In DNN, many hidden layers are used for solving complex problems such as speech and image recognition, toxicity prediction, drug discovery and many applications in the health domain, etc.

Convolutional neural network

It is a deep neural network and is mainly used for image analysis, text analysis, etc. The convolution layer is the main part of the architecture. In this layer, a filter is defined that convolved on the input across the height, width, and depth which help to produce the activation map. Then, the network learns filters and is activated when some specific type of feature at some spatial position in the input is detected by the network (Géron, 2019).

Boltzmann machine

It is an optimization technique used in ANN, which is capable of optimizing the problem-related weight and quantity (Salakhutdinov & Hinton, 2009). In BM, the recurrent structure is used. The neurons used in BM have two possible states, i.e., 0 or 1, and the state of the neurons is free or frozen state. BM can be obtained by applying simulated annealing on the Hopfield network. The variation of BM is a restricted Boltzmann machine having a layer of hidden binary variables that are used in this probabilistic model. It is mostly applied to high-dimensional data such as images and texts (Larochelle et al., 2012). In general, it is difficult to learn a fully connected BM. Hence, each layer in a multilayer deep Boltzmann machine (DBM) captures complicated and higher-order correlations between the activities of hidden features in the layer below it (Salakhutdinov & Hinton, 2009). A DBM has the potential of learning internal representations and is considered one of the most promising ways of solving complex problems such as object and speech recognition. The high-level representations are built from a large pool of unlabeled sensory inputs and then specific labeled data are used to fine-tune the model and are done for a specific task.

Autoencoders

An AE neural network is an unsupervised and semi-supervised ML algorithm that uses back propagation (Ranjan, 2020). AEs are used for reducing the size of the input into a smaller dimension. This model can be used to reconstruct the original data from the compressed data. It uses convolutional layers to learn and give better results for images, series data, and video. The representation of each layer is treated as the output. The pretrained layers from other models can be used to enhance the performance of encoder/decoder using transfer learning. The applications of AE include dimensionality reduction, deionizing images, detecting feature variation, watermark removal, image coloring, etc. The architecture of an AE consists of three layers, i.e., encoder, code, and decoder.

- *Encoder:* The function of this layer is to compress the input into a lower dimension. It means it helps to reduce the input image dimension. The compressed image formed in this layer is the distorted version of the original image.
- *Code:* This layer represents the compressed input and is considered the input to the decoder.
- *Decoder:* This layer helps in decoding the distorted image, which means it is reconstructed from the compressed image.

Deep learning approaches for diagnostic evaluation of HCC

In this section, a review of various DL approaches for the diagnostic evaluation of HCC is presented. The most recent studies carried out for a histopathological image analysis using DL models, radiomics and multiomics using DL, and four common DL models used for diagnosis of HCC are discussed. These four common DL models are deep ANN, CNN, DBM, and AE.

The credit for huge achievements documented in DL goes to the progressive evolution in graphics processing units (GPUs), which have a major function in expediting the computation requirements of DL (Tobore et al., 2019). These models are implemented in histopathology, medical imaging, mutation analysis, as well as survival prediction in HCC related to the precise diagnosis of the disease. A comprehensive summary of various DL-based models used in HCC diagnosis is outlined in Table 15.1.

TABLE 15.1 Deep learning (DL)-based models for hepatocellular carcinoma diagnosis.

Sl. no.	Title of the study	Diagnostic technology used	DL model used	Reference
1	Deep learning for accurate diagnosis of liver tumor based on magnetic resonance imaging and clinical data	MRI	CNN	Zhen et al. (2020)
2	Impact of a deep learning assistant on the histopathologic classification of liver cancer	Histopathology	DL-based model	Kiani et al. (2020)
3	Classification and mutation prediction based on open histopathology H&E images in liver cancer using deep learning	Histopathology	CNN	Chen et al. (2020)
4	Deep learning LI-RADS grading system based on contrast-enhanced multiphase MRI for differentiation between LR-3 and LR-4/LR-5 liver tumors	MRI	CNN	Wu et al. (2020)
5	Deep learning-based classification and mutation prediction from histopathological images of hepatocellular carcinoma	Histopathology	Deep CNN	Liao et al. (2020)
6	Liver cancer detection using hybridized fully convolutional neural network based on deep learning framework	CT	Hybridized fully CNN	Dong et al. (2020)
7	Prognostic analysis of histopathological images using pretrained convolutional neural networks: Application to hepatocellular carcinoma	Histopathology	CNN	Lu and Daigle (2020)
8	Application of artificial intelligence using a novel EUS-based convolutional neural network model to identify and distinguish benign from malignant hepatic masses	Endoscopic ultrasound	CNN	Marya et al. (2021)
9	Artificial neural network model for liver cirrhosis diagnosis in patients with hepatitis B virus-related hepatocellular carcinoma	Serological indicator screening	ANN	Mai et al. (2020)
10	Deep learning radiomics based on contrast-enhanced ultrasound might optimize curative treatments for very-early or early-stage hepatocellular carcinoma patients	Radiomics	Cox CNN	Liu et al. (2020)
11	Deep learning for liver tumor diagnosis part II: Convolutional neural network interpretation using radiologic imaging features	Radiology	CNN	Wang et al. (2019)
12	Automated classification of hepatocellular carcinoma differentiation using multiphoton microscopy and deep learning	Histopathology	CNN	Lin et al. (2019)
13	Grading of hepatocellular carcinoma based on diffusion-weighted images with multiple B-values using convolutional neural networks	Histology and MR imaging	CNN	Zhou et al. (2019)
14	Deep learning-based multiomics integration robustly predicts survival in liver cancer	Multiomics	DL-based model	Chaudhary et al., (2018)

TABLE 15.1 Deep learning (DL)-based models for hepatocellular carcinoma diagnosis—Cont'd

Sl. no.	Title of the study	Diagnostic technology used	DL model used	Reference
15	Artificial neural network as a classifier for the identification of hepatocellular carcinoma through prognostic gene signatures	DNA microarray	ANN	Jujjavarapu and Deshmukh (2018)
16	Deep learning with convolutional neural network for the differentiation of liver masses at dynamic contrast-enhanced CT	CT	CNN	Yasaka et al. (2018)
17	Joint multiple fully connected convolutional neural network with extreme learning machine for hepatocellular carcinoma nuclei grading	Histopathology	CNN	S. Li, Jiang, and Pang (2017)
18	Artificial neural network models for early diagnosis of hepatocellular carcinoma using serum levels of α-fetoprotein, α-fetoprotein-L3, des-γ-carboxy prothrombin, and Golgi protein 73	Serum biomarker screening	ANN	B. Li, Li, et al. (2017)
19	Restricted Boltzmann machines for the classification of hepatocellular carcinoma	ELISA, serum antigen mini array	Restricted Boltzmann machines	Koziol and Tan (2014)
20	Comprehensive proteomic profiling identifies serum proteomic signatures for the detection of hepatocellular carcinoma and its subtypes	Proteomics	ANN	Poon et al. (2003)
21	Application of classification tree and neural network algorithms to the identification of serological liver marker profiles for the diagnosis of hepatocellular carcinoma	Serum biomarker screening	DL-based neural networks	Poon et al. (2001)

DL-based histopathological image analysis for HCC diagnosis

Histopathological studies of tumors help in correlating the diagnosis of the disease with its prognosis and management, for personalized treatment. Phenotypic characteristics of a tumor, such as density of cell, mitotic index, nuclear atypia, and structure of the tissue, are the basic parameters that need to be examined from histopathological slides for staging and grading of the tumor. Presently, histopathological evaluation of the tumor is considered standard practice for determining the subtypes of HCC. The use of DL models in histopathological evaluation is now proven to lessen the burden of experts examining the whole slide images manually. In a study, researchers have designed a fully automated DL model pipeline using an R-program, named Breiman's "random forest" for quantitative analysis of histopathological slides for HCC patients, which could be trained further to distinguish HCC from adjacent or surrounding healthy tissues by considering 31 variants in test and external validation groups (Liao et al., 2020). This model could also evaluate the prognosis by determining the shorter and longer survival groups of patients by considering 46 histopathological parameters, thus predicting the patient's survival outcomes. In another study, a DL model has been proposed, which enables comparatively easier and cost-effective evaluation of whole slide images (WSI) than a direct analysis. Usually, one WSI comprises around >100,000 × 100,000 pixels, which is definitely strenuous and impractical to analyze each pixel directly (Sun et al., 2020). This proposed model constructs several groups of "patches" which can be independently analyzed. These patches are extracted cost effectively by a CNN model based on transfer learning and multiple instance learning. Chen et al. (2020) demonstrated another DL model based on a neural network named Inception V3 which could construct "tiles" cropped from WSIs. The tiles could categorize hepatic tissues as HCC vs normal and also well vs moderately vs poorly differentiated ones. The model was validated on the basis of Matthews correlation coefficient (MCC) and precision recall curves (PR curves) for its performance evaluation. The performance of this method was comparable (AUC = 0.961, 95% CI = 0.939–0.981) to 5-year experienced clinicians (Chen et al., 2020). A research group used three pretrained CNN models named Inception3, VGG16, and ResNet50 and extracted features from H&E-stained WSIs of HCC patients to classify subgroup and for HCC prognosis (Lu & Daigle, 2020). They could differentiate clearly between cancer and healthy cells. They have demonstrated a correlation between disease-free survival of patients with overall survival by analyzing with a univariate Cox regression method. They have also demonstrated a noteworthy correlation between features extracted from WSIs, which could be associated with several genetic and molecular pathways of HCC development.

DL-based radiomics and multiomics for HCC diagnosis

Radiomics is a new paradigm in precision medicine, which involves automated, computerized extraction of quantifiable features of radiological imaging phenotypes in a high-throughput manner and subsequent data mining to retrieve clinically relevant information such as tissue heterogeneity and compacity of normal and pathological tissue, that help further inaccurate disease diagnosis and prognosis (Dreher et al., 2020). Radiomics features can be morphological (e.g., size, volume, and shape of the volume of interest in tumors), histogram features (e.g., magnitude, asymmetry, dispersion, uniformity, etc., for the gray-level pixel image values), textural features (e.g., gray level co-occurrence and the run-length matrix), or high-order features (i.e., textural features extracted from Laplacian/Gaussian-filtered images (Park et al., 2020). Such radiomics approaches have been categorized as model-based, statistics-based, or transform-based, which are in practice for diagnosis, staging and grading, therapeutic selection, prognosis assessment, and surveillance of HCC. They have been analyzed either with CT, MRI, and PET images with many software packages such as Chang-Gung Image Texture Analysis, Radiomics, MaZda, and TexRAD (Hu et al., 2020). Although surgical resection (SR) is the first-line recommendation in patients with solitary HCC, postoperative recurrence is of high concern and no reliable prediction tools exist. Ji et al. designed an ML-based contrast-enhanced CT radiomics strategy to predict the postoperative recurrence of HCC after SR, in which a three-feature radiomics signature showed superior pre and postoperative prognostic performance with a C-index of 0.633–0.699, which was helpful in accurate prediction of individual HCC recurrence that may further assist in personalized HCC management (Ji et al., 2019). However, such conventional radiomics approaches are time-consuming and labor-intensive during feature extraction, machine learning, and modeling processes. Deep learning-based radiomics is advantageous over conventional radiomics because it relies on representation learning and can extract prognosis-relevant best quantitative indices by assessing morphological and textural changes in HCC during/after radiofrequency (RFA), liver transplantation, or surgical resection (SR) that may not be perceived accurately by radiologists, thus assisting in appropriate clinical decision-making. For the first time, Liu et al. developed DL-based radiomics signatures for RFA and SR to predict progression-free survival accuracy by analyzing contrast-enhanced ultrasound data, which further assisted in clinical decision-making for treatment selection in very-early to early-stage HCC (Liu et al., 2020). In another study, Zhen et al. designed a DL system using MRI-based radiomics approach ($N=31{,}608$ images) to classify hepatic tumors and the mined data from this DL network demonstrated accurate and high-level specificity and sensitivity compared to the performance of experienced radiologists (Zhen et al., 2020). These DL-based radiomics approaches for HCC diagnosis, prognosis assessment, staging, and therapeutic selection are emerging tools in precision medicine, but require rigorous clinical validation of deep learning radiomics algorithm to avoid overfitting problems before real-world clinical practice.

Multiomics approaches in the assessment of cancer progression generally involve correlated and integrative profiling and analysis of multiple bio-entities in a "big" data format. Such big data format in deep learning-based multiomics algorithms include not only primary omics data such as genomics, transcriptomics, proteomics, and metabolomics but also some relevant secondary omics data such as epigenomics, multilayer signature biomarkers, microbiome, immunomics, and multiplex family history data (Biswas & Chakrabarti, 2020). DL-based multiomics in HCC involves automated storing, processing, analysis, and interpretations of various types of patient-specific "omics" data and their complex interrelatedness to understand the underlying pathophysiological mechanism of HCC progression, identification of subtypes, disease prognosis, as well as therapeutic target identifications. In this regard, Chaudhary et al. developed a DL-based autoencoder framework model for multiomics datasets of HCC patients' RNA and miRNA sequencing as well as methylation data and validated the model in six different cohorts (Chaudhary et al., 2018). In their study, they integrated multiomics data using DL and predicted differential survival subpopulations of HCC patients robustly, with accurate identification of aggressive subtypes associated with certain mutational, stemness, and tumor markers, which might be helpful in HCC prognosis-associated precision medicine.

Deep learning model-based HCC diagnosis

CNN model for HCC diagnosis

CNNs are multilayered feed-forward networks with only a forward flow of information. In a recent study, Wang et al. (2019) have developed and validated a prototype CNN-based DL model that can classify common hepatic lesions on a multiphasic MRI. They have demonstrated accuracy and sensitivity of around 92% with a specificity of 98% for this DLS, and the result showed greater accuracy than two highly efficient and experienced radiologists. This CNN-based DL system could classify six classes of hepatic lesions (benign cyst, focal nodular hyperplasia, hepatocellular carcinoma (HCC), cavernous hemangioma, intrahepatic cholangiocarcinoma (CC), and colorectal carcinoma metastasis) along with hepatic lesions like benign, hepatocellular, and malignant non hepatocellular carcinoma lesions. In another study, Kiani et al. (2020) developed a CNN-based DenseNet architecture to differentiate HCC from cholangiocarcinoma using hematoxylin and eosin (H&E)-stained digital whole-slide

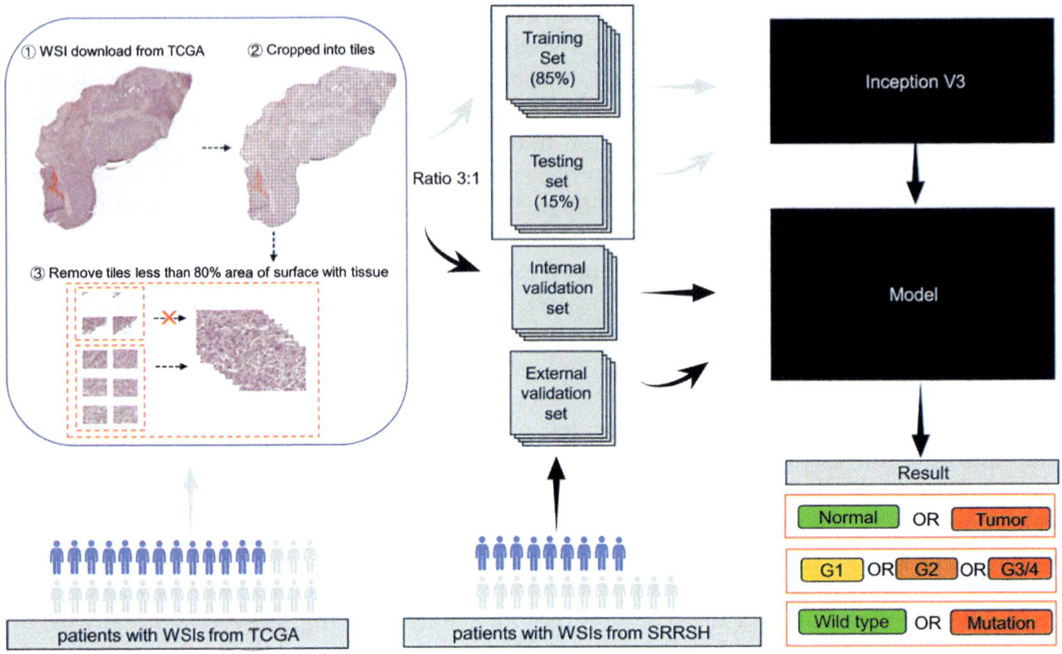

FIG. 15.2 Schematic representation of the DL framework based on convolutional neural network (CNN) proposed by Chen et al., for training the model to categorize and predict genetic mutations. In this study, Inception V3 model used whole-stained images (WSIs) from training cohorts for test and training, which was validated further with internal and external validation sets for learning, selection, and subsequent classification of HCC based on specific gene mutations. *Adapted with permission from Chen, M., Zhang, B., Win, T., Jiasheng, C., Hepan, Z., Sarun, J., Qijiang, M., Hong, Y., & Xiujun, C. (2020). Classification and mutation prediction based on histopathology H&E images in liver cancer using deep learning.* Npj Precision Oncology, *4(1). https://doi.org/10.1038/s41698-020-0120-3.*

images (WSI) of primary liver tumors. A diagnostic accuracy of around 95% was achieved in this model. In another interesting study, Chen et al. (2020) have developed and validated a CNN model named the InceptionV3 network using WSIs of liver tissue (Fig. 15.2). In this model, WSIs were collected from the Genomic Data Commons portal and also from Sir Run-Run Shaw Hospital (SRRSH), China, for the classification of liver tissue on the basis of some specific gene mutations. The model showed an accuracy of 96.0% for classification between benign and malignant lesions and an accuracy of 89.6% for a distinction between well, moderately, and poorly differentiated tumors of HCC. They have demonstrated that four crucial genes (TP53, CTNNB1, FMN2, and ZFX4) associated with HCC diagnosis could be potential prognostic biomarkers while assessing histopathological images using their proposed DL model. Nevertheless, it was a preliminary study, and more in-depth explorations might lead to improvised diagnostic methods. Furthermore, Wu et al. (2020) have developed an AlexNet CNN model assisting LI-RADS grading protocol for hepatic lesions, which has demonstrated an accuracy level comparable to experienced radiologists. This DL-based model proposed substantial clinical guidance to distinguish intermediate-stage liver lesions from late, malignant ones, for accurate HCC diagnosis.

Zhen et al. (2020) have developed a CNN-based model for HCC diagnosis using binary, three-way as well as seven-way malignancy classifiers, which is comparable to the performance shown by experienced radiologists both in sensitivity and specificity of prediction. Similarly, Wang et al. (2019) have established an archetype CNN-based DL system for clinical radiology to interpret and classify HCC. In another study, they have investigated two CNN-based models, model triphasic and model art/del, and their potential to differentiate different types of liver lesions including HCC using dynamic CT images. They could classify the liver lesions into five categories and could differentiate between benign masses and malignant tumors in HCC. Dong et al. (2020) have proposed a high-performance design based on hybridized fully convolutional neural network for HCC detection and efficient segmentation of liver cancer. An advanced CNN model with an extreme learning machine (MFC-CNN-ELM) was proposed by B. Li, Li, et al. (2017), for improved diagnosis and treatment of hepatocellular carcinoma by grading of nuclei. Marya et al. (2021) have designed a CNN model based on endoscopic ultrasound (EUS) images to identify and classify focal liver lesions for diagnosis of HCC.

ANN model for HCC diagnosis

ANN model works on basic principles like common statistical methodologies based on generalized linear models, nonparametric correlation and regression analysis, and cluster analysis. S. Li, Jiang, and Pang (2017) have studied on multilayer perceptron (MLP)-based ANN model, which can be used in combination with another ANN model adopting radial basis function (RBF) for the diagnosis of HCC and distinguishing early-stage HCC from liver cirrhosis by screening four potential serum biomarkers. ANN-based algorithms were adopted by Poon et al. (2001) to profile different serological markers for HCC diagnosis in patients having liver cirrhosis. In another study, Poon et al. (2003) developed an ANN utilizing the differential proteomic datasets for differentiating HCC with nondiagnostic AFP from chronic liver disease. This ANN-based model also enables clinicians to evaluate the advanced HCC subtypes with advanced metastasis stage from early HCC. Recently, Mai et al. (2020) have proposed a novel noninvasive ANN model for the diagnosis of liver cirrhosis in HBV-related HCC patients by screening the routine clinical serological biomarkers.

DBM model for HCC diagnosis

Restricted Boltzmann machine is a type of deep learning method which has a principle similar to the standard form of a neural network having a single hidden layer that gives a bipartite graph structure. This model consists of a symmetrically aligned connection between a visible layer and a hidden layer, but without any connection between the internal nodes of both layers. This model is quite in use for medical data analysis. Koziol and Tan (2014) have demonstrated the potential of such a model based on restricted Boltzmann machines using a panel of 12 antibody profiles for the diagnosis of HCC. They have proved the merits of restricted Boltzmann machines, a DLS for use in the classification of HCC.

AE model for HCC diagnosis

Autoencoder, unlike Boltzmann machine, is a triple-layered feed-forward neural network of visible and hidden layers, the former being more in number. In this autoencoder model, the number of neurons is the same in both input and output layers. Chaudhary et al. (2018) have designed an autoencoder-based DL system to understand the heterogeneity of HCC among patients for improved prognosis and classification of HCC subtypes using multiomics data. This multiomics data is based on genomics and clinical data. This model will ultimately help diagnose patients with precise HCC categories for their prognosis and treatment.

Challenges and future perspectives

The advancement in DL technology in the recent 5 years is substantial and appreciable though they are backed by many limitations for practical use in clinics. Transparency and intelligibility are the major issues against the practical integration of deep learning for the diagnosis of HCC through radiology and histopathology. Besides the sensitivity and specificity of a model, it is important whether and how the model is applicable for the diagnosis of HCC. Screening of serum AFP levels along with ultrasonography has been the gold standard biomarker for early diagnosis of subclinical HCC such as identification of small HCC cases (<3–5 cm), thereby presenting a better cumulative survival rate in such patients (Poon et al., 2001). However, it has been reported that if screening of AFP only is taken into consideration, it will be insufficient for precise detection of HCC, as more than 50% of the small HCC (<3 cm) would have been missed, leading to inaccuracy (Okuda et al., 1984). In his study, Okuda has strongly suggested that the conventional AFP test should be used in combination with advanced diagnostic techniques such as neural networks, which may enable an early and accurate diagnosis of HCC. Sera proteomic signatures of HCC patients are a potential tool for the early detection of the disease but future studies required less-biased exposition of the diagnostic accuracy (Poon et al., 2003). Although postoperative histopathological imaging of paraffin-embedded samples is still the gold standard for the detection of HCC grade, it consumes more time and requires highly skilled pathologists, making this method inconvenient in certain cases. In this context, CNN-based MRM imaging is convenient but it has the limitation of requiring a huge size of data. So, the development of DLS for automated diagnostic tools, which can give accurate output with fewer training images, is a challenge to AI researchers. Liu et al. (2020) have mentioned in their study two major limitations of their proposed model, a Cox CNN model using radiomics to detect very-early to early-stage HCC. The study has a limited population size and low interpretability. These limitations are to be overcome in future research in AI. Moreover, the use of DL for the evaluation of behavioral differences between HCC in cirrhotic and non-cirrhotic patients distinguishing primary liver lesions from metastatic ones, especially the differential diagnosis between HCC and cholangiocarcinoma, needs more in-depth explorations and optimizations.

Conclusion

In this chapter, we present a detailed study on conventional methods of the diagnostic evaluation of HCC. Subsequently, various DL models used in the literature for HCC diagnosis are studied. Also, the challenges involved in HCC diagnosis are identified. From this study, it is found that a number of potential DL models are suggested and validated for the clinical diagnosis of HCC. The integration of DL technology with conventional diagnostic techniques in medical science has advanced substantially in recent years. Though this progress is noteworthy, limitations are also unavoidable in the present scenario. Hence, larger studies are required to get a comparable yield of medical professionals with DL assistance vs professionals without DL assistance in order to gradually outperform human-based detection of HCC.

Acknowledgment

DKB acknowledges the funding support from the Science and Engineering Research Board (SERB), Department of Science and Technology, Govt. of India for the "Start-up Research Grant—2019" (SRG/2019/001995).

References

Befeler, A. S., & Di Bisceglie, A. M. (2002). Hepatocellular carcinoma: Diagnosis and treatment. *Gastroenterology, 122*(6), 1609–1619. https://doi.org/10.1053/gast.2002.33411.

Bialecki, E. S., & Di Bisceglie, A. M. (2005). Diagnosis of hepatocellular carcinoma. *HPB, 7*(1), 26–34. https://doi.org/10.1080/13651820410024049.

Biswas, N., & Chakrabarti, S. (2020). Artificial intelligence (AI)-based systems biology approaches in multi-omics data analysis of cancer. *Frontiers in Oncology, 10*. https://doi.org/10.3389/fonc.2020.588221.

Brancatelli, G., Baron, R. L., Peterson, M. S., & Marsh, W. (2003). Helical CT screening for hepatocellular carcinoma in patients with cirrhosis: Frequency and causes of false-positive interpretation. *American Journal of Roentgenology, 180*(4), 1007–1014. https://doi.org/10.2214/ajr.180.4.1801007.

Chaudhary, K., Poirion, O. B., Lu, L., & Garmire, L. X. (2018). Deep learning-based multi-omics integration robustly predicts survival in liver cancer. *Clinical Cancer Research, 24*(6), 1248–1259. https://doi.org/10.1158/1078-0432.CCR-17-0853.

Chen, M., Zhang, B., Win, T., Jiasheng, C., Hepan, Z., Sarun, J., Qijiang, M., Hong, Y., & Xiujun, C. (2020). Classification and mutation prediction based on histopathology H&E images in liver cancer using deep learning. *NPJ Precision Oncology, 4*(1). https://doi.org/10.1038/s41698-020-0120-3.

Chevret, S. (1999). A new prognostic classification for predicting survival in patients with hepatocellular carcinoma. *Journal of Hepatology, 31*(1), 80173–1. https://doi.org/10.1016/S0168-8278(99.

Dong, X., Zhou, Y., Wang, L., Peng, J., Lou, Y., & Fan, Y. (2020). Liver cancer detection using hybridized fully convolutional neural network based on deep learning framework. *IEEE Access, 8*, 129889–129898. https://doi.org/10.1109/ACCESS.2020.3006362.

Dreher, C., Linde, P., Boda-Heggemann, J., & Baessler, B. (2020). Radiomics for liver tumours. *Strahlentherapie und Onkologie, 196*(10), 888–899. https://doi.org/10.1007/s00066-020-01615-x.

El-Serag, H. B., & Rudolph, K. L. (2007). Hepatocellular carcinoma: Epidemiology and molecular carcinogenesis. *Gastroenterology, 132*(7), 2557–2576. https://doi.org/10.1053/j.gastro.2007.04.061.

Esteva, A., Robicquet, A., Ramsundar, B., Kuleshov, V., DePristo, M., Chou, K., Cui, C., Corrado, G., Thrun, S., & Dean, J. (2019). A guide to deep learning in healthcare. *Nature Medicine, 25*(1), 24–29. https://doi.org/10.1038/s41591-018-0316-z.

Farinati, F., Marino, D., De Giorgio, M., Baldan, A., Cantarini, M., Cursaro, C., Rapaccini, G., Del Poggio, P., Di Nolfo, M. A., Benvegnù, L., Zoli, M., Borzio, F., Bernardi, M., & Trevisani, F. (2006). Diagnostic and prognostic role of α-fetoprotein in hepatocellular carcinoma: Both or neither? *American Journal of Gastroenterology, 101*(3), 524–532. https://doi.org/10.1111/j.1572-0241.2006.00443.x.

Géron, A. (2019). *Hands-on machine learning with Scikit-Learn, Keras, and TensorFlow*. O'Reilly Media.

Hu, W., Yang, H., Xu, H., & Mao, Y. (2020). Radiomics based on artificial intelligence in liver diseases: Where we are? *Gastroenterology Report, 8*(2), 90–97. https://doi.org/10.1093/gastro/goaa011.

Huang, S., Yang, J., Fong, S., & Zhao, Q. (2020). Artificial intelligence in cancer diagnosis and prognosis: Opportunities and challenges. *Cancer Letters, 471*, 61–71. https://doi.org/10.1016/j.canlet.2019.12.007.

Ji, G. W., Zhu, F. P., Xu, Q., Wang, K., Wu, M. Y., Tang, W. W., Li, X. C., & Wang, X. H. (2019). Machine-learning analysis of contrast-enhanced CT radiomics predicts recurrence of hepatocellular carcinoma after resection: A multi-institutional study. *eBioMedicine*, *50*, 156–165. https://doi.org/10.1016/j.ebiom.2019.10.057.

Jujjavarapu, S. W., & Deshmukh, S. (2018). Artificial neural network as a classifier for the identification of hepatocellular carcinoma through prognostic gene signatures. *Current Genomics*, *19*(6), 483–490.

Kashyap, R., Jain, A., Nalesnik, M., Carr, B., Barnes, J., Vargas, H. E., Rakela, J., & Fung, J. (2001). Clinical significance of elevated α-fetoprotein in adults and children. *Digestive Diseases and Sciences*, *46*(8), 1709–1713. https://doi.org/10.1023/A:1010605621406.

Kiani, A., Uyumazturk, B., Rajpurkar, P., Alex, W., Rebecca, G., Erik, J., Yifan, Y., Langlotz, C. P., Ball, R. L., Montine, T. J., Martin, B. A., Berry, G. J., Ozawa, M. G., Hazard, F. K., Brown, R. A., Chen, S. B., Wood, M., Allard, L. S., Lourdes, Y., Jeanne, S. (2020). Impact of a deep learning assistant on the histopathologic classification of liver cancer. *NPJ Digital Medicine*, *3*(1). https://doi.org/10.1038/s41746-020-0232-8.

Kinoshita, A., Onoda, H., Fushiya, N., Koike, K., Nishino, H., & Tajiri, H. (2015). Staging systems for hepatocellular carcinoma: Current status and future perspectives. *World Journal of Hepatology*, *7*(3), 406–424. https://doi.org/10.4254/wjh.v7.i3.406.

Koziol, J. A., Tan, E. M., Dai, L., Ren P., & Zhang, J. (2014). Restricted Boltzmann machines for classification of hepatocellular carcinoma. *Computational Biology Journal*, 1–5. https://doi.org/10.1155/2014/418069.

Krinsky, G. A., Lee, V. S., Theise, N. D., Weinreb, J. C., Rofsky, N. M., Diflo, T., & Teperman, L. W. (2001). Hepatocellular carcinoma and dysplastic nodules in patients with cirrhosis: Prospective diagnosis with MR imaging and explantation correlation. *Radiology*, *219*(2), 445–454. https://doi.org/10.1148/radiology.219.2.r01ma40445.

Krinsky, G. A., Nguyen, M. T., Lee, V. S., Rosen, R. J., Goldenberg, A., Theise, N. D., Morgan, G., & Rofsky, N. M. (2000). Dysplastic nodules and hepatocellular carcinoma: Sensitivity of digital subtraction hepatic arteriography with whole liver explant correlation. *Journal of Computer Assisted Tomography*, *24*(4), 628–634. https://doi.org/10.1097/00004728-200007000-00022.

Larochelle, H., Mandel, M., Pascanu, R., & Bengio, Y. (2012). Learning algorithms for the classification restricted Boltzmann machine. *Journal of Machine Learning Research*, *13*, 643–669. http://jmlr.csail.mit.edu/papers/volume13/larochelle12a/larochelle12a.pdf.

Leung, T. W. T., Tang, A. M. Y., Zee, B., Lau, W. Y., Lai, P. B. S., Leung, K. L., Lau, J. T. F., Yu, S. C. H., & Johnson, P. J. (2002). Construction of the Chinese University Prognostic Index for hepatocellular carcinoma and comparison with the TNM staging system, the Okuda staging system, and the cancer of the Liver Italian Program staging system: A study based on 926 patients. *Cancer*, *94*(6), 1760–1769. https://doi.org/10.1002/cncr.10384.

Li, S., Jiang, H., & Pang, W. (2017). Joint multiple fully connected convolutional neural network with extreme learning machine for hepatocellular carcinoma nuclei grading. *Computers in Biology and Medicine*, *84*, 156–167. https://doi.org/10.1016/j.compbiomed.2017.03.017.

Li, B., Li, B., Guo, T., Sun, Z., Li, X., Li, X., Chen, L., Zhao, J., & Mao, Y. (2017). Artificial neural network models for early diagnosis of hepatocellular carcinoma using serum levels of α-fetoprotein, a-fetoprotein-L3, des-γ-carboxy prothrombin, and Golgi protein 73. *Oncotarget*, *8*(46), 80521–80530. https://doi.org/10.18632/oncotarget.19298.

Liao, H., Long, Y., Ruijiang, H., Wei, W., Lin, X., Mingheng, L., Zhen, Z., Zhenru, W., Xuequn, S., Xuefeng, L., Jiajie, P., Kefei, Y., & Zeng, Y. (2020). Deep learning-based classification and mutation prediction from histopathological images of hepatocellular carcinoma. *Clinical and Translational Medicine*. https://doi.org/10.1002/ctm2.102.

Lin, H., Wei, C., Wang, G., Chen, H., Lin, L., Ni, M., Zhuo, S. (2019). Automated classification of hepatocellular carcinoma differentiation using multiphoton microscopy and deep learning. *Journal of Biophotonics*, *12*(7), e201800435. https://doi.org/10.1002/jbio.201800435.

Liu, F., Liu, D., Wang, K., Xie, X., Su, L., Kuang, M., Huang, G., Peng, B., Wang, Y., Lin, M., Tian, J., & Xie, X. (2020). Deep learning radiomics based on contrast-enhanced ultrasound might optimize curative treatments for very-early or early-stage hepatocellular carcinoma patients. *Liver Cancer*, *9*(4), 397–413. https://doi.org/10.1159/000505694.

Llovet, J. M., Brú, C., & Bruix, J. (1999). Prognosis of hepatocellular carcinoma: The BCLC staging classification. *Seminars in Liver Disease*, *19*(3), 329–337. https://doi.org/10.1055/s-2007-1007122.

Lu, L., & Daigle, B. J. (2020). Prognostic analysis of histopathological images using pre-trained convolutional neural networks: Application to hepatocellular carcinoma. *PeerJ*, *2020*(3). https://doi.org/10.7717/peerj.8668.

Mai, R. Y., Zeng, J., Mo, Y. S., Liang, R., Lin, Y., Wu, S. S., Piao, X. M., Gao, X., Wu, G. B., Li, L. Q., & Ye, J. Z. (2020). Artificial neural network model for liver cirrhosis diagnosis in patients with hepatitis B virus-related hepatocellular carcinoma. *Therapeutics and Clinical Risk Management, 16*, 639–649. https://doi.org/10.2147/TCRM.S257218.

Manghisi, G., Elba, S., Mossa, A., Giorgio, A., Aloisio, V., Perrotta, A., Tardio, B., Del Naja, C., Caturelli, E., Calandra, M., Castellano, L., De Sio, I., Capuano, G., Pomponi, D., Castiglione, F., Cocchia, P., Farinati, F., Rinaldi, M., … Adinolfi, L. E. (1998). A new prognostic system for hepatocellular carcinoma: A retrospective study of 435 patients. *Hepatology, 28*(3), 751–755. https://doi.org/10.1002/hep.510280322.

Marya, N. B., Powers, P. D., Larissa, F.-L., Dayyeh, B. K. A., Gleeson, F. C., Chen, S., Long, Z., Hough, D. M., Vinay, C., Iyer, P. G., Elizabeth, R., William, S., Tarek, S., Storm, A. C., Wang, K. K., & Levy, M. J. (2021). Application of artificial intelligence using a novel EUS-based convolutional neural network model to identify and distinguish benign and malignant hepatic masses. *Gastrointestinal Endoscopy, 93*(5), 1121–1130. https://doi.org/10.1016/j.gie.2020.08.024.

Okuda, K., Obata, H., Yukio, N., Toshio, O., Nobuo, O., & Kunihiko, O. (1984). Prognosis of primary hepatocellular carcinoma. *Hepatology, 4*(S1), 3S–6S. https://doi.org/10.1002/hep.1840040703.

Park, H. J., Park, B., & Lee, S. S. (2020). Radiomics and deep learning: Hepatic applications. *Korean Journal of Radiology, 21*(4), 387–401. https://doi.org/10.3348/kjr.2019.0752.

Pazgan-Simon, M., Serafinska, S., Justyna, J.-L., Krzysztof, S., & Jolanta, Z.-J. (2015). Diagnostic challenges in primary hepatocellular carcinoma: Case reports and review of the literature. *Case Reports in Oncological Medicine*, 1–5. https://doi.org/10.1155/2015/878763.

Poon, T. C. W., Chan, A. T. C., Zee, B., Ho, S. K. W., Mok, T. S. K., Leung, T. W. T., & Johnson, P. J. (2001). Application of classification tree and neural network algorithms to the identification of serological liver marker profiles for the diagnosis of hepatocellular carcinoma. *Oncology, 61*(4), 275–283. https://doi.org/10.1159/000055334.

Poon, T. C. W., Yip, T. T., Chan, A. T. C., Yip, C., Yip, V., Mok, T. S. K., Lee, C. C. Y., Leung, T. W. T., Ho, S. K. W., & Johnson, P. J. (2003). Comprehensive proteomic profiling identifies serum proteomic signatures for detection of hepatocellular carcinoma and its subtypes. *Clinical Chemistry, 49*(5), 752–760. https://doi.org/10.1373/49.5.752.

Ranjan, C. (2020). *Understanding deep learning: Application in rare event prediction*. Connaissance Publishing.

Roberts, L. R., Sirlin, C. B., Zaiem, F., Almasri, J., Prokop, L. J., Heimbach, J. K., Murad, M. H., & Mohammed, K. (2018). Imaging for the diagnosis of hepatocellular carcinoma: A systematic review and meta-analysis. *Hepatology, 67*(1), 401–421. https://doi.org/10.1002/hep.29487.

Ryder, S. D. (2003). Guidelines for the diagnosis and treatment of hepatocellular carcinoma (HCC) in adults. *Gut, 52*(3), iii1–iii8. https://doi.org/10.1136/gut.52.suppl_3.iii1.

Salakhutdinov, R., & Hinton, G. (2009). Deep Boltzmann machines. In *Vol. 5. Proceedings of the twelth international conference on artificial intelligence and statistics. Proceedings of machine learning research* (pp. 448–455).

Sun, C., Xu, A., Liu, D., Xiong, Z., Zhao, F., & Ding, W. (2020). Deep learning-based classification of liver cancer histopathology images using only global labels. *IEEE Journal of Biomedical and Health Informatics, 24*(6), 1643–1651. https://doi.org/10.1109/JBHI.2019.2949837.

Szklaruk, J., Silverman, P. M., & Charnsangavej, C. (2003). Imaging in the diagnosis, staging, treatment, and surveillance of hepatocellular carcinoma. *American Journal of Roentgenology, 180*(2), 441–454. https://doi.org/10.2214/ajr.180.2.1800441.

Tang, A., Hallouch, O., Chernyak, V., Kamaya, A., & Sirlin, C. B. (2018). Epidemiology of hepatocellular carcinoma: Target population for surveillance and diagnosis. *Abdominal Radiology, 43*(1), 13–25. https://doi.org/10.1007/s00261-017-1209-1.

Tobore, I., Li, J., Yuhang, L., Al-Handarish, Y., Kandwal, A., Nie, Z., & Wang, L. (2019). Deep learning intervention for health care challenges: Some biomedical domain considerations. *JMIR mHealth and uHealth, 7*(8). https://doi.org/10.2196/11966.

Torzilli, G., Belghiti, J., & Makuuchi, M. (2004). Differences and similarities in the approach to hepatocellular carcinoma between Eastern and Western Institutions. *Liver Transplantation, 10*(2), S1–S2. https://doi.org/10.1002/lt.20032.

VanderLaan, P. A. (2016). Fine-needle aspiration and core needle biopsy: An update on 2 common minimally invasive tissue sampling modalities. *Cancer Cytopathology, 124*(12), 862–870. https://doi.org/10.1002/cncy.21742.

Vauthey, J. N., Lauwers, G. Y., Esnaola, N. F., Do, K. A., Belghiti, J., Mirza, N., Curley, S. A., Ellis, L. M., Jean-Marc, R., Asif, R., Cleary, K. R., & Nagorney, D. M. (2002). Simplified staging for hepatocellular carcinoma. *Journal of Clinical Oncology*, 1527–1536. https://doi.org/10.1200/JCO.2002.20.6.1527.

Villanueva, A. (2019). Hepatocellular carcinoma. *New England Journal of Medicine, 380*(15), 1450–1462. https://doi.org/10.1056/NEJMra1713263.

Wang, C. J., Hamm, C. A., Savic, L. J., Ferrante, M., Schobert, I., Schlachter, T., Lin, M. D., Weinreb, J. C., Duncan, J. S., Chapiro, J., & Letzen, B. (2019). Deep learning for liver tumor diagnosis part II: Convolutional neural network interpretation using radiologic imaging features. *European Radiology, 29*(7), 3348–3357. https://doi.org/10.1007/s00330-019-06214-8.

Wu, Y., White, G. M., Cornelius, T., Gowdar, I., Ansari, M. H., Supanich, M. P., & Deng, J. (2020). Deep learning LI-RADS grading system based on contrast enhanced multiphase MRI for differentiation between LR-3 and LR-4/LR-5 liver tumors. *Annals of Translational Medicine, 8*(11), 701. https://doi.org/10.21037/atm.2019.12.151.

Yang, J. D., Hainaut, P., Gores, G. J., Amadou, A., Plymoth, A., & Roberts, L. R. (2019). A global view of hepatocellular carcinoma: Trends, risk, prevention and management. *Nature Reviews Gastroenterology and Hepatology, 16*(10), 589–604. https://doi.org/10.1038/s41575-019-0186-y.

Yasaka, K., Akai, H., Abe, O., & Kiryu, S. (2018). Deep learning with convolutional neural network for differentiation of liver masses at dynamic contrast-enhanced CT: A preliminary study. Radiology, 286(3), 887–896. https://doi.org/10.1148/radiol.2017170706.

Zhen, S. H., Cheng, M., Tao, Y. B., Wang, Y. F., Juengpanich, S., Jiang, Z. Y., Jiang, Y. K., Yan, Y. Y., Lu, W., Lue, J. M., Qian, J. H., Wu, Z. Y., Sun, J. H., Lin, H., & Cai, X. J. (2020). Deep learning for accurate diagnosis of liver tumor based on magnetic resonance imaging and clinical data. *Frontiers in Oncology, 10*. https://doi.org/10.3389/fonc.2020.00680.

Zhou, W., Wang, G., Xie, G., & Zhang, L. (2019). Grading of hepatocellular carcinoma based on diffusion weighted images with multiple b-values using convolutional neural networks. *Medical Physics, 46*(9), 3951–3960. https://doi.org/10.1002/mp.13642.

Zimmermann, A. (2016). *Etiology and pathogenesis of hepatocellular carcinoma: Inflammatory and toxic causes* (pp. 2931–2959). Springer Science and Business Media LLC. https://doi.org/10.1007/978-3-319-26956-6_164.

Further reading

Okuda, K. (1986). Early recognition of hepatocellular carcinoma. *Hepatology, 6*(4), 729–738. https://doi.org/10.1002/hep.1840060432.

CHAPTER

16

Issues and future challenges in cancer prognosis: (Prostate cancer: A case study)

Dukka Karun Kumar Reddy[a], *H. Swapnarekha*[b], *H.S. Behera*[c], *S. Vimal*[d], *Asit Kumar Das*[e], *and Danilo Pelusi*[f]

[a]Dr. Lankapalli Bullayya College of Engineering, Visakhapatnam, Andhra Pradesh, India [b]Department of Information Technology, Aditya Institute of Technology and Management, Tekkali, Andhra Pradesh, India [c]Department of Information Technology, Veer Surendra Sai University of Technology, Burla, Odisha, India [d]Ramco Institute of Technology, Rajapalayam, Tamilnadu, India [e]IIEST Shibpur, Howrah, West Bengal, India [f]University of Teramo, Teramo, Italy

Introduction

Cancer disease can originate almost in any tissue or organ of the body. When abnormal cells rapidly multiply with uncontrollable expansion, they go beyond their unusual boundaries by infecting and spreading to other body parts or organs. Cancer is also known as a malignant or neoplasm tumor. A tumor's presence does not necessarily imply cancer; tumors can be non-cancerous (benign) and pre-cancerous (pre-malignant). Fig. 16.1 illustrates that after lung cancer, prostate cancer is the second most leading frequent cancer diagnosis and the fifth leading cause of death worldwide made in men, according to WHO 2018 statistics. The international cancer research agencies estimate that globally, 1 in 11 females and 1 in 8 males die from the disease. One in five people develops cancer during their lifetime. These statistics suggest that more than 50 million people live within 5 years of a past cancer diagnosis (International Agency for Research on Cancer, WHO, 2021).

DNA is found in all recognized cellular life with genetic instructions for living organisms' creation and work. The primary function of DNA in the cell is to store information since it provides the instructions for constructing other components of the cell, such as proteins

Computational Intelligence in Cancer Diagnosis
https://doi.org/10.1016/B978-0-323-85240-1.00001-8

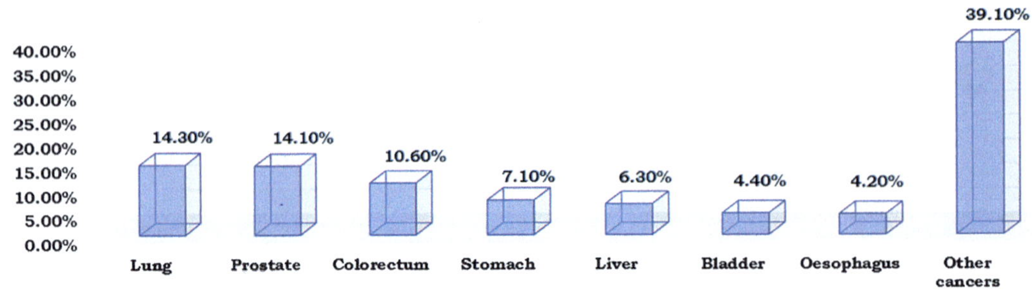

FIG. 16.1 Most frequent cancer diagnosis worldwide.

and RNA molecules. According to physicians, prostate cancer ascends to develop changes in the DNA of prostate cells. The DNA of cells contains the instructions for the cell to what to do. The differences in the cells cause them to divide and expand at a faster rate than normal cells. Where other cells die and dysfunctional cells continue to live. The abnormal cells clump together to form a tumor, which can spread and invade the surrounding tissue. Some abnormal cells may break away and spread to other parts of the body over time (metastasize).

Cancer that is related to or occurs in the prostate gland is known as prostate cancer. The prostate gland is a part of the male reproductive system located deep inside the groin, which comprises the penis, prostate, testicles, seminal vesicles, and rectum. The prostate is located between the rectum and the base of the penis just below the bladder. It is also one of the vital parts of the reproductive system. The prostate develops during puberty under the male hormone testosterone's influence and its by-product DHT (dihydrotestosterone). It provides semen (seminal fluid), which combines with the sperm from the testes. The seminal fluid with perfect consistency and environment aids the sperm to nourish and travel to fertilize the female's eggs. Around 30% of the seminal fluid comes from the prostate gland. Fibrolysin, protein-splitting enzymes, acid phosphatase, citric acid, fructose, potassium, calcium, sodium, and zinc are the critical components of its secretions. The seminal fluid provides sperm with the energy they need to travel to the egg. Many prostate cancers are slow-growing and limited to the prostate gland, which is unlikely to cause significant damage. While some forms of prostate cancer are slow-growing and need little or no medication, others are persistent and spread rapidly to the parts of the body. Table 16.1 shows the prostate cancer estimated number of cases from 2020 to 2040 in males of all ages region-wise (International Agency for Research on Cancer, WHO, 2021). Prostate cancer is often diagnosed in men using diagnostic tests such as a digital rectal exam or a Prostate-Specific Antigen Examination. Prostate cancer in its early stages typically has no symptoms. Symptoms of advanced prostate cancer, which are possible, are shown in Fig. 16.2.

The high global incidence of prostate cancer necessitates a strengthening of current resources for detecting patterns and developing preventive strategies to minimize the disease's potential public health effects. Prostate cancer registries, in particular, are important for the advancement of prostate cancer research to collect data on morbidity and prevalence. The disease characteristics presents, risk factors and trends, disparities in access to treatment, quality of care, and long-term data related to oncologic and health-related quality of life outcomes with management of the disease. As a consequence, improvements in data quality,

TABLE 16.1 Prostate cancer (males of all ages) estimated number of incident cases from 2020 to 2040.

Region	Estimated cases in 2020	Prediction in 2040	Increased cases by 2040	Percentage increase in cases (%)
Asia	371,225	720,659	349,434	+94.1
Europe	473,344	603,856	130,512	+27.6
Latin America and Caribbean	214,522	389,442	174,920	+81.5
Northern America	239,574	295,809	56,235	+23.5
Africa	93,173	192,695	99,522	+106.8
Oceania	22,421	33,107	10,686	+47.7
Total	1,414,259	821,309	2,235,568	+58.1

Estimated number of new cases (in millions).

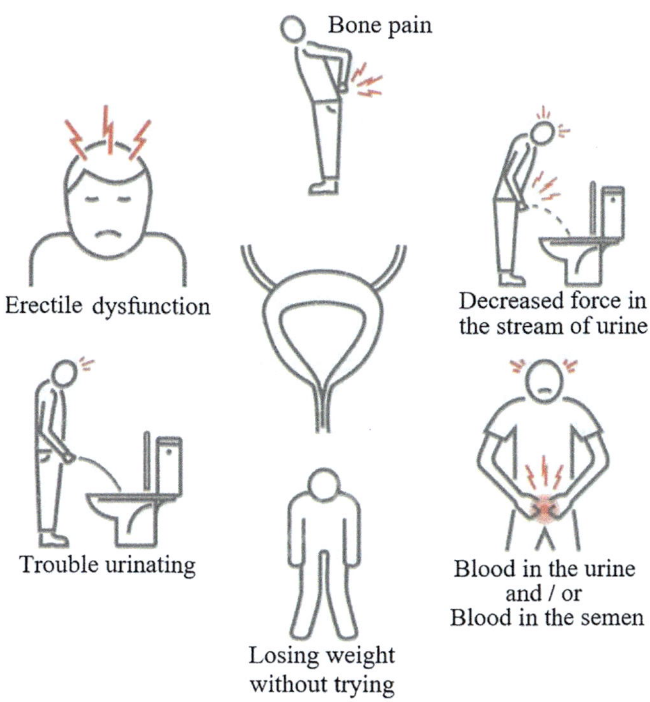

FIG. 16.2 Symptoms of prostate cancer advanced stage.

tissue acquisition, and data feedback to health care providers would increase the importance of epidemiological studies (Rawla, 2019).

ML algorithms key strength is their ability to analyze and use vast quantities of data more easily than humans using conventional statistical analysis. The rapidly growing area is quantitative data extraction from medical images like MRI (magnetic resonance imaging). These datasets are analyzed for clinical insights such as treatment outcomes, patient prognosis, and correlations with other biomarkers (Aerts et al., 2014). ML has been suggested for a variety of medical applications in recent years, the most important of which are the identification and characterization of malignant tumors in various body regions, reduced acquisition time, automatic segmentation of organs and lesions, and early detection of neurological disorders (Chang et al., 2018; Rawla, 2019). Unfortunately, machine learning algorithms are far from mainstream use in clinical practice, owing to the current rarity of large amounts of data needed for their implementation (Cuocolo et al., 2019).

The main contributions of this chapter are

- Design and implementation of the GB approach with PSO for hyperparameter optimization.
- Assessing the influencing features of prostate cancer disease from realistic and synthetic generated datasets to force the algorithms to learn the essential features and make it competent to evaluate clinically significant and insignificant prostate cancer.
- Study of performance analysis with the proposed method and other competent ML algorithms.
- Examining the performance metrics using precision, recall, FPR, AUC, and accuracy through cross-fold validation.
- Compared to ML classifiers, the predicted model has a robust impact on prostate cancer malignant and benign classification.

The chapter is systematized as follows: "Literature review" section presents a brief analysis of significant work in prostate cancer. "Proposed methodology" section describes the proposed GB with PSO-based technique with various ML algorithms for the data analysis part. "Materials and methods" section presents description of the dataset and synthetic dataset generation with performance metrics. "Experimentation and result analysis" section demonstrates the experimentation setup and result analysis. "Conclusion" section presents the conclusion.

Literature review

The early cancer prediction with the assessment of diagnosis is receiving more significance and attention. Subsequently, the periodicity of prostate cancer patients is rapidly increasing. The examination of the malignant, benign tumor will play a considerable challenge for the early assessment of cancer diagnosis. The ML technique has been a significant inspiration for numerous researchers because of its potential in characterizing and predicting the diagnosis of tumors at an early stage and symptom assessment of tumor risk (Table 16.2).

TABLE 16.2 Various ML models and extracted dataset with various parameters.

Reference	Approach	Type of study	Number of patients	Positive class	Number of lesions	Data origin	Data extraction	Classification problem	Metrics
Zhu et al. (2020)	ANN	Retrospective	153	64	116	Local	mpMRI	Gleason score ≥ 7 *vs.* ≤ 6	Specificity, sensitivity, and AUC
Zhong et al. (2019)	CNN	Retrospective	140	49	216	Local	mpMRI	Gleason score ≥ 7 *vs.* ≤ 6	Specificity, sensitivity, and AUC
Hussain et al. (2018)	SVM (with Gaussian, RBF, and polynomial approach), DT, and NB	Retrospective	20	–	–	Public	Scale-invariant transform, morphological, elliptic Fourier descriptors, and texture, entropy from MRI	Prostate and brachytherapy	NPV, PPV, FPR ROC, sensitivity, and specificity
Barlow et al. (2019)	KNN, SVM2 QD, DT, RF, MLP, and ADA	Retrospective	21,171	1130	–	Local	mpMRI	Gleason score ≥ 7(*PSA level* ≥ 20 ng/mL) and ≤ 6	Specificity, sensitivity, PPV, NPV, F1-Score, and AUC
Wang et al. (2017)	SVM (trained with radiomics, PI-RADS scores, and radiomics and PI-RADS scores)	Prospective	54	43	54	Local	mpMRI radiomics	Prostate cancer and normal (PZ, TZ)	Specificity, sensitivity, positive likelihood ratio, negative likelihood ratio, and AUC-ROC
Jović et al. (2017)	ELM, ANN, GP	Prospective	–	–	–	–	–	–	RMSE, R^2

Continued

TABLE 16.2 Various ML models and extracted dataset with various parameters—cont'd

Reference	Approach	Type of study	Number of patients	Positive class	Number of lesions	Data origin	Data extraction	Classification problem	Metrics
Toivonen et al. (2019)	LR	Prospective	62	80	100	Local	Multisequence mpMRI	Gleason score ≥ 7 *vs* ≤ 6	AUC
Nguyen et al. (2017)	RF and LR	Retrospective	144	–	–	Local	Morphological features with quantitative information from the glands and stroma	Gleason score ≥ 7 *vs* ≤ 6	AUC
Liu et al. (2017)	CNN and XGBoost	Retrospective	199	–	317	Local	3D mpMRI	Clinically significant or non-significant lesions	Sensitivity and specificity
Shafai-Erfani et al. (2019)	RF	Prospective	10	–	–	Local	MRI-based synthetic computed topographies	Dose evaluation	Accuracy
Cuocolo et al. (2020)	Deep learning	Retrospective	–	–	–	Public	MRI-based radical prostatectomy	Gleason score ≥ 7 *vs.* ≤ 6	AUC
Min et al. (2019)	LR	Retrospective	280	86	280	Local	MRI-based radiomics signature	Gleason score ≥ 7 *vs.* ≤ 6	Specificity, sensitivity, and AUC
Abbasi et al. (2020)	CNN	Prospective	20	–	–	Public	MRI images of T1-Weighted scans	Prostate and brachytherapy	PPV, NPV, accuracy and ROC

Transin et al. (2019)	Linear mix model	Retrospective	74	70	127	Local	mpMRI	Gleason score ≥ 7 *vs.* ≤ 6	Specificity, sensitivity, and AUC
Schelb et al. (2019)	CNN	Prospective	312	38	457	Local	mpMRI	Gleason score ≥ 7 *vs.* ≤ 6	Specificity, sensitivity, and AUC
Maas et al. (2019)	LR	Prospective	50	57	65	Local	mpMRI	Gleason score ≥ 7 *vs* ≤ 6	AUC
Li et al. (2019)	LDA	Retrospective	103	27	167	Local	mpMRI	Gleason score ≥ 7 *vs* ≤ 6	AUROC
Jensen et al. (2019)	k-NN	Retrospective	99	68	112	Public	mpMRI	ISUP 5 classes	AUC
Abraham & Nair (2018)	ANN	Retrospective	162	40	182	Public	MRI	ISUP 5 classes	Accuracy, PPV, kappa
Abraham & Nair (2019)	CNN	Retrospective	–	68	100	Public	mpMRI	ISUP 5 classes	Precision, recall, and AUC
Antonelli et al. (2019)	LR	Prospective	164	63	164	Local	3-T mpMRI	Gleason score 4 vs 3	Accuracy, sensitivity, specificity mean sensitivity and AUC

Proposed methodology

Ensemble learning is the ML concept with the progression of multiple models, predominantly used to perk up the function approximation, classification, prediction, etc. These multiple models like experts or classifiers are generated and combined through a strategic perspective in solving a specific computational intelligence problem by producing one optimal predictive model. Boosting is one of the most effectively utilized algorithmic calculations in data science, under ensemble learning techniques. Boosting is another effective way to enhance the predictive control of the regression tree and classical decision models, where a set of base learners (weak learners) alludes and combined to craft a strong learner that acquires better execution than a single one. The boosting techniques pull together base learners to shape into a strong rule. The base learning algorithm of ML is used with different distribution in sequential order to identify the weak rule and corrects its predecessor, which produces a new rule for weak prediction. The integration of all these weak rules is used to build a strong single prediction rule by applying the base learning algorithm every time. This entire iterative procedure followed by choosing the right distribution for each round with copious iterations is known as the general boosting technique shown in Fig. 16.3.

Gradient Boosting

The Gradient Boosting (GB) algorithm (Refer Table 16.3) is the most efficient and widespread implementation of the Adaptive Boosting algorithm. The derivation pursues a similar idea to the existing literature of AdaBoost. It comes under supervised learning, which is based on function approximation by optimizing a specific loss function. GB trains the models in a steady, stabilizer, and sequential manner. AdaBoost and GB's primary transformation is how these algorithms recognize weak learners' inadequacies (e.g., decision trees). Whereas the AdaBoost algorithm recognizes the inadequacies using high weight data points, GBA achieves the same by gradients in the loss function. The GB algorithm involves three significant steps for training the model into different sub-models, i.e., optimizing the loss function and base learners with predictions. Finally, to minimize the loss function, an adaptive model is incorporated with base learners. Each sub-model gradually reduces its loss function by

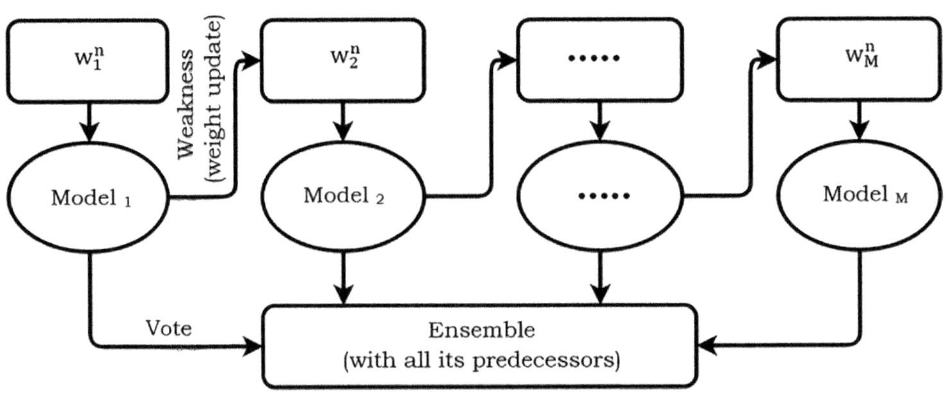

FIG. 16.3 General Boosting technique.

TABLE 16.3 Gradient Boosting algorithm.

Gradient Boosting algorithm	
Step 1:	Initialization of weights to $w : w_1\, w_2 \ldots w_n = \frac{1}{n}$ where n be the total instances of the dataset.
Step 2:	Initialize f with some constant value, where f ensemble function.
Step 3:	While $m < M$ do: where M is the number of models to be grown
	Step 3.1 Models are created and get the hypothesis is $H_m\,(x_n)$ for all data points, where x_n in the dataset and y_n corresponds labels.
	Step 3.2 Computing the negative gradient $\{g_m(H_m(x_n))\}_{n=1}^{T}$
	$g_m\,(H_m(x_n)) = \left(\frac{\partial\psi\,(y_n, f(H_m(x_n)))}{\partial f(H_m(x_n))} \Big\vert H_m(x_n) \right)_{f(H_m(x_n)) = f^{m-1}(H_m(x_n))}$
	Step 3.3 Fitting a new base learner function $h(H(x)\,\theta_m)$
	Step 3.4 Calculating the best gradient descent step size ρ_m
	$(\rho_m, \theta_m) = argmi\, n_{\rho,\theta} \sum_{n=1}^{T} \psi\left(\left(y_n, f^{m-1}(H_m(x_n)) \right) + \rho h(H_m(x_n), \theta_m) \right)$
	Step 3.5 Computing ψ_m which is a specific loss function $\psi(y,f)$
	Step 3.6 updating the function estimate for the model
	$f_m = f_{m-1} + \rho_m\, h(H_m(x_n)\theta_m)$
Step 4:	Following M iterations, the final regression functional output with: $f(x) = \sum_{m}^{M} f_m(x)$

using gradient descent (Friedman, 2001). This course of action integrates all the sub-models (trees) through successively learning and is provided with a highly accurate prediction for the given variables (Reddy et al., 2021). The final output estimation is derived from the integration of all the predictions from various base learners. In GB, the base learners are provided with DT. These DTs restrict every DT's nodes that take different features of the data subset for picking the best split. Every sub-model reflects on the preceding tree's errors and takes these considerations into account. Consecutive DTs built from preceding tree errors by ensemble, and all the preceding predecessors of DT's model GB integrates different base learners' predictions to generate the final performance prediction.

PSO

Optimization is defined as minimizing or maximizing a function in relation to a finite set, which often represents a range of options for a given situation. The function is determined with the best choice among comparisons with different choices. The optimization objective deals with a function application in deriving variational principles, optimal management, optimal design, minimal error, maximal profit, and minimal cost. Several studies on the social behavior of animal groups were developed in the early 1990s. These studies disclosed that certain animals like birds and fishes that belong to a certain group could exchange information among their group. PSO (Refer Table 16.4) was proposed by Dr. Kennedy and Dr.

TABLE 16.4 PSO algorithm.

PSO algorithm	
Step 1:	Initialization for each particle i in swarm population size P **Step 1.1** Initialize x_i *and* v_i randomly. **Step 1.2** Evaluate the fitness $f(X_i)$ and initialize $P_{best(i)}$ randomly. **Step 1.3** Initialize *gbest* from X_i best fitness.
Step 2:	Repeat until stopping criteria is reached **Step 2.1** For each particle i **Step 2.1.1** Update V_i^t *and* X_i^t /**where* V_i^t *is inertia** / $X_i^t(t+1)=X_i^t(t)+V_i^t(t+1)$ $V_i^t(t+1)=w*V_i^t+c_1*r_1*(P_{best(i)}^t-P_i^t)+c_2*r_2*(P_{gbest}^t-P_i^t)$ **Step 2.1.2** Evaluate the fitness $f(X_i)$ *if* $f(P_{best(i)})<f(X_i^t)$ *then* $P_{best(i)}=X_i^t$ *if* $f(P_{gbest})<f(X_i^t)$ *then* $P_{gbest}=X_i^t$
Step 3:	Return P_{gbest} as the best estimation as the global optimum solution.

Eberhart in the year 1995, inspired by bird flocking or fish schooling (Eberhart & Kennedy, 1995). Such capability grants that these have major survival advantages. It is a metaheuristic algorithm with population basis that is suitable to optimize non-linear continuous functions.

Implementation

The proposed work's objective is to finetune the hyperparameters of the GB algorithm with respect to PSO. The validation error of an ML model using a set of hyperparameters is the objective function in hyperparameter optimization to yield and identify the lowest error rate on the validation set with generalized optimum results. Since determining the objective function necessitates training, the ML model with a complex set of hyperparameters is costly. So, the proposed work explores and evaluates the search space through the PSO technique for minimizing the number of poor hyperparameter choices. PSO hyperparameter tuning uses a continuously updated particle model's velocity to concentrate on promising hyperparameters by reasoning from particle *pbest* and *gbest*. The hyperparameters of GB taken into consideration for the study are *n_estimators, max_depth, min_samples_leaf, and min_samples_split.* Five-fold cross-validation is performed on the dataset to observe the mean results. The PSO explores the learning model parameters within the search space through the number of evaluations that is considered for the study of PSO, which is 10.

Materials and methods

This section addresses the dataset collection and experimental details of the proposed framework, which comprises materials and the overall framework method for validating the performance measures used in the experiment. It is challenging for early detection and identification of malignant cancers and potentially crucial for early diagnosis strategies. In this study, a novel technique to support predicting prostate cancer was proposed, namely, GB+PSO. Accordingly, the GB machine was developed to predict cancer first; then, the PSO algorithm was applied to optimize the parameters of the GB model. The classical ML model was also developed and investigated to predict prostate cancer and compared with the proposed GB+PSO model. Using the prostate cancer disease data and taking calculated reference truth, the accuracy of model estimation was evaluated and the impact of parameter optimization on model estimation was tested. The results showed that the PSO algorithm could optimize the parameters of the GB model stably, the GB+PSO model could accurately estimate malignant cancers, and the estimation accuracy of the model increases with the optimized parameter tuning. Compared with other ML models, the GB+PSO model could obtain the best estimation performance of significant and insignificant prostate cancer. Fig. 16.4 shows the overall framework of the system. The dataset collection and observation are the initial processes of this framework. In this process, the collected dataset is observed strictly to discover the sorts of information (datatype). The learning algorithm uses the hyperparameter optimization technique to develop a final model using cross-fold validation. The ML algorithms were used to study the detailed analysis of the dataset. The following lists of algorithms that are taken into consideration for this study are Multi-layer Perceptron

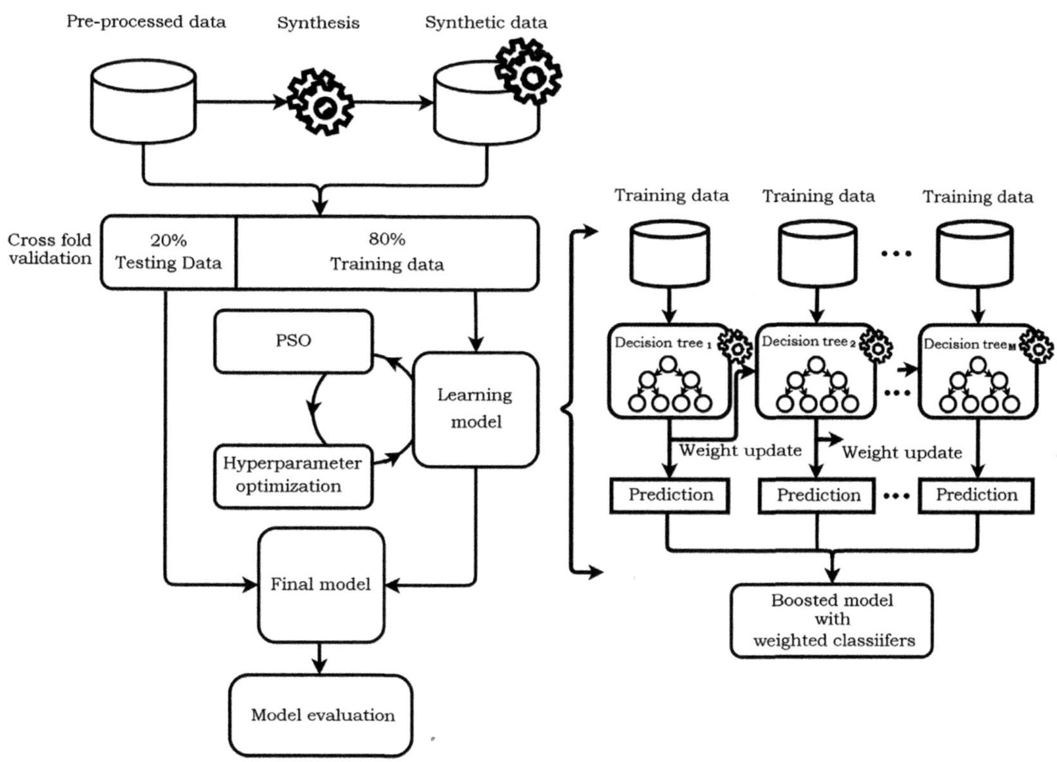

FIG. 16.4 Overall system framework of GB+PSO.

(MLP), Extra Trees Classifier (ETC), Decision Tree (DT), and Random Forest (RF). In addition to the above algorithms, GB is also considered for this study. The GB algorithm used PSO for finetuned hyperparameter and compared it with ML algorithms. The complete model was assessed by using multiple assessment metrics.

Dataset collection and description

The experiment dataset is collected from an open-source Kaggle (Saifi & Mahmood, 2019). The dataset consists of feature Id, radius, texture (significantly indices difference between homogeneous and heterogeneous lesions), area, perimeter, compactness, smoothness, fractal_dimension, symmetry, and diagnosis_result. The dataset is collected from solid prostate masses of clinical patients with and a simple computer graphical program, on the basis of a digital scan, which is capable of performing cytological feature analysis. The general features of malignancy in cytological features are based on digital scans of each cell nucleus. The features are high cellularity, cellular enlargement, nuclear hyperchromasia, increased nuclear ratio, discohesiveness of cells, abnormal distribution of nuclear chromatin, etc. The program computes 10 real-valued features from each cell nucleus in the sample by using a curve-fitting algorithm and then calculates the mean, extreme, and standard error value of each feature for the image, returning real-valued vectors. The diagnosis_result is the target class specified with labels as malignant (M) and benign (B). Malignant tumors are cancerous in which abnormal cells divide uncontrollably and destroy body tissue. A benign tumor is not the same as a malignant tumor. It does not infiltrate surrounding tissue or spread to other parts of the body, i.e., non-cancerous.

Synthetic data generation

As small-scale datasets provide hindrance to understand the patterns working with the ML algorithms significantly, a fixed number of instances will drive to limited samples, limited patterns, and a limited degree of the class difference between the samples. A synthetic dataset is generated programmatically with statistical distribution to the base dataset underlying the process of precisely controlled and tuned randomly. Its primary purpose is of being flexible for ML algorithms with desired properties as of the original dataset, which makes the degree of class separation controllable for ML algorithms to make the learning problem hard or easy. When it comes to modeling distributions of statistical counterparts, CTGANs have more modeling distributions (Xu et al., 2021). CTGAN is a GAN-based method for modeling and sampling rows from a tabular data distribution. A GAN is composed of two models with (i) a generator that learns to generate fake data and (ii) a discriminator that learns to differentiate the generator's fake data from the real data. The mode-specific normalization in CTGAN is used to cope with columns with complicated distributions. The complicated distributions contain non-Gaussian and multimodal distributions, while a conditional generator and training-by-sampling approaches are used to combat class imbalance problems. The generator in a GAN is fed with a vector sampled from a conventional multivariate normal distribution. By training together with critical neural networks, one eventually obtains a deterministic transformation that translates the conventional multivariate normal distribution into the distribution of the data. Since columns in a row do not have a local structure, fully connected networks in generator and critic are used to capture all conceivable correlations between columns. Generally, two fully connected hidden layers are found in both generator (batch-normalization and ReLU activation function is used) and critic (leaky ReLU function and dropout on each hidden layer). The synthetic row representation is generated using a mix of activation functions after the hidden layers. The synthetically produced data cause GANs to fall short of the baseline methods on various metrics such as machine learning efficacy and probability fitness. The synthetic dataset of 100 instances is created through CTGAN with the base dataset, and the evaluation metric gives a value of 0.219 to show how similar the two datasets are. The proposed work is carried out on the integration of these two datasets (Refer Fig. 16.5).

Proposed model

In this section, the GB + PSO model was proposed to predict the benign tumor and malignant tumor from the prognosis cancer dataset. For this stage, an initial GB model was developed first with fivefold cross-validation; then, its hyper-parameters were optimized by the PSO algorithm. In the initial GB model, default hyper-parameters were considered. For GB + PSO, four hyper-parameters were considered and optimized (n_estimators, max_depth, min_samples_leaf, and min_samples_split). For the determination of the optimal values of these parameters, the particles fly in the search space and exchange experiences. For each position, they calculate the fitness of particles via a fitness function, i.e., accuracy. For each value of hyper-parameters, a corresponding value was computed ("Evaluation criteria" section), and the best fit model is corresponding to the highest accuracy. The scheme of the development of the GB + PSO model for estimating the detection and identification of malignant cancers is shown in Fig. 16.4.

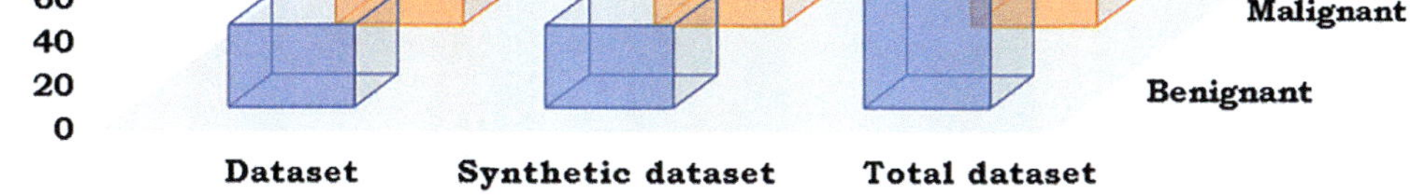

FIG. 16.5 The actual dataset and synthetic dataset with diagnosis result.

Evaluation criteria

Performance evaluation aims to assess the developed system's effectiveness and estimate the model's accuracy with an appropriate technique for this work on future data. For evaluating the output of a classification model, a confusion matrix is used. The confusion matrix offers a simple picture of classification by prediction outcomes, i.e., the ML errors, and types of errors made by the classification. The matrix compares the predicted ML values with the actual values by true negative (TN), true positive (TP), false negative (FN), and false-positive (FP). The model's exactness is given by precision, recall, false-positive rate (FPR), and accuracy (Reddy et al., 2020).

Experimentation and result analysis

The experimental design utilized IdeaPad 330 LENOVO laptop with an i5-8250U processor with Intel Core, 8GB RAM, and Windows 10 Operating system. The SDV library is used for synthetic data generation with statistical properties as the original dataset. Optunity library is used for hyperparameter tuning. For data observation and study, Pandas and NumPy libraries are used. For the data analysis of ML classifiers and metrics evaluations, Scikit-learn library was utilized through *Google colab interactive* environment.

The proposed GB + PSO algorithm outperformed well compared to conventional ML algorithms and GB in this portion of the resulting study. The algorithms mentioned above' confusion matrix is portrayed in Figs. 16.6–16.11. These figures show the confusion matrix at different folds (*k*-represents fold). The MLP model fails to predict FN instances, and the FP, i.e., wrongly predicted instances, are more. The MLP model performed well at $k=3$. The ETC performs well compared to MLP at every fold and performs well at $k=5$. The RF and DT performed well in classifying the malignant class at every fold. The RF at fold $k=1,2$ and 5 performed well in classifying the significant and insignificant prostate cancer compared to DT. The GB performed well in all the folds, except at $k=5$. At $k=1$, two wrongly classified instances are predicted.

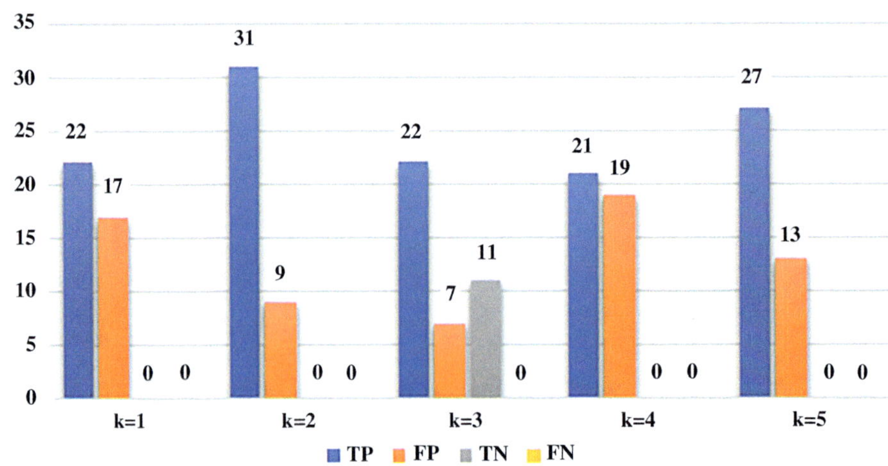

FIG. 16.6 Confusion matrix for cross fold validation (MLP).

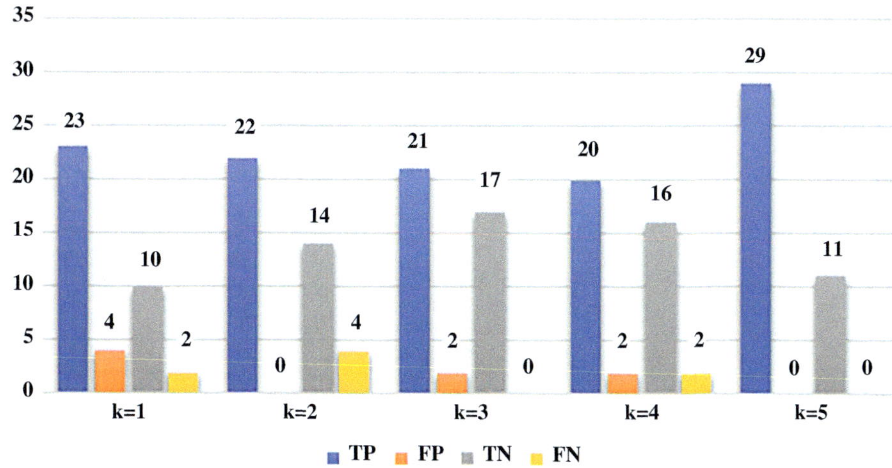

FIG. 16.7 Confusion matrix for cross fold validation (ETC).

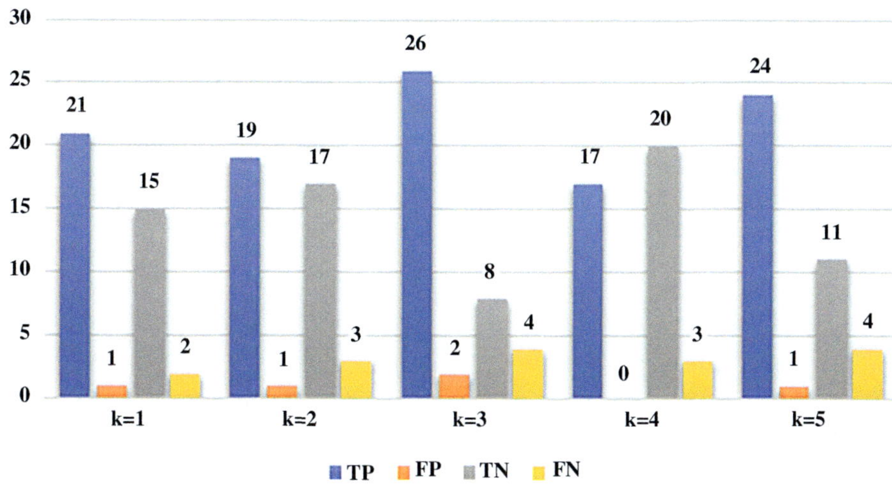

FIG. 16.8 Confusion matrix for cross fold validation (DT).

Similarly, at $k=2$, three instances are wrongly predicted. And at $k=3$ and 4, one instance is wrongly classified. The GB+PSO at $k=2$ have minimum misclassification as 0 and maximum misclassification of 3 at fold $k=1$, i.e., the proposed method has exactly classified the instances of significant and insignificant prostate cancer instances effectively. Tables 16.5–16.10 show the model's performance with metrics precision, recall, FPR, AUC, and accuracy. The proposed approach GB+POS gives an accuracy of 96% for classification to predict significant and insignificant prostate cancer, whereas the precision value for MLP is 0.65 and that for FPR with 0.87. The recall value of 1 is generated for MLP because TN and FN instances are classified as 0. The ETC produces a value of 0.93 and 0.1 for the precision, recall, and FPR. The RF has been evaluated with FPR with an area of 0.143.

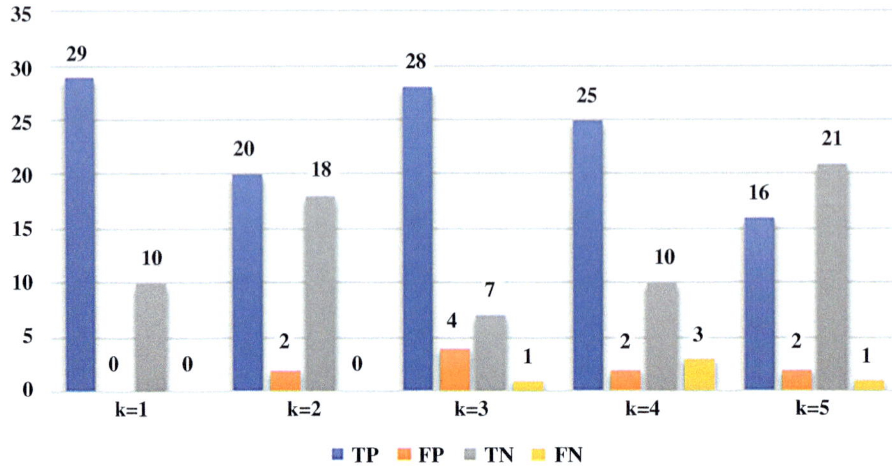

FIG. 16.9 Confusion matrix for cross fold validation (RF).

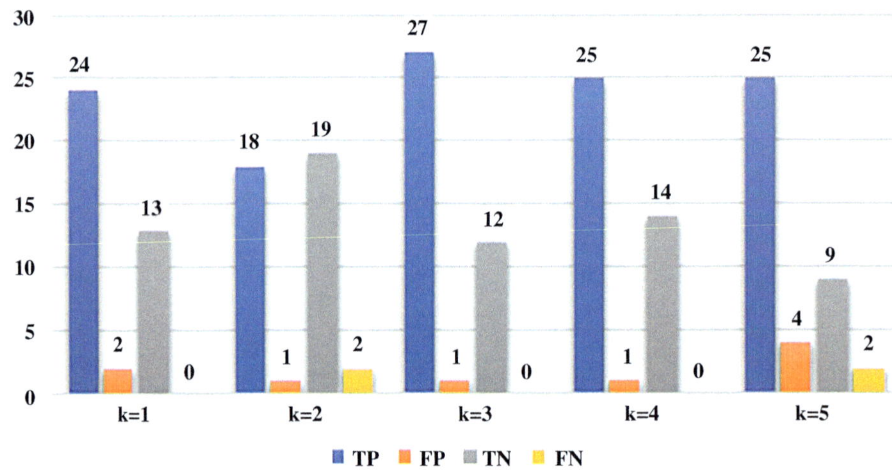

FIG. 16.10 Confusion matrix for cross fold validation (GB).

The precision and recall values are 0.91 and 0.95, respectively. The DT produces the least FPR comparative to GB, and in terms of precision, DT is evaluated with 0.95 and GB with 0.93. The GB + PSO classifies the malignant and benign class effectively when compared to the other models. The GB + PSO produced a mean FPR of 0.042, with an AUC of 0.96 and overall mean accuracy of 0.9646. Table 16.11 shows the GB + PSO with initial search space taken into consideration and the mean (average) explored search space of the evaluated hyperparameters.

Figs. 16.12 and 16.13 show the cross-fold validation metrics and confusion matrix of GB + PSO with a number of evaluations. The number of evaluations of PSO considered for this study is 10. So, at each evaluation, cross-validated results are illustrated in these figures.

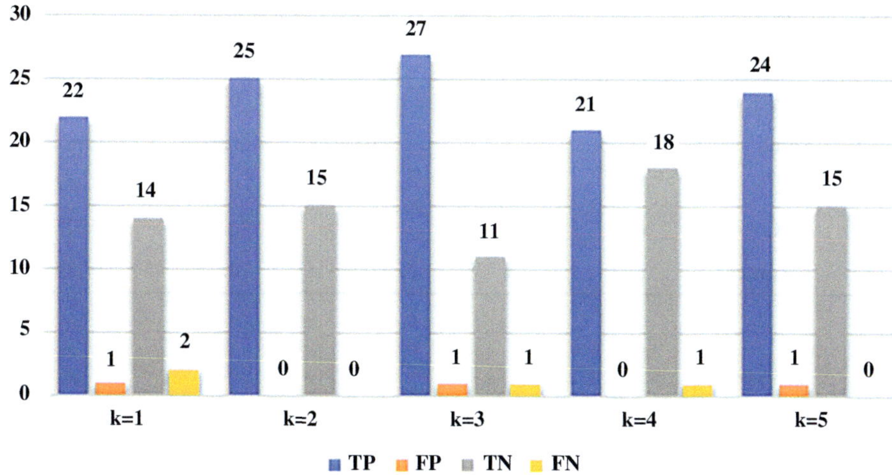

FIG. 16.11 Confusion matrix for cross fold validation (GB+PSO).

TABLE 16.5 MLP metrics.

ML model	Metrics	Fold-1	Fold-2	Fold-3	Fold-4	Fold-5	Mean
MLP	Precision	0.5641	0.775	0.7586	0.525	0.675	0.6595
	Recall	1	1	1	1	1	1
	FPR	1	1	0.3889	1	1	0.8778
	AUC	0.5	0.5	0.8056	0.5	0.5	0.5611
	Accuracy	0.5641	0.775	0.825	0.525	0.675	0.6728

TABLE 16.6 ETC metrics.

ML model	Metrics	Fold-1	Fold-2	Fold-3	Fold-4	Fold-5	Mean
ETC	Precision	0.8519	1	0.913	0.9091	1	0.9348
	Recall	0.92	0.8462	1	0.9091	1	0.935
	FPR	0.2857	0	0.1053	0.1111	0	0.1004
	AUC	0.8171	0.9231	0.9474	0.899	1	0.9173
	Accuracy	0.8462	0.9	0.95	0.9	1	0.9192

TABLE 16.7 RF metrics.

ML model	Metrics	Fold-1	Fold-2	Fold-3	Fold-4	Fold-5	Mean
RF	Precision	1	0.9091	0.875	0.9259	0.8889	0.9198
	Recall	1	1	0.9655	0.8929	0.9412	0.9599
	FPR	0	0.1	0.3636	0.1667	0.087	0.1435
	AUC	1	0.95	0.8009	0.8631	0.9271	0.9082
	Accuracy	1	0.95	0.875	0.875	0.925	0.925

TABLE 16.8 DT metrics.

ML model	Metrics	Fold-1	Fold-2	Fold-3	Fold-4	Fold-5	Mean
DT	Precision	0.9545	0.95	0.9286	1	0.96	0.9586
	Recall	0.913	0.8636	0.8667	0.85	0.8571	0.8701
	FPR	0.0625	0.0556	0.2	0	0.0833	0.0803
	AUC	0.9253	0.904	0.8333	0.925	0.8869	0.8949
	Accuracy	0.9231	0.9	0.85	0.925	0.875	0.8946

TABLE 16.9 GB metrics.

ML model	Metrics	Fold-1	Fold-2	Fold-3	Fold-4	Fold-5	Mean
GB	Precision	0.9231	0.9474	0.9643	0.9615	0.8621	0.9317
	Recall	1	0.9	1	1	0.9259	0.9652
	FPR	0.1333	0.05	0.0769	0.0667	0.3077	0.1269
	AUC	0.9333	0.925	0.9615	0.9667	0.8091	0.9191
	Accuracy	0.9487	0.925	0.975	0.975	0.85	0.9347

TABLE 16.10 GB + PSO metrics.

ML model	Metrics	Fold-1	Fold-2	Fold-3	Fold-4	Fold-5	Mean
GB + PSO	Precision	0.9565	1	0.9643	1	0.96	0.9762
	Recall	0.9167	1	0.9643	0.9545	1	0.9671
	FPR	0.0667	0	0.0833	0	0.0625	0.0425
	AUC	0.925	1	0.9405	0.9773	0.9688	0.9623
	Accuracy	0.9231	1	0.95	0.975	0.975	0.9646

TABLE 16.11 GB + PSO with optimized hyperparameters.

GB + PSO parameters	PSO optimized hyperparameters	
	Search space	**Mean explored search space**
n_estimators	[10, 50]	25.48
max_depth	[5,20]	13.76
min_samples_leaf	[1,2]	1.54
min_samples_split	[2,3]	2.06

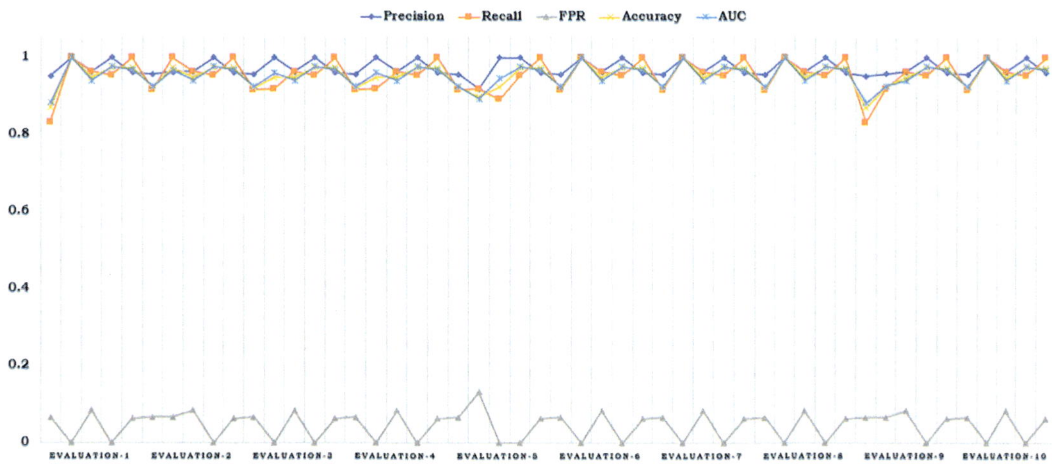

FIG. 16.12 GB + PSO metrics at different evaluations.

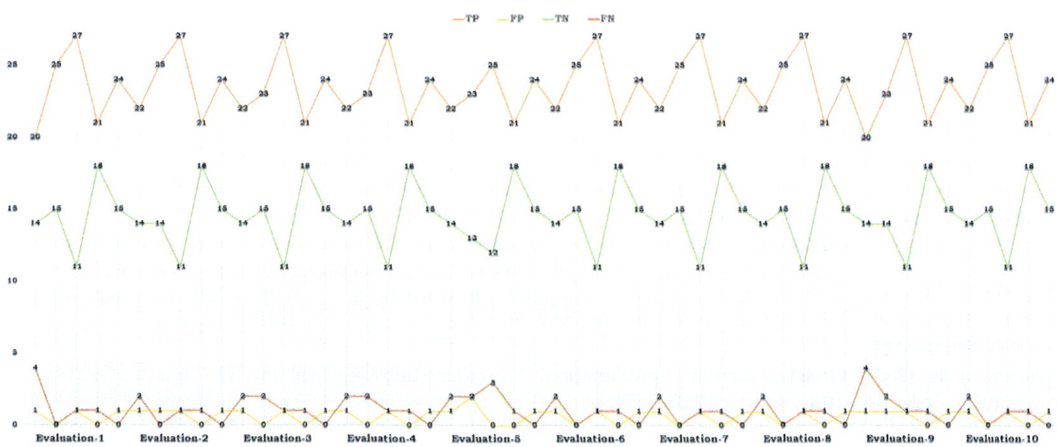

FIG. 16.13 GB + PSO confusion matrix at different evaluations.

Conclusion

With the advancement in technology, it is possible to thoroughly examine both common and less common signs of the tumor to avoid serious effects, which may be identified timely at different stages from initiation of the disease to diagnosis. ML algorithms confront major challenges in dealing with algorithm selection, data acquisition, time and resources, interpretation of results, and high error susceptibility. To assess the added value in decision-making, prospective studies and validation are required to unlock ML approaches' promising potential. Thus, to come back from the worst generalization outcomes, appropriate stopping methods and validation are required for the objective function minimization procedure. So, the proposed works deal with the validation error of the GB + PSO model using a set of hyperparameter tuning as the objective function. This analysis also reveals that methods need external validation, and no studies are discussing the actual use of optimization technique models in clinical workflows. Further research is required for an in-depth analysis by formulating a prototype with improved accuracy and reliability for new dimensions by data and individual clinical experiences.

References

Abbasi, A. A., et al. (2020). Detecting prostate cancer using deep learning convolution neural network with transfer learning approach. *Cognitive Neurodynamics*, *14*(4), 523–533. https://doi.org/10.1007/s11571-020-09587-5. Springer Netherlands.

Abraham, B., & Nair, M. S. (2018). 'Computer-aided classification of prostate cancer grade groups from MRI images using texture features and stacked sparse autoencoder. *Computerized Medical Imaging and Graphics*, *69*, 60–68. https://doi.org/10.1016/j.compmedimag.2018.08.006. Elsevier Ltd.

Abraham, B., & Nair, M. S. (2019). Automated grading of prostate cancer using convolutional neural network and ordinal class classifier. *Informatics in Medicine Unlocked*, *17*, 100256. https://doi.org/10.1016/j.imu.2019.100256. Elsevier B.V.

Aerts, H. J. W. L., et al. (2014). Decoding tumour phenotype by noninvasive imaging using a quantitative radiomics approach. *Nature Communications*, *5*(1), 4006. https://doi.org/10.1038/ncomms5006.

Antonelli, M., et al. (2019). Machine learning classifiers can predict Gleason pattern 4 prostate cancer with greater accuracy than experienced radiologists. *European Radiology*, *29*(9), 4754–4764. https://doi.org/10.1007/s00330-019-06244-2.

Barlow, H., Mao, S., & Khushi, M. (2019). Predicting high-risk prostate cancer using machine learning methods. *Data*, *4*(3). https://doi.org/10.3390/data4030129.

Chang, P., et al. (2018). Deep-learning convolutional neural networks accurately classify genetic mutations in gliomas. *American Journal of Neuroradiology*, *39*(7), 1201–1207. https://doi.org/10.3174/ajnr.A5667.

Cuocolo, R., et al. (2019). Machine learning applications in prostate cancer magnetic resonance imaging. *European Radiology Experimental*, *3*(1), 35. https://doi.org/10.1186/s41747-019-0109-2.

Cuocolo, R., et al. (2020). Machine learning for the identification of clinically significant prostate cancer on MRI: A meta-analysis. *European Radiology*, *30*(12), 6877–6887. https://doi.org/10.1007/s00330-020-07027-w.

Eberhart, R., & Kennedy, J. (1995). A new optimizer using particle swarm theory. In (Vol. #), *MHS'95. Proceedings of the sixth international symposium on micro machine and human science* (p. 39). IEEE. https://doi.org/10.1109/MHS.1995.494215.

Friedman, J. H. (2001). Greedy function approximation: A gradient boosting machine. *The Annals of Statistics*, *29* (5). https://doi.org/10.1214/aos/1013203451.

Hussain, L., et al. (2018). Prostate cancer detection using machine learning techniques by employing combination of features extracting strategies. *Cancer Biomarkers*, *21*(2), 393–413. https://doi.org/10.3233/CBM-170643.

International Agency for Research on Cancer, WHO. (2021). https://gco.iarc.fr/today/home.

Jensen, C., et al. (2019). Assessment of prostate cancer prognostic Gleason grade group using zonal-specific features extracted from biparametric MRI using a KNN classifier. *Journal of Applied Clinical Medical Physics, 20*(2), 146–153. https://doi.org/10.1002/acm2.12542.

Jović, S., et al. (2017). Prostate cancer probability prediction by machine learning technique. *Cancer Investigation, 35*(10), 647–651. https://doi.org/10.1080/07357907.2017.1406496.

Li, Q., et al. (2019). Radiological semantics discriminate clinically significant grade prostate cancer. *Cancer Imaging, 19*(1), 81. https://doi.org/10.1186/s40644-019-0272-y.

Liu, S., et al. (2017). Prostate cancer diagnosis using deep learning with 3D multiparametric MRI. In S. G. Armato, & N. A. Petrick (Eds.), *arXiv* (p. 1013428). https://doi.org/10.1117/12.2277121.

Maas, M. C., et al. (2019). A single-arm, multicenter validation study of prostate cancer localization and aggressiveness with a quantitative multiparametric magnetic resonance imaging approach. *Investigative Radiology, 54*(7), 437–447. https://doi.org/10.1097/RLI.0000000000000558.

Min, X., et al. (2019). Multi-parametric MRI-based radiomics signature for discriminating between clinically significant and insignificant prostate cancer: Cross-validation of a machine learning method. *European Journal of Radiology, 115*(March), 16–21. https://doi.org/10.1016/j.ejrad.2019.03.010.

Nguyen, T. H., et al. (2017). Automatic Gleason grading of prostate cancer using quantitative phase imaging and machine learning. *Journal of Biomedical Optics, 22*(3), 036015. https://doi.org/10.1117/1.JBO.22.3.036015.

Rawla, P. (2019). Epidemiology of prostate cancer. *World Journal of Oncology, 10*(2), 63–89. https://doi.org/10.14740/wjon1191.

Reddy, D. K. K., Behera, H. S., Nayak, J., Naik, B., Ghosh, U., & Sharma, P. K. (2021). Exact greedy algorithm based split finding approach for intrusion detection in fog-enabled IoT environment. *Journal of Information Security and Applications, 60.* https://doi.org/10.1016/j.jisa.2021.102866.

Reddy, D. K., et al. (2020). Deep neural network based anomaly detection in internet of things network traffic tracking for the applications of future smart cities. *Transactions on Emerging Telecommunications Technologies.* https://doi.org/10.1002/ett.4121.

Saifi, S., & Mahmood, D. S. (2019). *Prostate cancer.* Available at: https://www.kaggle.com/sajidsaifi/prostate-cancer/activity.

Schelb, P., et al. (2019). Classification of cancer at prostate MRI: Deep learning versus clinical PI-RADS assessment. *Radiology, 293*(3), 607–617. https://doi.org/10.1148/radiol.2019190938.

Shafai-Erfani, G., et al. (2019). Dose evaluation of MRI-based synthetic CT generated using a machine learning method for prostate cancer radiotherapy. *Medical Dosimetry, 44*(4), e64–e70. https://doi.org/10.1016/j.meddos.2019.01.002. Elsevier Inc.

Toivonen, J., et al. (2019). Radiomics and machine learning of multisequence multiparametric prostate MRI: Towards improved non-invasive prostate cancer characterization. *PLoS One, 14*(7), e0217702. https://doi.org/10.1371/journal.pone.0217702. Edited by K. L. Penney.

Transin, S., et al. (2019). Computer-aided diagnosis system for characterizing ISUP grade ≥ 2 prostate cancers at multiparametric MRI: A cross-vendor evaluation. *Diagnostic and Interventional Imaging, 100*(12), 801–811. https://doi.org/10.1016/j.diii.2019.06.012. Société française de radiologie.

Wang, J., et al. (2017). Machine learning-based analysis of MR radiomics can help to improve the diagnostic performance of PI-RADS v2 in clinically relevant prostate cancer. *European Radiology, 27*(10), 4082–4090. https://doi.org/10.1007/s00330-017-4800-5.

Xu, L., Skoularidou, M., Cuesta-Infante, A., & Veeramachaneni, K. (2021). *Modeling tabular data using conditional GAN.* ArXiv Preprint http://arxiv.org/abs/1907.00503.

Zhong, X., et al. (2019). Deep transfer learning-based prostate cancer classification using 3 tesla multi-parametric MRI. *Abdominal Radiology, 44*(6), 2030–2039. https://doi.org/10.1007/s00261-018-1824-5. Springer US.

Zhu, L., et al. (2020). Feasibility of integrating computer-aided diagnosis with structured reports of prostate multiparametric MRI. *Clinical Imaging, 60*(1), 123–130. https://doi.org/10.1016/j.clinimag.2019.12.010. Elsevier.

Further reading

International Agency for Research on Cancer, World Health Organization (n.d.). Available at https://gco.iarc.fr/today/home.

Liu, S., Zheng, H., Feng, Y., & Li, W. (2017). Prostate cancer diagnosis using deep learning with 3D multiparametric MRI. In *Vol. 10134. Progress in biomedical optics and imaging-proceedings of SPIE* SPIE. https://doi.org/10.1117/12.2277121.

Maas, M. C., Litjens, G. J. S., Wright, A. J., Attenberger, U. I., Haider, M. A., Helbich, T. H., Kiefer, B., Macura, K. J., Margolis, D. J. A., Padhani, A. R., Selnæs, K. M., Villeirs, G. M., Fütterer, J. J., & Scheenen, T. W. J. (2019). A single-arm, multicenter validation study of prostate cancer localization and aggressiveness with a quantitative multiparametric magnetic resonance imaging approach. In *Vol. 54. Investigative radiology* (pp. 437–447). Lippincott Williams and Wilkins. https://doi.org/10.1097/RLI.0000000000000558. Issue 7.

Reddy, D. K. K., & Behera, H. S. (2020). *Software effort estimation using particle swarm optimization: Advances and challenges. 1120*. https://doi.org/10.1007/978-981-15-2449-3_20.

Reddy, D. K. K., Behera, H. S., Nayak, J., Vijayakumar, P., Naik, B., & Singh, P. K. (2020). Deep neural network based anomaly detection in internet of things network traffic tracking for the applications of future smart cities. *Transactions on Emerging Telecommunications Technologies*. https://doi.org/10.1002/ett.4121.

Xu, L., et al. (2019). Modeling tabular data using conditional GAN. *arXiv*. (NeurIPS). Available at: http://arxiv.org/abs/1907.00503.

CHAPTER

17

A novel cancer drug target module mining approach using nonswarm intelligence

R. Gowri[a] *and R. Rathipriya*[b]

[a]Department of Computer Applications, Sona College of Arts and Science, Salem, Tamilnadu, India [b]Department of Computer Science, Periyar University, Salem, Tamilnadu, India

Nomenclature

Biological nomenclature

The various biological nomenclature related to this research is discussed in this sub-section for a better understanding.

Protein: The proteins are the linear chain of 22 amino acids (Pereira-Leal et al., 2004; Spirin & Mirny, 2013; Structures of Life, 2007), as in Fig. 17.1. It is used to represent the functionality of biological systems. They are responsible for various physical and chemical activities in the cell.

Protein interaction networks (PIN): They are the molecular networks formed by the proteins (Pereira-Leal et al., 2004; Spirin & Mirny, 2013; Structures of Life, 2007). These networks are responsible for signal transfer and commanding in various portions of the biological system. Fig. 17.3 is the sample PIN.

Drug targets: The drug targets are potential proteins that are responsible for transporting the drug to infected parts, boost the efficacy of the drug, and bind with the drug for effective treatment (Gashaw et al., 2011).

Drug target modules (DTM): The drug target modules are protein modules. The group of proteins are enriched with drug target characteristics that are responsible for a particular function of the cell. It is the dense part of the PIN, the suspicious region of infection or highly communicated region of PIN (Spirin & Mirny, 2013). Fig. 17.2 highlights the sample protein modules.

https://doi.org/10.1016/B978-0-323-85240-1.00019-5

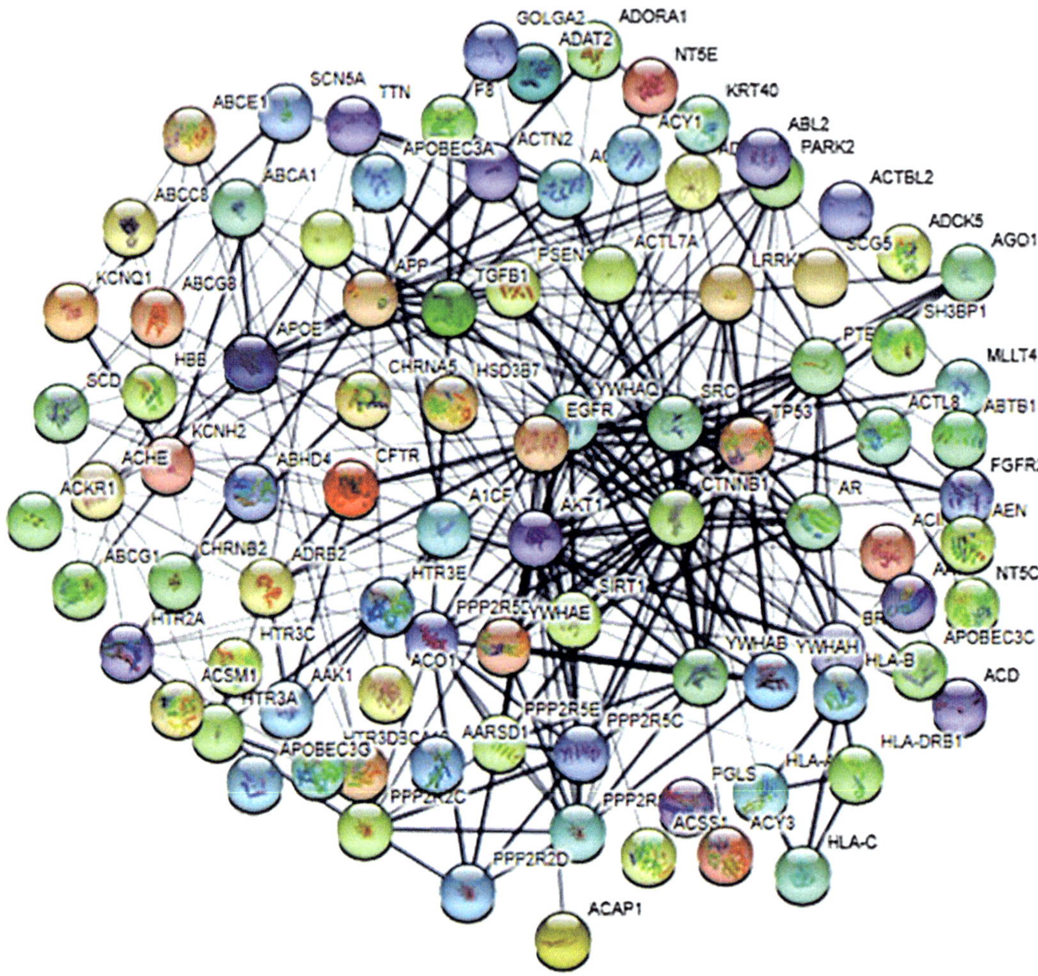

FIG. 17.1 Sample protein interaction network.

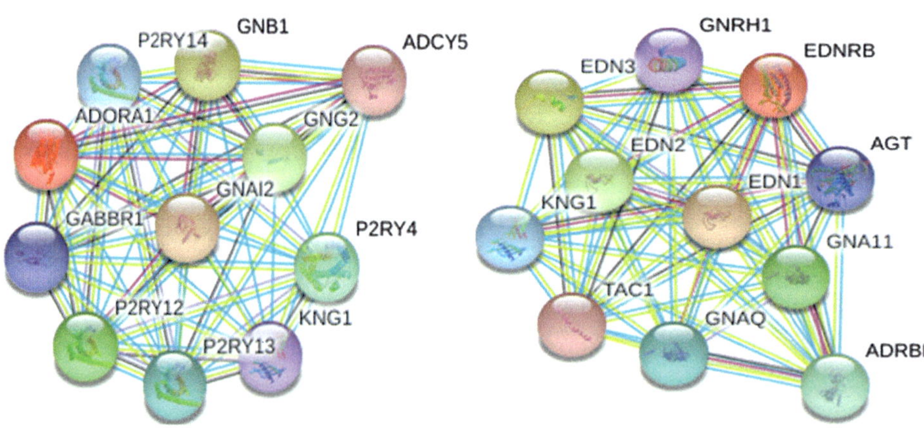

FIG. 17.2 Sample protein modules.

The biological significances of the proteins are represented using the functional annotations. The functional annotations are the de facto standard for describing the functional aspects of molecular products like p. These significances are categorized mainly under three types, and they are as follows:

Molecular function: There are activities like binding and catalysis that occur at the molecular level. They occurred as a single bio-product (protein) or a group of bio-products (protein complex) which depends on their location and environmental factors. The molecular functional annotations represent only the bio-product activity and not the bio-product combination responsible for this activity (Bult et al., 2019; Sussman & Prilusky, 2019).

Biological process: It represents the fundamental biological processes of the bio-products at the cellular level, such as metabolic activity and transportation. They are the combination of one or more molecular functions to execute a biological activity in the cell. They are performed by one or more bio-products (Bult et al., 2019; Sussman & Prilusky, 2019).

Cell component: It represents the relative location or cellular compartments of the bio-product when they perform any molecular functions. They are typically determined based on various aspects like relative cellular structures, cellular compartments, and stable macro-molecular complexes they located.

Mathematical nomenclature

The mathematical structures used to represent various concepts in this research are discussed here. The biological data are represented using the graph structure. This sub-section discusses the preliminaries related to graph theory.

Graph: The graph is a discrete structure representing a relationship between a pair or group of items (Bapat, 2010; Diestel, 2016; Gursoy & Keskin, 2008). It is a significant abstraction used more often in most of the research areas. Based on the direction of the relation, the graphs are categorized as either directed or undirected in nature. In this research, the undirected graphs are used to depict the Protein Interaction Networks (PINs).

Definition 17.1

Undirected graph: the undirected graph G is represented as a tuple (V, E), where V is set of Vertices and E is set of Edges in the graph.

The sample graph is shown in Fig. 17.3. The PIN is represented as G, which is the set of proteins in the PIN and their pair-wise interactions are represented as G: $V(G)=\{p_1, p_2, \dots p_n\}$ and G: $E(G)=\{(p_i, p_j) \mid p_i, p_j\}$.

Complete graph: The complete graph is a type of graph which is defined based on its connectivity between the vertices (Bapat, 2010; Diestel, 2016; Gursoy & Keskin, 2008).

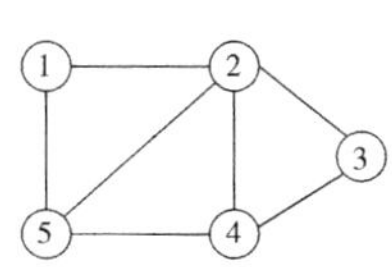

V= {1, 2, 3, 4, 5}, E = {(1,2), (1,5), (2,5), (2,4), (2,3), (3,4), (4,5)}

FIG. 17.3 A sample undirected graph G (V, E).

Definition 17.2

Complete Graph: A graph is said to be complete if there exists an edge between all the pairs of vertices in the graph.

The sample complete graph is in Fig. 17.4. The complete graphs represented as "K_n," where "n" is the number of vertices in that

Connected graph: The connected graphs also defined based on their connectivity between the vertices (Bapat, 2010; Diestel, 2016; Gursoy & Keskin, 2008).

Definition 17.3

Connected Graph: A graph is a connected graph, if there exists a path between all the pairs of vertices in the graph. i.e., each vertex is reachable from every other vertex. Otherwise, the graph is said to be a disconnected graph.

The sample connected and disconnected graphs are given in Fig. 17.5. In the second graph of the figure, the vertices <4> <5> are not reachable from other vertices $\{1, 2, 3\}$, so it is said to be disconnected.

Sub-Graph: It is a part of a graph which is explained in Definition 17.3 (Bapat, 2010; Diestel, 2016; Gursoy & Keskin, 2008). The expected results of this research are protein modules, which are the sub-graphs of the given PIN.

Definition 17.4

Sub-Graph: A sub-graph G_s of a graph G is also a graph whose set of vertices and edges are sub-set of vertices and Edges of G., i.e., $G_s(V_s, E_s)$ $G(V, E)$, where V_s V, E_s E.

The sample sub-graphs are in Fig. 17.6. The H_1, H_2, and H_3 are the sample sub-graphs of given Graph G. The sub-graphs are categorized into cliques and non-cliques based on their connectivity (Bapat, 2010; Diestel, 2016).

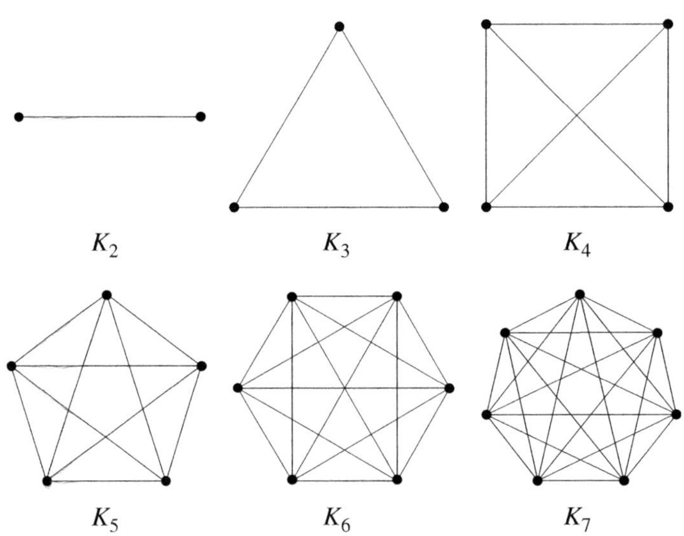

FIG. 17.4 Sample complete graphs.

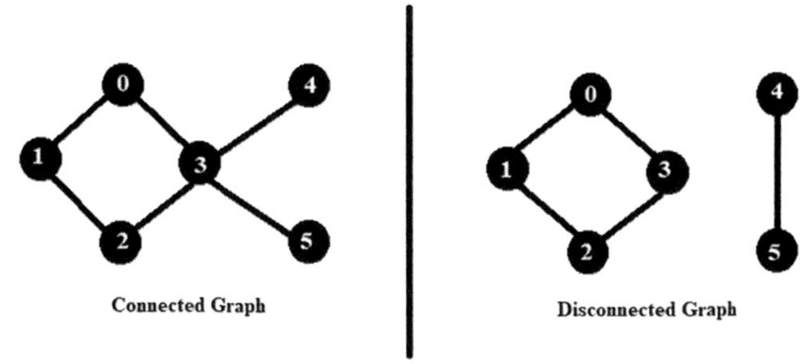

FIG. 17.5 A sample connected and disconnected graph.

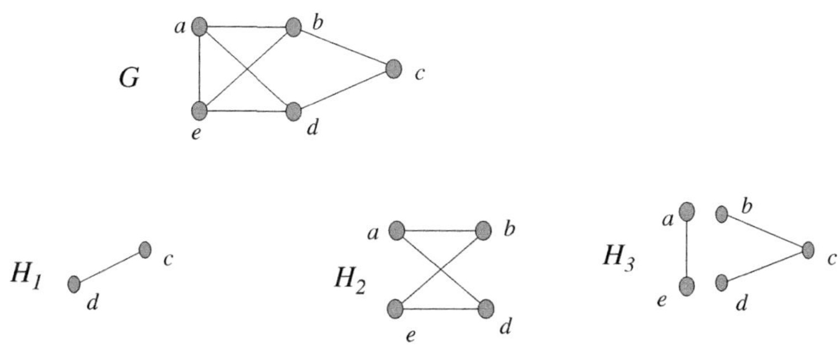

FIG. 17.6 Sample sub-graphs H_1, H_2, H_3 of Graph G.

Definition 17.5
a. Clique: It is a complete sub-graph. It has all possible edges between all the pairs of given vertices. b. Non-Clique: It is a connected sub-graph. It has paths between all the pairs of given vertices.

In Fig. 17.7, the H_1 is a clique, and H_2 is non-clique. The protein modules are in the form of either cliques or non-cliques in this research.

Adjacency matrix: The graphs are represented in the matrix format for computational purposes. They are adjacency matrix, path matrix, etc. (Bapat, 2010; Diestel, 2016; Gursoy & Keskin, 2008). It can be used based on the requirement of computational work. In this work, the adjacency matrix is used to represent the PIN.

Definition 17.6
Adjacency Matrix: It is a binary matrix used to represent the presence and absence of the edge between each pair of vertices.

A sample adjacency matrix (A) of G is in Fig. 17.7. The values of A (G) can be evaluated using Eq. (17.1)

$$A_{ij} = \begin{cases} 1, & (v_i, v_j) \in E \\ 0, & Otherwise \end{cases} \tag{17.1}$$

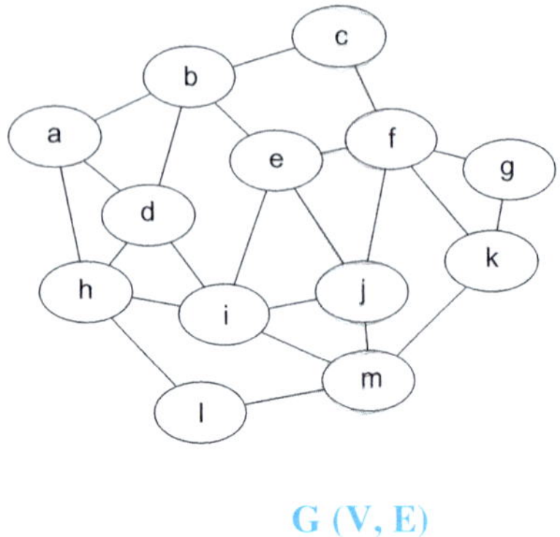

	a	b	c	d	e	f	g	h	i	j	k	l	m
a	0	1	0	1	0	0	0	1	0	0	0	0	0
b	1	0	1	1	1	0	0	0	0	0	0	0	0
c	0	1	0	0	0	1	0	0	0	0	0	0	0
d	1	1	0	0	0	0	0	1	1	0	0	0	0
e	0	1	0	0	0	1	0	0	1	1	0	0	0
f	0	0	1	0	1	0	1	0	0	1	1	0	0
g	0	0	0	0	0	1	0	0	0	0	1	0	0
h	1	0	0	1	0	0	0	0	1	0	0	1	0
i	0	0	0	1	1	0	0	1	0	1	0	0	1
j	0	0	0	0	1	1	0	0	1	0	0	0	1
k	0	0	0	0	0	1	1	0	0	0	0	0	1
l	0	0	0	0	0	0	0	1	0	0	0	0	1
m	0	0	0	0	0	0	0	0	1	1	1	1	0

Adjacency Matrix (A)

FIG. 17.7 Adjacency matrix of a simple graph *G*.

Introduction

Cancer is a lethal disease, which is the second most common cause of death in India. It is the uncontrollable growth of cells (mutated cells), which replaces the healthy cells with mutated cells (cancer cells) and affects the biochemistry of the body (MFMER, 2015). It spreads over a part of the organ first and continuously spreads all over the body, leading to lethal effects. Around 2.25 million people in India are affected by cancer. Every year, 11.50 lakh people are newly registered as cancer patients (ICMR, 2019). The incidence of cancer has doubled over the last 26 years in India. The year-wise prevalence of cancer in India is given in Fig. 17.8. Around 43% of the cancer patients are affected by breast, lung, oral, stomach, and cervical cancers (Fig. 17.9). The topmost cancer that affects both men and women is oral cancer (IndiaToday, 2018). In addition, the other prevailing cancers are ovarian, pancreatic, brain,

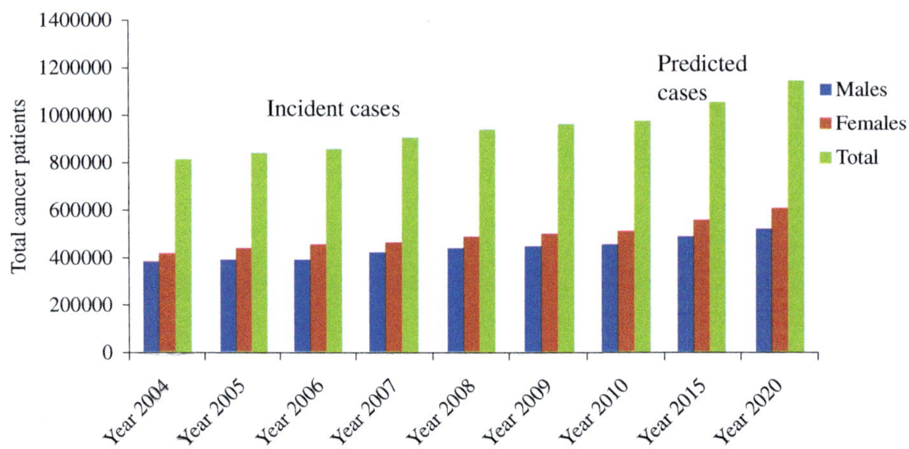

FIG. 17.8 Cancer affected patients count year-wise in India (Ali et al., 2011).

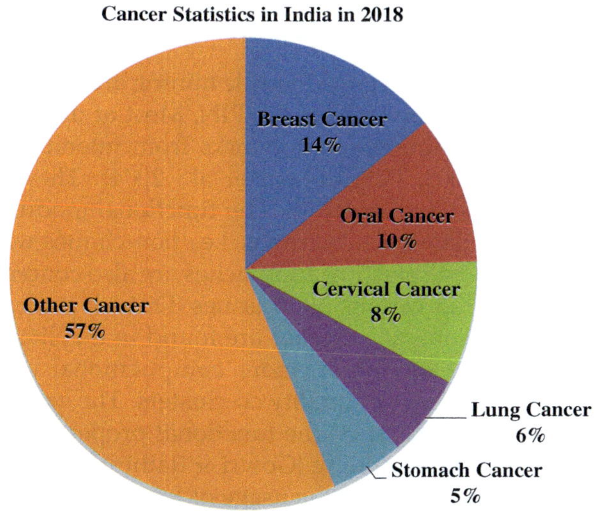

FIG. 17.9 Percentage incidence of different cancers in India by WHO (Pandey & Sengupta, 2018).

prostate, melanoma, colon, leukemia, lymphotic, uterine, kidney, sarcoma cancers, and so on. (ICMR, 2019). As per the health councils' reports, most of these cancers can be cured if the disease is detected early and treated adequately. It is the most challenging research in the medical field. There are several therapies for curing or (IndiaToday, 2018) reducing the severity of cancer, which depends on the cancer stage. The targeted therapy is one among them. Drug targeted therapy is a type of customized treatment for a particular disease. Currently, the targeted therapy, which is a rational drug design, emerged for cancer. This treatment is a target-specific treatment initially designed for cancer. These targeted therapies customized based on the potential target proteins. There are some limitations seen in this kind of therapy: (i) target proteins itself get mutated, leading to failure of the therapy; (ii) tumor cells will choose a different way to achieve cell multiplication. These limitations are due to the therapy depending on single target proteins. In such a case, the Drug Target Modules (DTM) consist of more than one target protein that can be chosen for more effective targeted therapy. Generally, the functional modules enriched with drug target proteins (Hase et al., 2014) and proteins with drug target characteristics are Drug Target Modules (DTM). These modules can be a potential drug target for specific diseases. These proteins and their interactions play an essential role in disease mechanism. They are used for targeted therapy development.

In this chapter, the DTM mining approach MR-CoC$_{VFO}$ is proposed for the *Homo sapiens* protein interaction network. It is the MR-CoC$_{multi}$ (Gowri & Rathipriya, 2018, 2020) approach using the VFO (Gowri & Rathipriya, 2021), an optimization technique for module-specific optimal functional score selection. The presence of drug targets and their characteristics are tested through their biological significance. From the results, few cancer drug target modules and their active pharmaceutical drugs are suggested. This paper further discusses the related research works; proposed Optimized Multi-Functional MR-CoC; experimental analysis and summarization of this research work.

Related works

The existing methodologies for functional module mining are studied over time. The functional modules are the dense sub-networks of the PIN. Most of the current methods use the graph-theoretic approaches for mining sub-networks from interaction networks (Bader & Hogue, 2003; Dittrich et al., 2008; Pereira-Leal et al., 2004). They focused on the fully connected sub-graphs called a clique structure. The functional modules may not be cliques all the time. They are densely cohesive sub-graphs, i.e., non-cliques with high network density (Pereira-Leal et al., 2004). Some existing approaches are also concentrated on these kinds of non-cliques, but they are facing various other issues (Dittrich et al., 2008). The most common problems faced by the existing approaches are found in this literature study. They are scalability issues, more time consumption, more computational overhead, ignorance of overlapped co-clusters, and ignorance of quality co-clusters. The existing approaches are focusing only on the network characteristics; the functional properties of the proteins are not considered for the functional module mining (Gowri & Rathipriya, 2018, 2020). The previous approach has satisfied all the issues except the usage of functional properties. The functional properties like Hydrophobic Residues Density Score, Sequence Length Score, Polar Residues Density Score, Amino Acid Density Score, and Molecular Weight Score are proposed by Gowri and Rathipriya (2020). These scores are chosen based on the drug target characteristics. All these scores may not suit all the functional modules, and the combination of these functional scores has to be selected for each functional module. The customized selection of functional scores is a tedious process.

Methods and material

This section discusses the various measures used in this research. It includes functional coherence, drug coherence, and network metrics which are evaluated in this research.

Functional coherence

It is a functional measure of the protein module to evaluate the consistency of each functional annotation within the protein module (Bastos et al., 2016). The functional annotations are the biological functional descriptors associated with each protein. It represents their roles in the biological system under various aspects. These functional annotations for each protein module are measured to project their characteristics. The commonality of these functional annotations is measured using the functional coherence measure (FC). It is the ratio between the number of proteins that possess a functional annotation and the proteins' count in the protein module as in Eq. (17.2).

$$FC_j(PM_i) = \frac{\left|p_{ij} \in PM_i\right|}{|p_i|} \tag{17.2}$$

The functional coherence (FC) of the jth annotation in the ith Protein Module (PM) is given in Eq. (17.1). The "p_{ij}" represents the proteins possessing jth annotation in the ith module, and

"p_i" is the proteins in the ith module. The value of FC_j lies between "1" and "0," where "1" represents all the proteins in the module has the jth annotation. Based on this FC value, the functionality of the protein module is described. This measure is used in this research to determine the functionalities of the resultant protein modules.

Drug coherence

The active drugs that correspond to each DTM are also extracted. The number of common drugs for each DTM is also highlighted along with these results. The commonality of the drugs is measured based on the drug coherence (DC) measure, which is the same as the functional coherence measure. The DC of each drug in the module is the ratio of the number of targets possessing that drug to the total number of targets in the module as in Eq. (17.3).

$$DC_j(PM_i) = \frac{|d_{ij} \in DTMM_i|}{|d_i|} \tag{17.3}$$

The drug coherence (DC) of the jth drug in the ith Drug Target Module (DTM) is given in Eq. (17.1). The "d_{ij}" represents the drug target possessing jth drug in the ith module and "d_i" is the drug target in the ith module. The value of DC_j lies between "1" and "0," where "1" represents all the proteins in the module has the jth annotation. Based on this DC value, the commonality of a particular drug to the DTM is described.

Network metrics

The various network metrics used in this research to describe the nature of the results (Franceschini et al., 2013). They are as follows:

Number of nodes: The interactors' count of the protein module

Number of edges: The number of communication links found among the proteins in the module. It shows the cohesiveness of the interactors.

Average node degree: It represents the possible presence of interactions for each node (Franceschini et al., 2013). It is the ratio between the sum of degrees of the nodes and the number of nodes in the protein module, as in Eq. (17.4).

$$avg\_deg(v) = \frac{\sum_{i=1}^{n} deg(v_i)}{number\ of\ nodes} \tag{17.4}$$

Expected number of edges: It represents the possible number of interaction that can be present in the given protein module (Franceschini et al., 2013). It is evaluated using Eq. (17.5).

$$\text{Expected no.of edges} = \frac{n(n-1)}{2} \tag{17.5}$$

Local clustering coefficient: It represents the ratio of the interactions between a vertex and its neighbors to the possible number of edges between them (Franceschini et al., 2013). For the cliques, the average local clustering coefficient (C) of a protein module will be "1," and for non-cliques, it will be less than "1." It is evaluated using Eq. (17.6), where "N_i" is the set of neighbors of "v_i," and "n_i" is the $|N_i|$. "E" represents the edges.

$$C = \frac{1}{n}\sum_{i=1}^{n} \frac{2\left|\{e_{jk} : v_j, v_k \in N_i, e_{jk} \in E\}\right|}{n_i(n_i - 1)} \tag{17.6}$$

PPI enrichment value: It represents the "*p*-value" of the randomness functional enrichments of the nodes (Franceschini et al., 2013). The smaller value indicates less randomness among the nodes. The functional enrichment of the nodes is high when the p-value is small.

Proposed MR-CoC$_{VFO}$

The MR-CoC$_{multi}$ is proposed by Gowri and Rathipriya (2018). Furthermore, this approach faced the limitation as discussed previously, and the optimal functional score selection is required for this purpose. This mechanism should have the capability of finding independent functional score sets. Because each DTM should have different characteristics and they possess different functional score measures, this can be identified using the independent search mechanism. So the NSI is better suited to this problem. So, the optimal functional score selection mechanism using VFO is proposed in this chapter for cancer-specific drug target module mining. The Venus Flytrap Optimization algorithm-based functional score selection (VFO_FSS) is discussed in this section. This approach is adapted with MR-CoC$_{multi}$ (Gowri & Rathipriya, 2020) for DTM mining in this chapter. This is the major contribution of this chapter.

VFO_FSS: The major works to be concentrated in this approach are defining population, fitness measure, and trigger function. A population is a group of individuals searching for a solution. For this problem, a binary string is chosen as individual, e.g., 110101. The 1's and 0's represent the presence and absence of a particular functional score in that individual, respectively. A population of 50 individuals is used for this research. The functional scores used in MR-CoC$_{multi}$ are Hydrophobic Residues Density Score (S_h), Sequence Length Score (S_{len}), Network Density Score (S_n), Polar Residues Density Score (S_p), Amino Acid Density Score (S_{aa}), and Molecular Weight Score (S_m) proposed by Gowri and Rathipriya (2020). Customized selection among these measures for each DTM privately is performed using this approach. For that, the individuals are designed as strings of flags representing it. The fitness function of this approach is defined as the sum of the percentage of each score which represents the weightage of the score set. This fitness function selects the score set (individual) that satisfies maximum scores instead of the minimum number of scores. It is defined in Eq. (17.7).

$$\text{fitness} = S_h + \frac{S_{len}}{\max\_len} + S_n + S_p + \frac{S_{aa}}{\max\_aa} + \frac{S_m}{900} \tag{17.7}$$

The next significant part that drives this approach is the trigger function. This function highlights the support criterion of the VFO approach. This function implements the support criterion which selects maximal feature scores rather than a minimal number of scores. The trigger function is defined in Eq. (17.8). The t_h (~0.7) and t_l (~0.4) are higher and lower priority constants, respectively, the weightage given to individuals to trigger more probably based on their protein weight. Protein weight is the number of proteins that possess a particular score in the module. Here, the score which is present in almost 50% of the proteins in the module is triggered. The pseudocode of VFO_FSS is in Algorithm 17.2

$$t = \begin{cases} t_h \times rand, pwt \geq n/2 \\ t_l \times rand, otherwise \end{cases} \quad (17.8)$$

MR-CoC$_{VFO}$: This approach is the same as MR-CoC$_{\text{multi}}$ using this VFO_FSS for functional score selection. The overview of MR-CoC$_{\text{VFO}}$ is shown in Fig. 17.10. It is a MapReduce-based approach where the VFO_FSS is used in the map phase. This is also clearly depicted in Algorithms 17.1–17.3.

Experimental analysis

The experiment is carried out in a standalone system with an I7 processor and 8GB RAM. These have experimented with in the MatLab environment. The default MapReduce framework in MatLab is used for this purpose. The datasets used in this research work are as follows:

PIN dataset: The Protein Interaction Networks (PIN) data have a high set of interactions that refer to the edge of two proteins. In this research, the *Homo sapiens* PIN is chosen from the STRING database, which consists of 19,427 proteins and 85,48,003 interactions.

Drug target dataset: The drug target proteins are used for analyzing the characteristics of protein modules by verifying the presence of drug target proteins in them. The standard therapeutic targets are taken from the Therapeutic Target Database (TTD) (Zheng et al., 2006).

The results are compared with those of the previous approaches for exploring its performance efficiency. The number of protein modules (cliques and non-cliques) extracted by each of the approaches MR-CoC$_{\text{VFO}}$, MR-CoC$_{\text{Multi}}$, and MR-CoC is in Table 17.1. This evidences that the performance of the proposed approach (*) is similar to that of other approaches.

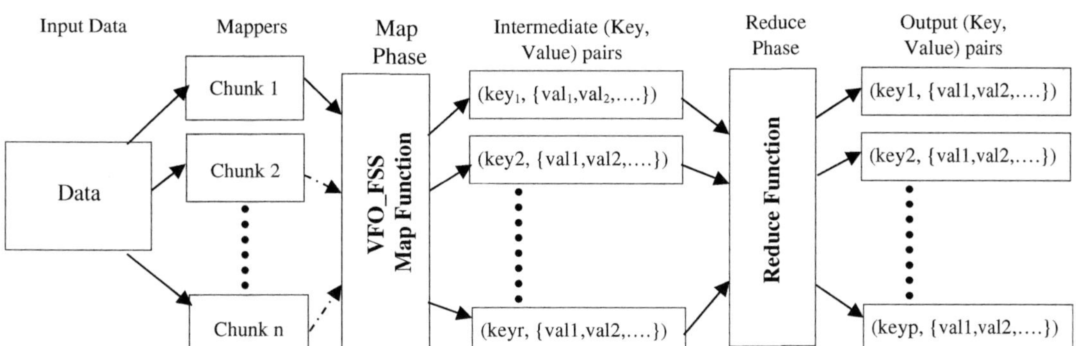

FIG. 17.10 Overview of the proposed MR-CoCVFO model.

Algorithm 17.1 MR-CoC$_{VFO}$

Input: Protein Interactions
Output : co-clusters(M = {C_1, C_2,...C_{ns} })
Step 1: *Fix the number of seeds (n_s) to be generated and size of the seed (len_s)*
Step 2: *Generate n_s seeds of size len_s*
Step 3: *Generate the Adjacency Matrix on PIN*
// Map Phase
Step 4: *For each seed s_i*
  a. *Apply VFO_FSS approach and return intermediate KV pairs {co-cluster, {count, functional_score vector}}*
// Reduce Phase
Step 5: *For each key generate output KV pair {co-cluster_size, {co-cluster, functional_score vector}}*
// After MR phases
Step 6: *Evaluate and visualize the obtained co-clusters.*

Algorithm 17.2 VFO-FSS

Input: Seed Vector
Output : co-cluster, multi-score

Step 1: *Evaluate the multi-score of C*
Step 2: *Select Optimal Scores for C using VFO*
Step 3: *If all selected scores satisfy their threshold conditions then*
  a. *Evaluate row (or column) multi-score of C*
  b. *Remove row and its corresponding column with low multi-score from C*
  c. *Evaluate multi-score of C*
  d. *Optimal Scores for C using VFO*
Step 4: *Return C and multi-score*

Algorithm 17.3 VFO

```
Input: Multi- Score of C
Output : optimal FSS
Step 1: Generate population ft_i (i = 1, 2, ..., n)
Step 2: Repeat until max_iteration

a. For each flytrap                //initial phase

i. At first stimulation (t=0), evaluate action potential (u_t) and charge
   accumulation (C)using f(ft)
ii. At second stimulation (t=rand()), if t<=T then
                          //trapping phase

➢ update ut, C and evaluate the fitness and object status

b. Rank the flytraps and find the current best  //sealing phase
c. seal the best flytrap until another best flytrap arrives

Step 3: Post process results and visualization
```

TABLE 17.1 Protein module counts of MR-CoC, MR-CoCMulti, and MR-CoCVFO.

Number of seeds (million)	Minimum module size = 3, score threshold = 0.9					
	Cliques			Non-cliques		
	MR-CoC	MR-CoC$_{Multi}$	*MR-CoC$_{VFO}$	MR-CoC	MR-CoC$_{Multi}$	*MR-CoC$_{VFO}$
1	831	549	768	3166	3693	3590
5	1129	983	1319	3578	4203	4844
10	2867	1992	2903	2090	4671	3991
50	3209	2201	3067	4672	5579	5933
100	3276	2890	3410	8901	9217	9328

As this approach uses biological measures, its performance is further studied based on their biological significance.

The Biological Process (BP)-, Molecular Function (MF)-, and Cell Component (CC)-based significances of the MR-CoC$_{VFO}$ results are obtained from the STRING database. Some of the modules are taken randomly from the results. The significances of the results are in Tables 17.2–17.5. These results are tabulated based on different combinations of significance are as follows:

- Presence of two drug targets proteins (Fig. 17.11, Table 17.2)

TABLE 17.2 Biological functionalities of functional modules with two drug target participants.

Participants		Biological significance	FC
FM13			
FLT1, **FLT4**, KDR, NRP1, NRP2, PGF, PIK3R1, LCG1, PTPN11, VEGFA, EGFB	BP	**Regulating, Signaling and Responding Activity:** kinase signaling pathway; regulating the localization and multicellular organismal process; responding to chemical	1
		Regulating, and Responding Activity: regulating the cell migration and developmental process; cellular responding to organic substance and stimulus	0.9
	MF	**Binding Activity:** ion and protein binding	1
		Catalytic and Binding Activity: anion, and carbohydrate derivative binding	0.7
	CC	**Membrane**, extracellular region	0.9
		Receptor Complex	0.3
FM21			
ADRBK1, **CHRM3**, **EDNRB**, GNAQ, GNB5, GNG7, GRM1, GRM5, OXT, PLCB1, TACR1	BP	**Receptor Signaling**	1
		Responding and Regulating Activity: responding to nitrogen compound, endogenous stimulus, oxygen-containing compound, and organic substance regulation of signaling	0.7
	MF	**Protein binding**	0.7
		Regulator and Receptor Activity	0.4
	CC	**Membrane, Plasma membrane**	0.9
		Plasma membrane bounded cell projection part	0.7
FM23			
F2, F2R, **F5**, FGA, FGB, FGG, GP1BA, PROC, SERPINC1, SERPIND1, THBD	BP	**Regulating the cellular processes:** blood coagulation,	0.9
		Regulating, Transporting and Responding Activity: regulating the cellular protein metabolic process and multicellular organismal process response to chemical	0.7
	MF	**Binding Activity:** metal ion binding	0.7
		Signaling receptor binding	0.4
	CC	**Cytoplasm**	1
FM24			
CLEC4E, FCAR, **FCER1A**, **FCER1G**, FCGR2B, FCGR3A, GP6, ITGB2, LYN, PLCG2, SYK	BP	**Signaling and Receptor Activity:** regulating cell surface receptor and immune response	1
		Localization, Responding and Regulation Activity: response to stress, immune, to stimulus regulating of response, cell activation and immune system process	0.8

TABLE 17.2 Biological functionalities of functional modules with two drug target participants—cont'd

Participants		Biological significance	FC
	MF	**Binding Activity:** protein-complex binding	1
		Receptor Activity: transmembrane signaling receptor activity	0.6
	CC	**Plasma membrane**	1
		Cell surface	0.7

TABLE 17.3 Biological functionalities of functional modules with receptors in membrane region.

Participants		Biological significances	FC
FM20			
ADRBK1, AGT, EDN1, EDN2, EDN3, **EDNRB**, GNA1, GNAQ, GNRH1, KNG1, TAC1	BP	**Receptor Activity:** G protein-coupled, multicellular organismal process	1
		response to external stimulus, regulation of biological quality, regulation of response to stimulus, and response to chemicals	0.9
	MF	**Receptor Binding**	1
		Receptor Activity: G protein-coupled, molecular function regulator, and hormone activity	0.5
	CC	**Extracellular space**	0.6
		Plasma membrane	0.4
FM27			
CD44, CDH1, CXCL1, MMP0.9, **MMP9**, PLG, SRC, STAT3, TIMP1, TIMP2, VEGFA	BP	**Regulating Activity:** regulation of cellular process and organization	1
		Regulating and Responding Activity: negative and positive regulation of cellular process, localization, and response to stimulus	0.9
	MF	**Binding Activity:** protein binding	1
		Receptor and Binding Activity: ion, signaling receptor, and protein-complex binding	0.6
	CC	**Cytoplasmic Part**	0.9
		Cytoplasmic vesicle and endomembrane system	0.8
FM31			
ADRA1D, ADRB2, CYSLTR1, CYSLTR2, EDNRA, KNG1, LTB4R, LTB4R2, **PTGER1**, PTGFR, UTS2R	BP	**Receptor Activity:** G protein-coupled signaling and multicellular organismal process	0.8

Continued

TABLE 17.3 Biological functionalities of functional modules with receptors in membrane region—cont'd

Participants		Biological significances	FC
		Receptor Activity: G protein-coupled signaling, responding to external stimulus, and homeostatic process	0.5
	MF	**Receptor Activity:** G protein-coupled signaling	0.9
		Receptor Activity: eicosanoid,G protein-coupled peptide, and leukotriene receptor activity	0.5
	CC	**Plasma Membrane**	1
		Integral Component of Plasma Membrane	0.9
FM32			
CD19, CD22, CD79A, CD79B, FCER2, IGHV3-1, IGLL5, **MS4A1**, MS4A3, MS4A7	BP	**Receptor and Responding Activity:** responding to immune system and regulation of immune; cell surface receptor signaling pathway	0.5
		Receptor signaling and Responding Activity: B cell, and immune response-activating cell surface receptor signaling pathways; responding to humoral immune and adaptive immune	0.3
	MF	**Antigen binding**	1
	CC	**Membrane**	0.9
		Plasma Membrane	0.8

TABLE 17.4 Biological functionalities of functional modules with transporters in membrane region.

Participants		Biological significance	FC
FM14			
ATF4, BSG, CD44, GCLM, KYNU, NFE2L2, SLC1A2, SLC1A5, SLC1A7, SLC3A2, **SLC7A1**	**BP**	**Responding to chemical**	1
		transport and response Activity: responding to drug, organic substance, nitrogen compound, and stress	0.7
	MF	**Transporter Activity:** carboxylic acid, amino acid, and secondary active transmembrane transportation	0.5
		Transporter Activity: drug, cation, and L-amino acid transmembrane transportation	0.4
	CC	**Plasma membrane**	0.8
		Integral Component of Plasma Membrane	0.6
FM18			
F0.9, **F2**, F3, **F5**, MCFD2, MMRN1, PLG, PROC, PROS1, SERPINA1, SERPINC1	**BP**	**Transport Activity:** blood coagulation	0.9

TABLE 17.4 Biological functionalities of functional modules with transporters in membrane region—cont'd

Participants		Biological significance	FC
		Transport Activity: vesicle-mediated and endoplasmic reticulum to Golgi vesicle-mediated **transport;** response to stimulus	0.6
	MF	**Binding**	0.9
		Binding Activity: ion, metal ion, and calcium ion binding	0.7
	CC	**Membrane-bounded organelle, intracellular organelle part**	0.9
		Endoplasmic Reticulum Part, Membrane	0.8
FM22			
BDNF, GABRG2, GABRR3, GAD2, GLRA2, GLRB, **SLC12A5**, SLC32A1, STK39, WNK1, WNK3	BP	**Signaling and Regulation Activity:** cell communication and biological regulation	1
		Signaling and Regulation Activity: regulation of biological quality and signal transduction; cell-cell signaling and localization	0.7
	MF	**Transporter and Binding Activity:** anion and inorganic molecular entity transmembrane transporter activity; drug, small-molecule, and anion binding	0.5
		Transporter Activity: chloride transmembrane and chloride channel activity	0.4
	CC	**Membrane, Plasma Membrane**	0.7
		Plasma membrane part, cell projection and junction	0.6
FM33			
ARG1, ARG2, ASL, ASS1, ATP2B4, CALM1, CTBP1, DLG4, GRIN2A, **NOS1**, NOS1AP	BP	**Regulating Activity:** regulating the cellular process, biological process, and cellular metabolic process	0.9
		Regulating Activity: regulating the multicellular organismal process, nitrogen compound and cellular metabolic process, biological quality, and cellular process	0.7
	MF	**Transport and Binding Activity:** protein binding	0.9
		Transport and Binding Activity: protein domain specific and enzyme binding; inorganic molecular entity, and ion transmembrane transporter activity	0.4
	CC	**Intracellular membrane-bounded organelle**	1
		Neuron projection, whole membrane	0.7

TABLE 17.5 Biological functionalities of functional modules with binding proteins in the membrane region.

Participants		Biological significance	FC
FM6			
ALOX12, ALOX15, ALOX15B, ALOX5, PTGDS, PTGES, PTGES2, PTGES3, PTGIS, **PTGS1**, TBXAS1	BP	**Metabolic Activity:** eicosanoid and long-chain fatty acid metabolic process	1
		Metabolic Activity: eicosanoid biosynthetic process, arachidonic acid metabolic process, and unsaturated fatty acid biosynthetic process	0.9
	MF	**Binding and Catalytic Activity**	1
		Binding Activity: ion, isomerase, oxidoreductase, and metal ion binding	0.7
	CC	**Membrane**	1
		Endomembrane System, Organelle Sub-compartment	0.8
FM10			
BMP1, **COL1A1**, COL1A2, COL3A1, COL5A1, COL5A2, COL6A1, COL6A2, COL6A3, CRTAP, SPARC	BP	Extracellular Matrix Organization	0.9
		Responding and Development Activity: multicellular organism development, protein-containing complex subunit organization, and response to organic substance	0.8
	MF	**Metal Ion Binding**	0.6
		Binding Activity: platelet-derived growth factor and protein-containing complex binding	0.4
	CC	**Extracellular Region, Endomembrane System**	1
		Protein Containing complex	0.8
FM11			
CDC5L, STMN1, TBCA, TBCD, TBCE, TUBA1A, TUBA1B, TUBA1C, TUBA3E, TUBA4A, **TUBB1**	BP	cellular component organization	0.9
		cytoskeleton organization, cell cycle	0.7
	MF	**Binding**	1
		Binding Activity: protein, heterocyclic compound, and organic cyclic compound binding	0.7
	CC	**Intracellular organelle part**	1
		Microtubule	0.9
FM26			
AKR1C3, CYP1A1, CYP1B1, CYP1B2, CYP17A1, CYP19A1, CYP21A2, HSD17B3, HSD17B6, **HSD3B1**, SRD5A1	BP	**Metabolic and Biosynthesis Process:** hormone and steroid biosynthetic process; cellular hormone metabolic process, and oxidation-reduction process	1
		Metabolic Process: organic hydroxy compound and small-molecule metabolic process	0.9

TABLE 17.5 Biological functionalities of functional modules with binding proteins in the membrane region—cont'd

Participants	Biological significance		FC
	MF	**Oxidoreductase and Binding Activity**	1
		Binding Activity: oxido-reductase activity; cofactor, heterocyclic compound, and iron ion binding	0.6
	CC	**Membrane, Intracellular Organelle Part**	1
		Endoplasmic Reticulum, Organelle Membrane	0.6

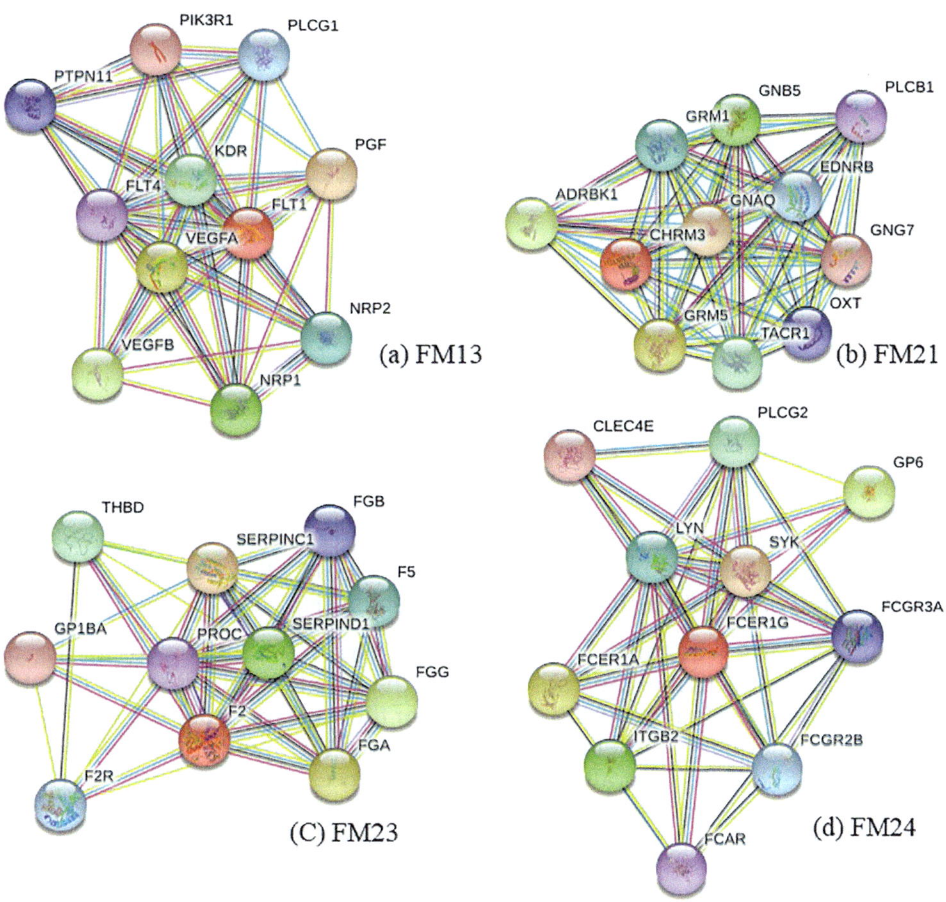

FIG. 17.11 Functional modules with two drug target participants.

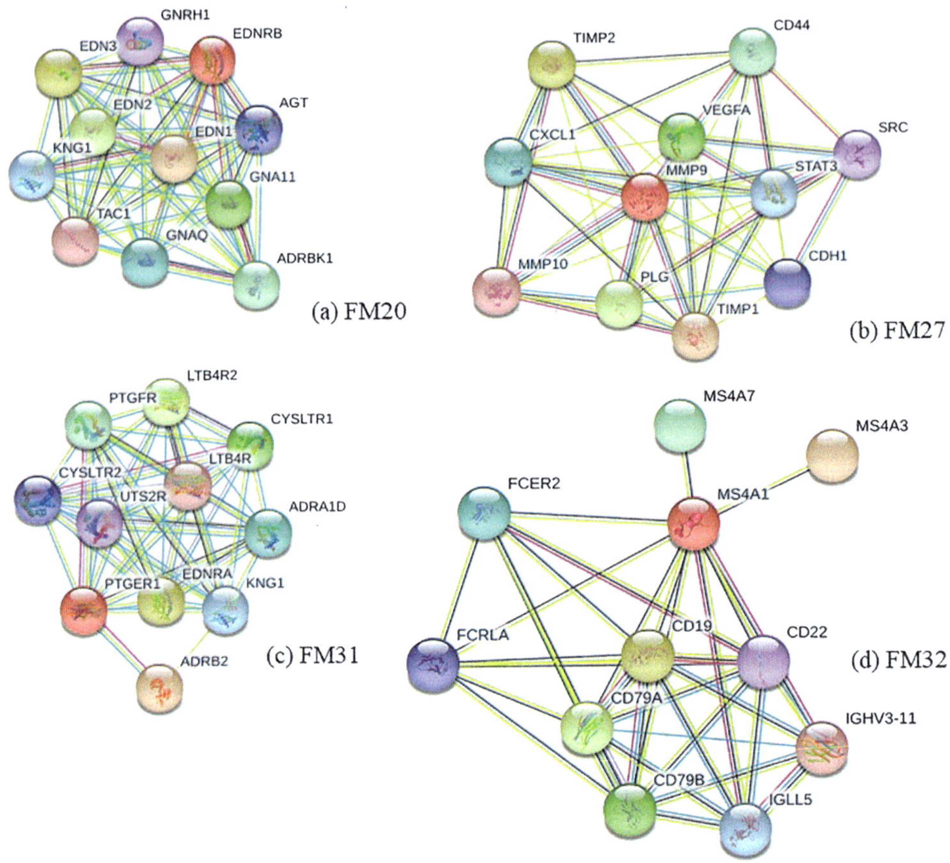

FIG. 17.12 Functional modules with receptors in membrane region.

- Presence of receptors in the membrane region (Fig. 17.12, Table 17.3)
- Presence of transporters in the membrane region (Fig. 17.13, Table 17.4)
- Presence of binding proteins in the membrane region (Fig. 17.14, Table 17.5)

As discussed earlier in this section, the drug targets are involved in the binding, transporting, receptor activities, and so forth, and they are located in the cell membrane regions. Initially, these resultant modules are tested for the occurrence of drug target proteins in them. These potential target proteins are taken from the therapeutic target database (TTD).

Tables 17.2–17.5 show the overall biological significances of 50 sample protein modules in the MR-CoCVFO results. In these tables, the first column lists the participants of the protein module where the highlighted proteins are potential drug targets, followed by the significant annotations under BP, MF, and CC with the two highest functional coherence (FC) in that protein module. The highlighted annotations in this column represent the dominating characteristic of that module.

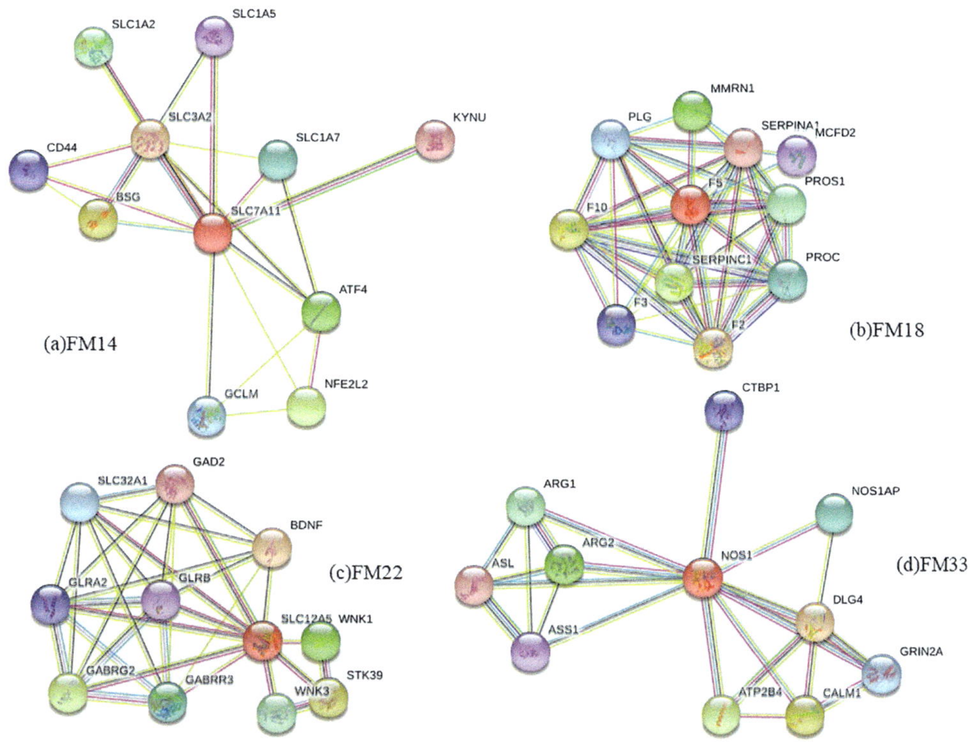

FIG. 17.13 Functional modules with transporters in membrane region.

Fig. 17.11 shows the network visualization of the protein modules with two drug targets proteins in them. They are FM13, FM21, FM23, and FM24. Their biological significances are also discussed in Table 17.2. The remaining Tables 17.3–17.5, contain the protein modules with one drug target protein. In Fig. 17.5, the drug target proteins are highlighted with red ellipses. These modules are uniquely named in the format, FM#, as in the figures.

Table 17.3 highlights the modules that have receptors in the membrane region, and their network structures are shown in Fig. 17.12. There are four functional modules of this type—FM20, FM27, FM31, and FM32—visualized in this figure. Fig. 17.13 visualizes the protein modules such as FM14, FM18, FM22, and FM33 that contain transporter proteins in the membrane region. The biological significances of these protein modules are given in Table 17.4.

Table 17.5 shows the binding proteins in the membrane region. The network visualization of these protein modules—FM6, FM10, FM11, and FM26—is shown in Fig. 17.14.

These biological significances of FMs evidenced the drug target functionalities in these modules. Further, three modules are suggested as Cancer DTMs that are enriched with cancer drug targets. The active pharmaceutical drugs of these DTMs are also highlighted in this chapter. These DTM figures include the network visualization of the module with highlighted

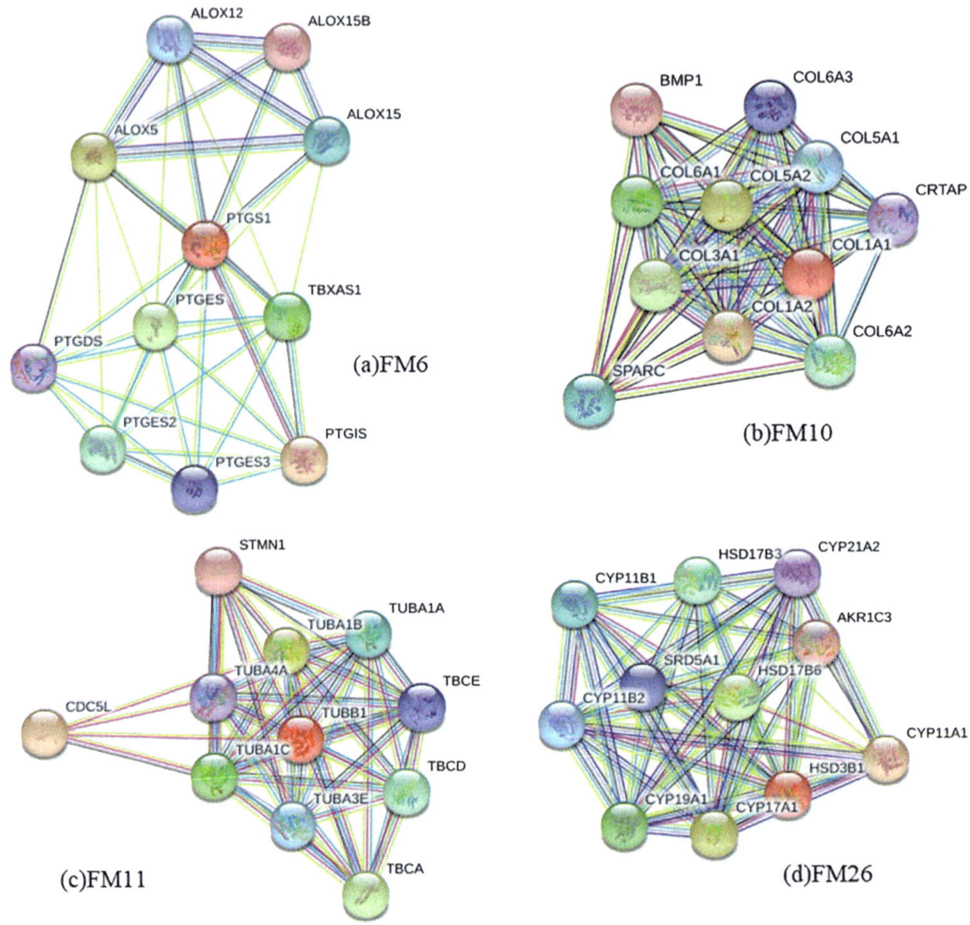

FIG. 17.14 Functional modules with binding proteins in the membrane region.

(red ellipse) drug targets, network metrics, and biological score measures of the module with selected scores (italicized) by VFO_FSS, biological significances, and KEGG pathways of the module with functional coherence. These DTMs are as follows:

DTM2: The DTM2 in Fig. 17.15 formed a non-clique with eight drug target proteins out of 13 participants. They are *CCNA2, CCND1, CCND3, CCNE1, CDK2, CDK4, CDK6, CDKN1A, E2F1, MAPK1, MAPK3, RAF1,* and RB1, and out of these, CCND1, CDK4, CDKN1A, E2F1, MAPK1, MAPK3, RAF1, and RB1 are cancer drug target proteins. They commonly perform the *regulation of metabolic processes* at the biological level and *protein binding process* at the molecular level. These proteins are present in the *protein complex and nucleoplasm region.* This

(a)

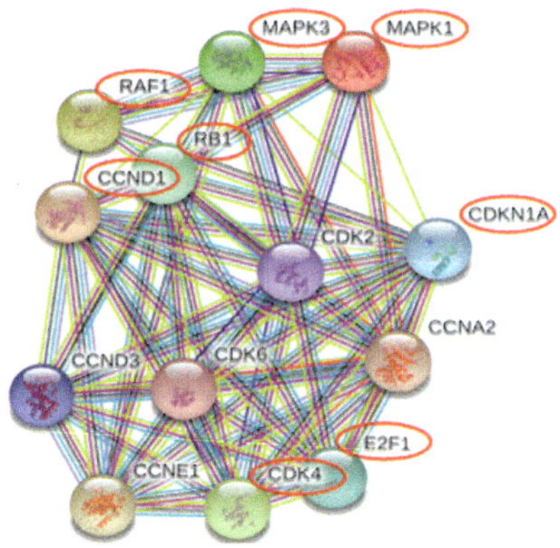

(b)

Score Measure	Value
Hydrophobic Density	0.5928
Sequence Length	3514
Network Density	0.895
Polar Residues Density	0.5274
Amino acid Density	16.625
Molecular weight	130.247
number of nodes	13
number of edges	63
average node degree	9.69
avg. local clustering coefficient	0.895
expected number of edges	16
PPI enrichment p-value	< 1.0e-16

*DTM2 → CCNA2, **CCND1**, CCND3, CCNE1, CDK2, **CDK4**, CDK6, **CDKN1A, E2F1, MAPK1, MAPK3, RAF1, RB1***

	Biological Significance	***FC***
BP	*regulation of gene expression, signal transduction, positive regulation of cellular process, regulation of cellular macromolecule biosynthetic process, regulation of nitrogen compound metabolic process, regulation of primary metabolic process regulation of cellular metabolic process, cellular macromolecule metabolic process, nitrogen compound metabolic process, primary metabolic process*	*1*
	cell cycle, positive regulation of nitrogen compound metabolic process, positive regulation of cellular metabolic process, positive regulation of macromolecule metabolic process	*0.92*
MF	*protein binding*	*1*
	Enzyme binding, protein serine/threonine kinase activity	*0.77*
CC	*Nucleoplasm, Protein Complex*	*1*
	cytoplasmic part, Cytoplasm, Cystosol	*0.84*

(c)

(d) KEGG pathways with:
FC=1: *Human papillomavirus infection, Cellular senescence*
FC=0.92: *Hepatitis B, Pathways in cancer, Viral carcinogenesis*
FC<0.9: *Non-small cell lung cancer, Glioma, Melanoma, Pancreatic cancer, Chronic myeloid leukemia, Cushing's syndrome, Prostate cancer, Breast cancer, Gastric cancer, Hepatocellular carcinoma, Kaposi's sarcoma-associated herpesvirus infection*

FIG. 17.15 Drug target module: DTM2.

DTM is less hydrophobic, i.e., hydrophobic residue density score is ignored by BWS. The proteins of DTM2 are related to human papillomavirus infection that causes cervical cancer. The other pathways like *Hepatitis B, pancreatic cancer, Non-small cell lung cancer, and melanoma* is in 90% of the proteins as in Fig. 17.15D.

The common drugs of DTM2 are shown in Fig. 17.16, where the drug *Curcumin* is common (DC = 1) to all the targets in DTM2. It is an anti-cancer drug that kills cancer cells in various

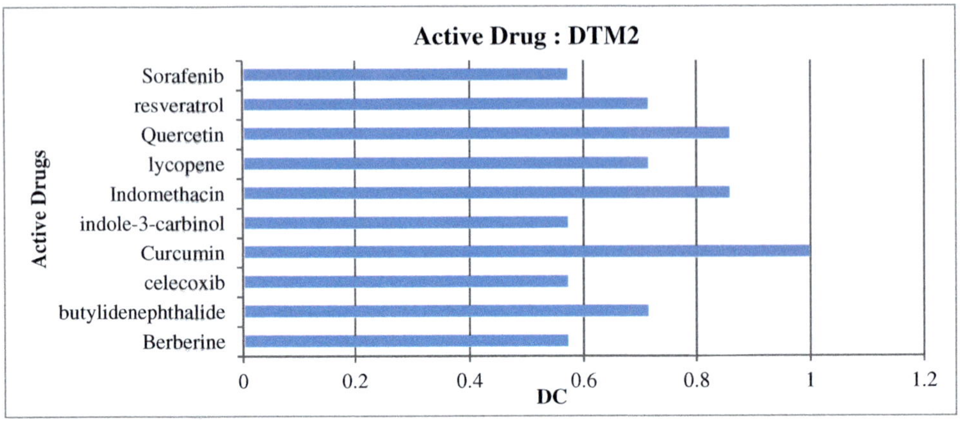

FIG. 17.16 Active drugs in DTM2.

cancer types (Shanmugam et al., 2019). In addition, *Quercetin* and *Indomethacin* have 85% DC, which are mainly used for breast cancer and colon cancer prevention (Jeong et al., 2009) (NCI Drug Dictionary, n.d.; Wang & Zhang, 2005).

DTM4: The DTM4 is a clique with three participants; all these interactors are drug target proteins, which are highlighted as in Fig. 17.17. This module is a high polar residue contented module. It is greater than its threshold; therefore, it is not selected by the BWS approach. *Regulating, signaling,* and *responding processes* are commonly seen under BP and MF, and they are located in the *membrane region* with maximal functional coherence. The cancer-related pathways like *Bladder cancer, EGFR tyrosine kinase inhibitor resistance, Colorectal cancer, Pancreatic cancer,* and *Melanoma* are common to these proteins as in Fig. 17.17D.

The active drugs of DTM4 are in Fig. 17.18. From the figure, the Allomaron (DC = 1) is found common to all the drug targets, which is used for treating hyperuricemia (Dincer et al., 2002). The other drugs are found in 66% of targets. It can be further analyzed for knowing their significance.

DTM6: The DTM6 has 14 proteins form a non-clique with four participants as cancer drug target proteins (DAPK1, DAPK3, MAPK1, and MAPK3), as in Fig. 17.19. *Regulation processes and catalytic activities* are in common in this module. They are commonly present in the *nucleus and cytoplasm* regions. Seventy-six percent of the proteins have common "*pathways in cancer.*" There are no common significant cancer types seen in this module. But, it has satisfied all the score measures. The presence of drug targets is minimal in this module when compared to other DTM.

The common drugs of DTM6 are listed in Fig. 17.20. In this figure, the *Decitabine* is common to all the targets. It is the anti-cancer ("antineoplastic" or "cytotoxic") chemotherapy drug, which is used for treating the myelodysplastic syndrome (MDS) (Cleveland, 2002).

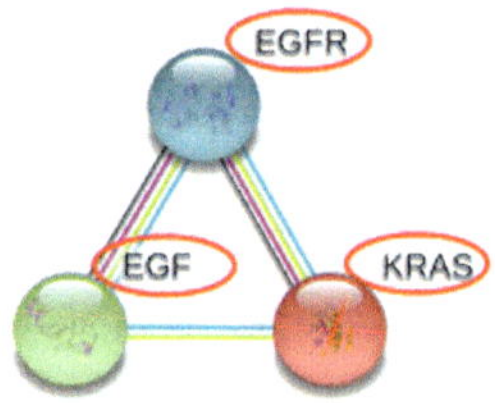

(a)

DTM4 (Clique)→***EGFR, EGF, KRAS***

Score Measure	Value
Hydrophobic Density	0.6185
Sequence Length	2606
Network Density	1
Polar Residues Density	0.5644
Amino acid Density	24.66
Molecular weight	130.019
number of nodes	3
number of edges	3
average node degree	2
avg. local clustering coefficient	1
expected number of edges	2
PPI enrichment p-value	< 0.161

(b)

	Biological Significance	*FC*
BP	*MAPK cascade, morphogenesis of an epithelium, positive regulation of cell population proliferation, animal organ morphogenesis, tube morphogenesis, positive regulation of MAP kinase activity, gland development, positive regulation of gene expression, cell surface receptor signaling pathway, regulation of developmental process, regulation of transcription, DNA-templated, regulation of RNA metabolic process, negative regulation of cellular process*	*1*
	positive regulation of transport, cell-cell signaling, forebrain development, learning or memory, response to light stimulus, camera-type eye development, positive regulation of defense response	*0.66*
MF	*phosphatidylinositol-4,5-bisphosphate 3-kinase activity, protein tyrosine kinase activity, Ras guanyl-nucleotide exchange factor activity*	*1*
CC	*whole membrane*	*1*
	clathrin-coated vesicle membrane, vesicle lumen, vesicle lumen, membrane raft	*0.66*

(c)

KEGG pathways with FC=1:
Bladder cancer, EGFR tyrosine kinase inhibitor resistance, ErbB signaling pathway, Gap junction, Colorectal cancer, Pancreatic cancer, Endometrial cancer, Glioma, Prostate cancer, Melanoma, Non-small cell lung cancer, Choline metabolism in cancer, FoxO signaling pathway, Hepatitis C, Phospholipase D signaling pathway, Breast cancer, Gastric cancer, Rap1 signaling pathway, Regulation of actin , cytoskeleton, Ras signaling pathway, MAPK signaling pathway, Human papillomavirus infection, PI3K-Akt signaling pathway, Pathways in cancer

FIG. 17.17 Drug target module: DTM4.

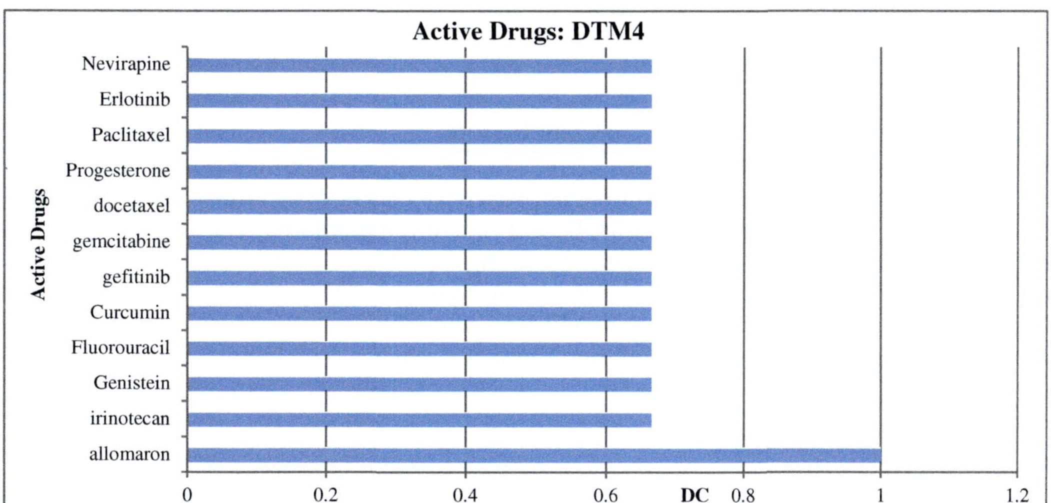

FIG. 17.18 Active drugs in DTM4.

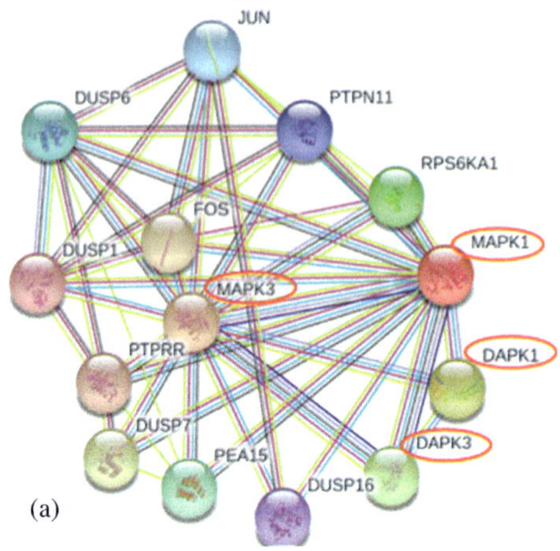

(a)

Score Measure	Value
Hydrophobic Density	0.6133
Sequence Length	2623
Network Density	0.836
Polar Residues Density	0.538
Amino acid Density	24.25
Molecular weight	131.937
number of nodes	14
number of edges	43
average node degree	6.45
avg. local clustering coefficient	0.836
expected number of edges	16
PPI enrichment p-value	1.25e-9

(b)

DTM6→ **DAPK1, DAPK3,** DUSP1, DUSP16, DUSP6, DUSP7, FOS, JUN, **MAPK1, MAPK3,** PEA15, PTPN11, PTPRR, RPS6KA1

	Biological Significance	***FC***
BP	*regulation of nitrogen compound metabolic process, regulation of primary metabolic process, regulation of cellular metabolic process, regulation of macromolecule metabolic process, cellular macromolecule metabolic process, nitrogen compound metabolic process, primary metabolic process*	*1*
	regulation of cellular protein metabolic process, macromolecule modification, regulation of response to stimulus, signal transduction	*0.92*
MF	*catalytic activity, acting on a protein, enzyme Binding*	*0.84*
CC	*Cytoplasm, nucleus*	*1*

(c)

(d) KEGG pathways with
FC= 0.76: *MAPK signaling pathway, Pathways in Cancer*

FIG. 17.19 Drug target module: DTM6.

Conclusion

This chapter proposed a computational approach for cancer drug target module mining in PIN. It is attempted on the *Homo sapiens* dataset. The implementation and application of this proposed optimized multi-functional co-clustering (MR-CoC$_{VFO}$) approach. It is the combination of the MR-CoC$_{multi}$ approach and the VFO algorithm. The results were compared with those of the previous approaches such as MR-CoC and MR-CoC$_{multi}$. This analysis evidenced the proposed approach performs well than these approaches. The biological significances of

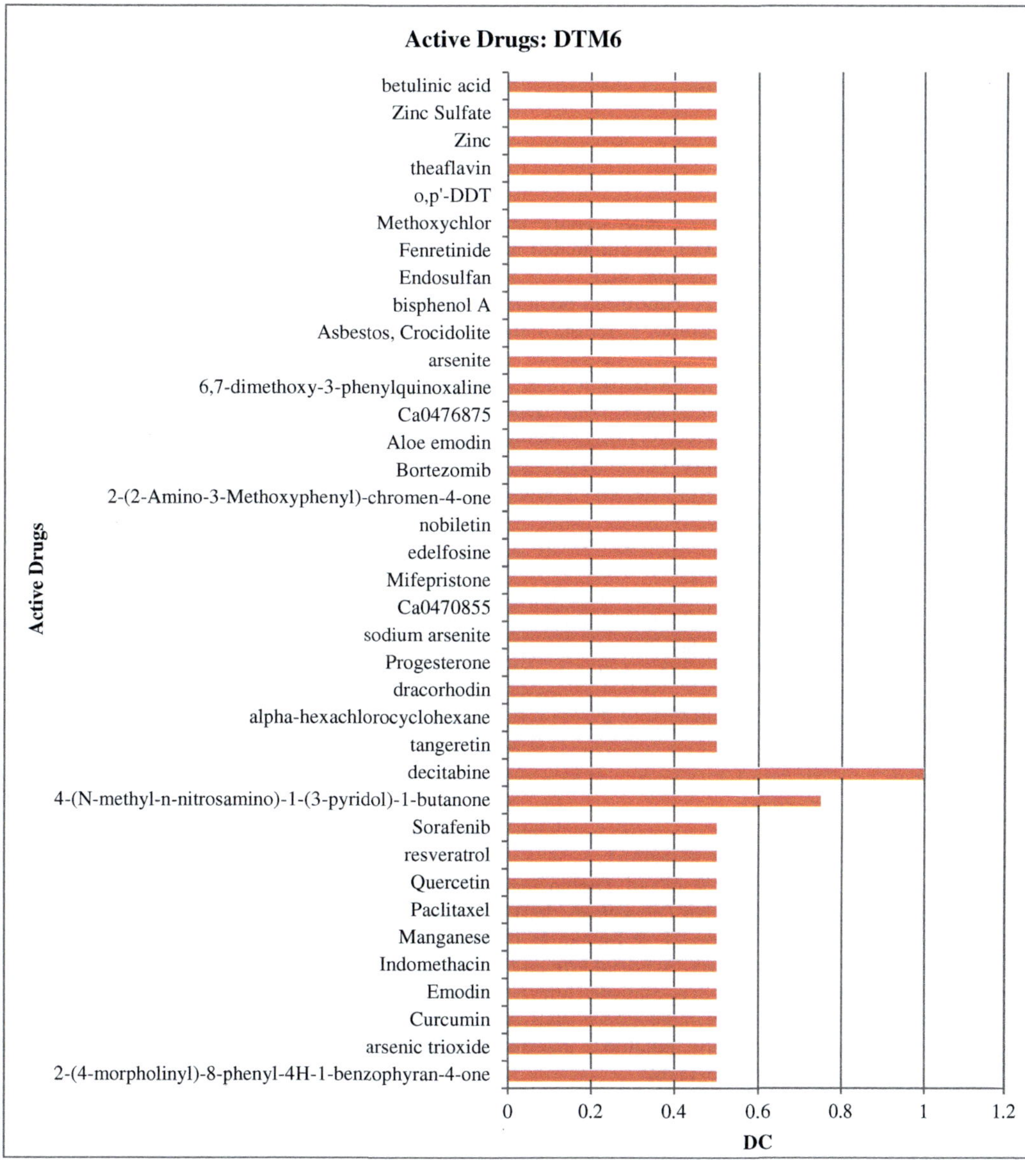

FIG. 17.20 Active drugs in DTM6.

the results were also analyzed for sample resultant modules, which were based on the drug target significances of the modules. Furthermore, the presence of cancer drug targets was analyzed. From this analysis, three protein modules are suggested and highlighted as cancer drug target modules with their active pharmaceutical drugs. These modules can be helpful for biologists in the development of targeted therapies for cancer.

Acknowledgment

The first author acknowledges the UGC for the financial support to her research under the UGC NET JRF (Student Id: 3384/(OBC) (NET JULY-2016)) Scheme.

References

Ali, I., Wani, W. A., & Saleem, K. (2011). Cancer scenario in India with future perspectives. Cancer Therapy, 8, 56–70.

Bader, G. D., & Hogue, C. W. V. (2003). An automated method for finding molecular complexes in large protein interaction networks. *BMC Bioinformatics*, *4*(2).

Bapat, R. B. (2010). *Graphs and matrices*. Hindustan Book Agency: Springer.

Bastos, H., Sousa, L., Clarke, L., & Couto, F. (2016). Functional coherence metrics in protein families. *Journal of Biomedical Semantics*, *7*(41), 1–11.

Bult, C., Blake, J., Smith, C., Kadin, J., Richardson, J., & Mouse Genome Database Group. (2019). Mouse genome database (MGD). *Mouse Genome Informatics*, D801–D806.

Cleveland, C. (2002). *Decitabine*. Chemocare. http://chemocare.com/chemotherapy/drug-info/decitabine.aspx.

Diestel, R. (2016). *Graph theory*. Springer.

Dincer, H., Dincer, A., & Levinson, D. (2002). Asymptomatic hyperuricemia: To treat or not to treat. *Cleveland Clinic Journal of Medicine*, *69*(8), 597–608.

Dittrich, M. T., Klau, G. W., Rosenwald, A., Dandekar, T., & Müller, T. (2008). Identifying functional modules in protein-protein interaction networks: An integrated exact approach. *Bioinformatics*, *24*, 223–231.

Franceschini, A., Szklarczyk, D., Frankild, S., Kuhn, M., Simonovic, M., Roth, A., Lin, J., Minguez, P., Bork, P., von Mering, C., & Jensen, L. J. (2013). STRING v9.1: Protein-protein interaction networks, with increased coverage and integration. *Nucleic Acids Research*, 808–815.

Gashaw, I., Ellinghaus, P., Sommer, A., & Asadullah, K. (2011). What makes a good drug target? *Drug Discovery Today - Elsevier*, 1037–1043.

Gowri, R., & Rathipriya, R. (2018). Cohesive sub-network mining in protein interaction network using score based co-clustering with MapReduce model. *Advances in BigData and Cloud Computing*, 227–236.

Gowri, R., & Rathipriya, R. (2020). *Functional module mining in protein interaction networks using co-clustering*. Apple Academic Press (in printing).

Gowri, R., & Rathipriya, R. (2021). Non-swarm intelligence algorithms: A case study. Computing, 103, 1815–1857. https://doi.org/10.1007/s00607-020-00870-1.

Gursoy, A., Keskin, O., & Nussinov, R. (2008). Topological properties of protein interaction networks from a structural perspective. *Biochemical Society Transactions*, *36*, 1398–1403.

Hase, T., Kikuchi, K., Ghosh, S., Kitano, H., & Tanaka, H. (2014). Identification of drug-target modules in the human protein-protein interaction network. Artificial Life and Robotics, 19, 406–413

ICMR. (2019). *India against cancer*. Cancer Statistics http://cancerindia.org.in/cancer-statistics/.

IndiaToday. (2018). *Cancer rate doubles in India: Facts, stats, cure and treatment of the most deadly disease in the world*. India Today Web Desk. https://www.indiatoday.in/education-today/gk-current-affairs/story/cancer-rate-india-stats-cure-treatment-1386739-2018-11-12.

Jeong, J.-H., An, J., Kwon, Y., Rhee, J., & Lee, Y. (2009). Effects of low dose quercetin: Cancer cell-specific inhibition of cell cycle progression. *Journal of Cellular Biochemistry*, *106*(1), 73–82.

MFMER, M. (2015). *Hospital for cancer surgery and treatment in India*. https://www.fortishealthcare.com/india/diseases/cancer-918.

NCI Drug Dictionary. (n.d.). National Cancer Institute. https://www.cancer.gov/publications/dictionaries/cancer-drug/def/indomethacin.

Pandey, K., & Sengupta, R. (2018, September 17). 9.6 million people will die of cancer this year. Retrieved from Down To Earth: https://www.downtoearth.org.in/news/health/9-6-million-people-will-die-of-cancer-this-year-61646.

Pereira-Leal, J. B., Enright, A. J., & Ouzounis, C. A. (2004). Detection of functional modules from protein interaction networks. *Proteins: Structure, Function, and Bioinformatics*, *54*, 49–57.

Shanmugam, M. K., Arfuso, F., Sng, J. C., Bishayee, A., PremKumar, A., & Sethi, G. (2019). Epigenetic effects of curcumin in cancer prevention. In *8. Epigenetics of cancer prevention* (pp. 107–128).

Spirin, V., & Mirny, L. (2013). Protein complexes and functional modules in the molecular networks. *PNAS*, *100*(21), 12123–12128.

Structures of Life. (2007). *US Department of Health and Human Services, National Institute of Health, National Institute of General Medical Services*.

Sussman, J., & Prilusky, J. (2019). *Proteopedia life in 3D*. http://www.proteopedia.org.

Wang, H.-M., & Zhang, G.-Y. (2005). Indomethacin suppresses growth of colon cancer via inhibition of angiogenesis in vivo. *World Journal of Gastroenterology*, *11*(3), 340–343.

Zheng, C., Han, L., Yap, C., Ji, Z., Cao, Z., & Chen, Y. (2006). Therapeutic targets: Progress of their exploration and investigation of their characteristics. *Pharmacological Reviews*, 259–279.

Index

Note: Page numbers followed by *f* indicate figures, *t* indicate tables, and *b* indicate boxes.

D

L

M

U

V

W

X

9780323852401

9780323852401